T0329556

Electrical Processes in Organic Thin Film Devices

Electrical Processes in Organic Thin Film Devices

From Bulk Materials to Nanoscale Architectures

Michael C. Petty
Department of Engineering
and
Centre for Molecular and Nanoscale Electronics
Durham University, UK

The right of Michael C. Petty to be identified as the author of this work has been asserted in accordance with law.

Registered Offices
John Wiley & Sons, Inc., 111 River Street, Hoboken, NJ 07030, USA
John Wiley & Sons Ltd, The Atrium, Southern Gate, Chichester, West Sussex, PO19 8SQ, UK

Editorial Office
The Atrium, Southern Gate, Chichester, West Sussex, PO19 8SQ, UK

For details of our global editorial offices, customer services, and more information about Wiley products visit us at www.wiley.com.

Wiley also publishes its books in a variety of electronic formats and by print-on-demand. Some content that appears in standard print versions of this book may not be available in other formats.

Library of Congress Cataloging-in-Publication Data

Name: Petty, Michael C., author.
Title: Electrical processes in organic thin film devices : from bulk
    materials to nanoscale architectures / Michael C. Petty.
Description: First edition. | Hoboken, NJ : John Wiley & Sons, Inc., 2022.
    | Includes bibliographical references and index.
Identifiers: LCCN 2021039428 (print) | LCCN 2021039429 (ebook) | ISBN
    9781119631279 (cloth) | ISBN 9781119631309 (adobe pdf) | ISBN
    9781119631347 (epub)
Subjects: LCSH: Organic semiconductors. | Organic thin films.
Classification: LCC TK7871.99.O74 P48 2022 (print) | LCC TK7871.99.O74
    (ebook) | DDC 621.3815/2 – dc23/eng/20211020
LC record available at https://lccn.loc.gov/2021039428
LC ebook record available at https://lccn.loc.gov/2021039429

Cover design by Wiley
Cover image: © NickSorl/Shutterstock,
artpartner-images/Getty, Ekely/Getty

Set in 9.5/12.5pt STIXTwoText by Straive, Chennai, India
Printed and bound by CPI Group (UK) Ltd, Croydon, CR0 4YY

C9781119631279_110122

For

Katie and Anna

Contents

Preface

Organic electronics has become a fast moving, interdisciplinary field of activity that is focused on the electronic and optoelectronic properties of organic materials. Over the past three decades, this has seen burgeoning scientific interest and significant commercial investment. The topic is wide-ranging, covering traditional solid-state devices that exploit organic semiconducting compounds to single-molecule architectures. In the former category, organic light-emitting displays are now firmly established in the marketplace while other products, such as transistors, memories, photovoltaic solar cells, and chemical sensors, are developing fast. Electronics technologies based on single molecules or small groups of molecules (molecular electronics) continue to be rooted in the research laboratory. However, the implications of operating future information processing systems with individual molecules are far-reaching. Single-molecule electronics has been recognized as a potential disruptive technology that could have a significant impact on society in the years to come.

Organic electronic devices are invariably based on thin films, using methods as diverse as thermal evaporation and molecular self-assembly. The aim of this book is to provide an understanding of the physical processes that determine their electrical characteristics. While band theory has been remarkably successful in accounting for the electrical behaviour of crystalline inorganic semiconductors and related devices, the ideas are not always relevant to organic materials, which exist in a variety of morphological forms. In some cases, band theory can be modified (as for amorphous inorganic semiconductors); in others different approaches are needed.

The first chapter introduces the important categories of electroactive organic materials and reviews the nature of their intra-molecular and inter-molecular chemical bonding, which naturally leads to a discussion of the applicability of the band model. The fundamental principles of electrical conduction (electric field, current density, charge carrier mobility, and excess carriers) and the constraints imposed by the thin-film morphology and dimensionality are then examined. The role of surfaces on device operation is becoming increasingly important as dimensions are reduced. Cartoon diagrams depicting molecules with abrupt interfaces within devices and are all very well, but these can seduce researchers into unrealistic physical models.

A critical, but often neglected, issue is the nature of the electrical contacts to a material or device. A separate chapter is assigned to this. In most cases, the electrical contacts should be Ohmic, but certain devices require the presence of a potential barrier or the ability to inject charge into the device. This is followed by an introduction to the field effect, which underpins the operation of many electronic components. Theories of low-frequency (DC) and high-frequency (AC) conduction processes in organic solids are considered separately but recognizing there is inevitably overlap.

I have included some details on how important quantities are measured and added a short section on electrical noise. Wherever possible, I have used published experimental data to illustrate particular points.

Application areas are described in the second half of this book. Chapters are devoted to devices currently in the marketplace (displays) and to those under development (transistors, solar cells, memories). A final chapter explores emerging organic thin film electronic components and circuits, such as single-molecule electronics, neuromorphic computing, and biologically inspired devices. Predicting our scientific and technological future is difficult, and this is very much a personal view. My suggestion is that new organic and biological devices will continue to emerge over the next decade, with many more finding their way into the commercial arena. Further on, organic (and perhaps biological) electronic information processing systems will begin to make an appearance. I submit that these will not, initially, complete with silicon and gallium arsenide technology, which will remain the mainstream for fast information processing. However, organic devices and systems will fulfill specific needs (for example, in the medical field), and may also provide solutions to some of the major technological energy and environmental challenges that lie ahead.

I have written this book for final-year undergraduates and post-graduate researchers, and assumed a basic knowledge of chemistry, materials science, electrical engineering, and physics topics such as quantum theory. Readers wishing to find more background information are directed to my book *Organic and Molecular Electronics: From Principles to Practice* (also published by Wiley).

Where would a scientific textbook be without some self-assessment problems? I have included a number of these at the end of each chapter. My aim is to consolidate and to extend the text. But some of the questions might require a bit of extra thought – and some research. Solutions may be found on the web pages for the book.

June 2021

Michael C. Petty
Durham University

Acknowledgements

I am grateful to current and former colleagues at Durham who have, over many years, provided invaluable input to this book: Jas Pal Badyal, David Bloor, Leon Bowen, Hajo Broersma, Martin Bryce, Mujeeb Chaudhry, Karl Coleman, Graham Cross, Ken Durose, Jim Feast, Andrew Gallant, John Gibson, Tariq Ginnai, Chris Groves, K. K. Kan, Ritu Kataky, Apostolos Kotsialos, Jonathan Lloyd, Phil Mars, Budhika Mendis, Andy Monkman, Boguslaw Obara, Chris Pearson, Ian Peterson, Graham Russell, Ifor Samuel, Brian Tanner, Richard Thompson, David Wood, Jack Yarwood, and Dagou Zeze. And to those elsewhere: Geoff Ashwell, André Barraud, Jeremy Burroughes, Horacio Cantiello, Jeon-Woo Choi, Michael Conrad, Danilo De Rossi, Panagiotis Dimitrakis, Stephen Evans, Leonid Goldenberg, Simon Harding, Felix Hong, Yulin Hua, Csaba Juhasz, Safa Kasap, Hans Kuhn, Yong Uk Lee, Yuri Lvov, Julian Miller, Dietmar Möbius, Jürgen Nagel, Claudio Nicolini, Pascal Normand, Asim Ray, Tim Richardson, Martin Taylor, Dimitris Tsoukalas, Gunnar Tufte, Ludovico Valli, Alexei Vitukhnovsky, Changsheng Wang, and Stephen Wood.

Some of the results described are, directly or indirectly, from students, post-docs, and visitors in my own research group: Napoleon Agbor, Munir Ahmad, Jin Ahn, Steve Baker, Paul Barker, John Batey, Zakia AL Busaidi, Duncan Cadd, Riccardo Cassalini, Pilar Cea, Phil Christie, John Cresswell, Gerald Davies, I. M. Dharmadasa, Ajaib Dhindsa, Nigel Evans, Alison Farrar, Mike Fowler, Vaughan Howarth, Yesul Jeong, Stécy Jombert, Gun-Young Jun, Nick Kalita, Mary Kilitziraki, Dan Kolb, Myung-Won Lee, Igor Lednev, Peter Lukes, Mohammed Mabrook, Kieran Massey, Bill Neal, Marco Palumbo, Jaehoon Park, Shashi Paul, Mike Petty (yes, there is another), Fawada Qaiser, Supachai Ritjareonwattu, Carole Stevens, Nick Thomas, John Tsibouklis, Eléonore Vissol-Gaudin, Diogo Volpati, Julian Warren, Nick Widdowson, Jason Wilde, Geoff Williams, and Youngjun Yun. It is a pleasure to also thank the numerous researchers worldwide whose data are reproduced in the book. And cheers to two perceptive LB chums, Paul Martin and Rodger Sykes, whose banter continues to befuddle, entertain, and enlighten.

Staff at Wiley provided an enormous amount of encouragement, help, and support, first to setup this project and then to see it through to completion: Jenny Cossham, Katrina Maceda, Elke Morice-Atkinson, Mahalakshmi Pitchai, Shirly Samuel, and Emma Strickland.

Last but not least, I shall always be indebted to Gareth Roberts, who steered me onto the organic electronics path. His only fault was that he supported a North London football team.

Constants

Velocity of light in free space	c	$2.998 \times 10^8 \text{ m s}^{-1}$
Permittivity of free space	ε_0	$8.854 \times 10^{-12} \text{ F m}^{-1}$
Electronic charge	e	$1.602 \times 10^{-19} \text{ C}$
Planck's constant	h	$6.626 \times 10^{-34} \text{ J s}^{-1}$
Boltzmann's constant	k_B	$1.381 \times 10^{-23} \text{ J K}^{-1}$
Avogadro's number	N_A	$6.022 \times 10^{26} \text{ kmole}^{-1} \ (= 6.022 \times 10^{23} \text{ mole}^{-1})$
Universal gas constant	R	$8.314 \times 10^3 \text{ J kmol}^{-1} \text{ K}^{-1}$
Faradays constant	F	$9.649 \times 10^7 \text{ C kmol}^{-1}$
Acceleration due to gravity	g	9.807 m s^{-2}

Useful Relationships

1 electronvolt (eV) $= 1.602 \times 10^{-19} \text{ J}$

For vacuum, energy in eV $= 1.240/(\text{wavelength in } \mu\text{m})$

$1 \text{ eV} = 8066 \text{ cm}^{-1}$

1 eV per particle $= 23\,060 \text{ kcal kilomole}^{-1} \ (= 23.06 \text{ kcal mol}^{-1})$

1 calorie $= 4.186 \text{ J}$

At 300 K, $k_B T = 0.026 \text{ eV}$

1 atmosphere $= 1.013 \times 10^5 \text{ Nm}^{-2}$

Symbols and Abbreviations

a	atomic spacing/monomer length (m)
a	distance between potential wells (m)
a	acceleration ($m\,s^{-2}$)
au	arbitrary units
A	acceptor
A	image force (N)
A	area (m^2)
A	(optical) absorbance
A	Hamaker constant (J)
A^*	effective Richardson constant ($A\,m^{-2}\,K^{-2}$)
$A^{-\cdot}$	radical anion
AC	alternating current
ADP	adenosine diphosphate
AFM	atomic force microscope/microscopy
Alq_3	tris-(8-hydroxyquinoline) aluminium
ALU	arithmetic logic unit
ANN	artificial neural network
ATP	adenosine triphosphate
AZO	aluminium-doped zinc oxide
B	magnetic field (T)
BCP	2,9-dimethyl-4,7-diphenyl-1,10-phenanthroline (or bathocuproine)
BHJ	bulk heterojunction
BSBCz	4,4′-bis[(N-carbazole)styryl]biphenyl
c	molecular concentration ($mol\,m^{-3}$)
C	Van der Waals constant ($J\,m^6$)
C	circumference (e.g. of nanotube) (m)
C	capacitance (F)
C	specific heat capacity ($J\,kg^{-1}\,K^{-1}$)
C_d	depletion layer capacitance per unit area ($F\,m^{-2}$)
C_d	drain capacitance (of SET) (F)
C_{fb}	flat-band capacitance per unit area ($F\,m^{-2}$)
C_g	gate capacitance (of SET) (F)
C_{hf}	high-frequency capacitance per unit area ($F\,m^{-2}$)

C_i	insulator (or gate) capacitance per unit area ($\mathrm{F\,m^{-2}}$)
C_{it}	capacitance associated with interface traps per unit area ($\mathrm{F\,m^{-2}}$)
C_{lf}	low-frequency capacitance per unit area ($\mathrm{F\,m^{-2}}$)
C_m	measured capacitance per unit area ($\mathrm{F\,m^{-2}}$)
C_{min}	minimum capacitance per unit area (e.g. of C–V curve) ($\mathrm{F\,m^{-2}}$)
C_p	parallel capacitance per unit area (capacitive element of parallel R–C circuit) ($\mathrm{F\,m^{-2}}$)
C_s	semiconductor capacitance per unit area ($\mathrm{F\,m^{-2}}$)
C_s	source capacitance (of SET) (F)
CB	conduction band
CD	compact disc
CGL	charge-generation layer
CIE	Commission International de l'Eclairage
CIO	indium-doped CdO
CMOS	complementary MOS
CRI	colour rendering index
CT	charge transfer
CTL	charge-transport layer
CuTTBPc	copper(II) 2,9,16,23-tetra-*tert*-butyl-29H,31H-phthalocyanine
d	distance or thickness (e.g. of thin film) (m)
D	donor
$D^{+\cdot}$	radical cation
D	diffusion coefficient ($\mathrm{m^2\,s^{-1}}$)
D	electric displacement ($\mathrm{C\,m^{-2}}$)
D	detectivity ($\mathrm{W^{-1}}$)
D^*	specific detectivity ($\mathrm{m\,Hz^{1/2}\,W^{-1}}$) or ($\mathrm{cm\,Hz^{1/2}\,W^{-1}}$ – Jones)
D_{it}	interface state charge density ($\mathrm{J^{-1}\,m^{-2}}$ or $\mathrm{eV^{-1}\,m^{-2}}$)
D_{itm}	interface state charge density in equilibrium with metal ($\mathrm{J^{-1}\,m^{-2}}$ or $\mathrm{eV^{-1}\,m^{-2}}$)
D_{its}	interface state charge density in equilibrium with semiconductor ($\mathrm{J^{-1}\,m^{-2}}$ or $\mathrm{eV^{-1}\,m^{-2}}$)
D_n	diffusion coefficient for electrons ($\mathrm{m^2\,s^{-1}}$)
D_p	diffusion coefficient for holes ($\mathrm{m^2\,s^{-1}}$)
DC	direct current
DDQ	2,3-dichloro-5,6-dicyano-1,4-benzoquinone
DLTS	deep level transient spectroscopy
DNA	deoxyribonucleic acid
DRAM	dynamic random access memory
DVD	digital video disc
e_p	emission rate for holes (from a trap) ($\mathrm{s^{-1}}$)
$\mathcal{E}$	electric field ($\mathrm{V\,m^{-1}}$)
$\mathcal{E}_m$	electric field at semiconductor/metal interface ($\mathrm{V\,m^{-1}}$)
$\mathcal{E}_b$	breakdown field ($\mathrm{V\,m^{-1}}$)
$\mathcal{E}_c$	coercive field ($\mathrm{V\,m^{-1}}$)

E	energy (J or eV)
E_0	energy band width (J or eV)
$E_a, \Delta E_a$	acceptor energy, acceptor ionization energy (J or eV)
E_a	electromigration activation energy (J or eV)
E_b	energy barrier (J or eV)
E_c	conduction band edge (J or eV)
$E_d, \Delta E_d$	donor energy, donor ionization energy (J or eV)
E_F	Fermi energy/level (J or eV)
E_{F0}	Fermi energy at 0 K (J or eV)
E_{Fi}	Fermi energy for intrinsic semiconductor (J or eV)
E_{Fm}	metal Fermi energy (J or eV)
E_{Fs}	semiconductor Fermi energy (J or eV)
E_{Fqn}	quasi-Fermi energy for electrons (J or eV)
E_{Fqp}	quasi-Fermi energy for holes (J or eV)
E_g	energy (band) gap (J or eV)
E_{HOMO}	energy of highest occupied molecular orbital (J or eV)
E_{LUMO}	energy of lowest unoccupied molecular orbital (J or eV)
E_{phot}	Illuminance (lux or $lm\,m^{-2}$)
E_n	quantized energy level (J or eV)
ΔE_{ST}	energy difference between singlet and triplet state (J or eV)
E_t	position of peak in Gaussian distribution of energy states (J or eV)
E_t	trap depth (from band edge) (J or eV)
E_v	valence band edge (J or eV)
EIL	electron-injection layer
EiM	evolution-in-materio
EL	electroluminescence (or electroluminescent)
EM	electromagnetic
EML	emissive layer
ENFET	enzyme field effect transistor
EPROM	electronically programmable read only memory
EEPROM or E^2PROM	electronically erasable/programmable read only memory
ETL	electron-transport layer
$f(E)$	Fermi–Dirac energy distribution function
$f_a(E)$	Fermi–Dirac energy distribution function for anode
$f_c(E)$	Fermi–Dirac energy distribution function for cathode
f	frequency (Hz)
$f_d(E_d)$	Fermi–Dirac function for donor atoms
f_{max}	maximum frequency (Hz)
f_T	cut-off frequency (for FET) (Hz)
F	force (N)
F	Fano factor
F	feature size (e.g. of crossed-bar resistive memory) (m)
F	fill-factor (of photovoltaic cell)

F_{field}	force due to electric field (N)
F_{wind}	force due to momentum transfer (N)
FET	field effect transistor
FeFET	ferroelectric field effect transistor
FLOPS	floating-point operations per second
FN	Fowler–Nordheim
FTO	fluorine-doped tin oxide
FPGA	field-programmable gate array
g	piezoelectric coefficient ($C\,N^{-1}$)
g_m	transconductance of FET ($A\,V^{-1}$)
G	conductance (S or Ω^{-1})
G	rate of electron–hole pair thermal generation ($m^{-3}\,s^{-1}$)
G	geometric gain
G^*	photoconductive gain
G_p	parallel conductance per unit area (conductive element of parallel R–C circuit) ($\Omega^{-1}\,m^{-2}$)
G_{ph}	rate of electron–hole pair photo generation ($m^{-3}\,s^{-1}$)
GMR	giant magnetoresistance ratio
HIL	hole-injection layer
HMDS	hexamethyldisilazane
HMTPD	N,N,N',N'-tetrakis(3-methylphenyl)-3,3'-dimethylbenzidine
HOMO	highest occupied molecular orbital
HTL	hole-transport layer
$i_{1/f}$	$1/f$ noise current (A)
i_{noise}	total noise current (A)
i_{RTN}	random telegraph noise current (A)
$i_{thermal}$	thermal noise current (A)
I	electric current (A)
I	spectral irradiance ($W\,m^{-2}\,m^{-1}$)
I_g	FET AC gate current (A)
I_d	FET drain current (A)
I_d	dark current (A)
I_{dd}	drain current in the dark (A)
I_{dl}	drain current in the light (A)
I_l	photocurrent (A)
I_{load}	load current (A)
I_{max}	current at maximum power point (A)
I_{phot}	luminous intensity (cd)
I_{sc}	short-circuit current (A)
IBSC	intermediate band solar cell
IDT	indacenodithiophene
IMFET	immunological field effect transistor
IPS	instructions per second

Ir(ppy)$_2$(acac)	bis[2-(2-pyridinyl-N)phenyl-C](2,4-pentanedionato-O^2,O^4) iridium(III)
ISC	intersystem crossing
ISFET	ion-sensitive field effect transistor
ITO	indium tin oxide
j	$\sqrt{-1}$
J	electric current density (A m^{-2})
J_0	reverse saturation current density (A m^{-2})
J_d	diffusion current density (A m^{-2})
J_d	dark current density (A m^{-2})
J_g	generation current density (A m^{-2})
J_n	current density of electrons (A m^{-2})
J_p	current density of holes (A m^{-2})
J_r	recombination current density (A m^{-2})
J_{te}	thermionic emission current density (A m^{-2})
$J_{te(max)}$	maximum thermionic emission current density (A m^{-2})
k	wavevector (m^{-1})
k	dielectric constant
K	scaling constant
K_{RISC}	RISC (reverse intersystem crossing) constant
l	mean free path (m)
l_{ph}	phase relaxation length (m)
L	length (m)
L	diffusion length (m)
L_D	Debye screening length (m)
L_{phot}	luminance (cd m^{-2})
LB	Langmuir–Blodgett
LC	liquid crystal
LCD	liquid crystal display
LEC	light-emitting electrochemical cell
LED	light-emitting diode
LSB	least significant bit
LSC	luminescent solar concentrator
LUMO	lowest unoccupied molecular orbital
m	mass (kg)
m_e^*	electron effective mass (kg)
M	molecular weight
M	number of electron sub-bands (or channels)
MDMO-PPV	poly(2-methoxy-5-(3′,7′-dimethyloctyloxy)-1,4-phenylenevinylene)
MEH-PPV	poly[2-methoxy-5-(2-ethylhexyloxy)-1,4-phenylenevinylene]
spiro-MeOTAD	2,2′,7,7′-tetrakis(N,N-di-4-methoxyphenylamino)-9,9′-spirobifluorene
MIS	metal/insulator/semiconductor
MOS	metal/oxide/semiconductor

MOSFET	metal/oxide/semiconductor field effect transistor
MPP	maximum power point (for a photovoltaic device)
MRAM	magnetic random access memory
MSB	most significant bit
MT	microtubule
MTF	mean time to failure
MWNT	multiwall (carbon) nanotube
n	quantum number
n	ideality factor (e.g. of Schottky diode or FET subthreshold swing)
n	refractive index
n	number (e.g. of charges, electron states) per unit volume, or concentration (m^{-3})
n_e	concentration of extrinsic carriers (m^{-3})
n_{eff}	effective refractive index
n_i	concentration of intrinsic carriers (m^{-3})
n_p	concentration of electrons in p-type semiconductor
n_{p0}	concentration of electrons at surface of p-type semiconductor (m^{-3})
n_{pb}	concentration of electrons in neutral (bulk) region of p-type semiconductor (m^{-3})
n_t	concentration of trapped charge carriers (m^{-3})
N	number (of dipoles, electron states etc.)
N_0	number of molecules per unit volume (m^{-3})
N_a	acceptor atom doping concentration (m^{-3})
N_c	effective density of states of conduction band (m^{-3})
N_d	donor atom doping concentration (m^{-3})
N_i	ionized impurity concentration (m^{-3})
N_t	trap concentration (bulk m^{-3}; surface m^{-2})
$N_t(E)$	number of surface traps per unit energy ($m^{-2}\,eV^{-1}$)
N_{t0}	surface trap concentration at band edge (m^{-2})
N_{te}	electron trap concentration (m^{-3})
N_{tp}	hole trap concentration (m^{-3})
N_r	recombination centre concentration (m^{-3})
N_v	effective density of states of valence band (bulk m^{-3}; surface m^{-2})
NAD	nicotinamide adenine dinucleotide
NADP	nicotinamide adenine dinucleotide phosphate
NDR	negative differential resistance
NEP	noise equivalent power (W)
NP	nanoparticle p-type semiconductor (m^{-3})
NT	naphtho[1,2-*c*:5,6-c]bis[1,2,5]-thiadiazole
OFET	organic field effect transistor
OLED	organic light-emitting device/diode
OLET	organic light-emitting transistor
OPBT	organic permeable base transistor
OPT	organic phototransistor

OPV	organic photovoltaic
OSIT	organic static induction transistor
OTS	n-octadecyltrichlorosilane
OXD-7	2-(4-*tert*-butylphenyl)-5-[3-[5-(4-*tert*-butylphenyl)-1,3,4-oxadiazol-2-yl]phenyl]-1,3,4-oxadiazole
p	momentum ($\mathrm{kg\,m\,s^{-1}}$)
p	electric dipole moment ($\mathrm{C\,m}$)
p	pyroelectric coefficient ($\mathrm{C\,m^{-2}\,K^{-1}}$)
p	hole concentration ($\mathrm{m^{-3}}$)
p_0	initial hole concentration ($\mathrm{m^{-3}}$)
p_D	hole concentration in dark ($\mathrm{m^{-3}}$)
p_p	concentration of holes in p-type semiconductor
p_{p0}	concentration of holes at surface of a p-type semiconductor ($\mathrm{m^{-3}}$)
p_{pb}	concentration of holes in neutral region of a p-type semiconductor ($\mathrm{m^{-3}}$)
p_t	concentration of trapped holes ($\mathrm{m^{-3}}$)
P	transmission probability (of an electron wave)
P	polarization ($\mathrm{C\,m^{-2}}$)
P_i	optical input power density ($\mathrm{W\,m^{-2}}$)
P_r	remanent or residual polarization ($\mathrm{C\,m^{-2}}$)
P3HT	poly(3-hexylthiophene)
PANi	polyaniline
$PC_{61}BM$	[6,6]-phenyl-C_{61}-butyric acid methyl ester
$PC_{71}BM$	[6,6]-phenyl-C_{71}-butyric acid methyl ester
PCPDTBT	poly[2,6-(4,4-bis-(2-ethylhexyl)-4*H*-cyclopenta[2,1-*b*;3,4-*b′*]-dithiophene)-*alt*-4,7-(2,1,3-benzothiadiazole)]
PDPPTzBT	diketopyrrolopyrrolethiazolothiazole copolymer
PDMS	polydimethylsiloxane
PEDOT	poly(3,4-ethylenedioxythiophene)
PEI	polyethylenimine
PEO	poly(ethylene oxide)
PET	poly(ethylene terephthalate)
PFO	poly(9,9-dioctylfluorene)
PMMA	poly(methyl methacrylate)
PNTT	copolymer based on naphtho[1,2-*c*:5,6-*c*]bis[1,2,5]-thiadiazole (NT) unit
PROM	programmable read only memory
PS	polystyrene (a-PS = amorphous polystryene)
PSD	phase-sensitive detection
PSQ	polysilsesquioxane
PSS	poly(styrene sulfonic acid)
PPP	poly(*p*-phenylene)
PPV	poly(*p*-phenylene vinylene)
PVA	poly(vinyl alcohol)

PVC	poly(vinyl chloride)
PVDF	poly(vinylidene difluoride)
PVK	poly(vinylcarbazole)
PVP	poly(4-vinyl phenol)
q	charge (C)
qubit	quantum bit
Q	charge per unit area (C m^{-2})
Q_{fg}	floating gate charge per unit area (C m^{-2})
Q_{fix}	fixed charge per unit area (C m^{-2})
Q_{it}	surface density of interface charge (C m^{-2})
Q_m	charge per unit area on metal surface (C m^{-2})
Q_{mob}	mobile charge per unit area (C m^{-2})
Q_n	charge per unit area in electron inversion region (C m^{-2})
Q_s	surface or bound charge per unit area (C m^{-2})
Q_{sc}	semiconductor space charge per unit area (C m^{-2})
r_{DA}	initial separation between charges in donor and acceptor (m)
R	rate of carrier recombination (m^{-3} s^{-1})
R	resistance (Ω)
R	radius (m)
R	hopping distance (m)
R_c	contact resistance (Ω)
R_d	drain resistance (of SET) (Ω)
R_{dg}	drain to gate leakage resistance (Ω)
R_{it}	resistance (reciprocal conductance) associated with unit area of interface traps (Ω m^2)
R_{load}	load resistance (Ω)
R_{series}	series resistance (Ω)
R_s	sheet resistance (Ω or Ω per square)
R_s	source resistance (of SET) (Ω)
R_{shunt}	shunt resistance (Ω)
R_t	resistance of tunnelling barrier (Ω)
RAM	random access memory
RF	radiofrequency
RFID	radiofrequency identification
R, G, B	red, green, blue (subpixel colour)
RISC	reverse intersystem crossing
RMS	root-mean-square
RNA	ribonucleic acid
ROM	read only memory
RRAM (or ReRAM)	resistive random access memory
RTD	resonant tunnelling diode
RTN	random telegraph noise
s	spacing (e.g. of probes) (m)
s	spin (of electron)

S	strain
S	subthreshold swing of FET (volts per decade)
S	singlet electronic state
S^*	specific sensitivity (of a photoconductor) ($m^2\,\Omega^{-1}\,W^{-1}$)
S_I	current noise power spectral density ($A^2\,Hz^{-1}$)
S_V	voltage noise power spectral density ($V^2\,Hz^{-1}$)
$S(E)$	density of states function ($J\,m^{-3}$ or $eV\,m^{-3}$)
$S(T)$	normalized DLTS signal
SAM	self-assembled monolayer
SCM	scanning capacitance microscope/microscopy
SET	single-electron transistor
SF	singlet fission
SNN	spiking neural network
S/N	signal-to-noise ratio (dB)
SKPM	scanning Kelvin-probe microscope/microscopy
SRAM	static random access memory
STM	scanning tunnelling microscope/microscopy
STP	standard temperature and pressure
STT-RAM	spin transfer torque random access memory
SWNT	single wall (carbon) nanotube
t	time (s)
t	thickness (m)
t	transfer integral (J or eV)
t_R	charge retention time (for memory) (s)
t_t	transit time (s)
T	mechanical stress ($N\,m^{-2}$)
T	temperature (K or °C)
T	triplet electronic state
T_{50} or $T_{1/2}$	OLED lifetime for luminance to drop to 50% of initial value (s)
T_{95}	OLED lifetime for luminance to drop to 95% of initial value (s)
T_c	characteristic temperature (K or °C)
T_C	Curie temperature (K or °C)
T_g	glass transition temperature (K or °C)
T_m	melting point (K or °C)
TADF	thermally activated delayed fluorescence
TAZ	3-(4-*tert*-butylphenyl)-4-phenyl-5-(4-phenylphenyl)-1,2,4-triazole
TCNQ	tetracyanoquinodimethane
TFE	tetrafluoroethylene
TIPS	triisopropylsilylethynl
TPBI	1,3,5-tris(*N*-phenylbenzimadazol-2-yl)benzene
TPD	*N,N'*-bis(3-methylphenyl)-*N,N'*-diphenylbenzidine
TrFE	trifluoroethylene
TSC	thermally stimulated conductivity

TTA	triplet-triplet annihilation
TTF	tetrathiafulvalene
U	net carrier transition rate ($m^{-3}\,s^{-1}$)
U_D	binding energy of donor exciton (J or eV)
UC	upconversion
USB	universal serial bus
v	velocity ($m\,s^{-1}$)
v	phase velocity of light in material ($m\,s^{-1}$)
v_d	drift velocity ($m\,s^{-1}$) (charge carrier velocity resulting from an applied electric field)
v_t	thermal velocity ($m\,s^{-1}$) (charge carrier velocity resulting from temperature)
V	potential energy (J or eV)
V	voltage (V)
$V(\lambda)$	eye sensitivity function
V_0	offset voltage (threshold or turn-on voltage) of an FET (V)
V_0, V_{00}	constants in quantum mechanical tunnelling equations (V)
V_{fb}	flat-band voltage (V)
V_{fg}	floating gate voltage (V)
V_g	gate voltage (V)
V_d	drain voltage (V)
V_d	diffusion voltage (or built-in potential) (V)
VDF	vinylidene difluoride
V_{DC}	DC voltage (V)
V_i	voltage across insulator or interfacial layer (V)
V_m	membrane potential (V)
V_{load}	load voltage (V)
V_{max}	voltage at maximum power point (V)
V_{oc}	open-circuit voltage (V)
V_t	turn-on or threshold voltage (e.g. for strong inversion) (V)
V_{TFL}	trap-filled limit voltage (V)
VB	valence band
VOFET	vertical organic field effect transistor
W	width (m)
W_d	depletion layer width (m)
W_{dmax}	maximum depletion layer width (m)
WOLED	white organic light emitting device/diode
WORM	write-once/read-many times memory
x_m	position of potential for maximum image force lowering of Schottky barrier (m)
Y	Young's modulus (Pa)
Y	electrical admittance (Ω^{-1} or S)
z	number of nearest neighbours
z	number of electron charges (valence)

Z	electrical impedance (Ω)
α	absorption coefficient (m^{-1})
α	inverse localization length (for an electron wave) (m^{-1})
α	temperature coefficient of resistivity $(\Omega\,\mathrm{m}\,\mathrm{K}^{-1})$
β	phase difference $(°)$
β	force (or spring) constant $(\mathrm{N}\,\mathrm{m}^{-1})$ or $(\mathrm{kg}\,\mathrm{s}^{-2})$
β	recombination constant $(\mathrm{m}^3\,\mathrm{s}^{-1})$
β	stretching factor (in OLED lifetime equation)
β_{S}	Schottky constant $(\mathrm{J}\,\mathrm{V}^{-1/2}\,\mathrm{m}^{1/2})$
β_{PF}	Poole–Frenkel constant $(\mathrm{J}\,\mathrm{V}^{-1/2}\,\mathrm{m}^{1/2})$
γ	Fowler–Nordheim constant $(\mathrm{V}\,\mathrm{m}^{-1})$
γ	minority carrier injection ratio
γ	activity coefficient
δ	insulator thickness (in MIS structure) (m)
$\tan\delta$	loss tangent
$\varepsilon_{\mathrm{hf}}$	high-frequency relative permittivity
ε_{i}	relative permittivity of insulator or interfacial layer
$\varepsilon_{\mathrm{lf}}$	low-frequency relative permittivity
ε_{r}	relative permittivity
$\varepsilon_{\mathrm{r}}'$	real part of relative permittivity
$\varepsilon_{\mathrm{r}}''$	imaginary part of relative permittivity
ε_{s}	relative permittivity of semiconductor
η	efficiency
η_{a}	photoabsorption efficiency
η_{c}	extraction efficiency (or coupling coefficient)
η_{cc}	charge collection efficiency
η_{ed}	exciton diffusion efficiency
η_{ext}	external quantum efficiency
η_{int}	internal quantum efficiency
η_{l}	luminous (or current) efficiency $(\mathrm{cd}\,\mathrm{A}^{-1})$
η_{p}	luminous power efficiency $(\mathrm{lm}\,\mathrm{W}^{-1})$
η_{PV}	power conversion efficiency of photovoltaic cell
θ	ratio of free to tapped carriers
θ	angle $(°)$
θ_{c}	critical angle $(°)$
λ	wavelength (m)
λ_{F}	Fermi wavelength (m)
Λ	periodicity of diffraction grating (m)
μ	charge carrier mobility $(\mathrm{m}^2\,\mathrm{V}^{-1}\,\mathrm{s}^{-1})$
μ_{D}	mobility in dark $(\mathrm{m}^2\,\mathrm{V}^{-1}\,\mathrm{s}^{-1})$
μ_{eff}	effective carrier mobility (in OFET) $(\mathrm{m}^2\,\mathrm{V}^{-1}\,\mathrm{s}^{-1})$
μ_{FET}	field effect mobility (as measured in an OFET) $(\mathrm{m}^2\,\mathrm{V}^{-1}\,\mathrm{s}^{-1})$
μ_{n}	electron mobility $(\mathrm{m}^2\,\mathrm{V}^{-1}\,\mathrm{s}^{-1})$

μ_0	reference carrier mobility or mobility within grain ($m^2\,V^{-1}\,s^{-1}$)
μ_p	hole mobility ($m^2\,V^{-1}\,s^{-1}$)
μ_i	carrier mobility resulting from ionized impurity scattering ($m^2\,V^{-1}\,s^{-1}$)
μ_r	relative permeability
ν_0	attempt to escape frequency (e.g. of carrier from trap) (s^{-1})
ν_{el}	electronic frequency (s^{-1})
ν_{ph}	phonon frequency (s^{-1})
ν_s	normalized surface potential
ρ	density ($kg\,m^{-3}$)
ρ	charge density ($C\,m^{-3}$)
ρ	resistivity ($\Omega\,m$)
σ	standard deviation of energy band width (J or eV)
σ	capture cross section (of a trapping centre) (m^2)
σ	electrical conductivity ($S\,m^{-1}$) or ($\Omega^{-1}\,m^{-1}$)
σ_{AC}	AC component of conductivity ($S\,m^{-1}$) or ($\Omega^{-1}\,m^{-1}$)
σ_D	conductivity in dark ($S\,m^{-1}$) or ($\Omega^{-1}\,m^{-1}$)
σ_{DC}	DC component of conductivity ($S\,m^{-1}$) or ($\Omega^{-1}\,m^{-1}$)
σ_e	extrinsic conductivity ($S\,m^{-1}$) or ($\Omega^{-1}\,m^{-1}$)
σ_i	intrinsic conductivity ($S\,m^{-1}$) or ($\Omega^{-1}\,m^{-1}$)
σ_n	conductivity due to electrons ($S\,m^{-1}$) or ($\Omega^{-1}\,m^{-1}$)
σ_n	capture cross section for electrons (m^2)
σ_L	conductivity in light ($S\,m^{-1}$) or ($\Omega^{-1}\,m^{-1}$)
σ_p	conductivity due to holes ($S\,m^{-1}$) or ($\Omega^{-1}\,m^{-1}$)
σ_p	capture cross section for holes (m^2)
σ_t	tunnelling conductivity ($S\,m^{-1}$) or ($\Omega^{-1}\,m^{-1}$)
τ	lifetime or relaxation time (s)
τ_b	carrier lifetime due to bulk scattering (s)
τ_f	carrier lifetime in thin film (s)
τ_{it}	interface trap lifetime or time constant (s)
τ_n	electron lifetime (s)
τ_p	hole lifetime (s)
τ_r	carrier (recombination) lifetime in depletion region (s)
τ_s	relaxation time for surface scattering (s)
ϕ	phase angle (radians)
ϕ_0	neutral level for trap distribution (J or eV)
ϕ_b	Schottky barrier height (J or eV)
ϕ_{b0}	Schottky barrier height at zero bias (J or eV)
ϕ_e	effective Schottky barrier height (J or eV)
ϕ_F	energy difference between semiconductor Fermi level and intrinsic Fermi level (J or eV)
ϕ_i	energy barrier due to insulator (J or eV)
ϕ_n	energy difference between Fermi energy and bottom of conduction band (for an n-type semiconductor) (J or eV)

ϕ_p	energy difference between Fermi energy and top of valence band (for a p-type semiconductor) (J or eV)
ϕ_t	effective energy barrier for tunnelling (J or eV)
$\Delta\phi$	image force lowering of Schottky barrier or energy offset (J or eV)
$\Delta\phi_m$	maximum image force lowering of Schottky barrier (J or eV)
Φ_m	metal work function (J or eV)
Φ_{phot}	luminous flux (lm)
Φ_{rad}	radiometric flux (W)
Φ_s	semiconductor work function (J or eV)
Φ_{ms}	difference between Φ_m and Φ_s (J or eV)
χ	electric susceptibility
X	electron affinity (J or eV)
X_A	electron affinity of acceptor molecule (J or eV)
X_D	electron affinity of donor molecule (J or eV)
X_i	electron affinity of insulator (J or eV)
X_s	semiconductor electron affinity (J or eV)
ψ	electron wave function
ψ_p	electrostatic potential (within a device) (V)
ψ_s	surface potential (V)
ω	angular frequency (radian s^{-1})
ω_c	cut-off frequency (radian s^{-1})

About the Companion Website

Electrical Processes in Organic Thin Film Devices is accompanied by a companion website:

www.wiley.com/go/petty/organic_thin_film_devices

The website includes:

- Solutions for the problems inside this book

1

Electronic and Vibrational States in Organic Solids

1.1 Introduction

Organic electronics concerns the electronic and optoelectronic properties of organic and biological materials [see, for example, 1, 2]. The field has attracted increasing attention over the last 30 years, from both academic researchers and commercial organizations. Many companies have now been established to exploit emerging ideas. Examples include light-emitting displays, domestic lighting, field effect transistors and memories, and organic solar cells. Moreover, electronics at the molecular level (electronics technologies based on single molecules or small groups of molecules) has been proposed as a disruptive technology that could have a significant impact on society as we move further into the twenty-first century.

Organic electronics relies on the key observation that metallic and semiconductive properties are not restricted to inorganic materials. For example, Figure 1.1 compares the electrical conductivity at room temperature for a variety of solid materials. The figures for organic polymers can extend over almost the entire spectrum of electrical conductivity, from insulating behaviour, as shown by diamond, to highly conducting, as for copper. An essential element in the academic and technological development of organic electronics is an understanding of the fundamental electrical behaviour of such materials and their associated device architectures. This book will address these issues and introduce the physical processes that play a role in the electrical conductivity of organic materials. Basic understandings of materials science (e.g. chemical bonding in solids), kinetic theory, and quantum mechanics are assumed.

Organic conductive materials comprise a wide range of different compounds that occur in various morphological forms. As well as the polymers noted in Figure 1.1, which can be semi-crystalline or amorphous, there are low molecular weight compounds such as organic dyes, charge-transfer complexes, and more 'exotic' substances, such as carbon nanotubes and graphene. Unfortunately, there is no single theory available that can account for the electrical conductivity in all these materials. The starting point for the discussion is therefore something that *is* reasonably well understood – band theory, which has been a great success story in terms of explaining the conductivity of inorganic crystalline materials. This approach will allow the introduction of fundamental concepts such as the density of electron energy states (energy levels), the population of these states, and chemical doping. The effects of bond disorder (as found in amorphous materials) are then discussed, and these ideas extended to encompass organic solids, which are held together by different types of chemical bonding.

The mechanical properties of materials can also play an important role in some electrical processes. A section on the physics of lattice vibrations (phonons) is therefore included in this introductory chapter.

Electrical Processes in Organic Thin Film Devices: From Bulk Materials to Nanoscale Architectures, First Edition. Michael C. Petty.
© 2022 John Wiley & Sons Ltd. Published 2022 by John Wiley & Sons Ltd.
Companion Website: www.wiley.com/go/petty/organic_thin_film_devices

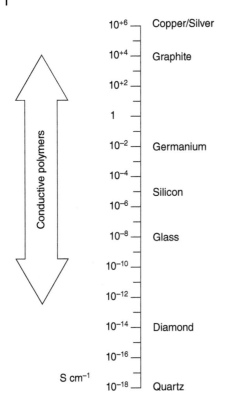

Figure 1.1 Range of conductivities for conductive polymers compared to various inorganic materials.

1.2 Band Theory for Inorganic Single Crystals

The classical approach to explaining the electrical behaviour of metals treats the electrons as a 'gas'. Kinetic theory is then applied to this electron gas. Accordingly, the electrons in a conductor possess a range of speeds, or energies, distributed about a mean value. The Maxwell–Boltzmann law describes the distribution of the electron energies. If the temperature is increased, the maximum and mean velocities of the electrons shift to higher values; at 0 K, the mean electron velocity becomes zero. As discussed in Chapter 2, this theory provides a useful insight into the phenomenon of electrical conductivity and provides order of magnitude values for the conductivity of some metals. However, the model fails to predict the correct temperature dependence of conductivity and does not explain why some materials are metals, others are good insulators, and some are of intermediate conductivity (semiconductors). One of the main problems with the classical approach is the postulate that the mean free path of the electrons is assumed to be of the same order as the lattice spacing. Mean free paths in metals are much larger ($\approx$50 nm in copper at 300 K). The band theory of solids, outlined in Section 1.2, addresses these issues.

When many isolated atoms are brought together to form a complete crystal, multiple splitting of the electron energy levels associated with the isolated atoms occurs. Figure 1.2 illustrates the

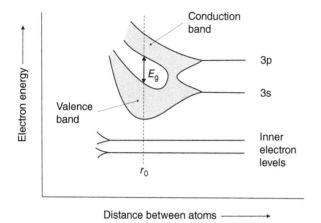

Figure 1.2 Energy levels in silicon as a function of inter-atomic spacing. The inner electron levels are completely filled with electrons. At the actual atomic spacing of the crystal r_0 the electrons in the 3s subshell and the electrons in the 3p subshell undergo sp^3 hybridization and all are accommodated in the lower valence band, while the upper conduction band is empty. The energy separation is the band gap E_g.

process of formation of a crystal of silicon and shows the electron energy levels as a function of the distance between the atoms; the equilibrium separation is shown as r_0. As the isolated atoms become closer together, their electron orbitals overlap, and discrete energy levels associated with the isolated atoms split to form bands of energies. A useful engineering analogy is that given by Kasap [3] for the resonant frequency in an resistor–inductor–capacitor (RLC) circuit. In isolation, the circuit will possess a characteristic resonance frequency (or energy). However, when two or more identical circuits are brought together, the circuits become coupled via mutual inductance, resulting in multiple resonant frequencies and a broad resonance band.

Isolated silicon atoms possess the electronic structure $1s^2 2s^2 2p^6 3s^2 3p^2$ in their ground state. As the s orbital is spherically symmetrical, it can form a bond in any direction. In contrast, the p orbitals are directed along mutually orthogonal axes and will tend to form bonds in these directions. Each atom has available two 1s states, two 2s states, six 2p states, two 3s states, six 3p states, and higher states. If N atoms are considered, there will be $2N$, $2N$, $6N$, $2N$, and $6N$ states of types 1s, 2s, 2p, 3s, and 3p, respectively. On decreasing the interatomic spacing decreases, these discrete energy levels split into bands beginning with the outer shell. As the 3s and 3p bands grow, these merge into a single band composed of a mixture of energy levels. This new band of 3s and 3p levels contains $8N$ available electron states. The chemical bonding in a silicon crystal consists of four sp^3 hybridized covalent bonds associated with each atom. As the distance between the atoms approaches r_0, the band splits into two, separated by a 'forbidden' energy gap, or band gap, E_g, shown in Figure 1.2. The upper band, called the conduction band (CB), contains $4N$ states, as does the lower-energy valence band (VB). In general, the total number of levels in a band must be the total number of levels in an individual atom multiplied by the total number of atoms in the solid. Thus, an s band will have $2N$ levels, a p band $6N$ levels, a d band $10N$ levels, and so on.

The $4N$ electrons in the original isolated silicon atoms ($2N$ in 3s states and $2N$ in 3p states) must occupy states in the valence or conduction bands. At 0 K, the electrons will occupy the lowest energy

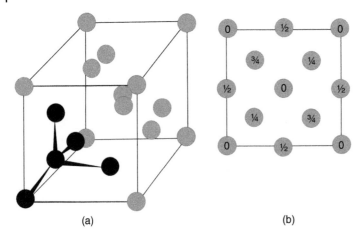

(a)   (b)

Figure 1.3 The sp^3 hybridized bonding in silicon results in a diamond crystallographic structure, based on two interpenetrating face-centred cubic (fcc) sub-lattices. (a) The highlighted bonds indicate the tetrahedral co-ordination for an individual silicon atom. (b) A plan view of the arrangement of silicon atoms in a unit cell; the numbers indicate the height of each atom above the base of the cube as a fraction of the cell dimension.

states available to them. For the silicon crystal, there are exactly $4N$ states in the valence band. Therefore, at 0 K, every state in the valence band will be filled, while the conduction band will be completely empty.

An estimate of the average energy level separation within a band can be obtained from the fact that there are approximately 10^{26} atoms (Avogadro's number, N_A, $= 6.02 \times 10^{26}$ kmol^{-1}) in a macroscopic (kg) quantity of a crystal. Energy bands are typically 1 eV wide, giving an energy separation of 10^{-26} eV. To a first approximation, therefore, the variation in electron energy across a band can be treated as being continuous.

The sp^3 hybridized bonding in silicon results in a diamond crystallographic structure shown in Figure 1.3, which is based on two interpenetrating face-centred cubic (fcc) sub-lattices; the cube side for silicon is 0.543 nm. In Figure 1.3a, the highlighted bonds indicate the nearest neighbour bonds and reveal the tetrahedral coordination for an individual silicon atom. The schematic diagram in Figure 1.3b shows the arrangement of silicon atoms in a unit cell, with the numbers giving the height of the atom above the base of the cube as a fraction of the cell dimension.

1.2.1 Schrödinger Wave Equation

Fundamental to quantum mechanics is the notion that electrons possess a wave nature. Erwin Schrödinger argued that it should be possible to represent these electron waves mathematically. Such a wavefunction will depend on position and time and can be represented in one dimension by $\psi(x, t)$. The probability of finding the electron per unit length at x at time t is $|\psi(x, t)^2|$, whereas the probability of finding the electron between $x + dx$ at time t is $|\psi(x, t)^2|dx$.

The Schrödinger wave equation is a key relationship involving the electron wavefunction. In essence, it is a statement of the conservation of energy, i.e. that the total of the kinetic energy, E, and potential energy, V, is a constant. It is useful to draw on an analogy with the classical case. Here, the appropriate equation can be written as:

$$\frac{1}{2m}p^2 + V = E \tag{1.1}$$

i.e.

kinetic energy + potential energy = total energy

where m, p are mass and momentum, respectively.

In the quantum mechanical formulation, for one dimension and for a potential energy that is only dependent on space, i.e. $V = V(x)$, the relevant equation becomes

$$\frac{d^2\psi}{dx^2} + \left(\frac{2m}{\hbar^2}\right)\psi(E - V(x)) = 0 \tag{1.2}$$

where $\hbar$ is the reduced Planck's constant, which equals Planck's constant (6.63×10^{-34} J s) divided by 2π. Essentially, the classical quantities of Eq. (1.1) have been replaced by quantum mechanical operators, which operate upon the wavefunction (representing a particular state of the system) to extract observable value quantities such as position, momentum, and energy. For example, the momentum p in Eq. (1.1) is replaced by the momentum operator $\frac{\hbar}{j}\frac{d}{dx}$, where $j = \sqrt{-1}$. Equation (1.2) is called the time-independent Schrödinger wave equation. Many of the problems in solid-state physics are concerned with solving this equation for various forms of the potential energy. Once the wavefunction has been determined, then the probability distribution and energy of the electron can be determined from ψ^2. The solutions for ψ are known as eigenfunctions (characteristic functions) and the corresponding energies are called eigenenergies.

1.2.2 Density of Electron States

Simple quantum mechanics can be used to predict how the number of electron energy levels, or energy states, varies with energy E. This is given by a function called the density of states, $S(E)$. To find out how many ways there are to obtain a particular energy in an incremental energy range dE, the quantum mechanical 'particle in a box' approach may be used to solve the Schrödinger wave equation [1–4]. This gives the energy of an electron in a three-dimensional cubic potential well of side L as follows:

$$E = \frac{h^2}{8m_eL^2}\left(n_x^2 + n_y^2 + n_z^2\right) \tag{1.3}$$

where m_e is the electron mass (9.11×10^{-31} kg) and n_x, n_y, and n_z are positive integers (quantum numbers) with values 1, 2, 3, ... According to the Pauli Exclusion Principle, each combination of n_x, n_y, and n_z will correspond to one electron orbital state. However, each of these orbital states can contain two electrons (one with spin 'up' and the other with spin 'down'). To work out the total number of electron states up to a defined energy E, consider the diagram shown in Figure 1.4, which depicts an octant of a sphere containing all possible positive values of n_x, n_y, and n_z. For large n values, the number of possible states is proportional to the volume of this octant. The radius of the sphere R in n-space is given by

$$R = \sqrt{\left(n_x^2 + n_y^2 + n_z^2\right)} \tag{1.4}$$

The total number of electron states N is therefore,

$$N = 2\left(\frac{1}{8}\right)\frac{4}{3}\pi R^3 \tag{1.5}$$

The factor of 2 at the beginning of Eq. (1.5) accounts for the fact that each electron orbital state contains two electrons, while the 1/8 arises as only positive n values (octant of the sphere) are considered.

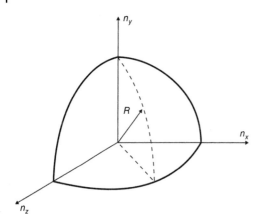

Figure 1.4 Octant of a sphere, radius R, containing all possible values of quantum numbers n_x, n_y, and n_z corresponding to a defined energy E. For large n values, the number of electron states will be proportion to the volume of this octant.

Using Eqs. (1.3) to (1.5) and dividing by volume gives an expression for the number of states per unit volume n:

$$n = \frac{N}{L^3} = \left(\frac{8\pi}{3}\right)\frac{(2m_e E)^{3/2}}{h^3} \tag{1.6}$$

The density of states function $S(E)$ can be obtained by differentiating Eq. (1.6) with respect to energy:

$$S(E) = \frac{dn}{dE} = (8\pi 2^{1/2})\left(\frac{m_e}{h^2}\right)^{3/2} E^{1/2} \tag{1.7}$$

The density of states function represents the number of electron states per unit volume per unit energy at energy E (Problems 1.1 and 1.4). Therefore, $S(E)dE$ represents the number of electron states (i.e. wavefunctions) in the energy interval E to $(E + dE)$ per unit volume. This expression can be applied to bulk three-dimensional materials and is independent of the dimension L. However, if the dimensionality is changed, the density of states function will also change. This is discussed further in Chapter 3.

1.2.3 Occupation of Energy States

The classical approach to electrical conductivity treats electrons as indistinguishable particles having velocities or energies governed by the Maxwell–Boltzmann distribution law. No restriction is placed on the number of electrons that can possess any value of energy. The quantum mechanical approach, on which band theory is based, postulates that the valence electrons in a metal obey Fermi–Dirac quantum statistics. In accordance with the Pauli Exclusion Principle, no two electrons in an atom can possess identical quantum numbers and must be regarded as indistinguishable particles. The Fermi–Dirac distribution function $f(E)$ governing the occupation of the electron states (or electron levels) over energy, at a temperature T, may be shown to be

$$f(E) = \frac{1}{\exp((E - E_F)/k_B T) + 1} \tag{1.8}$$

where k_B is the Boltzmann constant (1.38×10^{-23} J K^{-1}), $f(E)$ is the probability of a level at energy E being filled with an electron, and E_F, which has the dimensions of energy, is termed the Fermi level or Fermi energy (Problem 1.2).

When $E = E_F$, then from Eq. (1.8), $f(E) = 0.5$, i.e. the Fermi level is the energy at which there is a 50 : 50 chance of finding an electron. Figure 1.5 shows how the occupation of electron states

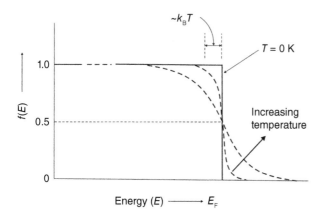

Figure 1.5 The Fermi–Dirac function $f(E)$ at $T = 0\,\text{K}$ and at $T > 0\,\text{K}$.

$f(E)$ varies with energy. At $T = 0\,\text{K}$, all energy levels below E_F are filled ($f(E) = 1$), while the levels are completely empty ($f(E) = 0$) above E_F. Thus, at absolute zero, the Fermi energy represents the demarcation between filled and empty states. At higher temperatures, the Fermi–Dirac distribution function becomes less step-like and some of the levels below E_F become depopulated and some above become populated. However, the effect is quite small except at high temperatures. The function $f(E)$ drops from unity to zero as the energy increases by a few $k_\text{B}T$. Since the value of $k_\text{B}T$ is only 0.026 eV at 300 K, the transition energy range is a very narrow one.

For values of energy that are more than a few $k_\text{B}T$ from the Fermi level (i.e. $(E - E_\text{F})/k_\text{B}T \gg 1$), Eq. (1.8) reduces to

$$f(E) = \exp\left(\frac{E_\text{F}}{k_\text{B}T}\right)\exp\left(-\frac{E}{k_\text{B}T}\right) \tag{1.9}$$

which is the Maxwell–Boltzmann law.

The number of occupied electron states within an incremental energy range can be found by multiplying the density of states function (Eq. (1.7)) by the probability of their occupancy:

$$N(E)dE = S(E)f(E)dE \tag{1.10}$$

where $N(E)dE$ is the number of occupied states in the energy range E to $E + dE$.

In the case of a metal, the Fermi energy at 0 K, denoted E_{F0}, can be shown to be [3]

$$E_{\text{F0}} = \left(\frac{h^2}{8m_\text{e}}\right)\left(\frac{3n}{\pi}\right)^{2/3} \tag{1.11}$$

Values of several electron volts per electron are typical, e.g. 7 eV for Cu, 5.53 for Au, 11.7 eV for Al. Furthermore, the Fermi energy at temperature T can be derived:

$$E_\text{F}(T) = E_{\text{F0}}\left(1 - \frac{\pi^2}{12}\left(\frac{k_\text{B}T}{E_{\text{F0}}}\right)^2\right) \tag{1.12}$$

As $E_{\text{F0}} \gg k_\text{B}T$, the Fermi energy will only weakly depend on temperature.

1.2.4 Conductors, Semiconductors, and Insulators

Every solid has its own characteristic energy band structure. After the allowed and forbidden energy regions have been determined, the occupation of the available energy levels is decided by

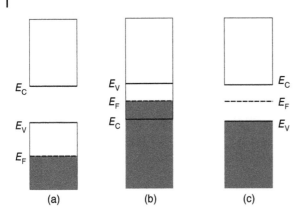

Figure 1.6 Possible energy band structures of crystalline solids. (a) Energy gap exists between the conduction and valence bands. The valence band is partly filled. (b) Overlapping valence and conduction bands. (c) Energy gap exists between the valence and conduction bands. The valence band is completely filled. Figures (a) and (b) represent the band structures of metals, while (c) is the band structure for an insulator. E_v = top of valence band. E_c = bottom of conduction band. E_F = Fermi energy.

the Fermi–Dirac distribution function (Eq. (1.8)). At absolute zero, all the states lying below the Fermi level are filled, and all those above are empty.

Either the energy bands overlap, or they do not, and the Fermi level may lie within one of the bands or in the forbidden energy region. This leads to essentially three cases, which are illustrated in Figure 1.6.

For conductivity to be possible, there must be available in the same band, energy states that are occupied and others that are empty. This is clearly the case for Figures 1.6a,b; these represent the energy band structures for metals. In Figure 1.6a, E_F lies within the valence band while the band arrangement depicted in Figure 1.6b represents the case in which the valence and conduction bands overlap, providing a continuum of electron energy levels. Metals generally have bands that are only half-filled. For example, in a piece of sodium comprising N atoms, there will be $2N$ available states in the uppermost, occupied energy band. As sodium has one valence electron (Group IA in the Periodic Table), there are only N electrons to be accommodated in the band, which is therefore half full.

In Figure 1.6c, the filled valence band (the highest energy of which is represented E_v in the diagram) is separated in energy from the next highest band, the conduction band (the lowest energy of which is represented E_c in the diagram). Conductivity is not possible, and the situation corresponds to that of an insulator. If a valence band electron were able to move because of an applied field, it would take energy from the field and change its energy state, i.e. move to a higher energy level in the valence band. This is not feasible since no higher energy levels are available for the electron to move into without violating the Pauli Exclusion Principle.

The forbidden energy gap in a crystal may be quite narrow. At a sufficiently high temperature, some of the valence electrons can gain enough thermal energy to transfer to states lying in the lowermost empty band, the conduction band. When an external electric field is applied, these electrons can gain additional energy from the field, and conduction is possible. The currents produced in such crystals are necessarily small, and these materials are distinguished from metallic conductors and true insulators by being called semiconductors. The energy gaps in inorganic semiconductors are mostly in the range 0.1–3 eV at room temperature (for Si, $E_g = 1.1$ eV; for GaAs, $E_g = 1.4$ eV).

Note that the conductivity of a semiconductor will increase as the temperature increases since more valence electrons can transfer to empty states lying in the conduction band. This is exactly opposite to the temperature dependence of conductivity in metals in which the resistivity increases with increasing temperature. The conductivity of a semiconductor will also be affected by radiation, i.e. it will be photoconductive. Incident radiation with energy equal to or greater than the band gap will be absorbed, thereby promoting electrons from the valence band to the conduction band and increasing the conductivity.

1.2.5 Electrons and Holes

The electrons in a crystal, such as silicon, are not completely free, but instead interact with the periodic potential of the lattice. As a result, their motion cannot be expected to be the same as for electrons in free space. In applying the usual equations of electrodynamics to charge carriers in a solid, it is reasonable to expect the electron mass will not be identical to that in free space. The classical relationship between electron kinetic energy, E, and momentum p may be modified by replacing the free electron mass m_e with an effective mass m_e^*:

$$E = \frac{p^2}{2m_e^*} \tag{1.13}$$

(The above equation can be derived using the relationships $E = \frac{1}{2}m_e v^2$ and $m_e v = p$.)

The effective mass will be related to the second derivative of the $E - p$ curve:

$$\frac{d^2 E}{dp^2} = \frac{1}{m_e^*} \tag{1.14}$$

An important parameter that is introduced in solid-state physics is the wavevector (or propagation constant) of the electrons, usually denoted by k (measured in units of m^{-1}). This is related to the wavelength λ of the electron wave by

$$k = \frac{2\pi}{\lambda} \tag{1.15}$$

The wavevector is also proportional to the electron momentum:

$$k = \frac{p}{\hbar} \tag{1.16}$$

The wave nature of particles such as electrons cannot be adequately described by a sinusoidal function with single angular frequency ω ($= 2\pi f$, where f is the frequency in Hz). Instead, a group or packet of many waves, spread over a small distance, represents the particle; each of the waves has a slightly different frequency and velocity to the other members of the group. The waves within the group interfere with one another to give a localized entity that moves through space with a wavelength λ and a group velocity, v_g, given by

$$v_g = \frac{d\omega}{dk} \tag{1.17}$$

The group velocity represents the speed at which the wave packet, or particle, propagates.

In contrast, the velocity of a single-frequency wave, known as the phase velocity, v, is simply:

$$v = \frac{\omega}{k} \tag{1.18}$$

This is the speed at which the crests (or troughs) of a wave move through space. In certain circumstances (e.g. high-frequency X-rays), it is possible for this velocity to exceed the speed of light. Only in the case of a linear connection between ω and k will $v = v_g$. The relationship between ω and k is known as a dispersion relation.

Using Eqs. (1.14) and (1.16), the electron effective mass may be defined:

$$m_e^* = \frac{\hbar^2}{d^2 E / dk^2} \tag{1.19}$$

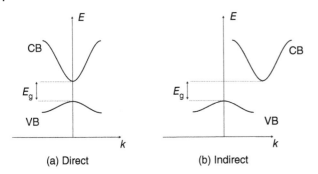

Figure 1.7 Electron energy E versus wavevector k diagrams. (a) Direct band gap semiconductor. (b) Indirect band gap semiconductor. Band gap $= E_g$. VB = valence band. CB = conduction band.

Therefore, the electron's mass is determined by the curvature of the energy versus wavevector curve. Figure 1.7 illustrates the energy versus wavevector diagrams for two important types of semiconductor energy bands. The curves are shown in the so-called 'reduced wavevector representation' in which the $E - k$ curves for higher bands are shown in same region of 'k-space' as the lowest energy band (this is possible as the electron energy within a particular band is a periodic function of the wavevector). The different curvatures of the $E - k$ curves close to the top of the valence band (VB in the figure) and near to bottom of the conduction band (CB) are evident. In Figure 1.7a, the conduction band minimum and the valence band maximum occur at the same value of k. This is termed a direct band gap semiconductor; GaAs is a typical example. In contrast, an indirect bandgap semiconductor arises if the minimum of the conduction band occurs at a different k value from the maximum of the valence band, Figure 1.7b. Silicon possesses an indirect energy gap. In this case, a transition of an electron from the valence band to the conduction band requires a change in wavevector (momentum).

Near the bottom of a band shown in Figure 1.7, $d^2 E/dk^2$ is a positive quantity and so is the effective mass. However, near the top of a band, the curvature is such that the effective mass of the electrons is now negative. A negative effective mass implies that if electrons near the top of the valence band have their energies increased by the application of an electric field, then this results in a reversal of their momenta (the momentum transfer from the crystal lattice to the electron is opposite and larger than the momentum transfer between the applied field and the electron).

For nearly filled bands, such as the valence band, it is much easier to deal with the behaviour of vacant sites rather than that of electrons. To achieve this, the electron effective mass, which is negative near the top of a band, is replaced by a hole effective mass, which is positive. If there is an electron missing from a given state, then the state is deemed to be occupied by a positive charge. Holes lower in the valence band possess a greater energy and it follows that representation of the hole energy in a band must be opposite to that of the electron energy. Valence band electrons with negative charge and negative mass move in an electric field in the same direction as holes with a positive charge and positive mass. Charge transport in the valence band can therefore be fully accounted for by considering hole motion.

Streetman provides a very useful analogy for the conduction by electrons and holes in their respective bands [4]. Suppose there are two bottles, one filled with water (valence band) and one empty (conduction band). When the bottles are tilted (electric field applied), will there be any net movement (transport) of water (electrical conduction)? In the case of the empty bottle, the answer is obviously no. For the filled bottle, there cannot be any motion because there is no empty space (no vacant electron energy levels) for the water to flow into. However, the situation changes if a small

Figure 1.8 Model for intrinsic conduction. Thermal excitation of electrons across the band gap E_g produces an increase in electron population of the conduction band and an increase in the hole population of the valence band.

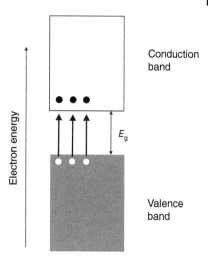

amount of water is transferred from the filled bottle to the empty bottle. Now when the bottles are tilted, water will flow downhill in one bottle and the air bubbles will move uphill in the other bottle. Similarly, a few electrons in an otherwise empty conduction band move in one direction on application of an electric field and the holes in an otherwise filled valence band move in the opposite direction. Figure 1.8 illustrates the process of electrons gaining energy and moving to the conduction band, leaving holes in the valence band.

1.2.6 Doping

As noted earlier, thermal or optical processes can affect the conductivity of a semiconductor. Heating a pure semiconductor such as Si from 0 K will produce equal numbers of electrons in the conduction band and holes in the valence band. This process is equivalent to breaking some of the covalent bonds in the silicon crystal. The vacant states or holes that are produced in the valence band correspond to incomplete bonds. On the application of an electric field, both the electrons and holes contribute to the resulting electric current. The conductivity is referred to as intrinsic. An intrinsic semiconductor is therefore a material in which the electronic properties (such as the conductivity) are not determined by (electrically active) impurities. The necessary level of purity depends upon the band gap and on the temperature. The semiconductor germanium, which has a band gap of 0.67 eV at room temperature, is intrinsic provided that the impurities are present at less than about 1 part in 10^{10}. On the other hand, at the same temperature, intrinsic silicon with the larger band gap of 1.1 eV would require a purity level of better than 1 part in 10^{13}.

At room temperature, the value of the thermal energy $k_B T$ is approximately 0.025 eV and the electrons in the crystal receive energies distributed about this mean value. Since the band gaps of semiconductors lie in the range 0.1–3 eV, the proportion of electrons excited into the conduction band will be very small. This will be particularly so for semiconductors having the larger energy gaps.

Whereas the conductivity of an intrinsic semiconductor is a function only of temperature, that of an extrinsic semiconductor is determined by the impurity content. The process of adding small amounts of impurity to control the electrical conductivity of semiconductors is called doping. In an extrinsic semiconductor, a specific impurity element is added in controlled amounts; its concentration determines the conductivity of the sample and the type of carriers present. When these are electrons, i.e. negative carriers, the sample is termed n-type; when these are holes, positive carriers, the sample is p-type.

A semiconductor such as silicon has tetrahedral covalent bonding in which each atom shares its four valence electrons with each of its four nearest neighbours. If a small amount of a Group V element, such as phosphorus, is incorporated into a silicon crystal, the phosphorus impurity atom is substituted for a silicon atom on the lattice so that it is tetrahedrally bound to its nearest neighbours. In the free atom state, phosphorus has two 3s electrons and three 3p electrons so that in the substitutional alloy the atom has one surplus electron over and above that necessary to form the covalent bonds. At very low temperatures, close to 0 K, this electron is loosely bound to its parent atom due to the Coulombic attraction of the nucleus. If the temperature of the semiconductor is increased, this electron is released from the attraction of the parent impurity atom by thermal energy and is then free to move throughout the crystal. Each impurity atom gives rise to one electron in the conduction band, and such atoms are called donors. With a suitable number of donor atoms in a sample, the electron concentration may be many orders of magnitude greater than that of an intrinsic sample.

The band representation of an n-type semiconductor is shown in Figure 1.9. The donor energy level, E_d, is located below and close to the bottom of the conduction band. At room temperature, each donor atom is ionized and provides an electron in the conduction band; the ionization energy is shown as ΔE_d. In addition to phosphorus, the other Group V elements, arsenic, antimony, and bismuth also behave as donor impurities in silicon.

The isolated atoms of the Group III elements – boron, aluminium, gallium, and indium – have outermost electronic shells of two electrons in s states and one electron in a p state. Consequently, when one of these elements is added to silicon, there is a deficiency of one electron for tetrahedral bonding of the impurity atom into the crystal lattice. At very low temperatures, this missing bond is localized in the Coulombic field of the impurity atom. At ordinary temperatures, however, the impurity atom, which is called an acceptor, receives an electron from another atom in the crystal and the missing bonding electron, or hole, becomes mobile. This p-type band representation is shown in Figure 1.10; the acceptor energy level is E_a and ionization energy ΔE_a.

A semiconductor may also contain simultaneously both donor- and acceptor-type impurity atoms. When the donor and acceptor concentrations are approximately equal, the material is referred to as being compensated. In other cases, the conductivity is governed by the net excess

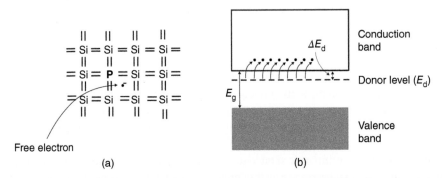

Figure 1.9 Donor impurity, such as phosphorus, in silicon. (a) A phosphorus atom substitutes for a silicon atom in the lattice, leaving a free electron. (b) Energy band diagram showing the formation of a donor level E_d with activation energy ΔE_d associated with the phosphorus impurity and located close to the conduction band of the silicon.

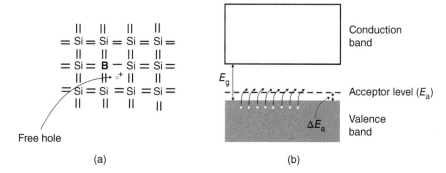

Figure 1.10 Acceptor impurity, such as boron, in silicon. (a) A boron atom substitutes for a silicon atom in the lattice, leaving a free hole. (b) Energy band diagram showing the formation of an acceptor level E_a with an activation energy ΔE_a associated with the boron impurity and located close to the valence band of the silicon.

impurity of one type. For silicon, when an impurity element from a group other than III or V is present it may give rise to one or more separate levels lying within the forbidden energy gap and such levels may lie deep within the gap.

The above principles may be applied to the doping of other inorganic semiconductors, including compound semiconductors such as GaAs. In such compounds, a departure from a strict stoichiometric composition (i.e. Ga:As = 1:1) leads to either an n-type or p-type sample depending on which element is present in the greater concentration. The III–V semiconductor gallium arsenide can also be doped n-type by adding tellurium from Group VI as a donor or p-type by adding zinc from group II as an acceptor.

The impurity concentrations in extrinsic inorganic semiconductors are minute, typically parts per million. Consequently, the impurity atoms are not sufficiently close for their electron orbitals to overlap. No energy bands are formed, and the impurities are associated with discrete or localized energy levels within the band gap of the semiconductor, E_D and E_A, in Figures 1.9 and 1.10, respectively. The probability of finding an electron is localized on the impurity site.

1.3 Lattice Vibrations

The electrical and mechanical properties of solids are intricately related. For example, it is well established that the increase in the resistivity of a metal with increasing temperature results from the scattering of electrons by vibrating atoms. Classically, the bonds between atoms can be thought of as mechanical springs. As the atoms vibrate about their equilibrium positions, these springs are being stretched and compressed, and the vibrations are coupled to neighbouring atoms. For a linear array of atoms, both longitudinal and transverse waves are possible. The displacement of the atoms from their mean positions as these waves propagate along the linear array with lattice spacing a are shown in Figure 1.11. For longitudinal waves (L wave in Figure 1.11b), the atomic vibrations are parallel to the direction of propagation. In contrast, the atoms are displaced sideways as a transverse wave propagates, i.e. the atom displacement is orthogonal to the propagation direction (T wave in Figure 1.11c).

The energy possessed by a lattice vibration of angular frequency ω will be quantized and the quantum $\hbar\omega$ is a known as a phonon. The angular frequency of the phonons and the wavevector

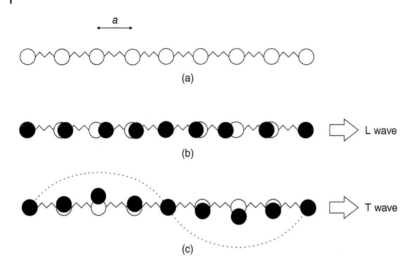

Figure 1.11 (a) A linear chain of atoms with spacing a in the absence of vibrations. (b) A snapshot of a travelling longitudinal (L) wave. Displacements are parallel to the atomic chain. (c) A snapshot of a transverse (T) wave propagating along the atomic chain. The displacements of the atoms are perpendicular to the atomic chain.

of the lattice k are related by

$$\omega = 2\left(\frac{\beta}{M}\right)^{1/2}\left|\sin\left(\frac{1}{2}\right)ka\right| \tag{1.20}$$

where β is a force (or spring) constant and M is the mass of an atom. This relationship between ω and k is the mechanical equivalent of the dispersion relation for electrons. Figure 1.12 shows the dispersion curves for both longitudinal and transverse lattice vibrations. At low values of lattice wavevector (long wavelength limit) both curves are linear, and the lattice waves will travel in the material with the velocity of sound appropriate for the wave. However, Eq. (1.20) reveals that there will be no frequencies higher than a maximum cut-off frequency, ω_c, which is given by

$$\omega_c = 2\left(\frac{\beta}{M}\right)^{1/2} \tag{1.21}$$

Figure 1.12 identifies different cut-off frequencies for the longitudinal, ω_{cl}, and transverse, ω_{ct}, phonons. These will depend on the precise nature of the interatomic bonding and the crystal structure (Problem 1.3). The cut-off frequencies will occur when $ka/2 = \pi/2$, i.e. when $k = \pi/a$. The dispersion relation of Eq. (1.20) is periodic, with a period $2\pi/a$. Only k values in the range $-\pi/a < k < \pi/a$ are physically meaningful.

If the linear atomic array depicted in Figure 1.11 is changed, so it contains two types of atoms with different masses arranged alternately along its length, additional vibrational modes are possible. Figure 1.13 shows that, as well as the normal transverse mode, there is a further vibrational mode in which the displacements of dissimilar atoms are always opposite to one another. Waves of this type, termed optical modes, can be excited by long wavelength infrared radiation in ionic crystals since the two dissimilar atoms in the crystal carry opposite electric charges and will therefore be displaced in opposite directions by the transverse electric field of the electromagnetic radiation.

Figure 1.14 shows the dispersion curves for both transverse acoustic and optical modes. The excitation of the optical mode is most pronounced for $k \to 0$, which corresponds to a nonzero angular frequency, ω_0. The optical and acoustic branches are separated by a forbidden band of frequencies.

Figure 1.12 Angular frequency ω versus wavevector k curves (dispersion curves) for longitudinal and transverse modes in a linear array of atoms. The different force constants for the two types of vibrations give rise to different cut-off frequencies ω_{cl} and ω_{ct} at $k = \pi/a$.

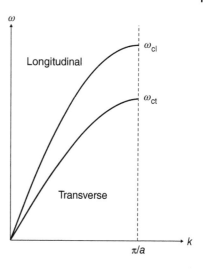

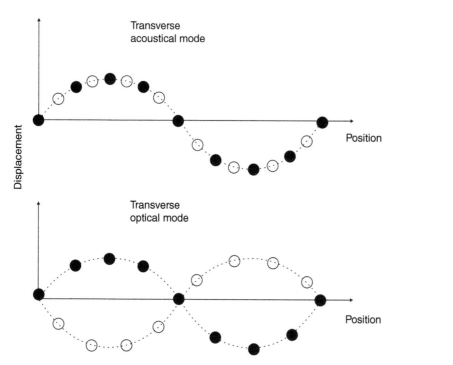

Figure 1.13 Snapshots of displacement of atoms in a linear array comprised of alternate atoms of different masses (depicted by the open and closed symbols) in transverse acoustical and transverse optical modes.

In this respect, Figure 1.14 bears some similarity to the energy versus wavevector curves for electrons shown in Figure. 1.7. Both are examples of a dispersion relationship: Figure 1.7 for electrons and Figure 1.14 for phonons. If the two atom masses are very different, the transverse optical mode is very flat, i.e. it covers a narrow frequency range. The introduction of the two types of atoms has also doubled the periodicity of the dispersion diagram and all possible frequencies are now contained for k over the range $-\pi/2a < k < \pi/2a$.

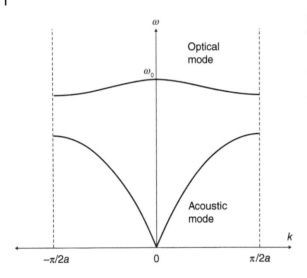

Figure 1.14 Dispersion (ω versus k) curves for the acoustical and optical modes for a linear array with alternate atoms of different masses. Note that the limiting cut-off frequencies of the waves now appear at $k = \pi/2a$ and that the limiting frequency ω_0 for the optical branch is not zero for $k = 0$.

An important phenomenon in inorganic semiconductors is that the drift velocity of the electrons (the electron velocity resulting from an applied electric field) is limited by scattering from lattice vibrations at high temperatures. In the case of organic semiconductor crystals, known as molecular crystals, two types of molecular oscillations may be distinguished. First, the individual molecules may oscillate with respect to one another – inter-molecular phonons. In contrast, within each molecule, the constituent atoms may oscillate. Such intra-molecular vibrations are known as vibrons. There can be a strong coupling between electronic excitations and these vibrons.

1.4 Amorphous Inorganic Semiconductors

The energy band model for semiconductors is a consequence of the ordered arrangement of the atoms on a lattice, i.e. a crystalline structure. Interesting questions then arise: what happens if the regular periodicity of the lattice is disrupted? Will the material still possess a recognizable band structure? The atoms in an amorphous solid are often arranged in a continuous random network. Amorphous pure silicon contains numerous dangling bonds (an unsatisfied bond in terms of its number of electrons), like those found at the surface of single crystal silicon. However, so long as the short-range order present in the crystalline phase is essentially unchanged (similar bond lengths, bond angles and local co-ordination) the main features of the density of states function is preserved. The overall result is that energy bands in amorphous materials are generally less defined than in their crystalline counterparts. Tails in the density of states function can extend, usually exponentially as a function of energy, into the band gap, as depicted in Figure 1.15a, and localized states can be found within the band gap [5]. For certain materials, the tails may spread so far into the energy gap that they partially overlap. The concept of mobility edges (carrier mobility is discussed in more detail in Chapter 2) within the band tails is introduced for amorphous materials. These are associated with the critical energies separating localized from extended states. The mobility edge is defined as the energy at which charge transport loses coherence. At energies below the mobility edge, the charge carriers are localized and need to be thermally excited to delocalized sates. Dangling bonds act as additional localized states that can influence the transport and recombination of charge carriers. The difference between the energies of the mobility edges is called the mobility

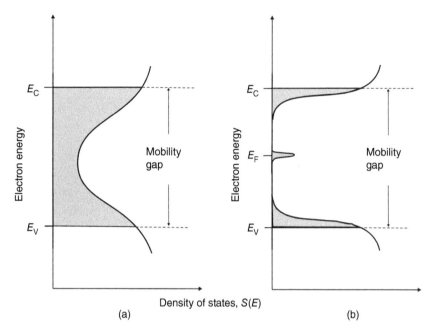

Figure 1.15 Electronic energy versus density of states diagrams for an amorphous semiconductor. (a) Overlapping conduction and valence band states. The shaded area represents localized states. A mobility gap separates the extended conduction and valence band states from the localized states. (b) Non-overlapping conduction and valence bands with localized states and localized defect states located near the Fermi energy. Source: Part (a) Mott and Davis [5].

gap, as depicted in Figure 1.15a. For amorphous inorganic semiconductors, this parameter is often used in preference to the energy gap.

Amorphous semiconductors, such as amorphous silicon (or α-Si), can also be doped to modify their electrical properties. This can lead to an additional band (or bands) of electron states within the mobility gap, as shown in Figure 1.15b. A high concentration of such levels can result in a Fermi level that is fixed, or pinned, to the defect energy levels.

1.5 Organic Semiconductors

While band theory might be considered as a reasonable starting point to explain the electrical behaviour of organic materials, this approach has several problems [2]. A key issue is that the permittivity – a macroscopic manifestation of atomic or molecular polarizability (Chapter 7) – is relatively low for organic materials. For instance, the relative permittivity for crystalline silicon is 12, compared to values for organic compounds, which are generally between 3 and 4. The permittivity is a measure of how a material responds to an applied electric field. For inorganic crystalline semiconductors, the dielectric screening provided by their higher permittivity is efficient so that the interactions between electrons are not important. However, these interactions cannot be neglected in organic semiconductors. This leads to the importance of entities known as excitons rather than free charge carriers (electrons and holes). Excitons may be approximated as a Coulomb-bound electron–hole pair.

Organic semiconductors can also exhibit very different morphologies to their inorganic counterparts – ranging from organic molecular single crystals, or molecular crystals, to semi-crystalline or amorphous polymers. Graphene, fullerenes, and carbon nanotubes are exceptional materials, in terms of their electronic behaviour and physical forms. Common groups of organic semiconductors are discussed in Sections 1.5.1–1.5.6 and the applicability, and limitations, of band theory are explored.

1.5.1 Bonding, Orbitals, and Energy Bands in Organic Compounds

Carbon has an atomic number of six and is found, directly above silicon, in Group IVB of the Periodic Table. Its electron configuration is $1s^2$, $2s^2$, $2p^2$, i.e. the inner s shell is filled and the four electrons available for bonding are distributed two in s orbitals and two in p orbitals. Carbon is a unique electronic material. It can be a good conductor in the form of graphite, an insulator in the form of diamond, or a flexible polymer (conductive or insulating) when reacted with hydrogen and other species. Carbon differs from other group IV elements, such as Si and Ge, that exhibit sp^3 hybridization. Carbon does not have any inner atomic orbitals except for the spherical 1s orbital, and the absence of nearby inner orbitals facilitates the formation of other hybridized orbitals – sp and sp^2 hybridizations involving only the valence (outer) s and p orbitals. The fact that such hybridizations do not readily occur in Si and Ge is related to the absence of 'organic materials' made from these elements.

In the simplest case, the carbon 2s orbital hybridizes with a single p orbital. Two sp hybrids result by taking the sum and difference of the two orbitals, shown in Figure 1.16, and two p orbitals remain. The sp orbitals are constructed from equal amounts of s and p orbitals; they are linear and 180° apart. In the case of three groups bonded to a central carbon atom, three equivalent sp^2 hybrids may be constructed from the 2s orbital and two p orbitals (e.g. a p_x and a p_y). Each orbital is 33.3% s and 66.7% p. The three hybrids, shown in Figure 1.17, lie in the xy plane (the same plane defined by the two p orbitals) directed 120° from each other, and the remaining p orbital is perpendicular to the sp^2 plane. As in the case of silicon, four sp^3 hybrids may be derived from an s orbital and three p orbitals. These are directed to the corners of a tetrahedron with an angle between the bonds of 109.5°; each orbital is 25% s and 75% p (Figure 1.3).

In the organic molecule ethylene, C_2H_4, each of the two carbons is attached to just three atoms. The apparent deficiency in the bonds is avoided by the joining of two carbon atoms by two bonds, or what is more properly called a double bond. The chemical structure is often noted as $CH_2=CH_2$. The bonding for hydrocarbons with carbon–carbon double bonds (alkenes) and triple bonds (alkynes) involve sp^2 and sp hybrids, respectively. In ethylene, two sp^2 hybrids on each carbon bond with the hydrogens. A third sp^2 hybrid on each carbon forms a $C(sp^2)$–$C(sp^2)$ single bond, leaving a p orbital 'left over' on each carbon. This orbital lies perpendicular to the plane of the six atoms. The two p orbitals are parallel to each other and have regions of overlap above and below the molecular plane. This type of bond in which there are two sideways bonding regions above and below a nodal plane

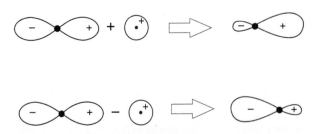

Figure 1.16   Mathematical combination of s and p orbitals to yield two sp hybridized orbitals.

Figure 1.17   Three sp^2 hybridized orbitals.

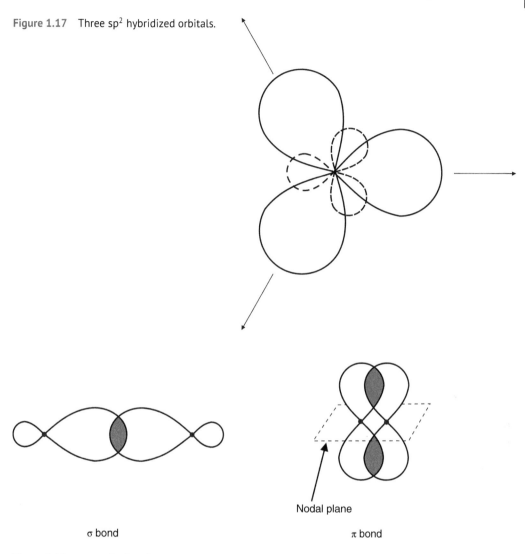

σ bond   π bond

Nodal plane

Figure 1.18   σ – and π-bonds.

is called a pi-or π-bond. In contrast, the bond formed by the head-on overlap of the two carbon sp^2 orbitals is called a σ-bond. Pi-bonds are the result of overlap of dumbbell-like p orbitals lying adjacent to each other, shown in Figure 1.18. One consequence of this structure is that the orbitals of the π-bond act like struts to stop the atoms at each end from rotating; the molecule stays flat.

The C=C double bond distance in ethylene is 0.133 nm, less than the value of 0.154 nm given above for a C—C single bond, while the C—H bond is 0.108 nm long. Strengths of the C=C and C—H bonds in ethylene are 6.61 eV (152 × 10^3 kcal kmol^{-1}) and 4.48 eV (103 × 10^3 kcal kmol^{-1}), respectively. The noncylindrical symmetric electron density about the C=C bond axis in ethylene, Figure 1.19, and related compounds leads to a barrier to rotation about this axis. Therefore, two isomers exist. (Isomeric compounds are those that possess identical chemical formulas but differ in the nature or sequence of bonding of their atoms or the arrangement of atoms in space.) These isomers are not easily interconverted and are called configurational isomers. These are known by

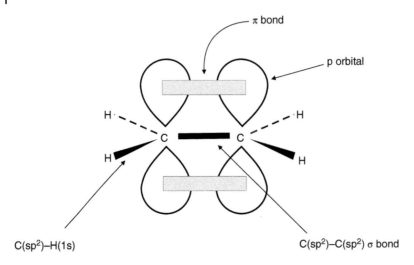

Figure 1.19 σ – and π-bonds in ethylene, $CH_2 = CH_2$.

the prefixes *cis-* (from the Latin for 'on this side' – both bonds are on the same side of the alkene plane) and *trans-* ('across' – the bonds are on opposite sides of the alkene).

As expected, double bonds are stronger than single ones (noted above for ethylene): to break the carbon–carbon link in ethylene requires more energy than is needed to do the same in ethane, C_2H_6, in which two CH_3 groups are linked by a single bond between carbons. The double bond is, however, considerably less than twice as strong; breaking open the π component of the double bond is easier than breaking a single bond. For this reason, ethylene reacts with other compounds more readily than does ethane. Carbon compounds that contain π-bonds are said to be unsaturated, meaning that the carbon atoms, while having formed the requisite number of bonds, are not fully saturated in terms of their number of potential neighbours. Saturated carbon molecules, on the other hand, contain only single bonds.

Atomic orbitals can be combined in different ways as these are simply mathematical functions. For example, the addition and subtraction of the wavefunctions associated with the two atoms in a hydrogen molecule are found to give rise to satisfactory molecular orbitals [1]. Addition (constructive interference between the atomic wavefunctions) leads to a bonding molecular orbital, while subtraction (destructive interference) gives rise to an anti-bonding orbital. These are generally differentiated by an asterisk: π* or σ*. The bonding combination corresponds to a decrease in energy (greater stability); the anti-bonding combination corresponds to an increase in energy (lower stability).

1.5.2 Molecular Crystals

Molecular crystals can possess a high degree of both short-range and long-range order. However, they differ from crystals of metals and inorganic materials because they are made up of discrete molecules. Pentacene, $C_{22}H_{14}$, comprising a linear arrangement of five interconnected benzene rings, is a good example of a molecular crystal. The molecule, depicted in Figure 1.20 may be regarded as a miniature lattice, with a precisely spaced series of atoms. The atoms are in proximity, giving rise to a good overlap of their atomic orbitals. Intra-molecular interactions between the atoms lead to a splitting of the carbon $2p_z$ orbitals and to a localization of the π electrons over the molecule. On a downward positive electron energy scale, there is a highest occupied molecular

Figure 1.20 Chemical structure of pentacene, $C_{22}H_{14}$.

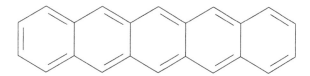

orbital, known by the acronym HOMO, and a lowest unoccupied molecular orbital, LUMO, with an energy separation of about 2.2 eV. In the simplest approximation of an energy-level description, internal electron–electron interactions are neglected, with the result that the LUMO describes both the lowest energetic position of an electron excited out of one of the occupied lower levels of a neutral molecule and the lowest possible energetic position of an additional electron brought in from outside the molecule.

The relatively strong intra-molecular forces in molecular crystals such as pentacene must be contrasted to the weaker van der Waals' inter-molecular forces that hold the molecules together in the solid crystalline state. Consequently, the qualitative description provided by the molecular orbital model is largely unaffected. In the organic crystalline solid, there is a moderate splitting of the molecular energy levels by these inter-molecular interactions into narrow bands. However, there is an important shift in these levels as neutral molecules become embedded in the solid-state environment. The resulting energy bands are very different from those found in inorganic semiconductors such as silicon. Their small width (typically a few 100 meV) and high associated effective mass lead to very small mobilities for the charge carriers (Chapter 2). The electronic energy levels in an isolated pentacene molecule (gas phase) and in the solid crystalline phase are contrasted in Figure 1.21 [1]. The energy required to remove an electron from the HOMO level in the crystal is I, the ionization energy, while the empty LUMO level can be similarly characterized by a binding energy or electron affinity, X, which, like I, is measured with respect to the vacuum level. Clearly, the HOMO–LUMO separation or energy gap E_g is given by

$$E_g = I - X \tag{1.22}$$

Figure 1.21 Electronic energy levels associated with pentacene. The levels on the left are for free molecules in the gaseous state, while those on the right are for the crystal. I = ionization energy. X = electron affinity. Source: Petty [1].

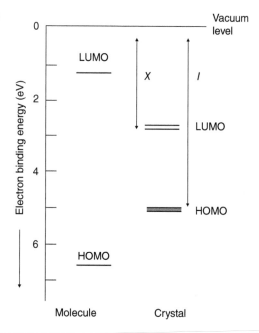

The value of E_g varies with the molecular weight of the organic compound. For example, for the acene chemical series from benzene to pentacene, this energy separation decreases with the number of benzene rings, and hence the size of the π-electron system.

At low temperatures, the charge carriers can move in the narrow energy bands, but they will possess high effective masses. At high temperatures, the electron waves become scattered on every molecular site by phonons or vibrons. The mean free path of the charge carriers will be of the same order as the lattice constant and band theory may no longer be adequate to account for the movement of charges through the crystal.

An important consideration, for the explanation of the operation of opto-electronic organic devices such as solar cells, is the absorption of light by a molecular crystal. As the electrons absorb the photon energy, three types of transitions may be distinguished. First, a transition may occur between the orbitals of the entire crystal, i.e. a transition between the HOMO and LUMO levels (valence and conduction bands). Second, the photon absorption may produce a transition between the orbitals of the constituent elements, thereby creating an excited atom or molecule. Lastly, the absorption can promote transitions between the orbitals of elements in adjacent crystal cells, creating charge-transfer transitions. The relative contribution of these three processes depends on the coupling between the constituent elements of the molecular crystal. Valence to conduction band transitions dominates in inorganic crystals because of the strong covalent bonding. In molecular crystals, photon absorption results predominantly in transitions between molecular orbitals, creating an excited state of the molecule. Charge-transfer is significant for particular types of molecular crystal, as described in Section 1.5.4.

The excitation of an electron in a molecular crystal produces an exciton. As noted above, this is a bound state of an electron and a hole, which are attracted to each other by the Coulomb force. The electrostatic attraction provides a stabilizing energy balance. Consequently, the exciton has slightly less energy than the unbound electron and hole and an exciton band forms with edges just below the LUMO and just above the HOMO bands. An exciton is an electrically neutral quasi-particle and can transport energy without transporting net electric charge.

Depending on the degree of delocalization, the excitons are identified as Frenkel, charge-transfer or Wannier–Mott. These two extreme cases are depicted in Figure 1.22. The Frenkel exciton, Figure 1.22a, corresponds to a correlated electron–hole pair localized on a single molecule. Its radius is therefore comparable to the size of the molecule (<0.5 nm) or is smaller than the inter-molecular distance. A Frenkel exciton can be considered as a neutral particle that can diffuse from site to site, perhaps moving hundreds of molecules away from its origin. In contrast, Wannier–Mott excitons, Figure 1.22b, occur in crystalline materials in which overlap between neighbouring lattice atoms reduces the Coulombic interaction between the electron and the hole of the exciton (leading to the greater permittivity values in such materials). This results in a large radius, 4–10 nm, many times the size of the lattice constant. This type of exciton is not found in van der Waals-bonded organic solids but is more typically a feature of inorganic semiconductors such as silicon or gallium arsenide. Figure 1.22 also indicates the energy levels associated with the different types of excitons. The Frenkel exciton can be considered as two distinct localized states, located above the HOMO and below the LUMO levels of the molecule, whereas the Wannier–Mott excitons are associated with the extended band structure of the crystal in which they are formed. Electronic excitation energy can be conducted within organic solids by excitons moving (hopping) from one location to another. This is an important and characteristic process.

Examples of organic molecular semiconductors include the acene materials, such as pentacene, noted above, the flat disc-shaped phthalocyanine molecules and the numerous low molecular

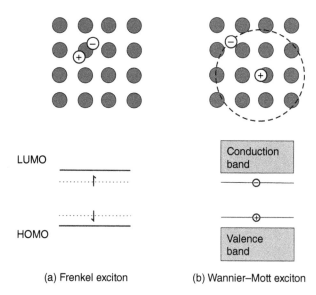

| LUMO | Conduction band |
| HOMO | Valence band |

(a) Frenkel exciton   (b) Wannier–Mott exciton

Figure 1.22 (a) Frenkel and (b) Wannier–Mott excitons with their respective electron energy diagrams.

Table 1.1 Comparison of electronic properties of inorganic semiconductor crystals and organic molecular semiconductor crystals.

Property	Inorganic semiconductor crystal	Molecular semiconductor crystal
Relative permittivity	≈12	≈3.5
Crystal basis	Atoms	Molecules
Chemical bonding	Covalent	Van der Waals
Width of energy bands	Few eV	50–500 meV
Band transport	Yes	Below 300 K
Light absorption at 300 K	Creates free carriers	Creates excitons (Coulomb-bound charges)
Exciton type	Wannier–Mott	Mostly Frenkel, sometimes charge-transfer

weight compounds that have been developed for electronic devices, such as transistors, photovoltaic cells, and light-emitting displays. Examples are provided in later chapters. Molecular crystals possess electronic band structures that differ markedly from their inorganic counterparts. Table 1.1 summarizes some of the key differences.

Disordered molecular crystals are distinct from the amorphous forms of crystalline inorganic semiconductors considered above in Section 1.4. No strong bonds are broken when the order in a molecular crystal is reduced, and there are no dangling bonds. The valence and conduction bands are simply decomposed into a narrow Gaussian distribution of localized states and charge carriers will hop between these [2, 6, 7]. The form of the density of the density of states can be approximated [7]:

$$S(E) = \frac{N_0}{\sqrt{2\pi}\sigma} \exp\left(-\frac{E^2}{2\sigma^2}\right) \tag{1.23}$$

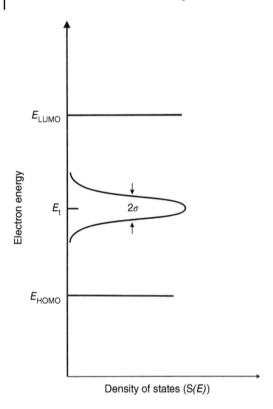

where σ is the energy width (standard deviation) of the distribution and N_0 is the molecular density. This is shown in Figure 1.23, where E_t is the position of the Gaussian distribution peak within the energy gap. The function differs from the parabolic density of states found in crystalline inorganic semiconductors (Eq. (1.7)) and from the exponential tails of states in, for example, amorphous silicon (Figure 1.15a). However, there is some similarity to the localized defect states that may occur in amorphous inorganic compounds (Figure 1.15b).

1.5.3 Polymers

An important feature of the band model is that the electrons are delocalized or spread over the lattice. The strength of the interaction between the overlapping orbitals determines the extent of delocalization that is possible for a given system. A solution to the Schrödinger wave equation is that if electrons are confined by an infinite potential energy barrier to a one dimensional 'box' of length L, the electron energies E_n are quantized and given:

$$E_n = \frac{\hbar^2 \pi^2}{2m_e L^2} n^2 \tag{1.24}$$

where n is a quantum number. For each value of n, there will be an electron wave with a characteristic energy. For many polymeric organic materials, the molecular orbitals responsible for bonding the carbon atoms of the chain together are the sp³ hybridized σ bonds, which do not give rise to extensive overlapping. The resulting band gap is large, as the electrons involved in the bonding are strongly localized on the carbon atoms and cannot contribute to the conduction process. Therefore, a simple saturated polymer such as polyethylene, $-(CH_2)_n-$, is an electrical insulator.

Figure 1.24 Overlapping π electron orbitals for a linear chain of atoms, separated by distance a.

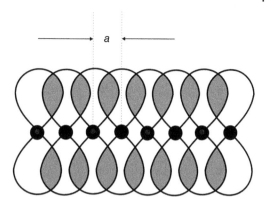

A significant increase in the degree of electron delocalization may be found in unsaturated polymers, i.e. those containing double and triple carbon–carbon bonds. If each carbon atom along the chain has only one other atom, e.g. hydrogen, attached to it, the spare electron in a p_z orbital of the carbon atom overlaps with those of carbon atoms on either side forming delocalised molecular orbitals of π-symmetry.

Suppose that such as a polymer is composed of a linear chain of atoms, with N atoms, each separated by a distance a, as shown in Figure 1.24. The total length of the chain is $(N-1)a$, which for many atoms approximates to Na. The energies of the electron orbitals will then be given:

$$E_n = \frac{\hbar^2 \pi^2 n^2}{2m_e (Na)^2} \tag{1.25}$$

If the π-electrons from the N p orbitals are available, with two electrons per molecular orbital (according to Pauli) the HOMO will be that given by $n = N/2$, and the corresponding energy will be

$$E_{\mathrm{HOMO}} = \left(\frac{N}{2}\right)^2 \left(\frac{\hbar^2 \pi^2}{2m_e (Na)^2}\right) \tag{1.26}$$

The LUMO has the energy:

$$E_{\mathrm{LUMO}} = \left(\frac{N}{2} + 1\right)^2 \left(\frac{\hbar^2 \pi^2}{2m_e (Na)^2}\right) \tag{1.27}$$

The energy required to excite an electron from the HOMO to the LUMO level is the band gap of the polymer, E_g:

$$E_g = E_{\mathrm{LUMO}} - E_{\mathrm{HOMO}} = (N+1)\left(\frac{\hbar^2 \pi^2}{2m_e (Na)^2}\right) \approx \left(\frac{\hbar^2 \pi^2}{2m_e a^2}\right)\left(\frac{1}{N}\right) \tag{1.28}$$

The band gap is therefore predicted to decrease with increasing length of the polymer chain and will practically vanish for macroscopic dimensions. For example, if $a = 0.3$ nm and $N = 100$, then $E_g = 42$ meV.

From the above, it might be expected that a linear polymer backbone consisting of many strongly interacting coplanar p_z orbitals, each of which contributes one electron to the resultant continuous π-electron system, would behave as a one-dimensional metal with a half-filled conduction band. In chemical terms, this is a conjugated chain and may be represented by a system of alternating single and double bonds. For one-dimensional systems, such a chain can more efficiently lower its energy by introducing bond alternation (alternating short and long bonds). The effect is known as Peierls distortion. This limits the extent of electronic delocalization that can take place along the backbone. The result is to open an energy gap in the electronic structure of the polymer. All

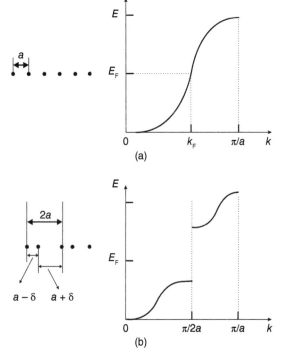

Figure 1.25 Peierls distortion in an isolated chain of equidistant monatomic sodium atoms, spacing a. Energy versus wavevector diagrams (a) without distortion and (b) with distortion. The distorted lattice, displacement δ from equilibrium, has a periodicity of one-half of that of the undistorted lattice.

conjugated polymers are large band gap semiconductors, with band gaps more than about 1.5 eV, rather than metals.

In 1955, Peierls showed that a monatomic metallic chain is unstable and will undergo a metal-to-insulator transition at low temperature. In Figure 1.25a, such a chain, for example sodium atoms, is shown. (This is simply a thought experiment because sodium atoms will not arrange in chains – they tend to form clusters.) An S-shaped energy versus wavevector relation, parabolic on both sides, will result. The band is half-filled because each sodium atom contributes one delocalized electron to the solid and two electrons – spin-up and spin-down – can be accommodated in each state. The Fermi energy E_F is at the band centre (the corresponding Fermi wavevector k_F is shown in Figure 1.25a).

If the arrangement of the atoms is now changed and every second atom is displaced by a small amount, δ, then the atoms are no longer equidistant; short spacings, $(a - \delta)$, and long spacings, $(a + \delta)$, alternate (Figure 1.25b). Again, the arrangement is periodic, but now with a repeat distance $2a$ instead of a. The effect can be contrasted to the doubling of the periodicity in the dispersion curves for phonons propagating in crystal lattices containing two different types of atoms (Figure 1.13). The system has thereby been transformed from a metal with no gap at the Fermi level into a semiconductor with a gap at $\pi/2a$. All states below the gap are filled at absolute zero; all above are empty.

The important question is what is the most stable (lowest energy) state for the one-dimensional array of atoms? It turns out that the answer is the distorted case. This is because the states in the gap have been accommodated above and below the gap. States cannot disappear, and there are as many states as atoms in the lattice (because the states are formed from atomic orbitals). Consequently, when summing the electronic energies from zero to E_F, it is evident that the creation of the gap had reduced the electronic energy. To achieve this, work must be done (against the interatomic forces) to displace the atoms. A full analysis reveals that the electronic energy is approximately linear in δ, whereas the elastic energy depends quadratically on δ. For sufficiently small displacements, the gain in electronic energy predominates over the elastic term. Consequently, under Peierls assumptions, there is always a gap at absolute zero. Electron-lattice coupling will drive the one-dimensional metal into an insulator.

The link between the preceding discussion for a linear array of sodium atoms and a chain of carbon atoms is perhaps unclear. The polymer polyethylene consists of long chains of carbon atoms with two hydrogen atoms per carbon atom. As noted above, the sp^3 hybridized σ bonds do not give rise to extensive overlapping, and the band gap is large. Removing one hydrogen atom from each carbon (to provide an unsaturated polymer chain) would leave unbonded electrons everywhere. The result is a chain of CH· *radicals*. The superscript dot denotes a chemical group carrying an odd number of electrons. It is these radicals that have some similarity with an alkali atom – both have an extra electron. Hence, the electrons in polyacetylene are not completely delocalized along the chain. There is an alternation of short and long bonds between the carbon atoms (where short bonds are drawn as double bonds and long bonds are single bonds), which leads to a semiconductive rather than a metallic band structure.

In summary, a completely delocalized electron system in one-dimension is expected to lead to the metallic state. However, as illustrated in Figure 1.26a, the Peierls transition leads to bond alternation, a doubling of the unit cell, and a semiconductive state. Figure 1.26b shows the electronic energy band structure for *trans*-polyacetylene (the band structure for the *cis* isomer is somewhat different). The valence and conduction bands of semiconductive polymers are often referred to as the π- and π*-bands (the π*-band represents the anti-bonding orbital – Section 1.5.1), respectively. In theory, the π-electron band structure extends over a band width E_0, given by

$$E_0 = 2zt \tag{1.29}$$

Figure 1.26 (a) The electrons associated with the CH· radicals in *trans*-polyactetyene will delocalize over the chain. The Peierls distortion leads to single/double bond alternation. (b) Resulting energy bands: the normally empty conduction (or π*) band and filled valence (π) band are indicated.

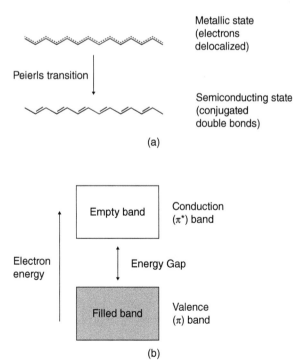

where z is the number of nearest neighbours and t is known as the transfer integral. This is a measure of the π-electron wavefunction overlap between neighbouring carbon sites along the backbone, i.e. the degree of delocalization. For linear polymers such as polyacetylene, $z = 2$ and $t \approx 2.5$ eV, giving $E_0 \approx 10$ eV.

The energy gap of *trans*-polyacetylene is about 1.4 eV, which is comparable to the values 1.1 eV for single crystal silicon and 1.4 eV for GaAs. However, the electrical conductivity in inorganic crystals and organic polymers is very different. This is because the conductivity also depends on the mobility of the charge carriers (Section 2.5). The band model of semiconductors predicts that the greater the degree of electron delocalization, the larger the width of the bands (in energy terms) and the higher the mobility of the carriers within the band. For inorganic semiconductors such as silicon or gallium arsenide, the three-dimensional crystallographic structure provides for extensive carrier delocalization throughout the solid, resulting in relatively high carrier mobilities.

The origin of the band gap in polyacetylene has been explained by a theory, developed originally by Su, Schrieffer, and Heeger [2, 8] that considers a hopping potential describing the energetic costs of moving an electron from one monomer to another.

The Su–Schrieffer–Heeger (SSH) theory is relatively simple compared to other models and predicts an energy for the electron given by

$$E = E_F \pm \sqrt{\left(\frac{E_0}{2}\right)^2 \cos^2(ka) + \left(\frac{E_g}{2}\right)^2 \sin^2(ka)} \tag{1.30}$$

where a is the monomer length, i.e. the distance between every second carbon atom on the polymer chain backbone. The Fermi level is located at the centre of the band gap. However, several experimental results have now been reported that cannot be reconciled with the SSH model and other models have subsequently been developed [2].

Finally, on an important practical point, it should be noted that polymers can comprise crystalline and amorphous regions, which can both influence the transport of charge. Such effects are explored in Section 3.7.

1.5.4 Charge-Transfer Complexes

A further important category of electrically conductive organic materials is charge-transfer compounds [9]. These materials are formed by the combination of two (or more) types of neutral molecules, one of which is an electron donor, D, i.e. it has a low ionization energy and can be easily oxidized (electron removal), and the other is an electron acceptor, A, i.e. has a high electron affinity and can easily be reduced (electron addition). The transfer of an electron from the donor molecule to the acceptor molecule can be represented as follows:

$$D + A \rightarrow [D^{+\cdot}][A^{-\cdot}] \tag{1.31}$$

The transfer process leaves behind an organic radical cation $[D^{+\cdot}]$ that has a 'free' electron, i.e. an electron that is not strongly involved in the chemical bonding (in addition to its charge, the radical ion has an unpaired electron spin). At the same time, the acceptor molecule gains an electron to become the anion radical $[A^{-\cdot}]$. Under certain electronic and structural criteria, these electrons can become conduction electrons, like those of traditional metals. In terms of the energy band diagram, Figure 1.27 reveals the energy-level arrangement in the separated donor and acceptor molecules compared to the D–A complex. In the latter, the LUMO and HOMO levels of both the D and A molecules split. Therefore, there is a smaller energy gap for excitation of an electron from the HOMO to the LUMO level in the complex.

Figure 1.27 Molecular orbitals formed in a charge-transfer complex. The HOMO and LUMO levels in the isolated donor D and acceptor A molecules interact to produce new energy levels in the D–A complex.

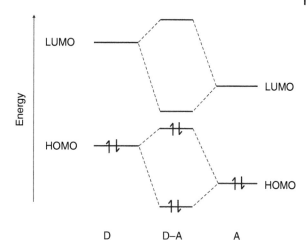

Figure 1.28 Charge-transfer reaction between pyrene, $C_{16}H_{10}$, and iodine to give a high-conductivity complex.

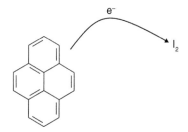

A good example of a charge-transfer process is the reaction between pyrene, $C_{16}H_{10}$ (conductivity $10^{12}\,S\,m^{-1}$) and iodine (conductivity $10^{-7}\,S\,m^{-1}$) to give a complex with a conductivity of about $1\,S\,m^{-1}$, as shown in Figure 1.28. In many cases, partial transfer of charge occurs between the donor and acceptor molecules, and say 6 electrons in every 10 donor atoms are transferred. This leads to mixed valence states in the complex. Charge-transfer complexes generally pack closely in their crystalline phase through the formation of rigid multi-sandwich stacks, resulting in a rather brittle solid.

The stacking can be of two types: mixed stacks in which the donors and acceptors stack alternately …ADADADAD… or segregated stacks in which the donors and acceptors form separate donor stacks (…DDDDDDDD…), and acceptor stacks (…AAAAAAAA…). These are illustrated in Figure 1.29. Molecular compounds with mixed stacks are not highly conductive because of electron delocalization on the acceptor species (Figure 1.29a). However, for segregated stacks, the π overlap and charge-transfer interaction between adjacent molecules in the stacking directions are strong, causing the unpaired electrons to delocalize partially along these one-dimensional molecular stacks resulting in a high conductivity in this direction (Figure 1.29b). This overlap is different from the p orbital overlap forming the π-bands in conjugated polymers. Overlapping in conjugated polymers occurs sideways, in the direction of the polymer axis, and leads to very wide bands, $\approx10\,eV$. Overlapping in charge-transfer salts is top to bottom, along the stacking axis and leads to rather narrow bands, with widths of the order of $1\,eV$. Conjugated polymers are *intra-molecular* one-dimensional conductors, whereas charge-transfer salts are *inter-molecular* conductors (there is also an inter-molecular contribution in polymers, for example inter-chain overlapping with band width $<1\,eV$, and an inter-stack overlap in charge-transfer salts, with a band width $\ll1\,eV$).

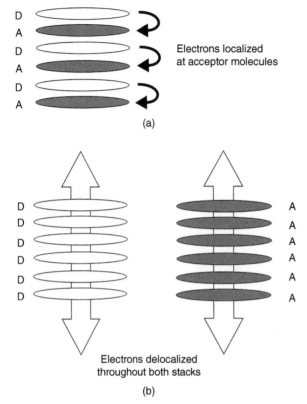

D
A
D
A
D
A

Electrons localized
at acceptor molecules

(a)

D D D D D D

A A A A A A

Electrons delocalized
throughout both stacks

(b)

Figure 1.29 Possible stacking sequences for donor D and acceptor A molecules. (a) In mixed stacks, the electrons become localized on the acceptor molecules and are unable to move through the stack. (b) For segregated stacks, the electrons are delocalized throughout both stacks resulting in high conductivity.

Well-known donor and acceptor molecules are tetrathiafulvalene, $C_6H_4S_4$ (TTF), and tetra-cyanoquinodimethane, $C_{12}H_4N_4$ (TCNQ), Figure 1.30a. The latter compound is a very strong acceptor forming first the radical anion and then the dianion, Figure 1.30b. The stability of the semi-reduced radical ion with respect to the neutral molecule mainly arises from the change from the relatively unstable quinoid structure (a benzene nucleus containing two instead of three double bonds within the nucleus) to the aromatic one allowing extensive delocalization of the π-electrons over the carbon skeleton. Consequently, TCNQ not only forms typical charge-transfer complexes but is also able to form true radical-ion salts, incurring complete one-electron transfer. On addition of lithium iodide to a solution of TCNQ, the simple lithium TCNQ salt is formed:

$$TCNQ + LiI \leftrightarrow Li^+TCNQ^{-\cdot} + \frac{1}{2}I_2 \tag{1.32}$$

Following removal of the free iodine precipitate, the TCNQ salt may be crystallized. The crystals show an electronic conductivity of about $10^{-5}\,S\,cm^{-1}$. A 1:1 TCNQ:TTF salt exhibits a high room temperature conductivity ($5 \times 10^2\,S\,cm^{-1}$), and metallic behaviour is observed as the temperature is reduced to 54 K. Superconductivity can also be observed in some charge transfer salts, usually at low temperature and high pressure.

1.5.5 Graphene

Important electroactive compounds that may find application in organic and molecular electronics are based on forms of carbon. Graphite consists of vast carbon sheets, which are stacked one on top of another like a sheaf of papers. In pure graphite, these layers are about 0.335 nm apart, but

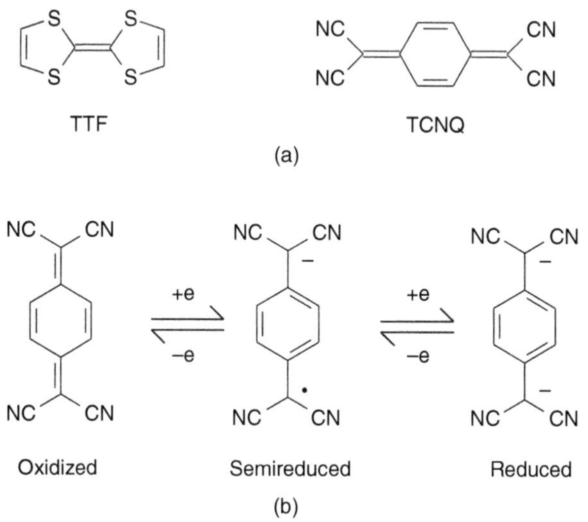

(a)

| Oxidized | Semireduced | Reduced |

(b)

Figure 1.30 (a) Charge-transfer compounds tetrathiafulvalene, $C_6H_4S_4$ (TTF), and tetracyanoquinodimethane, $C_{12}H_4N_4$ (TCNQ). (b) Oxidized, semireduced and reduced forms of TCNQ.

intercalating various molecules can separate them further. The bonding between the carbon atoms in the planes is mainly sp^2 hybridizations consisting of a network of single and double bonds. Weak interactions between the delocalized electron orbitals hold adjacent sheets together. Graphene is the name given to a flat monolayer of carbon atoms tightly packed into a two-dimensional (2D) honeycomb lattice and is a basic building block for graphitic materials of all other dimensionalities.

Graphene was isolated in 2004 by two researchers at The University of Manchester, UK, Andre Geim and Konstantin Novoselov. One Friday, the two scientists removed some flakes from a lump of bulk graphite with sticky tape (mechanical exfoliation). They noticed some flakes were thinner than others. By separating the graphite fragments repeatedly they managed to create flakes that were just one atom thick [10, 11]. Six years after their ground-breaking isolation of graphene, Geim and Novoselov were awarded the 2010 Nobel Prize for Physics.

A graphene sheet is depicted in Figure 1.31. This shows how the unit vectors $\mathbf{a}_1$ and $\mathbf{a}_2$ generate the hexagonal graphene lattice (note that these unit vectors are not orthogonal in the hexagonal lattice). If a (= 0.142 nm) is the carbon–carbon bond length, then $|\mathbf{a}_1| = a(\sqrt{3}, 0)$ and $|\mathbf{a}_2| = a(\sqrt{3}/2, 3/2)$ (Problem 1.5). This graphene sheet can be wrapped into 0D fullerenes, rolled into 1D nanotubes, or stacked into 3D graphite.

Graphene is the thinnest material ever made. It is one hundred times stronger than steel, a better electrical conductor and heat conductor than copper, flexible and optically transparent.

The electronic properties of graphene depend very much on the number of layers. Architectures consisting of more than 10 graphene layers can be considered as a graphite thin film since these essentially exhibit the electronic properties of graphite. Graphene (single layers and bilayers) is best described as a zero band gap semiconductor or a semi-metal. In intrinsic (undoped) graphene each carbon atom contributes one electron completely filling the valence band and leaving the conduction band empty. As such, the Fermi level is situated precisely at the energies where the conduction and valence bands meet [12, 13]. These are known as the Dirac or charge neutrality points. Unfortunately, the lack of an intrinsic band gap in graphene is a fundamental obstacle to its use as the channel layer in a field effect transistor. This is further discussed in Section 8.6.

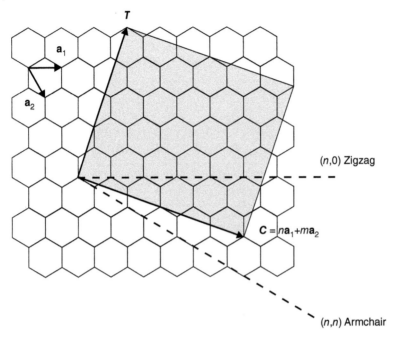

Figure 1.31 Graphene sheet showing the unit vectors $\mathbf{a_1}$ and $\mathbf{a_2}$ of the two-dimensional unit cell. In cutting a rectangular sheet, shown as the shaded region, a circumference vector $\mathbf{C} = n\mathbf{a_1} + m\mathbf{a_2}$ is defined. The direction of the axis of the nanotube is shown by the vector $\mathbf{T}$. Zigzag and armchair structures are defined by the depicted directions of $\mathbf{C}$.

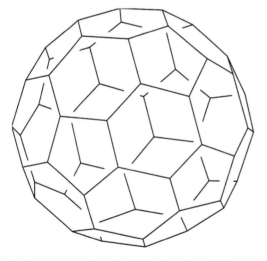

Figure 1.32 Structure of the fullerene C_{60}. Carbon atoms are positioned at the junctions of the bonds.

1.5.6 Fullerenes and Carbon Nanotubes

Under certain conditions, carbon forms regular clusters of 60, 70, 84, etc., atoms. A C_{60} cluster, shown in Figure 1.32, is composed of 20 hexagons and 12 pentagons and resembles a football. The diameter of the ball is about 1 nm. As with graphite, each carbon atom in C_{60} is bonded to three other carbon atoms. Thus, C_{60} can be considered as a rolled-up layer of a single graphene sheet. The term 'buckminsterfullerene' was given originally to the C_{60} molecule because of the resemblance to the geodesic domes designed and built by Richard Buckminster Fuller. However, this term (or simply fullerene or buckyball) is used quite generally to describe C_{60} and related compounds. For example, a fullerene molecule with the formula C_{70} can be formed by inserting an extra ring of hexagons around the equator of the sphere, producing an elongated shell more like a rugby ball.

The electronic structure of C_{60} is unique for a π-bonded hydrocarbon in that the molecule is a strong electron acceptor with an electron affinity of 2.65 eV. This is a consequence of its geometric structure, which influences the electronic energy levels in two simple ways. First, as noted above, the structure may be regarded as 20 six-membered rings with 12 five-membered rings. Conjugated five-membered rings always lead to higher electron affinity because of the aromatic stability associated with the $C_5H_5^-$ (cyclopentadienyl) anion. The non-planarity of the structure also means that the π-electrons are no longer pure p in character. The slight pyramidalization of each carbon atom (referring to the downward deflection of the three atoms surrounding each carbon from the plane in which they would all lie in the graphite structure) induces a small re-hybridization in which some s-character is introduced into the π-orbitals. The molecular orbital energy diagram for C_{60} reveals that the molecule might be expected to accept at least 6 electrons, and possibly up to 12.

Larger fullerenes such as C_{76}, C_{80}, and C_{84} have been found. Some smaller fullerenes also exist. For example, a solid phase of C_{22} has been identified in which the lattice consists of C_{20} molecules bonded together by an intermediate carbon atom. When suitably doped, such smaller fullerenes may exhibit higher superconducting transition temperatures than the C_{60} materials. The existence of the carbon fullerenes has stimulated some discussion about similar clusters of other atoms, such as silicon or nitrogen. Theory has shown that the N_{20} cluster should be stable. This material has also been predicted to be a powerful explosive. However, it has yet to be synthesized.

In addition to the spherical-shaped fullerenes, it is possible to synthesize tubular variations – carbon nanotubes. Such tubes are comprised of graphene sheets, curled into a cylinder [13, 14]. Each tube may contain several cylinders nested inside each other. The tubes are capped at the end by cones or faceted hemispheres. Because of their very small diameters (down to around 0.7 nm) carbon nanotubes are prototype one-dimensional nanostructures.

An important feature of a carbon nanotube is the orientation of the six-membered carbon ring in the honeycomb lattice relative to the axis of the nanotube. In cutting a rectangular sheet, shown as the shaded region in Figure 1.31, a circumference vector $\mathbf{C} = n\mathbf{a_1} + m\mathbf{a_2}$ is defined. The integers n and m denote the number of unit vectors along the crystallographic axes. The direction of the axis of the nanotube is shown by the vector $\mathbf{T}$ in Figure 1.31 (in a direction orthogonal to $\mathbf{C}$). The radius of the resulting nanotube R is then given by

$$R = \frac{C}{2\pi} = \left(\frac{\sqrt{3}}{2\pi} \right) a \sqrt{n^2 + m^2 + nm} \tag{1.33}$$

The primary classification of a carbon nanotube is as either being chiral or achiral. An achiral nanotube is one whose mirror image has an identical structure to the original. There are only two cases of achiral nanotubes: armchair and zigzag (these names arise from the shape of the cross-sectional ring). When the circumference vector lies along one of the two basis vectors, the nanotube is said to be of the zigzag type, for which $m = 0$. For an armchair nanotube, the circumference vector is along the direction exactly between the two basis vectors: in this case $n = m$. All other (n, m) indexes correspond to chiral nanotubes. Three examples of single-wall nanotubes (SWNTs) are shown in Figure 1.33.

The electronic structure of a SWNT is either metallic or semiconducting, depending on its diameter and chirality. Each carbon atom in the hexagonal lattice of the graphene sheet possesses six electrons. The inner 1s orbital contains two electrons, while three electrons in $2sp^2$ hybridized

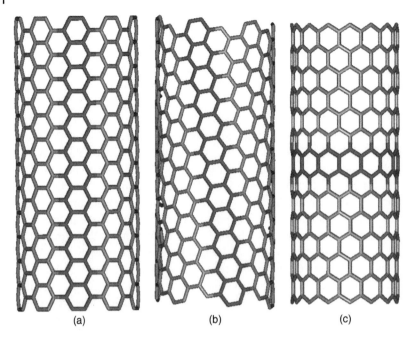

Figure 1.33   Three classes of single-wall carbon nanotube (SWNT) (a) (10,10) armchair SWNT, (b) (12,7) chiral SWNT and (c) (15,0) zigzag SWNT.

orbitals form three bonds in the plane of the graphene sheet. This leaves the final electron in a 2p orbital, perpendicular to the graphene sheet and to the nanotube surface. A delocalized π-electron network is therefore formed across the nanotube surface, responsible for its electronic properties. The band structure of a carbon nanotube can be derived from that of graphene, which, as noted above, is semi-metal with the valence and conduction bands meeting at several points in 'k' space. A full analysis leads to the following condition for carbon nanotubes to be metallic [15]:

$$|n - m| = 3I \tag{1.34}$$

where I is an integer. Nanotubes for which this condition does not hold are semi-conducting. Furthermore, it can be shown that the band gap of semi-conducting nanotubes decreases inversely with an increase in the tube radius, as depicted in Figure 1.34 for zigzag nanotubes. The relationship is approximately that the band gap $E_g = 0.45/R$ eV, where R is the radius of the tube in nanometres.

Some deviations in the electronic properties of nanotubes (from the simple π-electron network model of graphene) arise due to the curvature of the tube. As a result, nanotubes satisfying Eq. (1.34) develop a small curvature-induced band gap and hence are semi-metallic. Armchair nanotubes are an exception because of their special symmetry and remain metallic for all diameters. The band gap of semi-metallic nanotubes is small and varies inversely as the square of the nanotube diameter. For example, a semi-metallic nanotube with a diameter of 1 nm has a band gap of about 40 meV [15].

Nanotubes are found in a variety of forms and shapes other than the single wall tube. Bundles of SWNTs are frequently observed. The individual tubes in the bundle are attracted to their nearest neighbours via van der Waals interactions, with typical distances between the tubes being comparable to the inter-planar spacings in graphite, 0.31 nm. Multi-wall nanotubes (MWNTs) consist of SWNTs nested inside one another, like a Russian doll. Carbon nanotubes also occur in more interesting shapes such as junctions between nanotubes of different chiralities and three terminal

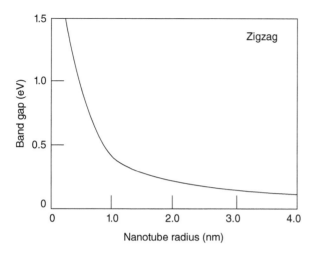

Figure 1.34 Band gap versus radius for zigzag carbon nanotubes. Source: Anantram and Léonard [15].

junctions. Such junctions are atomically precise in that each carbon atom is bonded primarily to its three nearest neighbours, and there are no dangling bonds.

At low temperature, a single-wall carbon nanotube is a quantum wire in which the electrons in the wire move without being scattered (ballistic transport – Section 2.1). In addition to their unique electronic properties, single-wall carbon nanotubes are also some of the strongest materials that are known, exhibiting very high tensile strengths, around 45×10^9 Pa (20 times stronger than steel) and large values of Young's modulus, 1.3 to 1.8×10^9 Pa (almost 10 times that of steel). When carbon nanotubes are bent, they buckle like straws but do not break and can be straightened back without damage. This results from the sp^2 bonds, which can rehybridize as they are bent. MWNTs also possess excellent mechanical properties, but they are not quite as good in this respect as their single-walled counterparts.

1.5.7 Doping of Organic Semiconductors

The principle of doping in organic semiconductors is like that for their inorganic counterparts. Impurities are added which either transfer an electron to the electron conducting (LUMO or π^*) states (n-type doping) or remove an electron from the hole conducting (HOMO or π) states to generate a free hole (p-type doping). However, the term 'doping' can be a misnomer as it tends to imply the use of minute quantities, parts per million or less, of impurities introduced into a crystal lattice, as is the case for inorganic semiconductors. For conductive polymers, typically 1–50% by weight of chemically oxidizing (electron withdrawing) or reducing (electron donating) agents are used to alter physically the number of π-electrons on the polymer backbone, leaving oppositely charged counter ions alongside the polymer chain. These processes are redox (reduction–oxidation) chemistry. For example, the halogen doping process that transforms polyacetylene, $-(CH)_n-$, to a good conductor is oxidation (or p-doping):

$$(CH)_n + \frac{3ny}{2}I_2 \rightarrow \left((CH)^{y+}(I_3^-)_y\right)_n \tag{1.35}$$

Reductive doping (n-doping) is also possible, e.g. using an alkali metal:

$$(CH)_n + nyNa \rightarrow \left((CH)^{y-}yNa^+\right)_n \tag{1.36}$$

In both cases, the doped polymer is a salt. The counter ions, I_3^- or Na^+, are fixed in position while the charges on the polymer backbone are mobile and contribute to the conductivity. The doping effect can be achieved because a π-electron can be removed (or added) without destroying the σ backbone of the polymer. In this way, the charged polymer remains intact. The resulting increase in conductivity can be many orders of magnitude.

A completely different type of doping is possible with the polymer polyaniline (PANi) [1]. This material can be considered as being derived from a polymer, which consists of alternating reduced repeat units. The average oxidation state can be varied continuously from the completely reduced polymer to the fully oxidized material. Complete protonation of the imine nitrogen atoms in the emeraldine base form of polyaniline by aqueous HCl, for example, leads to a structural change with one unpaired spin per repeat unit, but with no change in the number of electrons. The result is a half-filled band and a conductive state, where there is a positive charge in each repeat unit (from protonation) and an associated negative charge (e.g. Cl^-). The protonation is accompanied by an increase in conductivity of about around 11 orders of magnitude.

Doping of molecular crystals can be achieved in a similar way to that described above. For example, high conductivities can be achieved when organic dyes with a weak donor character (such as phthalocyanine) are exposed to strongly oxidizing gases such as iodine or bromine. Alternatively, relatively large aromatic molecules, which are strong π-electron donors or π-electron acceptors, can be used.

There are several important derivatives of graphite and graphene that can be formed by chemical modification. Graphite oxide (formerly called graphitic oxide or graphitic acid) is a compound of carbon, oxygen, and hydrogen in variable ratios, obtained by treating graphite with strong oxidizing agents. The maximally oxidized bulk product is a yellow solid with C:O ratio between 2.0 and 2.9. In contrast to hydrophobic graphite, graphite oxide is a highly hydrophilic layered material and can easily be exfoliated by sonication in water, yielding stable dispersions that consist mostly of single-layer sheets. These monomolecular sheets are also referred to as graphene oxide by analogy to graphene. The structure and properties of graphite (or graphene) oxide depend on the synthesis method and degree of oxidation. The material typically preserves the layer structure of the parent graphite, but the layers are buckled and the inter-layer spacing is about twice (~ 0.7 nm) that of graphite. Chemical reduction to reduced graphene oxide removes a significant amount of oxygen. While graphene oxide is an insulating material, reduced graphene oxide is conductive (but exhibits strongly reduced conductivity as compared to graphene).

The atomic structures of graphene oxide and reduced graphene oxide remain a matter of debate due to a rather random functionalization of each layer and compositional variations depending on the method of preparation. Both materials lend themselves to covalent functionalization due to the presence of defects in the graphene lattice that act as sited for reactivity. The resulting chemically modified graphenes could potentially be much more adaptable for a lot of applications.

The molecular orbital energy diagram for C_{60} reveals that the molecule might be expected to accept at least six electrons, and possibly up to 12. In the solid state, the C_{60} molecules form a crystal lattice with a face-centred cubic structure and a molecular spacing of 1 nm. Alkali atoms can easily fit into the empty spaces in the fcc lattice. Consequently, when C_{60} crystals are heated with potassium, the metal vapour diffuses into the fullerene lattice to form the compound $K_3 C_{60}$. The molecule C_{60} is an insulator, but when it is doped with an alkali metal it becomes electrically conducting. In the case of $K_3 C_{60}$, the potassium atoms are ionized to form K^+, and their electrons are associated with the C_{60}, which becomes a C_{60}^{3-} triply negative ion. Each C_{60} molecule has three electrons that are loosely bonded to the fullerene molecule and can move through the lattice. The $K_3 C_{60}$ has a superconducting transition temperature (Section 2.8) of 18 K. Higher transition temperatures are found for other alkali atoms, e.g. 33 K at ambient pressure and 38 K at 0.8 GPa for Cs_3C_{60} [16, 17]. The transition temperature increases with the radius of the dopant alkali metal ion.

Carbon nanotubes can also be doped either by electron donors or electron acceptors. After reaction with the host materials, the dopants are intercalated in the inter-shell spaces of the multi-walled nanotubes, and, in the case of single-walled nanotubes, either in between the individual tubes or inside the tubes.

The doping of organic semiconductive materials can be achieved in two further ways: photo-doping and charge-injection. In the former case, the material is locally oxidized and reduced by photo-absorption. In the second case, electrons and holes can be injected from suitable metal contacts directly into the π^*- and π-bands, respectively.

Problems

1.1 The width of an energy band in a solid is typically 10 eV. Calculate (i) the number of states per unit volume at the mid-point of the band and within a small energy range $k_B T$ and (ii) the number of states per unit volume within the energy range $k_B T$ to $2k_B T$ from the bottom of the band.

1.2 Using Eq. (1.10) for the number of occupied states $N(E)dE$ in the energy range E to $E + dE$, obtain an expression for the Fermi energy of a metal at 0 K, $E_F(0)$. Assuming that sodium contributes one free electron per atom, calculate the value of $E_F(0)$. Hence, calculate the electron velocity at the Fermi energy at 0 K. How does this result differ from that expected for a classical Maxwell–Boltzmann of electron energies?

1.3 Differentiate Eq. (1.18) to obtain an equation for the group velocity of lattice vibrations. If the maximum value of this expression (long wavelength limit) represents the velocity of sound waves in a material, calculate the force constant for aluminium. Look up the appropriate material constants. Hence, calculate the 'sound velocity' at wavelengths of (i) 1 mm, (ii) 1 μm, and (iii) 1 nm. Comment on your answers.

1.4 Derive expressions for the density of states $S(E)$ for potential wells of one dimension and two dimensions. Comment on your answers. Give examples of organic materials that might be considered as one-dimensional and two-dimensional.

1.5 Download three copies of a graphene sheet and cut along the edges of the outermost hexagons. Starting from an origin carbon atom in the top left hexagon, label the corresponding carbons in the other hexagons (0,0), (1,0), (2,0)...(1,1) ...(6,6). By rolling up the two-dimensional sheets, create the following nanotubes: zig-zag (7,0), armchair (6,6), and chiral (9,2). Calculate the radius of each tube.

References

1 Petty, M.C. (2019). *Organic and Molecular Electronics*, 2e. Chichester: Wiley.
2 Köhler, A. and Bässler, H. (2015). *Electronic Processes in Organic Semiconductors*. Weinheim: Wiley-VCH.

3 Kasap, S.O. (2008). *Principles of Electrical Engineering Materials and Devices*, 3e. Boston: McGraw-Hill.

4 Streetman, B.G. and Banerjee, S. (2000). *Solid State Electronic Devices*, 5e. Hoboken, NJ: Prentice Hall.

5 Mott, N.F. and Davis, E.A. (1979). *Electronic Properties in Non-crystalline Materials*, 2e. Oxford: Clarendon Press.

6 Cápek, V. and Muzikante, I. (2001). Electronic states in organic molecular crystals. In: *Organic Electronic Materials* (eds. R. Farchioni and G. Grosso), 241–282. Berlin: Springer.

7 Horowitz, G. (2015). Validity of the concept of band edge in organic semiconductors. *J. Appl. Phys.* 118: 115502.

8 Geoghegan, M. and Hadziioannou, G. (2013). *Polymer Electronics*. Oxford: Oxford.

9 Bryce, M.R. Conductive charge-transfer complexes. In: *An Introduction to Molecular Electronics* (eds. M.C. Petty, M.R. Bryce and D. Bloor), 168–184. London: Edward Arnold.

10 Novoselov, K.S., Geim, A.K., Morozov, S.V. et al. (2004). Electric field effect in atomically thin carbon films. *Science* 306: 666–669.

11 Geim, A.K. and Novoselov, K.S. (2007). The rise of graphene. *Nat. Mater.* 6: 183–191.

12 Warner, J.H., Schäffel, F., Bachmatiuk, A., and Rümmeli, M.H. (2013). *Graphene: Fundamentals and Emergent Applications*. Waltham, MA: Elsevier.

13 Foa Torres, L.E.F., Roche, S., and Charlier, J.-C. (2020). *Introduction to Graphene-Based Nanomaterials: From Electronic Structure to Quantum Transport*, 2e. Cambridge: Cambridge University Press.

14 Saito, R., Dresselhaus, G., and Dresselhaus, M.S. (1998). *Physical Properties of Carbon Nanotubes*. London: Imperial College Press.

15 Anantram, M.P. and Léonard, F. (2006). Physics of carbon nanotube electronic devices. *Rep. Prog. Phys.* 69: 507–561.

16 Saito, G. and Yoshida, Y. (2011). Organic superconductors. *Chem. Rec.* 11: 124–145.

17 Ganin, A.Y., Takabayashi, Y., Khimyak, Y.Z. et al. (2008). Bulk superconductivity at 38 K in a molecular system. *Nat. Mater.* 7: 367–371.

Further Reading

Blakemore, J.S. (1969). *Solid State Physics*. Philadelphia: Saunders.

Blythe, T. and Bloor, D. (2005). *Electrical Properties of Polymers*. Cambridge: Cambridge University Press.

Canatore, E. (ed.) (2013). *Applications of Organic and Printed Electronics*. New York: Springer.

Cicoira, F. and Santato, C. (eds.) (2013). *Organic Electronics – Emerging Concepts and Technologies*. Weinheim: Wiley.

Cuevas, J.C. and Scheer, E. (2017). *Molecular Electronics: An Introduction to Theory and Experiment*, 2e, 2017. Singapore: World Scientific.

D'Souza, F. and Kadish, K.M. (eds.) (2016). *Handbook of Carbon Nano Materials*, World Scientific Series on Carbon Nanoscience. Vols. 7 and 8. Singapore: World Scientific.

Forrest, S.R. (2020). *Organic Electronics: Foundations to Applications*. Oxford: Oxford University Press.

Heeger, A.J. (2002). Semiconducting and metallic polymers: the fourth generation of polymeric materials. *Synth. Met.* 125: 23–42.

Karl, N. (2001). Low molecular weight organic solids. Introduction. In: *Organic Electronic Materials* (eds. R. Farchioni and G. Grosso), 215–239. Berlin: Springer.

Katsnelson, M.I. (2012). *Graphene: Carbon in Two Dimensions*. Cambridge: Cambridge University Press.

Klauk, H. (ed.) (2012). *Organic Electronics II: More Materials and Applications*. Weinheim: Wiley-VCH.

MacDiarmid, A.G. (2002). Synthetic metals: a novel role for organic polymers. *Synth. Met.* 125: 11–22.

Mody, C.C.A. (2017). *The Long Arm of Moore's Law*. Cambridge, MA: MIT Press.

Pauling, L. (1960). *The Nature of the Chemical Bond*, 3e. New York: Cornell University Press.

Roth, S. (1995). *One-dimensional Metals*. VCH: Weinheim.

Schwoerer, M. and Wolf, H.C. (2007). *Organic Molecular Solids*. Weinheim: Wiley-VCH.

Streitwieser, A. and Heathcock, C.H. (1976). *Introduction to Organic Chemistry*. New York: Macmillan.

Wolf, E.L. (2013). *Graphene: A New Paradigm in Condensed Matter*. Oxford: Oxford University Press.

2

Electrical Conductivity: Fundamental Principles

2.1 Introduction

In Chapter 1, the energy distribution of electrons in a solid was described. The statistics that govern the occupation of the energy levels, or states, was also introduced. Attention now turns to what happens when an electric field is applied. In general, the electrons will acquire additional momentum (the effect of an electric field on ionic species is considered in Section 6.2). However, there are important questions. Are all the electrons in the solid affected? How is equilibrium achieved? The chapter begins with the traditional approach, based on Newton's laws and kinetic theory.

2.2 Classical Model

The idea of electrons in a metal being free to move about as if they constituted an 'electron gas' was first put forward by Drude in 1900, soon after the discovery of the electron. According to this theory, the behaviour of free electrons in a metal is, in many respects, analogous to that of gaseous molecules. There is one significant difference between the two cases. In the free electron theory, collisions between electrons are neglected and the electrons only undergo collisions with the metal ions. Drude also assumed that the distance travelled by an electron between collisions, the mean free path (see later), is governed by the lattice spacing in the crystal and is independent of the electron's speed. Between collisions, the electrons move in straight line (in the absence of any electric field), and they emerge after collision in a random direction with speeds appropriate to the temperature of the region where collision occurred, the hotter the region, the higher the speed of the emerging electron.

For an electric field $\mathscr{E}$ applied in one dimension, say the x-direction, the component of the field is given by

$$\mathscr{E}_x = -\frac{dV}{dx} \tag{2.1}$$

where V is the potential in Volts. The units of $\mathscr{E}$ are, therefore, $V\,m^{-1}$. (The use of bold text in this book denotes a vector quantity.) The negative sign in Eq. (2.1) arises because the potential difference is the work done per unit charge *against* the electrostatic force to move a unit charge between different points in the field.

The electrical conductivity (measured in units of Siemens per metre, $S\,m^{-1}$, or reciprocal Ohms per metre, $\Omega^{-1}\,m^{-1}$) of a solid is the rate at which charge is transported across unit area as a result

Electrical Processes in Organic Thin Film Devices: From Bulk Materials to Nanoscale Architectures, First Edition. Michael C. Petty.
© 2022 John Wiley & Sons Ltd. Published 2022 by John Wiley & Sons Ltd.
Companion Website: www.wiley.com/go/petty/organic_thin_film_devices

of a unit applied electric field. If the current per unit area, or current density, in the x-direction is J_x (units A m^{-2}), the conductivity, σ, can be written as

$$\sigma = \frac{J_x}{\mathcal{E}_x} \tag{2.2}$$

The basic Drude model is that of a random motion of the electrons before any electric field is applied, i.e. as many electrons are moving in any one direction as in the opposite direction, so that there is no net flow in any direction and the current is zero. Drude supposed that each atom in the metal would contribute one or more electrons to the electron gas and assumed that each electron possessed a kinetic energy corresponding to three classical degrees of freedom, i.e. that each electron has a kinetic energy $\frac{3}{2}k_B T$. The magnitude of the average thermal velocity of an electron, $\bar{v}_t$, can, therefore, be estimated:

$$\frac{1}{2}m_e \bar{v}_t^2 = \frac{3}{2}k_B T \tag{2.3}$$

The value of $k_B T$ at 300 K is 4.14×10^{-21} J (0.026 eV) and Eq. (2.3) predicts an average thermal velocity for all the electrons of about 1×10^5 m s^{-1}. However, one important consequence of the quantum mechanical arguments outlined in Chapter 1 is that the electric field only affects the electrons close to the Fermi energy. The Fermi energy values for metals are several eV (Eq. (1.11)). The corresponding electron velocity is, therefore, significantly greater than that predicted by Eq. (2.3).

The direction of electron motion may be altered when it collides with a metal ion. The average distance travelled by an electron before it experiences such a collision is the mean free path, l. In some special cases, the electrons are not subject to scattering events. This situation is referred to as ballistic transport and occurs, for example, when electrons are confined in ultra-small regions in certain semiconductor structures or in carbon nanotubes at low temperatures. Ballistic transport is determined by the electronic structure of semiconductors and allows ultra-fast devices to be fabricated (Section 3.8.2).

With the electric field applied across the sample, the electrons will be accelerated during their free periods between collisions. This acceleration is generally taken to be in a direction opposite to that of the field since the charge on the electron is negative. Simultaneously, collisions of the electrons with the ions will tend to restore the condition in which all the electron velocities are random. The situation is illustrated in Figure 2.1: Figure 2.1a shows the case with no field applied while Figure 2.1b indicates the effect of an applied electric field. In equilibrium, the situation is equivalent to the valence electrons in the metal all possessing a common drift velocity due to the applied electric field. This electron drift constitutes the electric current, which, in the normal convention will be in the same direction of the electric field. It is important to realize that this drift velocity, v_d, is superimposed on the random electron velocities due to thermal motion and *its value will be many orders of magnitude less than* v_t.

A simple relationship for the conductivity of a material may be evaluated using Newton's laws. With an electric field applied, the force acting on an individual electron of charge, e (1.60×10^{-19} C), will be proportional to its charge and to the applied field. Therefore, the acceleration, **a**, experienced by the electron is (Newton's Second Law):

$$\mathbf{a}m_e = e\mathcal{E} \tag{2.4}$$

The average time of flight between collisions is denoted by τ, an appropriate time constant or relaxation time. This parameter is related to the average electron thermal velocity by

$$l = \bar{v}_t \tau \tag{2.5}$$

Figure 2.1 Model for electron drift. (a) In the absence of an applied electric field, a conduction electron moves about in a metal being frequently and randomly scattered by thermal vibrations of the atoms. There is no net electron drift in any direction. (b) With an applied field, there is a net drift of the electrons in the direction of the field. After many scattering events, the electron has been displaced a small distance from its original position towards the positive electrode.

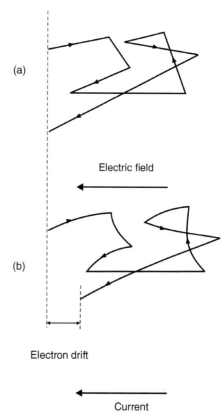

(a)

Electric field

(b)

Electron drift

Current

If all the extra energy gained from the applied field is lost upon collision, then the average drift velocity in the electric field, $\bar{\mathbf{v}}_d$, is given by $\mathbf{a}\tau$. Hence,

$$\bar{\mathbf{v}}_d = \frac{e}{m_e}\mathbf{\mathscr{E}}\tau \tag{2.6}$$

If there are n electrons per unit volume taking part in the conduction, then the current density will be the product of the charge density and the drift velocity:

$$\mathbf{J} = ne\bar{\mathbf{v}}_d \tag{2.7}$$

2.3 Boltzmann Transport Equation

A major simplification made by Drude was the assumption that all the electrons possessed identical thermal speeds (at the same temperature). An improvement to this model was subsequently developed by H. A. Lorentz (1905) who applied classical kinetic theory of gases to the free electron gas in a metal. According to this hypothesis, these electrons would possess a range of speeds (or energies) distributed about a mean velocity, $\bar{v}_t$, as depicted in Figure 2.2. The resulting curve is known as the Maxwell–Boltzmann distribution and has the mathematical form:

$$n(v_t) = Av_t^2 T^{-3/2} \exp\left(-Bv_t^2/T\right)\, dv_t \tag{2.8}$$

where $n(v_t)$ is the number of electrons having a velocity in the range v_t to $v_t + dv_t$, T is the absolute temperature, and A and B are constants. If the temperature is increased, the maximum and mean

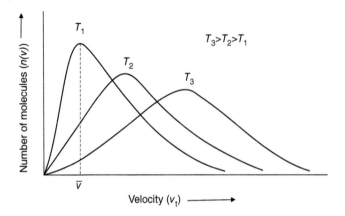

Figure 2.2 Maxwell–Boltzmann distribution of the electron velocities in a metal at different temperatures. Mean velocity at temperature $T_1 = \bar{v}$.

velocities shift to higher values and the Maxwell–Boltzmann curve flattens out. If the temperature is reduced to zero, $\bar{v}_t$ becomes zero. A slightly different form of Eq. (2.8) has previously been provided by Eq. (1.9).

The effect of the application of an electric field to the electron population can be determined by the Boltzmann transport equation, which describes the statistical behaviour of a system not in thermodynamic equilibrium. This is a better approach than the classical one, as it considers how the system of electrons relaxes toward its equilibrium distribution once it has been disturbed (the classical approach assumes that the electron's drift velocity falls to zero following a collision).

When an electric field (and/or a temperature gradient) is applied to a conductor, this results in general in a flow of charge (and/or heat). It is assumed that any momentum that is transferred from the electrons to the lattice is soon lost in random thermal vibrations, or phonons. Consequently, the applied electric field alters the distribution of the electrons in such a way that their average velocity is no longer zero. However, there is no effect on the phonon (lattice vibration) distribution. At any instant, and at any location, the rate at which the electron distribution function f changes is the sum of two kinds of contribution:

$$\left(\frac{\partial f}{\partial t}\right)_{\text{total}} = \left(\frac{\partial f}{\partial t}\right)_{\text{field}} + \left(\frac{\partial f}{\partial t}\right)_{\text{scat}} \tag{2.9}$$

The first term on the right-hand side of Eq. (2.9) represents the influence of the electric field in changing the distribution, while the second is the effect of scattering in the attempt to restore the equilibrium distribution.

Since force is the rate of change of momentum and an electron with velocity $\mathbf{v}$ has a momentum $m\mathbf{v}$ and experiences a force $-e\mathcal{E}$ in an electric field $\mathcal{E}$, then the field term in Eq. (2.9) is

$$\left(\frac{\partial f}{\partial t}\right)_{\text{field}} = \left(\frac{\partial \mathbf{v}}{\partial t}\right)\left(\frac{\partial f}{\partial \mathbf{v}}\right) = \left(\frac{-e\mathcal{E}}{m_e}\right)\left(\frac{\partial f}{\partial \mathbf{v}}\right) \tag{2.10}$$

Hence, the effect of an electric field in changing the distribution depends on the derivative of the distribution function with respect to velocity (which is related to the derivative with respect to energy).

In considering the influence of collisions in minimizing any departure from the equilibrium distribution, it is assumed that the scattering term in Eq. (2.9) is directly proportional to $f - f_0$, where f_0 is the distribution function under thermal equilibrium conditions. This is equivalent to assuming that a relaxation time τ can be defined:

$$\left(\frac{\partial f}{\partial t}\right)_{scat} = \frac{f_0 - f}{\tau} \tag{2.11}$$

This relaxation time τ is related in an integral manner to the mean free time introduced in Eq. (2.5). Using Eqs. (2.9)–(2.11), the Boltzmann transport equation can be expressed as

$$\left(\frac{\partial f}{\partial t}\right)_{total} = \frac{f_0 - f}{\tau} - \frac{e\mathscr{E}}{m_e}\left(\frac{\partial f}{\partial \mathbf{v}}\right) \tag{2.12}$$

When a constant electric field is applied for a time that is long compared to τ, a steady state is reached. The term on the left-hand side of Eq. (2.12) becomes zero and the two terms on the right-hand side must be equal. A suitable integration over the displaced distribution then demonstrates a non-zero velocity for all the electrons, and a finite electrical conductivity. In order to carry out such an integration, it is necessary to make the simplifying assumption that the relaxation time depends only on the magnitude of the electronic speed and not on the direction of the electron velocity. The usual assumption in solving the Boltzmann equation is that a simple power law can express the energy dependence of electrons:

$$\tau \propto E^\lambda \tag{2.13}$$

where λ is a constant depending on the scattering process. For carriers of a given energy, the relaxation time is independent of temperature if the scattering is due to static imperfections. Otherwise, τ becomes a function of temperature as well as energy. Solutions to this equation for various external influences on the electron velocity distribution can be predicted from the Boltzmann transport equation and standard integrals are used to evaluate from properties such as the electrical conductivity, thermal conductivity, and Seebeck coefficient [1, 2]. However, assuming τ is a constant, then the result of Eq. (2.7) is obtained for the current density resulting from an applied field.

2.4 Ohm's Law

Substituting the value of $\bar{v}_d$ from Eq. (2.6) into Eq. (2.7) gives

$$\mathbf{J} = \frac{ne^2\tau}{m_e}\mathscr{E} \tag{2.14}$$

which can be recognized as a form of Ohm's Law – the current density is proportional to the electric field. The electrical conductivity of the material can then be calculated by noting Eq. (2.2):

$$\sigma = \frac{ne^2\tau}{m_e} \tag{2.15}$$

The electrical behaviour of a material may also be characterized in terms of its resistance, R, measured in Ohms (Ω) or its resistivity, ρ, measured in Ω m. The latter quantity provides a useful parameter that is independent of the dimensions of the sample and is simply the reciprocal of the conductivity:

$$\rho = \frac{1}{\sigma} = \frac{m_e}{ne^2\tau} \tag{2.16}$$

For a wire of cross-sectional area, A, length, L, with a resistance, R, as shown in Figure 2.3, the resistivity is given by (Problem 2.1):

$$\rho = \frac{RA}{L} \tag{2.17}$$

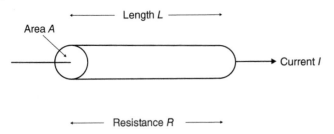

Figure 2.3 Sample of a solid wire of length L, cross-sectional area A, with current I. The resistance of the wire is R.

Noting that

$$\mathcal{E} = \frac{V}{L} \text{ and } J = \frac{I}{A}$$

it is evident that Eq. (2.14) can also be written as

$$V = IR \tag{2.18}$$

which is the most familiar form of Ohm's law. In some situations (for example, when dealing with electrical conductivity at high frequencies, Chapter 7) a parameter called the conductance, G, is used; this is simply the reciprocal of the resistance ($G = 1/R$) with units Ω^{-1} or S.

For many applications in organic electronics, the conductive materials are in the form of thin films, and it is the sheet resistance (or surface resistance), R_s, that is important. This is because the sheet resistance is easier to measure than the resistivity. The sheet resistance of a layer with a thickness, t, is given by

$$R_s = \frac{\rho}{t} \tag{2.19}$$

Strictly speaking, the units of sheet resistance are Ohms. It is common to quote the sheet resistance in units of Ohms per square, although this can give rise to some confusion. This nomenclature is particularly useful when the resistance of a rectangular piece of material, with length L (between the electrodes) and width W is needed. The resistance is simply the product of the sheet resistance and the number of squares:

$$R = R_s \frac{L}{W} \tag{2.20}$$

2.5 Charge Carrier Mobility

Equation (2.15) can also be written as follows:

$$\sigma = ne\mu \tag{2.21}$$

where

$$\mu = \frac{e\tau}{m_e} \tag{2.22}$$

The parameter μ is called the mobility of the electrons and is equal to the average drift velocity per unit applied electric field (Eq. (2.6)):

$$\mu = \frac{\bar{v}_d}{\mathcal{E}} \tag{2.23}$$

The mobility will have dimensions of dimensions $m^2\ V^{-1}\ s^{-1}$ (common units are $cm^2\ V^{-1}\ s^{-1}$). This parameter will also apply to holes, but the electron and hole mobilities in a particular solid will generally be different. As noted above, the velocity that the electrons acquire because of their finite temperature (thermal velocity) is normally much larger than the carrier drift velocity (Problems 2.2 and 2.3). In terms of the mean free path, l, the mobility may be written as follows:

$$\mu = \frac{el}{m_e \bar{v}_t} \tag{2.24}$$

Using Eq. (2.3) for the average thermal velocity of the electrons and noting that the mean free path in crystalline semiconductors is normally greater than the atomic spacing, a, gives the result:

$$\mu > \frac{ea}{(3k_B T m_e)^{1/2}} \sim 1\ cm^2\ V^{-1}\ s^{-1} \tag{2.25}$$

The concept of the mobility of a charge carrier is an important one in both inorganic and organic semiconductor device physics. It is a measure of how quickly the carriers (electrons or holes) will respond to an applied electric field and provides an indication of the upper frequency limit of the material if it is used in a device such as a transistor (Chapter 8). Table 2.1 contrasts the room temperature carrier mobility values (in the commonly used units of $cm^2\ V^{-1}\ s^{-1}$) for Si, GaAs, pentacene, polyacetylene, and a few other organic compounds [3–8]. Most conductive organic compounds are predominantly p-type conductors, and the mobility values will refer to holes. The mobility of

Table 2.1 Room temperature carrier mobilities for field effect transistors based on organic semiconductors.

Material	Carrier mobility ($cm^2\ V^{-1}\ s^{-1}$)
Si single crystal (electrons)	1500
GaAs single crystal (electrons)	8500
Rubrene single crystal	20
Tetracene single crystal	2.4
TCNQ single crystal	1.6
Pentacene	10^{-3}–1
Polyacetylene	10^{-4}
Polythiophene	10^{-5}
Phthalocyanine	10^{-4}–10^{-2}
Thiophene oligomers	10^{-4}–10^{-1}
Organometallic dmit complex	0.2
Benzothiophene derivative	7
C_{60}	0.3
Carbon nanotube	10^3–2×10^4
Graphene	2.5×10^5

After Facchetti [3], Dimitrakopoulos and Mascaro [4], Pearson et al. [5], Abe et al. [6], Novoselov et al. [7] and McEuen and Park [8]. The electron mobilities in single crystal silicon and gallium arsenide are also given.
Sources: Adapted from Dimitrakopoulos and Mascaro [4]; Pearson et al. [5]; Novoselov et al. [7].

charge carriers in organic molecular and polymeric compounds is generally lower than in inorganic semiconductors, as evident from Table 2.1. The highest values for the charge carrier mobilities for organic compounds are invariably reported for materials in the form of single crystals, indicating the important link between structural order and the ability of a material to transport electric charge. For example, the electrical conduction in polymers not only requires carrier transport along the polymer chains but also transfer, or 'hopping', between these chains, which tend to lie tangled up like a plate of spaghetti. Some improvement in the carrier mobility can be achieved by both increasing the degree of order of the polymer chains and by improving the purity of the material. A useful discussion on the mobility of charge carriers in inorganic semiconductors and the factors that influence it can be found in the review article by Groves [9].

A high mobility in a material does not necessarily imply a high electrical conductivity; Eq. (2.21) reveals that σ also depends on the concentration of the charge carriers. Although the mobility values for the organic materials listed in Table 2.1 are quite low, other features make these attractive for certain types of electronic device, as discussed in Chapters 8 and 10.

For inorganic semiconductors, the temperature dependence of the carrier relaxation time and hence the temperature dependence of the carrier mobility can be predicted for various scattering processes (using the ideas of the Boltzmann Transport Equation [1]). For example, for scattering of carriers by lattice vibrations, the mobility μ_l is given

$$\mu_l \propto T^{-3/2} \tag{2.26}$$

In contrast, if the main source of carrier scattering is by ionized impurities, the dependence of the mobility μ_i on temperature takes the following form [2]:

$$\mu_i \propto \frac{T^{3/2}}{N_i} \tag{2.27}$$

where N_i is a concentration of all ionized impurities.

If there are several scattering processes at work, leading to mobilities μ_1, μ_2, μ_3, etc., and the energy dependence of the relaxation time is the same for all the scattering processes, the effective mobility μ_{eff} can be calculated:

$$\frac{1}{\mu_{\text{eff}}} = \frac{1}{\mu_1} + \frac{1}{\mu_2} + \frac{1}{\mu_3} + \dots \tag{2.28}$$

Equation (2.28) is often used as an approximate formula even when the energy dependence of the relaxation time is not the same for all the scattering mechanisms. Equation (2.28) is a form of Matthiessssen's rule and shows that the resistivities arising from different scattering mechanisms are additive.

Equation (2.23) suggests that the carrier drift velocity will continue to increase as the applied electric field increases. At very high electric fields, the carrier energy can become larger than the normal thermal energy. The carriers are referred to as 'hot', and this situation leads to a reduction in the carrier mobility and a saturation of the drift velocity, at a value v_{sat}, as shown in Figure 2.4 for silicon [10]. The value of v_{sat} for both electrons and holes is approximately 10^5 m s^{-1}, corresponding to an applied field of 5×10^6 V m^{-1}. This is of the same order as the thermal velocity calculated from Eq. (2.3). For certain semiconductors (e.g. GaAs, InP), it is possible for the carrier drift velocity to *decrease* with increasing electric field. The effect is shown in Figure 2.5. For low applied fields, v_d increases with $\mathcal{E}$ as expected. However, with increasing electric field the electrons are transferred to a higher energy (indirect) conduction band in which the carriers have a much higher effective mass. The result is an overall decrease in the electrons' drift velocity and hence their mobility. The region where the carrier velocity decreases with increasing electric field is one of negative

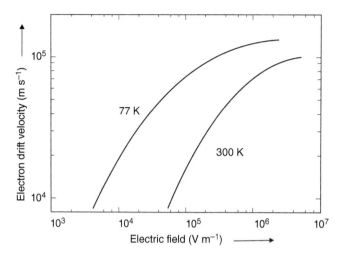

Figure 2.4 Electron drift velocity versus electric field for silicon showing saturation at high values of field. Source: With permission from Jacoboni et al. [10].

Figure 2.5 Characteristic of carrier drift velocity versus electric field for a semiconductor exhibiting the transferred electron mechanism. A negative *differential* mobility (or resistance) region is evident at high fields.

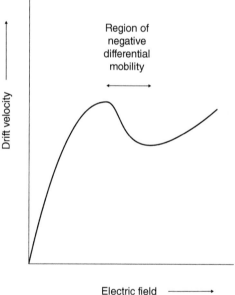

differential mobility or negative differential resistance. The actual material resistance is, of course, not negative only for the *change* of resistance with electric field. The phenomenon is known at the transferred electron effect and can be exploited in high-frequency oscillators.

For low mobility solids, such as organic materials, the conductivity principles outlined above require some modification [1]. For example, a carrier mobility of $1 \text{ cm}^2 \text{ V}^{-1} \text{ s}^{-1}$ implies a relaxation time of about 6×10^{-16} s from Eq. (2.22) (and even less if the electron mass is smaller than that of a free electron). The Heisenberg Uncertainty Principle provides an insight into the consequences of such a small value of τ:

$$\Delta E \Delta t \geq \frac{\hbar}{2} \tag{2.29}$$

where ΔE, Δt are the uncertainties in energy and time, respectively. Taking $\Delta t = \tau = 6 \times 10^{-16}$ s gives $\Delta E \approx 0.5$ eV. The allocation of energies in a band therefore becomes meaningless if the energy spread in the semiconductor is only of the order of $k_B T$. Band transport can occur only if the energy bands are wider than the uncertainty of the charge carrier's energy. Theories which involve the jumping or hopping of carriers from one site to another become preferable to band conduction for such cases. Such transport processes are generally thermally activated and can lead to an Arrhenius exponential law for carrier mobility:

$$\mu = \mu_0 \exp\left(-\frac{\Delta E}{k_B T}\right) \tag{2.30}$$

where μ_0 is a constant and ΔE is an appropriate activation energy for the transport mechanism.

2.6 Equilibrium Carrier Statistics

The density of the charge carriers and their mobility controls the conductivity of a semiconductor (whether it is in the form of an organic compound or an inorganic material). The temperature will determine the number of electrons, holes, and phonons within the solid. For example, the conductivity will increase when the value of $k_B T$ is large enough to permit some thermal excitation of electrons from the upper part of an otherwise filled band to the lower portion of the next higher (normally empty) band. When thermodynamic equilibrium is attained, the number of electrons being promoted to higher energies will equal the number falling back into lower energy levels. The Fermi distribution function (Eq. (1.8)) can be used to calculate the concentrations of electrons and holes in the semiconductor if the densities of available states in the conduction bands are known. These calculations are relatively straightforward for single crystal inorganic semiconductors with well-defined band structures and are outlined below for the cases of intrinsic and extrinsic conduction.

2.6.1 Intrinsic Conduction

As noted in Chapter 1, the conductivity of a semiconductor can be affected by thermal or optical processes. Heating an intrinsic semiconductor such as Si from 0 K will produce equal numbers of electrons in the conduction band and holes in the valence band (Figure 1.8). The number of electrons in the conduction band at any temperature can be obtained by integrating Eq. (1.10) from the bottom of the conduction band (E_c) to its top ($E_c + X$), where X is the electron affinity:

$$n = \int_{E_c}^{E_c + X} S(E)f(E)dE \tag{2.31}$$

This integration can be performed by making a few assumptions and using Eq. (1.7) for $S(E)$, the density of states function, with the electron mass replaced by the effective mass for electrons m_e^*. First, it is assumed that $(E_c - E_F) \gg k_B T$, enabling the Fermi–Dirac function (Eq. (1.8)) to be replaced by the Boltzmann distribution (Eq. (1.9)). The upper limit of the conduction band can be taken to be infinity rather than $(E_c + X)$, as the Boltzmann function decays exponentially with energy. Furthermore, $S(E)f(E)$ is only significant above and close to E_c. Hence, Eq. (2.31) becomes

$$n \approx \frac{(8\pi 2^{1/2})\, m_e^{*3/2}}{h^3} \int_{E_c}^{\infty} (E - E_c)^{1/2} \exp\left(-\frac{(E - E_F)}{k_B T}\right) dE \tag{2.32}$$

which gives

$$n = N_c \exp\left(-\frac{E_c - E_F}{k_B T}\right) \tag{2.33}$$

where

$$N_c = 2\left(\frac{2\pi m_e^* k_B T}{h^2}\right)^{3/2} \tag{2.34}$$

N_c is a temperature-dependent constant called the effective density of states at the conduction band edge (as an aside, a useful exercise is to confirm that N_c, as given by Eq. (2.34), does have the units of m^{-3}). If all the states in the conduction band are replaced with an *effective* concentration N_c (number of states per unit volume) located at the energy corresponding to the conduction band minimum E_c and this is multiplied by the Boltzmann function, then the number of electrons in the conduction band is obtained. A similar analysis can be undertaken to find the concentration of holes in the valence band p. The result is

$$p = N_v \exp\left(-\frac{(E_F - E_v)}{k_B T}\right) \tag{2.35}$$

where N_v is the effective density of states at the valence band edge and is given

$$N_v = 2\left(\frac{2\pi m_h^* k_B T}{h^2}\right)^{3/2} \tag{2.36}$$

The hole effective mass is denoted $m_h{}^*$ in Eq. (2.36) to distinguish it from the electron effective mass. By multiplying Eqs. (2.33) and (2.35), the following important result is obtained:

$$np = N_c N_v \exp\left(-\frac{E_g}{k_B T}\right) \tag{2.37}$$

where the band gap $E_g = E_c - E_v$. Thus, the product of the electron and hole concentrations in a semiconductor is not dependent on the Fermi-level position, only on the temperature and material properties. This is an example of the law of mass action and applies to intrinsic material and to material containing impurities (see below). For the case of intrinsic material, $n = p$, and the intrinsic carrier concentration is

$$np = n_i^2 = N_c N_v \exp\left(-\frac{E_g}{k_B T}\right) \tag{2.38}$$

or

$$n_i = (N_c N_v)^{1/2} \exp\left(-\frac{E_g}{2k_B T}\right) \tag{2.39}$$

This variation of n_i with temperature will be dominated by the exponential term in the above equation. A plot of $\ln(n_i)$ versus $1/T$ will therefore be almost linear and will provide an approximate value for E_g. For organic molecular semiconductors, the effective density of states can be taken to be the density of the molecules.

The carrier concentration can be substituted into Eq. (2.21) to provide an expression for the temperature dependence of the conductivity. As both electrons and holes are present in the semiconductor, the more general form of this relationship is

$$\sigma = ne\mu_n + pe\mu_p \tag{2.40}$$

where n, p, and μ_n, μ_p are the carrier concentrations and mobilities of the electrons and holes, respectively. The carrier with the greater concentration is termed the majority carrier, giving rise to a majority carrier current, while the carrier with the smaller concentration is the minority carrier, associated with a minority carrier current. Of course, for intrinsic material $n = p = n_i$ and so the intrinsic conductivity σ_i can be written as

$$\sigma_i = n_i e(\mu_n + \mu_p) \tag{2.41}$$

If the temperature dependences of the carrier mobilities are slowly varying functions of temperature, then the conductivity will exhibit an almost exponential increase with temperature:

$$\sigma_i = \sigma_0 \exp\left(-\frac{E_g}{2k_B T}\right) \tag{2.42}$$

where σ_0 is a constant.

Equations (2.35) and (2.38) can be used to evaluate the position of the Fermi level E_{Fi} in intrinsic material:

$$N_v \exp\left(-\frac{(E_{Fi} - E_v)}{k_B T}\right) = (N_c N_v)^{1/2} \exp\left(-\frac{E_g}{2k_B T}\right) \tag{2.43}$$

which gives

$$E_{Fi} = E_v + \frac{E_g}{2} - \frac{1}{2}k_B T \ln\frac{N_c}{N_v} \tag{2.44}$$

If $N_c \approx N_v$ (i.e. the electron and hole effective masses are similar), then the Fermi level will be located at a position at approximately mid-point of the band gap.

2.6.2 Carrier Generation and Recombination

The carrier concentrations described in the previous section will be equilibrium values. Electron–hole pairs are constantly not only being generated but also being removed. When an electron and hole 'meet' in a material, they can recombine, and the electron–hole pair is annihilated. The rate of recombination R (units m^{-3} s^{-1}) will be proportional to the number of electrons and also to the number of holes:

$$R \propto np \tag{2.45}$$

The rate of electron–hole pair generation G will depend on how many electrons are available for excitation from the valence band, N_v, the number of empty states in the conduction band, N_c, and the probability that the electron will make the transition, $\exp(-E_g/k_B T)$:

$$G \propto N_c N_v \exp\left(-\frac{E_g}{k_B T}\right) \tag{2.46}$$

In thermal equilibrium, the rate of generation will equal the rate of recombination, i.e. $G = R$. This provides an expression for the product of the electron and hole concentrations that is identical to Eq. (2.37).

2.6.3 Extrinsic Conduction

The band representation of an n-type semiconductor has been shown in Figure 1.9. At room temperature, each donor atom is ionized and provides an electron in the conduction band. The ionization energy ΔE_d represents the energy separation between the donor level and the bottom of the conduction band and is generally around 0.05 eV. Values of ΔE_d may be estimated by using the standard equation for the first ionization energy of a hydrogen atom but using the electron effective mass m_e^* rather than the free electron mass m_e and using the permittivity of the semiconductor rather than the permittivity of free space ε_0. Hence,

$$\Delta E_d = \frac{m_e^* e^4}{8\varepsilon_s^2 \varepsilon_0^2 h^2} = (13.6 \ \text{eV}) \left(\frac{m_e^*}{m_e}\right) \left(\frac{1}{\varepsilon_s^2}\right) \tag{2.47}$$

where ε_s is the relative permittivity (dielectric constant) of the semiconductor.

Using $\varepsilon_s = 11.9$ and $m_e^* \approx m_e/3$ for silicon, $\Delta E_d = 0.032 \ \text{eV}$.

The relationship between the electron and hole concentrations in extrinsic material is also provided by Eq. (2.38). If the electron concentration n has been increased by n-type doping, then the hole concentration p must have been reduced to maintain the condition $np = n_i^2$.

The temperature dependence of the carrier concentration of an extrinsic semiconductor n_e depends on the relationship between the magnitude of the ionization energies for donors (or acceptors) and the absolute temperature. For example, for an n-type material at sufficiently high temperature, all the donor atoms will be ionized and:

$$n_e = N_d \tag{2.48}$$

where N_d is the concentration of donor atoms. At lower temperatures, not all the donors will be ionized and the probability of finding an electron in a state at an energy corresponding to the donor energy must be known. This probability function $f_d(E_d)$ is like the Fermi–Dirac function (Eq. (1.8)) except that it has a factor of $\frac{1}{2}$ – the spin degeneracy weighting factor – multiplying the exponential term:

$$f_d(E_d) = \frac{1}{1 + \dfrac{1}{2}\exp\left(\dfrac{(E_d - E_F)}{k_B T}\right)} \tag{2.49}$$

The degeneracy factor arises as the electron state at the donor can take an electron with spin up or spin down. The Coulomb repulsion of two localized electrons (with opposite spins) raises the energy of the doubly occupied level so that double occupation is essentially prohibited.

In this case, the number of extrinsic carriers (equal to the number of ionized donors) will be given by the number of donors multiplied by the probability of *NOT* finding an electron at E_d:

$$n_e = N_d(1 - f_d(E_d)) = \frac{N_d}{1 + 0.5\exp((E_d - E_F)/k_B T)} \tag{2.50}$$

Under the conditions, $k_B T \ll \Delta E_d$ (low temperature), the extrinsic electron concentration can be shown to depend on temperature by the following expression:

$$n_e = \left(\frac{N_c N_d}{2}\right)^{1/2} \exp\left(\frac{-\Delta E_d}{2k_B T}\right) \tag{2.51}$$

A plot of $\ln n_e$ versus $1/T$ will be approximately linear can be used to estimate the ionization energy of the donors.

A similar approach to that above can be used to obtain an expression for the number of ionized acceptors. However, in this case, the spin degeneracy factor is 2. This results from the fact that an acceptor level, viewed as an *electron* level, can be singly or doubly occupied, but not empty. The latter situation would result in a situation, where two holes are localized in the presence of the acceptor impurity, which has a very high energy due to the mutual repulsion of the holes.

Figure 2.6 shows how the carrier concentration in an n-type semiconductor is expected to vary with temperature. In the figure, $\log(n)$ is plotted as a function of T^{-1}. At low temperatures (and

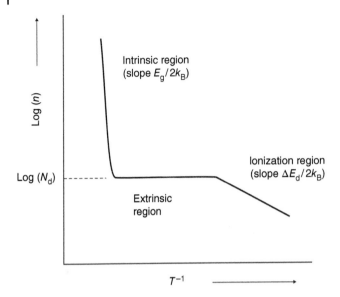

Figure 2.6 The variation of the carrier concentration, n, in an n-type semiconductor with temperature, T. Energy gap $= E_g$. Concentration of donor atoms $= N_d$. Ionization energy of donors $= \Delta E_d$.

above 0 K), there is sufficient thermal energy to ionize the donor atoms and the carrier concentration versus temperature relationship is that of Eq. (2.51). This is the ionization region. At intermediate temperatures, all the donor atoms are ionized, and the carrier concentration is constant (Eq. (2.48)). A further temperature increase will take the semiconductor into the intrinsic conduction region (Eq. (2.39)), where there is sufficient thermal energy available to promote electrons from the valence band to the conduction band.

As noted in Chapter 1, a semiconductor may also contain simultaneously both donor and acceptor type impurity atoms. In the case of compensated material (for which the donor and acceptor concentrations are approximately equal), the extrinsic carrier concentration is given by a modified version of Eq. (2.51) in which the exponential term is $-\Delta E_d/k_B T$, rather than $-\Delta E_d/2k_B T$ [2]. Impurity elements (e.g. transition metals in silicon) may also give rise to more than one separate level lying deep within the forbidden energy gap.

2.6.4 Fermi-Level Position

There is an interdependence of the position of the Fermi level and the carrier concentration for both intrinsic and extrinsic semiconductors. At 0 K, the Fermi level represents the energy below which all electron states are occupied and above which all electron states are filled. In a metal, E_F coincides with the top occupied level in a partially filled band. When the temperature rises, some electrons will move from energy band levels below E_F to levels above E_F, thereby conforming to the distribution shown in Figure 1.5. The Fermi energy no longer separates filled and unfilled states but is still a useful reference level. For intrinsic semiconductors (and insulators) at 0 K, the Fermi level can still be positioned somewhere between the valence band (all states filled) and the conduction band (all states empty). Equation (2.44) reveals that the intrinsic Fermi level is located very nearly midway between the valence and conduction bands. As the temperature rises, E_F remains positioned approximately at the midpoint in the energy gap, at energy E_{Fi}.

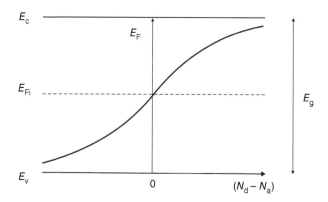

Figure 2.7 The variation in the position of the Fermi level E_F with net donor or acceptor concentration. For $N_d = N_a$, the position of the Fermi level lies close to the middle of the band gap, at energy E_{Fi}.

In the case of extrinsic material, the Fermi level will move towards the conduction band for n-type material and towards the valence band for p-type, as indicated in Figure 2.7. For samples containing both donor and acceptor impurities, the position of the Fermi level will be determined by the net donor or acceptor concentration ($N_d - N_a$). As an extrinsic material is heated and it moves into the region of intrinsic conductivity, E_F will again move towards the middle of the band gap, to E_{Fi}.

The equilibrium electron and hole concentrations can be related to the position of the Fermi level by

$$n = n_i \exp\left(\frac{E_F - E_{Fi}}{k_B T}\right) \tag{2.52}$$

$$p = n_i \exp\left(\frac{E_{Fi} - E_F}{k_B T}\right) \tag{2.53}$$

These equations are valid whether the material is intrinsic or doped. The equations show that when the Fermi level is located at the intrinsic Fermi level, the electron concentration (= hole concentration) is n_i. However, n will increase exponentially as the Fermi level moves away from E_{Fi} and towards the conduction band. Similarly, the hole concentration in the valence band will increase rapidly as E_F moves from E_{Fi} towards the valence band.

Under conditions of very heavy doping, the density of conduction band electrons (or valence band holes) can exceed the effective density of states. The dopant impurities become so close together that they can no longer be considered as being composed of discrete, non-interacting energy states. Instead, the donor (acceptor) states can form a band, which may overlap with the bottom of the conduction band (top of valence band). In such cases, the semiconductor is termed degenerate, and the Fermi level lies within the conduction (valence) band.

An important concept in semiconductor (inorganic or organic) device physics is that there can be no discontinuity or gradient in the equilibrium Fermi level across the material:

$$\frac{dE_F}{dx} = 0 \tag{2.54}$$

From a thermodynamic perspective, the Fermi level represents the electrochemical potential energy, or the free energy per electron, in the material, i.e. E_F represents the potential of an electron to do electrical work ($e \times V$) or nonmechanical work through chemical or physical processes. When two materials, with different Fermi energies, are brought into intimate contact, electrons are transferred from one material to the other until the free energy per electron for the entire system is minimized and E_F is constant throughout the materials. This principle is used in Chapters 4 and 8 where the physics behind various electronic devices is examined.

2.6.5 Meyer–Neldel Rule

The exponential relationship between conductivity and reciprocal temperature that is observed in many undoped and doped materials (Eqs. (2.39) and (2.51)), indicates that the pre-exponential factor and exponent are independent quantities. However, for many disordered materials, including amorphous inorganic semiconductors and organic compounds, there is experimental evidence to suggest that these quantities are linked. For example, if the conductivity versus temperature relationship for a material can be written (cf Eq. (2.42)):

$$\sigma = A \exp\left(-\frac{\Delta E}{k_B T}\right) \tag{2.55}$$

where ΔE is an activation energy and A is a constant, then, there exists a link between σ and ΔE of the form:

$$\log \sigma = \alpha \Delta E + \beta \tag{2.56}$$

where α and β are constants. This empirical relationship (sometimes called the compensation rule) is referred to as the Meyer–Neldel Rule [11, 12]. Different activation energies can be obtained in the same material depending on its method of preparation or applied bias. The experimental data are usually presented as a series of straight-line activation energy plots which intersect at, or extrapolate to, a common focal point at a characteristic temperature T_0. The conductivity may then be more conveniently described by the equation:

$$\sigma = B \exp\left(-\frac{\Delta E}{k_B(T - T_0)}\right) \tag{2.57}$$

The Meyer–Neldel rule now follows with $\alpha = k_B/T_0$ and $\beta = \log B$. Figure 2.8 shows a generalized representation of this relationship. The different lines are obtained by changing a variable V_n

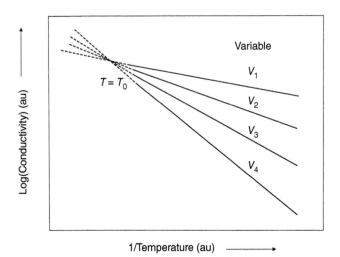

Figure 2.8 General form of the Meyer–Meldel Rule in Arrhenius plots. The activation energy for conductivity depends on a process or variable V_n (e.g. doping, bias) in such a way that the Arrenhius plots all pass through (or can be extrapolated to) a common temperature, T_0.

($n = 1, 2, 3$, etc.,) which might represent a different preparation method (e.g. doping) or a different applied bias.

The Meyer–Neldel Rule has also been demonstrated for the charge carrier mobilities in electronic organic devices such as field effect transistors other structures [13, 14]. It is possible that this law has many different explanations, each valid for only a limited class of solids. But in view of the generality of the phenomenon it seems more likely that there is one explanation. If a single model of the Meyer–Neldel rule is to be universally valid, it cannot relate to the microscopic details of the conduction process; it must be a phenomenological model. One attempt to define this, in terms of an exponential probability distribution of energy barriers, has been made by Dyre [15]. Although this theory was developed to explain the direct current (DC) electrical behaviour of materials, it also predicts a correlation between the Meyer–Neldel rule and the alternating current (AC) properties (Chapter 7), in which there is a power-law frequency dependence of the AC conductivity of the form $\sigma(\omega) \propto \omega^s$, where $s = 1 - (T/T_0)$.

2.7 Excess Carriers

Most semiconductor devices, whether based on inorganic or organic materials, operate by the creation of charge carriers in excess of thermal equilibrium values. Excess carriers can be produced optically, by exposing the device to light, or electrically, by the injection of carriers from an electrode. When such carriers are created non-uniformly in a semi-conductor, the electron and hole concentrations vary with position in the sample. Any such carrier gradient will result in the net motion of carriers from regions of high carrier concentration to regions of low carrier concentration. The origin of this diffusion lies in the random motion of particles. This process is a type of transport phenomenon and can also apply to the spontaneous spreading of other physical properties, such as heat, momentum, or light. The different forms of diffusion can be modeled quantitatively using the diffusion equation. In all cases, the net flux of the transported quantity (atoms, energy, or electrons) is equal to a physical property (diffusivity, thermal conductivity, and electrical conductivity) multiplied by a gradient (concentration, thermal, or electric field gradient). For transport in a single (x) direction, the flux of particles, J_x (the number of particles diffusing through and perpendicular to a unit cross-sectional are per unit time) is given by

$$J_x = -D\frac{dC}{dx} \tag{2.58}$$

where C is the particle concentration. The constant of proportionality D is called the diffusion coefficient, which has units of $m^2\ s^{-1}$. The negative sign in Eq. (2.58) indicates that the direction of diffusion is down the concentration gradient. This expression is called Fick's first law.

2.7.1 Quasi–Fermi Level

The Fermi level E_F, as discussed in Section 2.6.4, is meaningful only when no excess carriers are present. However, it is often desirable to refer to the steady-state electron and hole concentrations in terms of Fermi levels, which can be included in energy band diagrams for various devices. Expressions can be written for the steady-state concentrations in the same form as the equilibrium equations by defining separate quasi-Fermi levels E_{Fqn} and E_{Fqp} for the electrons and holes (NB sometimes these levels are referred to as IMREFs – Fermi spelt backwards). The quasi-Fermi level for either carrier is defined as that quantity which, when substituted into the place of the Fermi level, gives the concentration of that carrier under non-equilibrium conditions. For example, the resulting carrier concentrations of electrons and holes under non-equilibrium conditions are now given:

$$n = n_i \exp \left(\frac{E_{Fqn} - E_{Fi}}{k_B T} \right) \tag{2.59}$$

and

$$p = n_i \exp \left(\frac{E_{Fi} - E_{Fqp}}{k_B T} \right) \tag{2.60}$$

There is no longer a simple relationship between the electron and hole concentrations such as $np = n_i^2$. However, Eqs. (2.59) and (2.60) give the result:

$$np = n_i^2 \exp \left(\frac{E_{Fqn} - E_{Fqp}}{k_B T} \right) \tag{2.61}$$

So the difference in the electron and hole quasi-Fermi levels gives a measure of the departure of the product np from its equilibrium value.

2.7.2 Diffusion and Drift

If an electric field is present in addition to the carrier gradient, the current densities for electrons and holes, J_n and J_p, will each possess a drift and a diffusion component:

$$J_n(x) = e\mu_n n(x)\mathcal{E}(x) + eD_n \frac{dn(x)}{dx} \tag{2.62}$$

$$J_p(x) = e\mu_p p(x)\mathcal{E}(x) - eD_p \frac{dp(x)}{dx} \tag{2.63}$$

where D_n and D_p are the diffusion coefficients for electrons and holes, respectively. The first term in each equation above represents the drift process while the second term reflects diffusion. The total current density is the sum of the contributions due to electrons and holes:

$$J(x) = J_n(x) + J_p(x) \tag{2.64}$$

The situation is best visualized by the diagram shown in Figure 2.9. Here, the electric field is assumed to be in the x-direction, along with the carrier distributions $n(x)$ and $p(x)$, which both decrease with increasing x. Thus, diffusion takes place in the +x-direction. The resulting electron and hole diffusion currents are in opposite directions. Holes drift in the direction of the electric field, whereas electrons drift in the opposite direction because of their negative charge. However, the resulting drift current is in the +x-direction in each case. Note that the drift and diffusion components of the current are additive for holes when the field is in the direction of decreasing hole concentration, whereas the two components are subtractive for electrons under similar conditions.

An important result from Eqs. (2.62) and (2.63) is that the minority carriers can contribute significantly to the current through diffusion. Since the drift terms are proportional to carrier concentration, minority carriers seldom provide much drift current. In contrast, diffusion current is proportional to the *gradient* of concentration. For example, in n-type material, the minority hole concentration p may be many orders of magnitude smaller than the electron concentration n, but the gradient dp/dx can be significant. As a result, minority carrier currents through diffusion can sometimes be as large as majority carrier currents.

In discussing the motion of carriers in an electric field, the influence of the field on the energies of electrons can be indicated in the band diagrams. Assuming an electric field $\mathcal{E}(x)$ in the x-direction, the energy bands as depicted in Figure 2.10 show the change in potential energy of electrons in

Figure 2.9 Drift and diffusion directions for electrons and holes in a carrier gradient and an electric field. Current directions are indicated by the solid arrows.

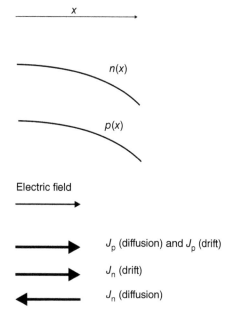

Figure 2.10 Energy band diagram for a semiconductor in an electric field. Conduction band minimum = E_c; valence band maximum = E_v; Fermi-level position in intrinsic material (approx. mid-gap) = E_{Fi}.

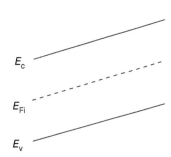

the field. Since electrons drift in a direction opposite to the field, the potential energy for electrons increases in the direction of the field, as shown in the figure. The electrostatic potential $V(x)$ varies in the opposite direction, since it is defined in terms of positive charges and is therefore related to the electron potential energy $E(x)$ by:

$$V(x) = -\frac{E(x)}{e} \tag{2.65}$$

From the definition of electric field (Eq. (2.1)), $\mathcal{E}(x)$ can be related to the electron potential energy in the band diagrams by choosing some reference in the band for the electrostatic potential. Choosing E_{Fi}, the mid-point in the band gap, as a convenient reference, the electric field can be related to this reference:

$$\mathcal{E}(x) = -\frac{dV(x)}{dx} = -\frac{d}{dx}\left(\frac{E_{Fi}}{-e}\right) = \frac{1}{e}\frac{dE_{Fi}}{dx} \tag{2.66}$$

The slope of the bands shown in Figure 2.10 is such that electrons drift 'downhill' in the field, which points 'uphill' in the band diagram.

At equilibrium, no net current flows in a semiconductor. Any fluctuation which would begin a diffusion current also sets up an electric field, which redistributes carriers by drift. An examination of the requirements for equilibrium indicates that the diffusion coefficient and mobility must be related. Assuming that the current is due to electrons and setting Eq. (2.62) to zero for equilibrium conditions:

$$\mathcal{E}(x) = -\frac{D_n}{\mu_n} \frac{1}{n(x)} \frac{dn(x)}{dx} \tag{2.67}$$

Using Eq. (2.59) for $n(x)$:

$$\mathcal{E}(x) = \frac{D_n}{\mu_n} \frac{1}{n(x)} \frac{1}{k_B T} \frac{d}{dx} \left(n_i \exp \left(\frac{E_{Fqe} - E_{Fi}}{k_B T} \right) \right) = \frac{D_n}{\mu_n} \frac{1}{k_B T} \left(\frac{dE_F}{dx} - \frac{dE_{Fi}}{dx} \right) \tag{2.68}$$

The equilibrium Fermi level does not vary with x, and the derivative of E_{Fi} is given by Eq. (2.66). Hence,

$$\frac{D}{\mu} = \frac{k_B T}{e} \tag{2.69}$$

This relationship between the carrier mobility and diffusion coefficient is known as the Einstein relation and is obtained for either carrier type. This allows either D or μ to be calculated if the other quantity is known (Problem 2.4).

2.7.3 Gradients in the Quasi-Fermi Levels

In equilibrium, the position of the Fermi level is constant throughout a sample (Eq. (2.54)). For non-equilibrium conditions with electron drift and diffusion, the current density is given by Eq. (2.62), where the gradient in the electron concentration is

$$\frac{dn(x)}{dx} = \frac{d}{dx} \left(n_i \exp \left(\frac{E_{Fqn} - E_{Fi}}{k_B T} \right) \right) = \frac{n(x)}{k_B T} \left(\frac{dE_{Fqn}}{dx} - \frac{dE_{Fi}}{dx} \right) \tag{2.70}$$

Using the Einstein relation (Eq. (2.69)), the total electron current becomes

$$J_n(x) = e\mu_n n(x)\mathcal{E}(x) + \mu_n n(x) \left(\frac{dE_{Fqn}}{dx} - \frac{dE_{Fi}}{dx} \right) \tag{2.71}$$

From Eq. (2.65), the second term in the brackets in Eq. (2.71) will cancel with the term outside the brackets, leaving

$$J_n(x) = \mu_n n(x) \frac{dE_{Fqn}}{dx} \tag{2.72}$$

Consequently, the process of drift and diffusion are taken into account by the spatial variation of the quasi-Fermi level. The current due to drift and diffusion can therefore be allowed for by a modified Ohm's law:

$$J_n(x) = e\mu_n n(x) \frac{d(E_{Fqn}/e)}{dx} = \sigma_n(x) \frac{d(E_{Fqn})}{dx} \tag{2.73}$$

A similar equation can be derived for the hole current. Therefore, any electron (hole) current resulting from drift or diffusion (or a combination) will be proportional to the gradient of the electron (hole) quasi-Fermi level.

2.7.4 Carrier Lifetime

Carrier recombination occurs continuously in all semiconductors (inorganic or organic). If this were not the case, the thermal generation of carriers would increase steadily. At some temperature, the rates of thermal generation and recombination must be equal (Section 2.6.2). When an excess of carriers is created, the average time that an electron, having been excited, will remain in the conduction band is defined as the electron lifetime. A similar definition applies to the hole lifetime.

If light impinges on a semiconductor with wavelength larger than the energy gap (or LUMO–HOMO separation), the concentration of both electrons and holes will increase and no longer reflect the equilibrium values. However, the deviation from equilibrium will be much more pronounced for the minority carriers. For example, suppose that the concentration of electrons is 0.1% of the hole concentration. An increase in the hole concentration of 0.1% due to light absorption would increase the electron concentration by 100% (since both electrons and holes will be created equally). In the presence of a predominance of majority carriers, the rate of approach to equilibrium is therefore governed by the recombination of the excess of minority carriers.

For an n-type semiconductor, assume that the excess concentration of minority carriers (holes) produced by photogeneration is Δp_n. At any instant, the rate of increase in minority carrier concentration will equal the rate of photogeneration, G_{ph}, minus the rate of recombination of minority carriers:

$$\frac{d\Delta p_n}{dt} = G_{ph} - \frac{\Delta p_n}{\tau_p} \tag{2.74}$$

where τ_p is the hole (minority carrier) lifetime. This is a general expression that describes the time evolution of the excess minority carrier concentration under the assumption that the excess carrier density is not unduly large. It can be shown that the carrier lifetime (electrons or holes) may be related to a diffusion length L [16]:

$$L = \sqrt{D\tau} \tag{2.75}$$

where D is the diffusion coefficient for the charge carrier (Problem 2.5).

2.8 Superconductivity

For particular materials and under certain conditions, zero electrical resistance is observed. Such materials exhibit superconductivity. Only a decade after the discovery of the electron (by J. J. Thompson in 1897) and the introduction of Drude's model of the electron gas (1900), Onnes discovered that the resistivity of mercury suddenly vanishes when the sample is cooled slightly below the boiling point of liquid helium. Many superconductors have now been identified, including elements (e.g. Pb, Sn, Ta), alloys and intermetallic compounds, organic charge-transfer salts, fullerenes, and oxides. However, conductors with some of the highest conductivities, such as silver, gold, and copper, do not exhibit superconductivity. The resistivity of these normal conductors at low temperatures is limited by scattering from impurities and crystal defects and saturates at a finite value determined by the residual resistivity. The highest superconducting transition temperatures have been achieved with the ceramic-based oxides (e.g. Hg–Ba–Ca–Cu–O) (around 135 K at atmospheric pressure).

The Bardeen, Cooper, and Schrieffer (BCS) theory for superconductivity formulated in 1957 assumes an attractive interaction between the electrons. This is provided by the exchange of phonons, which bind two electrons together to form entities called Cooper pairs. These pairs of electrons are correlated over quite large distances (100 nm–1 μm) compared to the average distance between electrons. Consequently, the Cooper pairs interpenetrate highly. The two electrons involved in a Cooper pair have opposite spin and the quasi-particle representing the pair has no net spin. Hence, the Cooper pairs do not obey Fermi–Dirac statistics. They can therefore all 'condense' to the lowest energy state and possess one single wavefunction that describes the entire assembly of Cooper pairs. Since the electrons operate as a pair, an individual electron cannot gain or lose small amounts of energy from an applied field or through collisions. Cooper pairs can move through the lattice without any energy exchange in collisions, which of course is the origin of electrical resistance.

A superconductor below its critical temperature expels all the magnetic field from the bulk of the sample as if it were a perfectly diamagnetic substance. This phenomenon is called the Meissner effect and is often used as a test for superconductivity.

Organic superconductors can be subdivided into one-, two- and three-dimensional solids and encompass the charge-transfer salts and the fullerenes [17, 18]. Perhaps the 'Holy Grail' in superconducting materials research is to identify a compound that exhibits the superconducting state at room temperature. In 1964, Little envisaged a superconducting polymer with a conjugated backbone and dye groups attached regularly to the chain, as shown in Figure 2.11a [19]. The dye groups, Figure 2.11b, are characteristically polarizable (Section 7.2). Little suggested that formation of Cooper pairs amongst the π-electrons of the chain might occur through such polarization of the side groups. Considering that the electronic mass is much smaller than any atomic mass, the coupling might be expected to be very large indeed in comparison to that in conventional semiconductors. As a consequence, Little suggested a critical temperature for the transition from the superconducting state to the normal state of about 2000 K. This dramatic prediction – a room temperature organic superconductor – met with considerable criticism and has not yet been realized in practice. Organic superconductors do exist, but with transition temperatures considerably below room temperature. The highest transition temperature for a charge-transfer salt is about 13 K and about 38 K for fullerenes [17]. However, there is a suggestion that a transition temperature of 123 K is achievable [18].

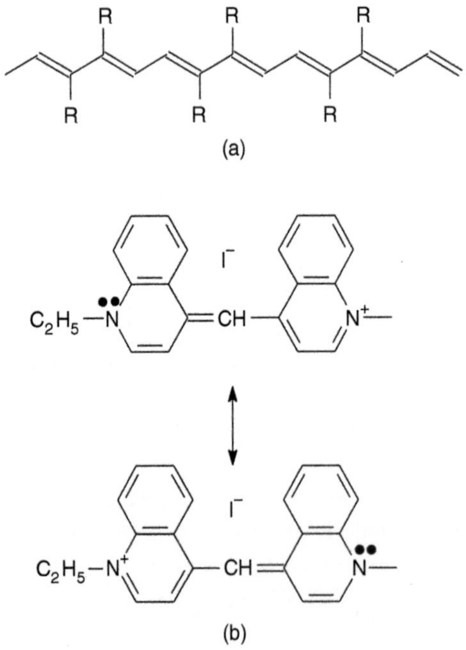

(a)

Figure 2.11 (a) Proposed structure of a room temperature polymeric superconductor by Little [19]. Side groups, designated R are attached to a polyacetylene backbone. (b) Suggestion for R substituent. These side groups are resonating hybrids of the two extreme structures depicted. Source: Based on Little [19].

(b)

Problems

2.1 If the maximum current density is $10^8\,A\,m^{-2}$ in a thin aluminium film, calculate the maximum current that can be carried by an Al integrated circuit connection 100 nm thick and 1 μm wide. Contrast this with the current that could be carried by a single carbon nanotube of diameter 1 nm if the maximum tolerable current density is $10^{15}\,A\,m^{-2}$. Discuss the problems associated with the practicalities of using carbon nanotubes as interconnects in nanoelectronic circuitry.

2.2 Copper has an electrical conductivity at 300 K of $6.0 \times 10^7\,\Omega^{-1}\,m^{-1}$, a density of $8.96 \times 10^3\,kg\,m^{-3}$ and a relative atomic mass of 63.5. Determine the drift velocity of electrons in the metal for an applied field of $100\,V\,m^{-1}$. Compare this figure to the average thermal velocity of electrons calculated (i) using simple kinetic theory and (ii) from quantum mechanical considerations. Assume that the Fermi energy of Cu is 7 eV and temperature independent. Comment on your answers.

2.3 A 100 nm thin film of a conductive polymer is deposited on an insulating glass plate of dimensions 1 cm long and 1 mm wide. The carrier (hole) density is $2.0 \times 10^{20}\,m^{-3}$ and the carrier mobility is $6 \times 10^{-8}\,m^2\,V^{-1}\,s^{-1}$. If a voltage of 10 V is applied across the length of the polymer, calculate (i) the current flowing in the film, (ii) the current density, and (iii) the sheet resistance of the conductive polymer.

2.4 The hole concentration in thin film of polycrystalline pentacene, thickness 200 nm, length 10 μm, and width 1 μm, varies linearly from $10^{15}\,cm^{-3}$ to $10^{10}\,cm^{-3}$ along its length. Calculate the hole diffusion current density and the current in the film at 300 K. Take the hole mobility of the pentacene to be $1\,cm^2\,V^{-1}\,s^{-1}$.

2.5 A semiconductor, with a band gap of 1 eV, absorbs 10 mJ of radiation in which the photons are just sufficiently energetic to create electron–hole pairs. Calculate the excess carrier concentration. If the excess carrier concentration falls to 20% of its peak value in 0.1 ms, calculate the carrier lifetime. If the electron diffusion coefficient is $3.5 \times 10^{-3}\,m\,s^{-1}$, what is the diffusion length?

References

1 Smith, R.A. (1969). *Wave Mechanics of Crystalline Solids*, 2e. London: Chapman and Hall.

2 Blakemore, J.S. (1969). *Solid State Physics*. Philadelphia: Saunders.

3 Facchetti, A. (2008). Molecular semiconductors for organic field-effect transistors. In: *Introduction to Organic and Optoelectronic Materials and Devices* (eds. S.-S. Sun and L.R. Dalton), 287–318. Boca Raton: CRC Press.

4 Dimitrakopoulos, C.D. and Mascaro, D.J. (2001). Organic thin-film transistors: a review of recent advances. *IBM J. Res. Dev.* 45: 11–27.

5 Pearson, C., Moore, A.J., Gibson, J.E. et al. (1994). A field effect transistor based on Langmuir–Blodgett films of an Ni(dmit)$_2$ charge transfer complex. *Thin Solid Films* 244: 932–935.

6 Abe, M., Mori, T., Osaka, I. et al. (2015). Thermally, operationally, and environmentally stable organic thin-film transistors based on bis[1]benzothieno[2,3-*d*:2′,3′−*d*′]naphtho[2,3-*b*:6,7-*b*′]

dithiophene derivatives: effective synthesis, electronic structures, and structure-property relationship. *Chem. Mater.* 27: 5049–5057.

7 Novoselov, K.S., Fal'ko, V.I., Colombo, L. et al. (2012). A roadmap for graphene. *Nature* 490: 192–200.

8 McEuen, P.L. and Park, J.Y. (2004). Electron transport in single-walled carbon nanotubes. *MRS Bull.* 29: 272–275.

9 Groves, C. (2017). Simulating charge transport in organic semiconductors and devices: a review. *Rep. Prog. Phys.* 80: 026502.

10 Jacoboni, C., Canali, C., Ottaviani, G., and Qauranta, A.A. (1977). A review of some charge transport properties of silicon. *Solid State Electron.* 20: 77–89.

11 Rosenberg, A., Bhowmik, B.B., Harder, H.C., and Postow, E. (1968). Pre-exponential factor in semiconducting organic substances. *J. Chem. Phys* 49: 4108–4114.

12 Stallinga, P. (2009). *Electrical Characterization of Organic Electronic Materials and Devices*. Chichester: Wiley.

13 Meijer, E.J., Matters, M., Herwig, P.T. et al. (2000). The Meyer–Neldel rule in organic thin-film transistors. *Appl. Phys. Lett.* 76: 3433–3435.

14 Pivrikas, A., Ullah, M., Singh, T.B. et al. (2011). Meyer–Neldel rule for charge carrier transport in fullerene devices: a comparative study. *Org. Electron.* 12: 161–168.

15 Dyre, J.C. (1986). A phenomenological model for the Meyer–Neldel rule. *J. Phys. C: Solid State Phys.* 19: 5655–5664.

16 Streetman, B.G. and Banerjee, S. (2000). *Solid State Electronic Devices*, 5e. Hoboken, NJ: Prentice Hall.

17 Saito, G. and Yoshida, Y. (2011). Organic superconductors. *Chem. Rec.* 11: 124–145.

18 Geilhufe, R.M., Borysov, S.S., Kalpakchi, D., and Balatsky, A.V. (2018). Towards novel organic high-T_c superconductors: data mining using density of states similarity search. *Phys. Rev. Mat.* 2: 024802.

19 Little, W.A. (1964). Possibility of synthesizing an organic superconductor. *Phys. Rev. A* 134: 1416–1424.

Further Reading

Blythe, T. and Bloor, D. (2005). *Electrical Properties of Polymers*. Cambridge: Cambridge University Press.

Kasap, S.O. (2008). *Principles of Electrical Engineering Materials and Devices*, 3e. Boston: McGraw-Hill.

Sze, S.M. and Ng, K.K. (2007). *Physics of Semiconductor Devices*, 3e. Hoboken, NJ: Wiley-Interscience.

3

Defects and Nanoscale Phenomena

3.1 Introduction

When the periodic and repeated arrangement of atoms in a solid is perfect and extends throughout the entirety of the sample without interruption, the result is a single crystal. As noted in Chapter 1, the energy band model for single crystal semiconductors, such as silicon, results directly from this ordered arrangement of atoms on a lattice. Bands are also present for organic materials in single crystal form; however, their relatively small width and high associated effective mass provide small carrier mobilities. The lower permittivity for organic materials also leads to the importance of excitons as charge carriers. Disorder in the regular atomic arrangement – for both inorganic and organic compounds – will result in a disruption of the band structure and produce localized energy levels within the band gap (Figures 1.15 and 1.23).

Materials are rarely used in their bulk form in electronic and optoelectronic device architectures. The physical nature of a typical organic thin film, which may form the basis of a transistor, is a long way removed from that of a single crystal. If the semiconductor is an evaporated molecular solid, such as pentacene, then it will almost certainly be polycrystalline. The grain boundaries will generally possess different electrical properties to those exhibited by the material within the individual grains. The latter will also be far from perfect containing defects that can act as carrier traps and recombination centres. In the case of polymers, the thin films produced using solution processing methods such as spin-coating and inkjet printing will only be partly crystalline. Defects, such as solitons and polarons, will occur along the polymer chains. For all types of thin films, dimensional effects will come into play as the film thickness is decreased. The properties of a film may also be dominated by its surface(s). In this chapter, all such effects are examined and their consequences on the electrical behaviour are explored. First, the practical (but often forgotten) issue of material purity is examined.

3.2 Material Purity

Ultra-high material purity is a fundamental requirement for ensuring high performance and reliability in all types of electronic devices. The relationships between the nature and concentration of impurities in inorganic semiconductor devices, such as those based on Si and GaAs, are well established. For example, substitutional lattice impurities such as phosphorus or boron can change the conductivity of a host silicon semiconductor by several orders of magnitude if introduced at the parts per billion level. This is because each impurity atom disrupts the valence states of the

Electrical Processes in Organic Thin Film Devices: From Bulk Materials to Nanoscale Architectures, First Edition. Michael C. Petty.
© 2022 John Wiley & Sons Ltd. Published 2022 by John Wiley & Sons Ltd.
Companion Website: www.wiley.com/go/petty/organic_thin_film_devices

neighbouring lattice atoms (Section 1.2.6). Consequently, the level of chemical impurities in single crystal silicon used in the electronic device industry must be less than one part per billion.

In organic electronics, the level and nature of impurities and their effect on performance may differ significantly from those affecting inorganic semiconductor devices. For example, no equivalent-ordered lattice exists in most organic solids. Impurity doping can strongly influence the conductive properties of an organic material, although the doping levels are usually higher those required for inorganic semiconductors (Section 1.5.7). Impurities may also act as traps in all types of semiconductor materials, extracting charge and limiting their performance in electronic devices.

Organic electronic materials can generally be classified into two groups: 'small molecules', and polymers (although a third group – complex biological molecules – is making an impact in some research areas). Small-molecule materials have well-defined molecular weights, allowing for straightforward separation of the host from the impurities. One common means for accomplishing this is via thermal gradient sublimation, whereby the organic source material is heated in a vacuum furnace, and then allowed to condense downstream in a cooler region of the furnace. The low- and high-molecular-weight impurities each condense in a different temperature zone from the desired source material, making separation of these components possible. Using sublimation techniques, fractional impurity concentrations as low as 10^{-4} are potentially achievable, although it remains an important challenge to measure this quantity precisely because of the complex role that impurities play in affecting the properties of the host material.

Because polymer chains in solution have a dispersity of molecular weights, there are a few strategies for purifying the material based on molecular weight alone. Chromatography and other 'distillation' processes are commonly used to achieve the highest level of purity, although attaining <1% impurity concentrations remains a challenge.

3.3 Point and Line Defects

No real crystal is perfect. Thermodynamic equilibrium requires that certain defects are always present in crystals. The surface of a crystal also sets an upper limit on its perfection. Macroscopic samples of most solids comprise many crystallites, randomly oriented, with grain boundaries (Section 3.5) separating one crystallite from the next. Such grain boundaries can be locations for many forms of impurity, some of which may be mobile. Within each crystalline region, there will certainly be finite concentrations of point and line defects.

The point defects in a solid can be grouped into two principal categories. These originate either from foreign atoms (impurities) or from native atoms. Certain atomic sites may not be occupied. This results in point defects known as vacancies, illustrated for a cubic lattice in Figure 3.1. Figure 3.1a shows a perfect lattice. Removal of an atom from one of the lattice sites produces a vacancy, Figure 3.1b. If the atom is relocated into a non-regular atomic site, this produces an interstitial defect, Figure 3.1c. These two types of defects formed by native atoms are also referred to as Schottky (vacancy) and Frenkel (interstitial) defects. In both cases, the defects produce a deformation of the regular lattice. All crystalline solids contain vacancies, and it is not possible to create such a material that is completely free of these defects. The fractional concentration of defects varies enormously from one material to another – around 10^{-4} for vacancies in aluminium just below its melting point (660 °C) to about 10^{-12} for vacancies in germanium just below its melting point (938 °C). The laws of thermodynamics explain the necessity of the existence of vacancies. In essence, the presence of vacancies increases the entropy (randomness) of the crystal,

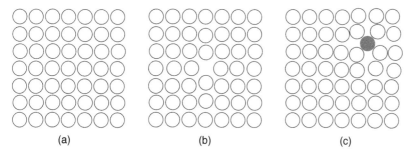

Figure 3.1 (a) Perfect cubic lattice. (b) Vacancy or Schottky defect. (c) Interstitial or Frenkel defect. The defects cause a deformation of the regular lattice.

Figure 3.2 Substitutional and impurity atoms (shown in dark) in a cubic lattice.

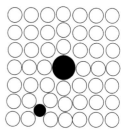

thereby lowering its free energy. The number of defects present at any temperature will then depend on their energy of formation.

Impurity atoms in otherwise pure crystals can be also considered as point defects, and they play an important role in the electronic (e.g. by doping) and mechanical properties of materials. There is, of course, no thermodynamic requirement that these defects must be present. Such atoms may either occupy a normal atomic site in the host lattice (substitutional impurity) or a non-regular atomic site (interstitial impurity). Examples are shown in Figure 3.2. As for the case of the native defects described above, impurity point defects also produce a local distortion in the otherwise perfect lattice, the degree of distortion depending on the crystal structure, parent atom size, impurity atom size, and crystal bonding. These local distortions act as additional scattering centres to the flow of charge carriers through the crystal, thereby increasing its electrical resistance.

In the case of ionic crystals, the removal of an ion produces a local charge as well as distortion of the lattice. To conserve overall charge neutrality, the vacancies occur either in pairs of oppositely charged ions or in association with interstitials of the same sign (e.g. in Na^+Cl^-, an Na^+ ion may move into an interstitial site). Ionized impurities can act as significant scattering centres of charge carriers and can have a direct effect on their mobility, as noted previously in Chapter 2. The temperature dependence of mobility μ_i was given by Eq. (2.27). The full relationship is [1, 2]

$$\mu_i = \frac{64\sqrt{\pi}(\varepsilon_s\varepsilon_0)^2(2k_BT)^{3/2}}{N_i e^3 m_e^{*1/2}} \left(\ln\left[1 + \left(\frac{12\pi\varepsilon_s\varepsilon_0 k_B T}{e^2 N_i^{1/3}} \right)^2 \right] \right)^{-1} \propto \frac{T^{3/2}}{N_i} \qquad (3.1)$$

where N_i is the ionized impurity density and ε_s is the relative permittivity of the semiconductor. As the impurity concentration increases, the mobility decreases, as predicted by the above equation.

Line defects in a crystalline solid are called dislocations. There are two basic types of dislocation, edge and screw dislocations, depicted in Figure 3.3. The edge dislocation, Figure 3.3a, can be visualized as an extra plane of atoms inserted part way into the crystal lattice. The edge of this extra plane is the actual dislocation. There is severe distortion in the region around the dislocation and

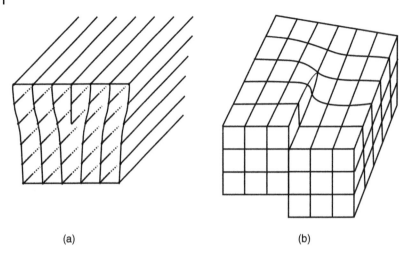

(a)   (b)

Figure 3.3 Two idealized forms of dislocation. (a) Edge dislocation. (b) Screw dislocation. In general, a dislocation follows a curved path, which has varying edge and screw components.

the lattice planes are bent. The presence of the edge dislocation greatly facilitates slip in the crystal when a shear force is applied, the slip occurring normal to the line marking the end of the extra plane of atoms. In a screw dislocation, Figure 3.3b, part of the lattice is displaced with respect to another point, and the displacement is parallel to the direction of the dislocation. A dislocation in a real crystal is likely to have both edge and screw character.

The dislocation density is an indication of crystal perfection. For single crystals of highly perfect solids such as silicon, it is possible to create crystals in which the dislocation density is less than $100\,\mathrm{cm^{-2}}$. A more typical value for other crystals is around $10^4\,\mathrm{cm^{-2}}$, while dislocation densities for metallic crystals are usually at least $10^7\,\mathrm{cm^{-2}}$. Unlike vacancies, thermal equilibrium does not require the presence of dislocations in a single crystal. The application of mechanical stress encourages dislocations to sweep through a crystal, generating point defects until the movement is pinned either by impurities or by the intersection with the path of other dislocations (a process known as work hardening). For liquid crystals, a related imperfection is found – called a disclination. This is a defect in the orientation of director (a dimensionless unit vector called used to represent the direction of preferred orientation of molecules), whereas a dislocation is a defect in positional order.

3.4 Traps and Recombination Centres

Impurities such as those described above may give rise to one or more localized energy levels within band gap of the semiconductor. The effect of chemical doping to control the conductivity of semiconductors has already been introduced in Section 1.2.6. The relatively shallow (i.e. close to the conduction/valence band edge) energy levels produced by dopants are generally fully ionized at room temperature. Deeper localized states can also exchange charge with the conduction and valence bands. For example, such states can temporarily remove electrons or holes from the conduction or valence bands, respectively, acting as carrier traps. Alternatively, the localized level can attract first an electron (hole) and subsequently a hole (electron). This is the process of recombination, and the localized level is referred to as a recombination centre. Trapping involves the temporary removal of a carrier (which is subsequently excited back into the valence or conduction

Figure 3.4 Major transitions within the band gap of a semiconductor. (a) Carrier recombination between conduction and valence bands, (b) trapping and (c) recombination at a localized impurity level, (d) trapping and detrapping via conduction or valence bands. Source: Based on Haneef et al. [3].

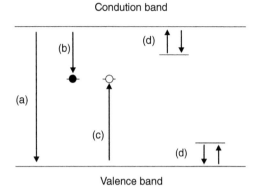

band) while, in the case of recombination, the carriers are permanently removed from the bands. Traps can be the limiting factor on the performance of many semiconductor (both organic and inorganic) devices [3]. Figure 3.4 illustrates the various trapping/recombination processes that may occur – (a) represents band-to-band recombination, (b) and (c) are trapping/recombination processes at localized impurity levels, and (d) represents trapping and detrapping via the conduction and valence bands.

3.4.1 Direct Recombination

For direct band gap semiconductors (Figure 1.7a), the direct transition of an electron from the conduction band to the valence band (Figure 3.4a) is possible by the emission of a photon (radiative process) or by the transfer of the energy to another free electron or hole (Auger process). The latter is the reverse of impact ionization, while the former is the inverse of direct optical transitions (i.e. the wavevector is unchanged). This type of band-to-band recombination is referred to as bimolecular recombination. The carrier recombination rate R will be proportional to the number of electrons and to the number of holes (Eq. (2.45)). The constant of proportionality is β (units $m^3\ s^{-1}$), a recombination coefficient (also referred to as the bimolecular recombination constant):

$$R = \beta np \tag{3.2}$$

In equilibrium, the thermal generation rate G is equal to R and $pn = n_i^2$. The recombination coefficient will therefore be related to G by

$$\beta = \frac{G}{n_i^2} \tag{3.3}$$

The net transition rate $U\ (= R - G)$ then equals zero. If excess carriers are generated (e.g. by optical means), a simple model for generation-recombination is that U is proportional to the excess carrier density. Under low-level injection, defined as the case where the excess carriers $\Delta p = \Delta n$ are fewer than the majority carriers, for an n-type semiconductor, $p = p_0 + \Delta p$, where p_0 is the hole density prior to injection, and $n \approx N_d$, the net recombination rate is given as

$$U = R - G = \beta\left(np - n_i^2\right) \approx \beta\Delta p N_d \equiv \frac{\Delta p}{\tau_p} \tag{3.4}$$

where the carrier lifetime for holes τ_p is

$$\tau_p = \frac{1}{\beta N_d} \tag{3.5}$$

In the case of p-type material, the electron lifetime τ_n is

$$\tau_n = \frac{1}{\beta N_a} \tag{3.6}$$

As noted in Section 2.7.4, the recombination rate of the majority carriers depends on the excess minority carrier density, as these limit the recombination rate.

For high-level injection ($\Delta n = \Delta p > n$ and p), the carrier lifetime becomes

$$\tau_n = \tau_p = \frac{1}{\beta \Delta n} \tag{3.7}$$

3.4.2 Recombination Via Traps

In the case of indirect band gap semiconductors (Figure 1.7b), the dominant transitions are recombination/generation processes via trapping centres. A free electron may be captured at a defect state (Figure 3.4b) or a free hole may be captured at a defect state (Figure 3.4c). The capture process can be described through a capture coefficient β (rather than the recombination constant in Eq. (3.2)) such that the rate of capture R of a species with density n (e.g. electrons in the conduction band) by a species of density N_t (trap density) is given as

$$R = \beta n N_t \tag{3.8}$$

The capture coefficient β is often expressed as the product of a capture cross section σ (units m^2) and the average thermal velocity of the free carriers $\overline{v_t}$:

$$\beta = \sigma \overline{v_t} \tag{3.9}$$

Capture cross sections of trapping centres can range from about 10^{-12} cm^2 for Coulombic attractive capture to less than 10^{-20} cm^2 for Coulombic repulsive capture (a trapping centre surrounded by an energy barrier) [4, 5]. Figure 3.5 shows potential energy diagrams for both types of centres. In the case of an attractive Coulombic centre, Figure 3.5a, the largest value for a capture cross section can be estimated by using simple models based on Coulomb's law or on carrier drift versus diffusion (asking at what radius a free electron will diffuse into such a centre rather than away from it) (Problem 3.1) [4]. The value of σ for a centre surrounded by a repulsive Coulombic centre will depend on the height of the potential energy barrier – ΔE in Figure 3.5b – and its radius.

A captured carrier at a defect site may do one of two things: (i) recombine with a carrier of the opposite type, or (ii) be thermally re-excited to the nearest energy band before recombination

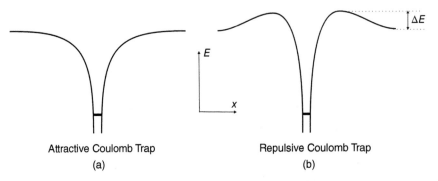

Attractive Coulomb Trap

(a)

Repulsive Coulomb Trap

(b)

Figure 3.5 Electron capture by traps. Potential energy versus distance diagrams for (a) an attractive Coulombic centre and (b) a repulsive Coulombic centre, ΔE = potential energy barrier. Sources: Adapted from Rose [4] and Bube [5].

occurs. In the latter case, the defect is referred to as a trap and the capture and release processes are called trapping and de-trapping; these are illustrated in Figure 3.5d.

The single-level recombination can be described by two processes – electron capture and hole capture. The net transition rate U can be described by the Shockley–Read–Hall statistics [6–8]:

$$U = R - G = \frac{\sigma_n \sigma_p \overline{v_t} N_t \left(pn - n_i^2\right)}{\sigma_n \left(n + n_i \exp\left(\frac{E_t - E_{Fi}}{k_B T}\right)\right) + \sigma_p \left(p + n_i \exp\left(\frac{E_t - E_{Fi}}{k_B T}\right)\right)} \tag{3.10}$$

where E_t is the energy level of the trap, E_{Fi} is the intrinsic Fermi energy (approximately at mid-gap), and σ_n and σ_p are the electron and hole capture cross sections, respectively. It is evident from this equation that the net transition rate is proportional to $pn - n_i^2$. The transition rate will also be a maximum when $E_t = E_{Fi}$, indicating that traps near to mid-gap will be most effective as recombination or generation centres. Considering only such traps:

$$U = \frac{\sigma_n \sigma_p \overline{v_t} N_t \left(pn - n_i^2\right)}{\sigma_n(n + n_i) + \sigma_p(p + n_i)} \tag{3.11}$$

Again, for low-level injection in an n-type semiconductor, the net recombination rate becomes

$$U = \frac{\sigma_n \sigma_p \overline{v_t} N_t \left((p_0 + \Delta p)n - n_i^2\right)}{\sigma_n n}$$

$$\approx \sigma_p \overline{v_t} N_t \Delta p \equiv \frac{\Delta p}{\tau_p} \tag{3.12}$$

where the hole lifetime τ_p is given by

$$\tau_p = \frac{1}{\sigma_p \overline{v_t} N_t} \tag{3.13}$$

Similarly, for a p-type semiconductor, the electron lifetime τ_n is

$$\tau_n = \frac{1}{\sigma_n \overline{v_t} N_t} \tag{3.14}$$

For low-level injection, the lifetime arising from indirect transitions is inversely proportional to the trap density N_t, while for direct band-to-band transitions, the lifetime is inversely proportional to the doping concentration.

In the case of high-level injection ($\Delta n = \Delta p > n$ and p), the carrier lifetime resulting from traps is

$$\tau_n = \tau_p = \frac{\sigma_n + \sigma_p}{\sigma_n \sigma_p \overline{v_t} N_t} \tag{3.15}$$

Comparison of Eq. (3.15) with Eqs. (3.13) and (3.14) reveals that the carrier lifetime is higher with high-level injection. It is also interesting to note that the lifetime due to band-to-band recombination decreases with injection ratio, while that due to trap recombination increases with injection ratio.

3.5 Grain Boundaries and Surfaces

The predominant plane defect in a crystalline material is the grain boundary, illustrated in Figure 3.6. Most crystalline solids do not consist of large single crystals (exceptions are perhaps the semiconductors Si, Ge, GaAs, and InP, the crystal growth of which has been perfected over the

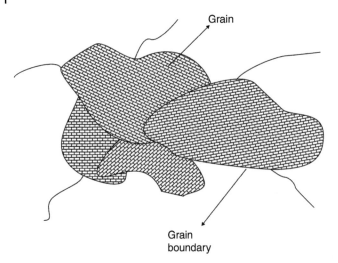

Grain

Grain
boundary

Figure 3.6 Individual grains in a polycrystalline material. The grains are separated by grain boundaries.

latter half of the twentieth century) but of many randomly oriented crystallites. This is certainly the case for thin films of organic molecular semiconductors, such as pentacene. The junction of these crystallites or grains results in grain boundaries, which represent mismatches in the rows and planes in the adjoining crystallites. Each grain itself is an individual crystal and certainly contains the point and line defects described above. Various degrees of crystallographic misalignment between adjacent grains are possible. When this orientation mismatch is slight, on the order of a few degrees, then the term 'low-angle' grain boundary is used. These boundaries can be described in terms of dislocation arrays.

The atoms are bonded less regularly along a grain boundary, and there is an interfacial or grain boundary energy (similar to surface energy) as a consequence of this boundary energy. Grain boundaries are more chemically reactive than the grains themselves and impurity atoms often preferentially segregate in these regions.

A further plane imperfection is the stacking fault, which occurs when mistakes are made in the sequence of stacking the crystal planes. The plane separating two incorrectly juxtaposed layers is the stacking fault. These defects occur most readily in certain types of crystals, for example the face-centred cubic structure.

Crystal surfaces represent a special type of plane defect. Much of the understanding of solids is based on the fact that these are perfectly periodic in three dimensions; for example the electronic and vibrational behaviour can be described in great detail using methods that rely on this periodicity (Chapter 1). The introduction of a surface breaks this periodicity in one direction and can lead to structural and to subsequent electronic, changes. As technology moves further in nanoscale dimensions, surfaces are becoming increasingly important in the determination of the behaviour of related devices. In the limiting cases, devices built using nanotechnology will be dominated by surface effects.

When the crystal lattice is abruptly terminated by a surface, the atoms cannot fulfil their bonding requirements, as illustrated for a silicon surface in Figure 3.7. Each silicon atom in the bulk of the crystal has four covalent bonds, and each bond involves two electrons. The Si atoms at the surface are left with dangling bonds, that is, bonds that are half full, having only one electron. These dangling bonds look for other atoms to which they can interact.

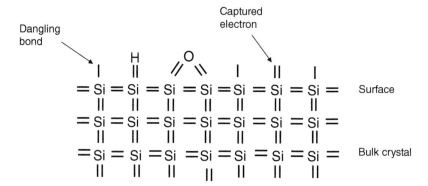

Figure 3.7 Schematic representation of possible defects occurring at the surface of silicon.

Atoms from the environment can therefore bond with the Si atoms on the surface. For example, a hydrogen atom can be captured by a dangling bond and hence become absorbed, or chemisorbed. A foreign atom or molecule is absorbed if it chemically bonds with the atoms on the surface. The H atom in Figure 3.7 forms a covalent bond with a silicon atom and hence becomes absorbed. However, an H_2 molecule cannot form a covalent bond, but due to hydrogen bonding, it can form a secondary bond with a surface Si atom. Molecules that only form secondary bonds with surface atoms are said to be *ad*sorbed. Similarly, a water molecule in the air can readily become adsorbed at the surface of the crystal.

The difference between chemisorption and physisorption (adsorption) therefore lies in the form of the electronic bond between the adsorbate and the substrate. If an adsorbed molecule or atom suffers significant electron modifications relative to its state in the gas phase to form a chemical bond with the surface (covalent or ionic), it is said to be chemisorbed. If on the other hand, it is held on the surface only by van-der-Waals forces, relying on the polarizability of the otherwise undisturbed molecule, it is said to be physisorbed.

3.5.1 Interface States

The dangling bonds represent surface or interface traps (also called interface states, surface states, or fast states) for carriers and can affect the performance of electronic devices (see, for example, Chapters 5, 8, and 10). This interface probably exists over several atomic layers. For Si/SiO$_2$ devices, most of the interface-trapped charge can be neutralized by low-temperature (450 °C) annealing in a hydrogen atmosphere. The total surface charge can be as low as 10^{10} cm^{-2}, which amounts to about one interface trap per 10^5 surface atoms.

Organic compounds, such as polymers and molecular crystals, are much less likely to have the bonding disruptions at their surfaces than inorganic materials. For example, the energy band structure of pentacene (Figure 1.21) is determined by the arrangement of the individual atoms in the pentacene molecule, and the HOMO and LUMO levels are likely to be similar at the surface of the crystal to those in the bulk. However, the effects of polymer surfaces and interfaces cannot be neglected. These can be regions when excitons can recombine thereby having a major effect on the performance of devices such as solar cells (Chapter 11).

When left unprocessed, the surface of any material will have absorbed and adsorbed atoms and molecules from the environment. This is one of the reasons that the surface of a silicon wafer in microelectronics technology is first etched and then oxidized to form SiO$_2$, a passivating layer on the crystal surface. Passivation layers are also continually sought for organic films that are exploited

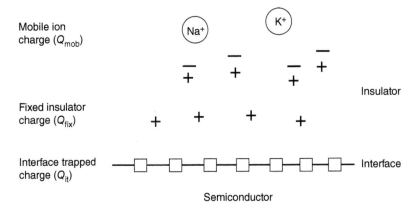

Figure 3.8 Possible mobile and fixed charges associated with an insulator/semiconductor structure. Q_{it} = interface trapped charge. Q_{fix} = fixed (immobile) charge in insulator. Q_{mob} = mobile charge in insulator.

in electronic device architectures. In many instances, such layers are also used as physical barriers, for example to prevent the ingress of moisture.

Figure 3.8 shows the type of charges that can exist at, or close to, the interface between a semiconductor and an insulator (as may be used in a metal/semiconductor/insulator device configuration, Chapter 5). Interface state traps of density N_{it} and charge per unit area Q_{it}, which are located at the semiconductor/insulator interface with energy states within the semiconductor band gap, can exchange charges with the semiconductor in a relatively short time (the reason the term 'fast state' is sometimes used). The semiconductor Fermi level will generally determine their occupancy, unless the insulator is very thin. Therefore, the amount of charge will depend on the presence of an external electric field. In contrast, fixed insulator charge per unit area, Q_{fix}, will be immobile under an electric field. Finally, mobile ionic charges per unit area, Q_{mob}, such as sodium or potassium ions, may be present. These can move within the insulator under bias-temperature stress conditions. Fixed and mobile charges can arise during the processing of the insulating layer, which may involve solvents with small levels of ionic impurities or can be created following the film deposition.

Like bulk impurities, an interface trap is considered a donor if it is neutral and can become positively charged by donating an electron. An acceptor interface trap is neutral and becomes negatively charged by accepting an electron. The distribution functions for interface traps are similar to those for bulk impurity levels discussed in Chapter 2.

A useful notation is to assume that an interface will possess both acceptor and donor traps and that the overall trap distribution will possess an energy level, the neutral level ϕ_0, above which the states are of an acceptor type, and below which are of donor type, as shown in Figure 3.9 [9]. To calculate the trapped charge Q_{it}, it can also be assumed that at room temperature, the trap occupancy takes on the value of 0 above the Fermi level and 1 below E_F. The interface charge can then be calculated by

$$Q_{it} = -e \int_{\phi_0}^{E_F} D_{it} dE \qquad E_F \text{ above } \phi_0$$

$$Q_{it} = +e \int_{E_F}^{\phi_0} D_{it} dE \qquad E_F \text{ below } \phi_0 \tag{3.16}$$

The surface charges will represent the effective net charges per unit area. Because the interface trap levels are distributed across the band gap, these are characterized by an interface trap density

Figure 3.9 Any interface state distribution at a semiconductor/insulator interface can be modelled by the equivalent distribution depicted. ϕ_0 represents a neutral level: above ϕ_0 the interface states are acceptor type and below ϕ_0 these are donor type. The occupancy (and charge) of the traps is determined by the position of the Fermi level, E_F. Source: Sze and Ng [9].

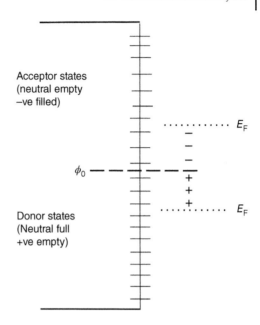

distribution D_{it} given by

$$D_{it} = \frac{1}{e}\frac{dQ_{it}}{dE} \qquad (3.17)$$

D_{it} is the number of traps per unit area per unit energy, in SI units of m^{-2} J^{-1}. However, common units for D_{it} are cm^{-2} eV^{-1}. The presence of surface states can influence the band structure of the semiconductor at the interface.

In the absence of any interface charges, the energy bands will continue undisrupted to the interface, as shown in Figure 3.10 for an n-type semiconductor interface with vacuum (provided, of course, that there is no external electric field). In this diagram, χ_s represents the electron affinity of the semiconductor, the energy difference between the lowest lying state in the conduction band and the vacuum level while Φ_s, which is the energy difference between the Fermi level and vacuum, is called the semiconductor work function. If acceptor-type surface states are now introduced below the Fermi level, they will not be in equilibrium with the semiconductor if they remain

Figure 3.10 Flat energy bands at the interface between an n-type semiconductor and insulator. No external electric field is applied. E_{vac} = vacuum energy level. E_c, E_v are edges of conduction and valence bands, respectively. E_F = Fermi level. X_s = semiconductor electron affinity. Φ_s = semiconductor work function.

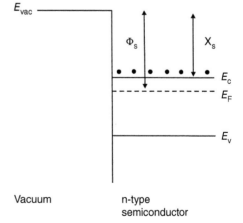

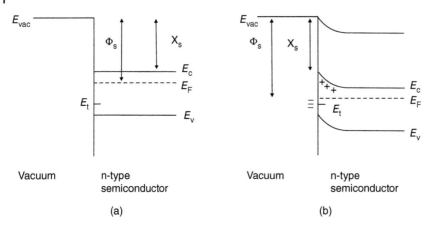

Vacuum   n-type   Vacuum   n-type
           semiconductor   semiconductor

(a)   (b)

Figure 3.11 Energy bands at the interface between an n-type semiconductor and insulator in the presence of an acceptor trap located at energy E_t. No external electric field is applied. E_{vac} = vacuum energy level. E_c, E_v are edges of conduction and valence bands, respectively. E_F = Fermi level. X_s = semiconductor electron affinity. Φ_s = semiconductor work function. (a) Non-equilibrium situation with trap uncharged. (b) In equilibrium, trap is negatively charged and compensated by a positive space charge in the semiconductor. Semiconductor bands bend upwards.

uncharged, Figure 3.11a. However, since the states are empty and below E_F, some of the electrons in the conduction band will fall into them. In this process, the semiconductor surface becomes negatively charged. To compensate, a positive space-charge layer forms below the surface. Hence, the energy bands at the surface bend upwards with respect to the Fermi level, which remains constant throughout the semiconductor (Section 2.6.4). The equilibrium situation, in which the positive charge in the space-charge region just balances the negative charge in the surface states, is shown in Figure 3.11b. The electron deficient space-charge region (E_F is further from E_c than it is in the semiconductor bulk) is referred to as the depletion region. An increase in the density of the surface states will produce a greater band bending in the semiconductor. An analogous situation will occur for donor-like surface states if these are located above E_F. In this case, a positive charge in these states will produce a downward band bending in the semiconductor. The introduction of acceptor-like surface states above the Fermi level or of donor-like states below the Fermi level will, of course, has no influence on the band structure.

An indication of the distance over which the semiconductor bands are bent (i.e. depletion layer width) is given by a parameter called the Debye screening length (or Debye length), L_D, which can be defined as

$$L_D = \sqrt{\frac{\varepsilon_s \varepsilon_0 k_B T}{e^2 N}} \tag{3.18}$$

where N is N_d or N_a, depending on which is the larger. The Debye length is an important concept in semiconductor device physics. This provides an idea of the distance scale in which charge imbalances are screened or smeared out. If a positively charged sphere is inserted into an n-type semiconductor, mobile electrons will be attracted to the sphere. At several Debye lengths from the charge centre, the positively charged sphere and the negative electron cloud will appear as a neutral entity. The Debye length depends inversely on the doping density because, the higher the concentration, the more easily screening takes place.

Interface states may also exist at grain boundaries. The charge that resides in these can produce band bending within the individual grains. Both surface states and interface states can be affected by external factors such as exposure to different gas ambient. For example, Figure 3.12 illustrates

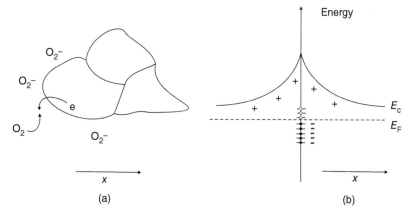

Figure 3.12 (a) Grain structure in polycrystalline material showing the chemisorption of oxygen at the surface of the grains. (b) The negatively charged oxygen molecules produce an upward curvature of the electronic energy bands at the grain boundary, providing a depletion region.

the processes that might occur if an n-type polycrystalline semiconductor is exposed to oxygen gas. The gas will first become chemisorbed, Figure 3.12a. The molecules act as surface acceptors, removing electrons from the conduction band of the semiconductor and giving rise to an increase in the hole concentration at the surface. The conduction band bends upwards in the energy diagram of Figure 3.12b to provide a depletion region with positive fixed charge (due to fixed donor atoms) that compensate for the negative charge trapped at the grain surface. This region will become less n-type than the bulk semiconductor.

3.6 Polymer Defects

So far the defects that have been discussed are mostly associated with single crystal and polycrystalline materials. There are also several specific defects associated with polymers that can influence their electrical and optical properties. An overview of these is provided in this section.

3.6.1 Solitons

First, consider the bonding structure in the *trans* form (Section 1.5.1) of polyacetylene. The double and single bonds can be interchanged without affecting the overall electron energy. Therefore, there are two degenerate (identical energy) lowest energy states A and B, shown in Figure 3.13, possessing distinct bonding structures. A simple defect can occur when these two degenerate forms of the polymer are joined, as shown in Figure 3.14. In this instance, a bond mismatch occurs, i.e. a carbon site with one too few π-electrons so that it cannot form a double bond. This produces a dangling bond. The defect results in one unpaired π-electron, but as the entire system is electrically neutral, it has the same number of protons as electrons.

The topological defects described above are called solitons due to their non-dispersive nature (an everyday manifestation of a soliton, or solitary wave, is the bow wave of a boat). The soliton will be mobile due to the translational symmetry of the chain. Although the soliton illustrated in Figure 3.14 is shown to occur as an abrupt change from the A to the B form of polyacetylene, the evidence is that the defect extends over a few carbon atoms. Solitons can become positively or

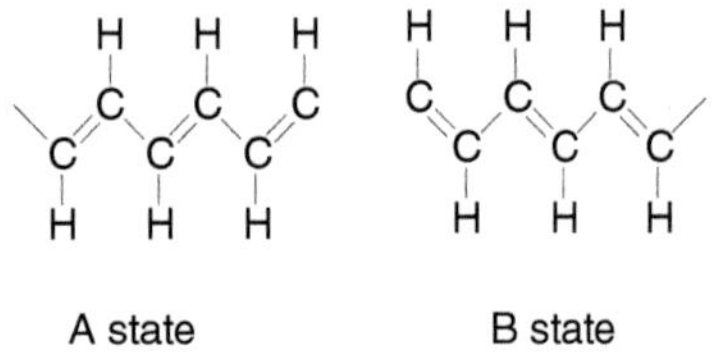

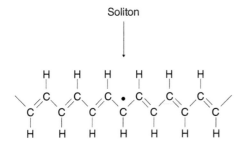

Figure 3.13 Two degenerate (equal energy) forms (A and B states) of *trans*-polyacetylene.

A state   B state

Figure 3.14 Soliton formation in a *trans*-polyacetylene chain.

negatively charged. If the localized state contains one electron, the soliton is neutral and can be associated with an energy level halfway between the valence and conduction bands, as shown in Figure 3.15a. The charge q and the spin s are indicated in the diagram. The unpaired electron will have a spin of $\frac{1}{2}$ and can be detected by the technique of electron spin resonance, ESR. When the electron in the localized state is removed, for example by p-doping, the soliton is positively charged with spin $= 0$, Figure 3.15b, and is no longer detectable by ESR. Similarly, if n-doping occurs, a negative soliton is obtained with spin $= 0$, Figure 3.15c. The ESR studies reveal that neutral solitons are highly mobile, whereas positive and negative solitons are believed to be localized over a number of carbon atoms.

3.6.2 Polarons and Bipolarons

Polyacetylene is a conjugated polymer with degenerate ground states. However, most conjugated polymers have non-degenerate ground states. In such materials, solitons are not stable and double-conjugational defects are found instead. Such a species is termed a polaron if it is singly charged or a bipolaron if it is doubly charged. A polaron can be thought of as the bound state of a charged soliton and a neutral soliton, whose mid-gap energy states hybridize to form bonding and anti-bonding levels (a 'combination' of the levels depicted in Figure 3.15a,c). Consequently, a polaron is characterized by two states in the gap. Polarons are also encountered in inorganic semiconductor physics: an electron moves through a lattice and affects the constituent ions. However, the resulting distortion of the lattice in inorganic semiconductors is small compared to the polaron defect in conjugated polymers.

The polaron and bipolaron are illustrated schematically for poly(*p*-phenylene) (PPP) in Figure 3.16. The positive (negative) polaron is a radical cation (anion), an entity consisting of a single electronic charge associated with a local geometrical relaxation of the bond lengths. Similarly, a bipolaron is a bound state of two charged solitons of like charges (or two polarons whose neutral solitons annihilate each other) with two corresponding mid-gap levels. The formation of these defect states leads to new localized energy levels in the band gap; consequently, characteristic optical absorption signatures will be observed. For further information, the reader is referred to the excellent book by Roth, which contains a more detailed description of the 'menagerie' of conjugational defects found in conductive polymers [10].

Figure 3.15 Energy levels associated with (a) a neutral soliton, (b) a positively charged soliton, and (c) a negatively charged soliton. The charge q and spin s of the defects are indicated.

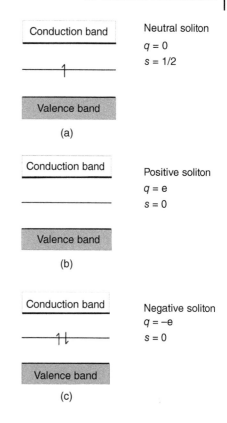

Soliton and polaron states play an important role in the conductivity of conjugated polymers. For example, it is generally believed that electrical conduction in doped polyacetylene proceeds by a hopping mechanism that may involve the capture of a mobile electron from a neutral soliton in an adjacent chain. However, due to disorder, solitons and polarons lose many of their characteristic features in real polymer samples.

3.7 Disordered Semiconductors

The effects of disorder on the energy band structure for both inorganic and organic semiconductors have been discussed in Chapter 1. Here, some other consequences of disorder on the morphology of materials and their electronic transport behaviour are explored.

Non-crystalline or amorphous inorganic materials are often formed by rapidly cooling a liquid so that the random arrangement of the atoms or molecules in the liquid phase is frozen-in. These materials are called glasses. The amorphous structure may have short-range order, as the bonding requirement of the individual molecules must be satisfied. The structure is a continuous random network of atoms. Because of the lack of long-range order, amorphous materials do not possess such crystalline imperfections as grain boundaries and dislocations, a distinct advantage in some applications. Some materials can exist in both the crystalline and amorphous forms. Figure 3.17 contrasts these two forms of silicon dioxide, SiO_2. The ordered, crystalline phase is depicted in Figure 3.17a. In contrast, Figure 3.17b represents the amorphous phase; this is vitreous silica and is formed by rapidly cooling molten SiO_2.

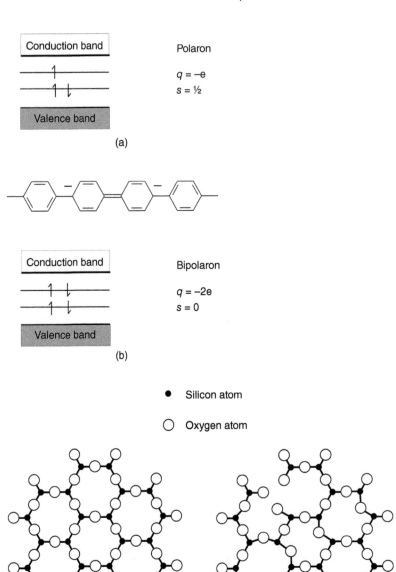

Figure 3.16 Schematic diagram of (a) a polaron and (b) a bipolaron in poly(*p*-phenylene) (PPP). The charge *q* and spin *s* of the defects are indicated.

Conduction band

Polaron

$q = -e$
$s = ½$

Valence band

(a)

Conduction band

Bipolaron

$q = -2e$
$s = 0$

Valence band

(b)

● Silicon atom

○ Oxygen atom

(a)   (b)

Figure 3.17 Silicon dioxide. (a) Crystalline form of SiO_2. (b) Amorphous SiO_2.

Amorphous pure silicon contains numerous dangling bonds like those found at the surface of single crystal silicon. However, so long as the short-range order present in the crystalline phase is essentially unchanged (similar bond lengths, bond angles, and local co-ordination) the main features of the density of states function is preserved. The overall result is that energy bands in amorphous materials are generally less defined than in their crystalline counterparts. Tails in the density of states function can extend into the band gap, as depicted in Figure 1.15; moreover, localized states can be found within the band gap.

In the case of organic molecular crystals, the transition from the crystalline to the amorphous state only involves the re-arrangement of relatively weak van der Waals bonds and narrow bands of localized states are formed (Figure 1.23). No dangling bonds are formed at surfaces or interfaces, which give organic molecular semiconductors an advantage over their inorganic counterparts. The amorphous state is favoured by the atomic makeup of many organic molecular materials, as close packing of the individual molecules is difficult to achieve. Material processing can also influence the degree of crystallinity. For example, when a thin film of an organic semiconductor is formed from a solution, the evaporation rate of the solvent will determine the film's morphology. If a low boiling point solvent is chosen, then the solvent evaporation rate can occur faster than the process of crystallization and an amorphous film is formed. Higher boiling point solvents can be used to increase deliberately the film drying time. This so-called 'solvent annealing' can result in improved thin film morphology and enhanced carrier mobilities.

The crystalline state may exist in polymeric materials. If the chains of polymeric molecules were all the same length and completely stretched out or if the chains were folded back and forth in a symmetrical manner, as depicted in Figure 3.18, then these molecules would form completely crystalline structures. However, most polymeric materials comprise of regions that are both crystalline and amorphous. The density of a crystalline polymer will be greater than an amorphous one of the same material and molecular weight, since the chains are more closely packed together for the crystalline structure. The degree of crystallinity may therefore be determined from accurate density measurements.

The molecular chemistry as well as chain configuration influences the ability of a polymer to crystallize. Crystallization is not favoured in polymers that are composed of chemically complex mer structures (the basic chemical repeat unit of the polymer). On the other hand, crystallization is not easily prevented in chemically simple polymers such as polyethylene, even for very rapid cooling rates. When a polymer is stretched or drawn, the internal shearing action tends to align the long molecules preferentially in the stretch direction. Not surprisingly, such one-dimensional ordering tends to induce crystallization. Charge transport in polymeric materials will consist of both carrier movements along the chains and carrier transfer, a hopping process, between the chains.

Figure 3.18 Chain-folded lamella example of polymer crystallinity.

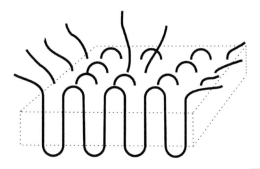

3.8 Electron Transport in Low-Dimensional Systems

As the size associated with materials and devices moves into the nanoscale, dimensional effects become increasingly important. Electron transport on the nanoscale depends on the relationship between the sample dimensions and three important characteristic lengths:

(i) The mean free path, l, which represents the average distance that an electron travels before it collides inelastically with impurities or lattice vibrations;

(ii) The phase relaxation length, l_{ph}, which is the distance after which the phase memory of electrons (electron coherence) is lost; and

(iii) The electron Fermi wavelength, λ_F, which is the wavelength of electrons that dominate electrical transport.

The electron transport is diffusive if the relevant sample dimension $L > l$. The transport is ballistic if $L \ll l$, l_{ph}, then the electron does not scatter and the electron wavefunction is coherent. In some high-mobility inorganic semiconductor heterostructures, l and l_{ph} can be tens of micrometres, whereas for polycrystalline metal films, l is only a few tens of nanometres. The conductance is quantized when $L \sim \lambda_F$. Diffusive transport involves electrons with a wide energy distribution, but ballistic transport involves only those electrons with energies close to the Fermi energy.

Other peculiarities of low-dimensional systems relate to the density of states function $S(E)$. (Section 1.2.2 and Problem 1.4). This is shown in Figure 3.19 for different dimensionalities. In the case of a bulk, three-dimensional (3D) semiconductor, the dependence of $S(E)$ on energy is given by the square-root relationship derived in Eq. (1.7). An equivalent expression for a 2D structure, for example a quantum well, can be obtained by counting all the possible states in a plane. The result is that $S(E)$ becomes independent of energy. For a 1D quantum wire, such as a carbon nanotube, the states are counted along a line. The results shows that $S(E) \propto E^{-1/2}$. For a 0D structure, such as quantum dot, all the available states exist only at discrete energies and can be represented by a delta function. In practice, the size distribution of an array of quantum dots leads to a broadening of this delta function.

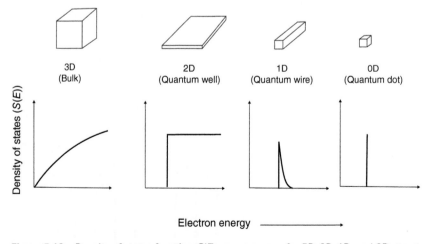

Figure 3.19 Density of states function $S(E)$ versus energy for 3D, 2D, 1D, and 0D structures.

3.8.1 Two-Dimensional Transport

Electrons in a 3D block of material are free to travel in any direction, forming an electron 'gas'. In the case of a thin slab of material, the electrons can still travel freely in the plane of the slab, but their motion in the third dimension is restricted. The wave function of an electron in this dimension is represented by a standing wave – the 'electron in a box model', often used in quantum mechanics. The motion of the electron in the third dimension is quantized and can be represented by a ladder of levels of increasing energy, with the energy separation between the levels becoming larger as the slab is made thinner. Two-dimensional electron gases can be artificially created using inorganic crystalline semiconductors. A widely exploited system is the gallium arsenide/aluminium gallium arsenide (GaAs/AlGaAs) hetero-structure which can be grown to near-epitaxial perfection using the technique of molecular beam epitaxy. This contains a 2D electron gas at the interface between the two materials. There is little disorder in this region so that the electrons are scattered much less than they are in silicon and possess high values of mobility. The 2D electron gas in a GaAs/AlGaAs hetero-junction has a Fermi wavelength which is a hundred times larger than in a metal. This makes it possible to study a constriction with an opening comparable to the wavelength and much smaller than the mean free path for impurity scattering. Such a constriction is called a quantum point contact and has been used to discover a sequence of steps in the conductance of a constriction of a 2D electron gas as its width is varied. Examples of two-dimensional organic solids include graphene sheets and the phthalocyanine compounds.

Generally, the carrier mobility in a thin film (<1 μm) of a semiconductor will differ from that in a bulk material. There are two distinct ways in which a surface can behave towards free carriers. First, the carriers can be scattered in a diffuse manner. This implies that the carriers emerge from the surface having lost all memory of their velocities prior to the collision. This type of scattering leads to a reduction in mobility of the carriers drifting within a mean free path of the surface. Specular scattering, on the other hand, requires that only the momentum component normal to the surface changes (the parallel components remaining constant) and does not lead to a mobility reduction. This type of scattering is expected from an ideal surface. The disorder present on a non-ideal (real) surface will result in a measure of diffusivity.

For mathematical convenience, rigorous theories taking into account surface scattering assume a constant relaxation time [11]. For example, consider a one-carrier system, corresponding to an extrinsic n-type semiconductor in the form of a thin film of thickness $2d$ comparable to the mean free path of the carriers. The band edges are assumed to continue flat up to the surface, as depicted in Figure 3.10. The electron density is taken to be uniform throughout the thin film and equal to the bulk value n_b. The effect of surface scattering is introduced in the form of some average collision time τ_s, just as bulk scattering is characterized by the relaxation time τ_b. If it is assumed that the bulk and surface-scattering processes are additive (Matthiessen's rule, Eq. (2.28)), then the average relaxation time for electrons in the thin film τ_f is given by

$$\frac{1}{\tau_f} = \frac{1}{\tau_s} + \frac{1}{\tau_b} \tag{3.19}$$

Another way of expressing Eq. (3.19) is to say that the probability per unit time that an electron is scattered is given by the sum of the scattering probabilities for the bulk and the surface taken separately. An estimate of τ_s (the mean distance d of a carrier from the surface divided by the mean velocity of the electrons in a direction perpendicular to the plane of the thin film, $\overline{v_z}$) is taken as follows:

$$\tau_s \approx \frac{d}{v_z} = \left(\frac{d}{l}\right) \tau_b \qquad (3.20)$$

where l is the mean free path defined by $l = \tau_b \overline{v_z}$. The average electron mobility in the thin film μ_f is given in Eq. (2.22), i.e. $\mu_f = e\tau_f/m_e^*$. Using Eqs. (3.19) and (3.20):

$$\frac{\mu_f}{\mu_b} = \frac{1}{1 + l/d} \qquad (3.21)$$

Thus, the carrier mobility in the thin film decreases with decreasing film thickness, while for $d \gg l$, it approaches the value in the bulk (Problem 3.2). Equation (3.21) will be valid not only for thick films in which the energy bands remain flat but also for thin films such that d is small compared with the Debye length given by Eq. (3.18).

Using the above approach, estimates can also be made for the surface mobility of thick films ($d \gg L_D$) [11]. An interesting example is the thin accumulation layer of majority carriers that can be formed at the surface of a semiconductor by the field-effect (Section 5.2). In this case, the semiconductor bands will be bend at the surface, compared to the bulk. The charge carriers in this layer can be considered as moving in a thin film with one surface (the physical surface) a diffuse scatterer and the other a specular reflector. The surface mobility μ_s will be determined by the amount of band bending at the surface and can be expressed as follows:

$$\frac{\mu_s}{\mu_b} = \frac{1}{1 + \left(1/L_c\right)\sqrt{1 + \frac{e\psi_s}{k_B T}}} \qquad (3.22)$$

where L_c is a parameter related to the Debye length. For small amounts of band bending (approaching flat bands conditions), L_c approaches L_D. In Eq. (3.22), ψ_s is known as the surface potential and is a measure of the surface band bending (for flat bands, $\psi_s = 0$). This parameter (units of V) is formally defined in Section 5.2.

3.8.2 One-Dimensional Transport

If the electrons in a thin metal wire can be considered by the electron in a box model, and if the size of the box is just the Fermi wavelength, only the first energy state is occupied. If the energy difference to the next level is much larger than the thermal energy ($\gg k_B T$), there are only completely occupied and completely empty levels, and the system is an insulator.

A thin wire represents a small box for electronic motion perpendicular to the axis, but it is a very large box for motion along the wire. Hence, in two-dimensions (radially), it is an insulator, and in one-dimension (axially), it is a metal. If there are only a few electrons, the Fermi energy is small, and the Fermi wavelength is large. This is the situation for semiconductors at low doping concentrations. Wires of such semiconductors are already one-dimensional if their diameter is around 10 nm.

In a 1D wire, the electrons are now quantum mechanically confined in two dimensions. The electrical conductance of such a wire, where the carriers travel ballistically (no scattering) is given by

$$G = \frac{2e^2}{h} \qquad (3.23)$$

One-dimensional electron transport is quantized. The quantum of resistance is e^2/h ($= 25.8\,k\Omega$) and is known as the von Klitzing constant. In a ballistic conductor, there are often a finite number of electron modes M, or parallel 1D sub bands. For example, metallic armchair nanotubes (Section 1.5.7) possess two conduction sub bands ($M = 2$) at the Fermi energy. Furthermore, not all electrons injected at one contact arrive at the other contact, and the electron wave function can be likened to tunnelling though a barrier with transmission probability T. Hence, the conductance of a ballistic conductor between two reflectionless contacts at $0\,\text{K}$ is given by the Landauer formula:

$$G = \frac{2e^2}{h}MT \tag{3.24}$$

In the absence of collisions, the resistance can only originate from the conductor/contact interface, and hence R is often referred to as the contact resistance (Problem 3.3). The conductance of such ballistic structures is therefore independent of the sample length. One-dimensional (and even zero-dimensional) systems may also be achieved by selecting the anisotropy of the material. For example, certain materials (conductive polymer chains and carbon nanotubes) may be regarded as natural (quasi) one-dimensional. Quantized conductance has been reported in individual multi-wall carbon nanotubes [12].

The quantization of an electron's resistance can be understood in semi-classical terms. When a voltage V is applied between two contacts, it generates a current $I \propto e\overline{v_d}S(E)V$. However, since $\overline{v_d} \propto E^{1/2}$ and $S(E) \propto E^{-1/2}$ for a 1D structure, the two terms cancel, and the resistance depends only on constants.

Finally, it should be noted that ballistic conduction differs from superconductivity due to the absence of the Meissner effect in the material (Section 2.8). A ballistic conductor would stop conducting if the driving force is turned off, whereas in a superconductor current would continue to flow after the driving supply is disconnected.

3.8.3 Zero-Dimensional Transport

Semiconductor quantum dots may be considered as zero-dimensional objects. These structures are small discs of material that are small compared to the Fermi wavelength so that the electrons are restricted in all three dimensions. The quantum dots of the organic world are the fullerenes, regular clusters of carbon atoms (Section 1.5.6). A C_{60} cluster (sixty carbon atoms) has a diameter of about 1 nm.

Whenever an electron is added to a conductor, for instance when current is flowing through a device, the additional charge will cause the existing charges to re-arrange slightly. This has the effect of modifying the potential energy of the conductor. In most situations, the effect is not observable (expect for some types of noise, such as shot noise) as the perturbation is too small. However, in nanoscale and molecular devices, this phenomenon can become appreciable and can have a marked effect on the transport characteristics.

Single-electron devices are those that can control the motion of even a single electron and consist of quantum dots or nanoparticles associated with tunnel junctions [13]. However, the terminology can be a little confusing [14]. The name suggests that the devices work with only one electron, which is not quite correct. Even a single atom contains a number of electrons and a metal- or semiconductor-based 'single-electron' device will have many electrons, which have relevance to its operation.

The principle of Coulomb blockade enables electrons to be localized on an isolated island or transported to one so that a precise number are transferred. In the single-electron transistor based on this effect, a capacitively coupled gate electrode is used to control this transfer (Section 8.7). The simplest device is the arrangement depicted in Figure 3.20. Electrons can tunnel between a

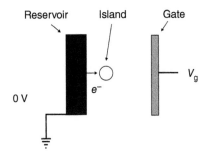

Reservoir   Island   Gate

0 V

V_g

e^-

Figure 3.20 Principle of Coulomb blockade. Electronic charge is isolated on an 'island'. Electrons are transferred from a charge reservoir to the island by the application of a voltage V_g to a gate electrode.

reservoir and an island. The gate voltage controls this electron motion. Moving an electron from the reservoir to the island is accomplished by applying a voltage to the gate V_g, where

$$V_g = \frac{e}{C} \tag{3.25}$$

C is the capacitance of the island due to the gate electrode. The charging energy E is given by the expression for the energy associated with a capacitor:

$$E = \frac{e^2}{2C} \tag{3.26}$$

The capacitance of a sphere imbedded in an insulator of relative permittivity ε_i scales with its radius, R:

$$C = 4\pi\varepsilon_i\varepsilon_0 R \tag{3.27}$$

(Students of electrostatics might think it unusual that an isolated sphere has a finite value of capacitance; however, deriving the capacitance between two concentric spheres and then allowing the radius of the outer sphere to become infinitely large can easily demonstrate the above formula.)

For a 1 μm island (sphere) in free space, $C \sim 5 \times 10^{-17}$ F and $E \sim 2 \times 10^{-22}$ J or about 1.4 meV. At 300 K, the thermal energy $k_B T$ is about 25 meV, an order of magnitude larger. However, if the island is very small (<10 nm) or the temperature is very low, the charging energy becomes greater than the thermal energy and single electrons may be isolated on the island. Strong Coulomb repulsion – Coulomb blockade – will block the transfer of a second electron (Problem 3.4).

A second requirement necessary for the observation of Coulomb blockade is that the quantum fluctuations in the electron number n on the island are sufficiently small so that the charge is localized on the island. This translates to a lower limit for the resistance of the tunnelling barrier, R_t.

If the typical time to charge or discharge the island is Δt, then

$$\Delta t = R_t C \tag{3.28}$$

The Heisenberg Uncertainty principle (Eq. (2.29)):

$$\Delta E \Delta t \approx \frac{e^2}{2C} R_t C \geq \frac{\hbar}{2} \tag{3.29}$$

implies a lower limit for R_t (related to the resistance quantum discussed above) for the energy uncertainty to be much smaller than the charging energy:

$$R_t \geq \frac{\hbar}{e^2} \tag{3.30}$$

This is the von Klitzing constant, encountered above in Section 3.8.2.

3.9 Nanosystems

The development of molecular electronics, as with all other areas of scientific endeavour, is constrained by the laws of physics; in most cases, these are the laws of classical or Newtonian physics. However, on the nanoscale, other principles become more important, even crucial. Quantum mechanical tunnelling (Section 6.3) becomes a significant electrical conduction process at dimensions less than 5 nm and therefore will have a key role as molecular scale electronics develops. In the following sections, some other concepts are introduced that might be significant for electronic and/or optoelectronic devices operating at nanometre dimensions.

3.9.1 Scaling Laws

The magnitudes of physical quantities characterizing nanoscale systems differ considerably from those familiar from the macro-scale world. Some of these quantities can be estimated by applying scaling laws to the values for macro-scale configurations [14, 15]. For example, the strength of a structure and the force it exerts can be assumed, in the first instance, to scale with its cross-sectional area. Nanoscale devices accordingly exert only small forces: a stress of 10^{10} N m^{-2} equates to 10^{-8} N nm^{-2} or 10 nN nm^{-2}.

Of relevance to this chapter is the scaling of classical (macroscopic) electromagnetic systems. Here, it is convenient to assume that electrostatic field strengths (hence electrostatic stresses) are independent of scale. The onset of strong field-emission currents from conductors limits the electrostatic field strength permissible at the electrodes of nanoscale systems; values of 10^9 V m^{-1} can readily be tolerated. At this field strength, one nanometre corresponds to a potential difference of 1 V.

If all the dimensions of a material are reduced by a constant K, then the effects on various important electrical parameters can be calculated. For example, electrical resistance R will scale with K, while the capacitance C (for a parallel plate capacitor) will scale as K^{-1} (assuming of course that the resistivity and permittivity of the material remain unchanged as its dimensions are reduced). An important consequence is that the time constant of a resistor capacitor combination, the RC product, will remain unchanged. This is also the case for the current density J (J = current/area, and both the current and area decrease as K^{-2} with scaling). Current densities in aluminium interconnections in microelectronic circuitry are limited to 10^{10} A m^{-2} or less by electromigration, which is a diffusive process that redistributes metal atoms under the influence of electrical forces and eventually interrupts circuit continuity (Section 6.9). This current density equates to 10 nA nm^{-2}.

The above represent general scaling laws. Similar arguments can be applied to electronic devices, such as the field effect transistor, as described in Chapter 8 [16]. In some models, as the device dimensions are reduced by K, the doping density is increased by K to keep the resistance constant. In this case, the RC time constant varies as K^{-1}, thereby providing faster switching times. A disadvantage, however, is that the current density now varies as K. The assumption of a constant electric field also provides problems as this leads to a reduction in operating voltages, which means that compatibility with standard voltage levels used in other parts of the electronic system will be lost (Problem 3.5).

The diffusion of matter (the net transport of particles from a region of higher concentration to a region of lower concentration) is a macroscopic manifestation of Brownian motion. Brownian motion can be described by a random walk, in which a particle moves a fixed distance, hits something, and then sets out in a totally different (random) direction. The effect is that the distance travelled is proportional to the square root of the number of steps (in contrast to a normal walk in which, for travel in a straight line, the distance travelled is proportional to the number of steps). This has important consequences on the nanoscale. The diffusion coefficient of oxygen in water is 18×10^{-10} m^2 s^{-1}. The time taken for a molecule to diffuse a given distance is approximately the (distance)^2 divided by the diffusion coefficient. Hence, an oxygen molecule will move 10 nm in about 50 ns but will take around 90 minutes to move 1 cm. Bigger molecules diffuse somewhat more slowly than smaller ones. The diffusion coefficient for a sugar molecule (sucrose) in water is about 5×10^{-10} m^2 s^{-1}, while for a macro-molecule like a protein, the diffusion coefficient is likely to be less than 10^{-10} m^2 s^{-1}.

3.9.2 Interatomic Forces

Solids are held together by strong ionic or covalent forces in the macro-world. Other bonds, such the van der Waals bond and the hydrogen bond are much weaker, but these can become important as nano-dimensions are approached. The classical van der Waals bond is a very short-range interaction with its associated energy varying as r^{-6}, where r is the distance between

(a) Atom-surface

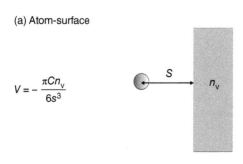

$$V = -\frac{\pi C n_v}{6s^3}$$

Figure 3.21 Potential energy of the van der Waals attractions V for three different extended geometries. (a) atom-surface, (b) sphere-surface, (c) two parallel chains. n_v is the number of atoms per unit volume. A is the Hamaker constant, a tabulated material parameter and C is the van der Waals constant. Sources: Adapted from Drexler [14] and Reitman [17].

(b) Sphere-surface

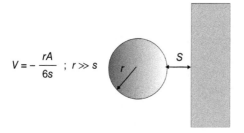

$$V = -\frac{rA}{6s} \quad ; \quad r \gg s$$

(c) Two parallel chains

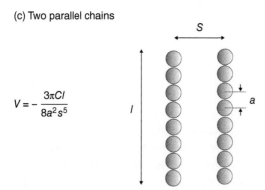

$$V = -\frac{3\pi C l}{8a^2 s^5}$$

the interacting molecules. However, if a single molecule or atom is interacting with a large body, such as a plane or a large sphere, the dependence of the interaction on the spacing from the point to the plane or sphere will become r^{-3}. Other possibilities are shown in Figure 3.21 [14, 17].

A further force, the Casimir force, operates between metallic surfaces, forcing them together. This is associated with the modes of oscillation of the electromagnetic field in an enclosed region. The force is vanishingly small except for extremely closely spaced surfaces, those that are relevant to molecular scale devices. For example, for two metal surfaces spaced by 10 nm, there is an attractive force of about 100 kPa (one atmosphere) arising from the Casimir effect.

Problems

3.1 A simple model for a Coulombic attractive trap is to equate the Coulomb potential energy equal to the average electron energy for a critical distance r_c. If the distance from the trap $r < r_c$, then the electron is captured.

Using the above approach, derive an expression for the capture cross section σ of a Coulombic attractive trap. Hence, estimate a value for σ at 300 K for a semiconductor with a relative permittivity of 10. How does this figure compare with the geometric cross-sectional area of the trap?

3.2 The bulk, room temperature, carrier mobilities of an inorganic semiconductor and an organic molecular crystalline solid are 1000 and 10 cm^2 V^{-1} s^{-1}, respectively. Estimate the effective carrier mobilities at room temperature for thin films of these materials with the following thicknesses: (i) 2 μm, (ii) 200 nm, and (iii) 20 nm.

3.3 Assuming that a carbon nanotube can be considered as a ballistic conductor, with two channels and perfect electrical contacts at each end, calculate its expected resistance. There will be an energy (power) associated with the current passing through the ballistic conductor with a finite value of resistance. Where is this power dissipated?

3.4 A single electron transistor operates by transferring electrons from a charge reservoir to a small spherical 'dot'. If the device is to operate satisfactorily at room temperature, estimate a maximum radius for the dot.

3.5 In a 1974 paper (*Chem. Phys. Lett.* 1974; 29 : 277–283), Aviram and Ratner proposed that a simple electronic rectifying device could be based on an organic molecule containing a donor and an acceptor group separated by a short σ-bonded bridge. Look up the original reference and estimate the molecular weight of this rectifying molecule. Hence, estimate the number of 'devices' that are contained in 100 g of the solid compound. How does this figure compare to (i) the number of transistors contained in a typical computer chip and (ii) the number of transistors manufactured, worldwide, per annum? (Again, some research will be needed to find these answers). What are the practical problems in exploiting the potential of the molecular rectifier in a computational architecture?

References

1 Conwell, E. and Weisskopf, V.F. (1950). Theory of impurity scattering in semiconductors. *Physiol. Rev.* 77: 388–390.

2 Smith, R.A. (1969). *Wave Mechanics of Crystalline Solids*, 2e. London: Chapman and Hall.

3 Haneef, H.F., Zeidell, A.M., and Jurchescu, O.D. (2020). Charge carrier traps in organic semiconductors: a review on the underlying physics and impact on electronic devices. *J. Mater. Chem. C* 8: 759–787.

4 Rose, A. (1963). *Concepts in Photoconductivity and Allied Problems*. New York: Robert E. Krieger Publishing Co.

5 Bube, R.D. (1992). *Photoelectronic Properties of Semiconductors*. Cambridge: Cambridge University Press.

6 Sah, C.T., Noyce, R.N., and Shockley, W. (1957). Carrier generation and recombination in *p-n* junction and *p-n* junction characteristics. *Proc. IRE* 45: 1228–1243.

7 Hall, R.N. (1952). Electron-hole recombination in germanium. *Physiol. Rev.* 87: 387.

8 Shockley, W. and Read, W.T. (1952). Statistics of recombination of holes and electrons. *Physiol. Rev.* 87: 835–842.

9 Sze, S.M. and Ng, K.K. (2007). *Physics of Semiconductor Devices*, 3e. Hoboken, NJ: Wiley-Interscience.

10 Roth, S. (1995). *One-Dimensional Metals*. Weinheim: VCH.

11 Many, A., Goldstein, Y., and Grover, N.B. (1965). *Semiconductor Surfaces*. Amsterdam: North-Holland.

12 Uchida, K. (2005). Single-electron devices for logic applications. In: *Nanoelectronics and Information Technology*, 2e (ed. R. Waser), 423–441. Weinheim: Wiley-VCH.

13 Yano, K., Ishii, T., Sano, T. et al. (1999). Single-electron memory for giga-to-tera bit storage. *Proc. IEEE* 87: 633–650.

14 Drexler, E.K. (1992). *Nanosystems*. New York: Wiley.

15 Rogers, B., Adams, J., and Pennathur, S. (2015). *Nanotechnology: Understanding Small Systems*. Boca Raton, FL: CRC Press.

16 Kasap, S.O. (2017). *Principles of Electrical Engineering Materials and Devices*, 4e. Boston: McGraw-Hill.

17 Reitman, E.A. (2001). *Molecular Engineering of Nanosystems*. New York: Springer.

Further Reading

Blakemore, J.S. (1969). *Solid State Physics*. Philadelphia: Saunders.

Chopra, K.L. (1969). *Thin Film Phenomena*. New York: McGraw-Hill.

Mataré, H.F. (1971). *Defect Electronics in Semiconductors*. New York: Wiley-Interscience.

Nicollian, E.H. and Brews, J.R. (1982). *MOS (Metal Oxide Semiconductor) Physics and Technology*. New York: Wiley-Interscience.

Petty, M.C. (2019). *Molecular Electronics: From Principles to Practice*, 2e. Chichester: Wiley.

Shong, C.W., Haur, S.C., and Wee, A.T.S. (2010). *Science at the Nanoscale*. Singapore: Pan Stanford.

Streetman, B.G. and Banerjee, S. (2000). *Solid State Electronic Devices*, 5e. Hoboken, NJ: Prentice Hall.

Sun, S.-M. and Dalton, L.R. (eds.) (2008). *Introduction to Organic Electronic and Optoelectronic Materials and Devices*. Boca Raton: CRC Press.

4

Electrical Contacts: Ohmic and Rectifying Behaviour

4.1 Introduction

Electrical measurements on semiconductor thin films (either inorganic or organic) are traditionally made by first establishing metallic electrical contacts to the sample. In this chapter, the physical nature of these contacts and their effects on the measured current versus voltage characteristics are discussed. First, the experimental challenges in obtaining and interpreting electrical conductivity data for thin films are considered.

4.2 Practical Considerations

Over many years, reliable equipment has been developed by the silicon industry to measure the DC resistivity of inorganic semiconductors, both in bulk and thin film form. The experimental methods are mostly based on probe techniques, in which an array of metallic probes is placed in direct contact with the semiconductor surface. Figure 4.1 shows an example of a four-point probe. The probes take the form of a linear array, with spacing s (other geometrical configurations are possible). A constant current, I, is passed through the outer probes and a voltage, V, is measured across the inner probes. The advantage of this scheme is that the effect of instrument lead resistances can be eliminated.

For a slab of material with infinite dimensions, the semiconductor resistivity, ρ, is given by

$$\rho = \frac{V}{I} 2\pi s \tag{4.1}$$

Thus, the four-point probe measures the material resistivity directly. In the case of a thin semiconductor wafer, where the wafer thickness $t \ll s$, then it is the sample sheet resistance R_s (Section 2.4) that is measured:

$$\rho = \frac{V}{I} \frac{\pi}{\ln 2} t = R_s \frac{\pi}{\ln 2} t \tag{4.2}$$

Correction factors are introduced to consider other sample geometries [1].

Probe methods are not particularly suited to organic thin films as the metallic probes can easily damage the relatively soft material. Moreover, organic semiconductors invariably possess lower conductivities than their inorganic counterparts and contact resistances can cause problems. Two-terminal methods are generally used in either a sandwich or an in-plane configuration. However, the experimental data can be difficult to interpret. One reason is that the solid metallic contacting electrodes may possess attendant oxide layers, which will play a role in the electronic

Electrical Processes in Organic Thin Film Devices: From Bulk Materials to Nanoscale Architectures, First Edition. Michael C. Petty.
© 2022 John Wiley & Sons Ltd. Published 2022 by John Wiley & Sons Ltd.
Companion Website: www.wiley.com/go/petty/organic_thin_film_devices

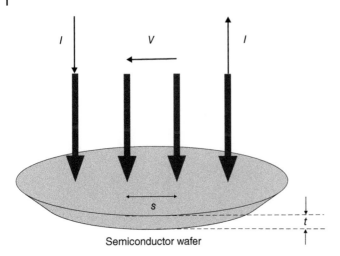

Figure 4.1 Four-point measurement of conductivity of a semiconductor wafer with an in-line arrangement of electrode probes. A constant current *I* passes through the outer probes and voltage *V* measured across the inner probes. Probe separation = *s*. Wafer thickness = *t*.

transport though the sample. The thicknesses of such oxides may, in some cases, be comparable with those of the thin organic film under scrutiny. Moreover, the electrical conductivity of the surface layers may be very low, leading to large contact resistances. The physical and chemical nature of the oxide layers will depend on the method of preparation. For example, the effective work function of an aluminium layer thermally evaporated onto a glass substrate, exposed to air, and subsequently coated with an organic film will differ from that of an aluminium layer deposited onto glass at reduced pressure and then coated, without breaking the vacuum, with the same organic film. The electrical conductivity of many organic materials is also affected by the presence of moisture and/or oxygen; certain chemical sensors exploit this fact. Stringent precautions are taken to encapsulate electronic devices, such as field effect transistors and displays, to improve their lifetimes. It is therefore imperative that all experiments on organic thin films and their associated devices are undertaken in a dry gas or under vacuum and, in some cases, in a dark environment. A further complication (which is common to all areas of microelectronics) is that of structural order in the organic film (Chapter 3). Defects, such as pinholes and grain boundaries, may act as nucleation points for the growth of metallic filaments.

For crystalline inorganic semiconductors, 'ideal' metal/semiconductor contacts may be established by cleaving the semiconductor in an ultra-high vacuum (<10^{-9} mbar) to create a fresh surface, and then immediately evaporating the metal. The contact is therefore formed before there is time for the surface of the semiconductor to become contaminated by the residual gases in the vacuum chamber. The resulting junction is ideal from a chemical point of view because of the lack of any insulating (or other) layer between the metal and the semiconductor. There may, however, be some physical damage to the semiconductor surface because of the cleaving process.

For organic materials, electrical contacts are usually made under vacuum, in air, or under an inert gas, using solid metallic electrodes. However, this is not always the case. It is possible to use liquid metals (e.g. Hg) or even electrolytes; the use of a scanning probe microscope for studies is another useful experimental technique. Such approaches are described later in this chapter. Typical contacting arrangements with solid electrodes are depicted in Figure 4.2a for investigations of out-of-plane conductivity and Figure 4.2b in the case of in-plane conductivity. The structure in Figure 4.2a facilitates easy contacting arrangements to both the top and bottom electrodes without damaging the

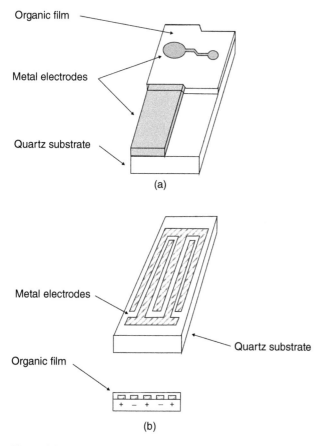

Organic film

Metal electrodes

Quartz substrate

(a)

Metal electrodes

Quartz substrate

Organic film

(b)

Figure 4.2 Electrode arrangements for measurements of organic thin film conductivity: (a) out-of-plane; (b) in-plane.

organic film. It is fabricated by first thermally evaporating a metal, such as aluminium, onto part of a glass substrate. The organic film is then deposited (e.g. by thermal evaporation or spin coating) on top of this. Finally, the top contact is established by evaporating another metal through a contact mask to provide the 'lollipop' shaped electrode. The latter process must be carried out very carefully to avoid affecting the organic film. It is desirable to use a metal that can be deposited at a low temperature (Al, Ag, Bi, or Mg are all suitable). A relatively slow evaporation rate is preferred, although it is important that the organic film is not exposed to the radiant heat from the evaporating source for an excessive time. Electrical contacts to the top and bottom electrodes may be made subsequently by soldering to the metal pads or by using an air-drying conductive (e.g. Ag) paste. Variations on this approach use a crossed-bar architecture (as shown in Figure 9.6). Photolithography can also be exploited to define the metallic electrodes. However, the various solvents used in the process need to be carefully chosen so that these do not affect the organic film.

Neglecting any influence of electrode interfacial layers, the out-of-plane electrical conductivity of the organic film sandwiched two electrodes, as shown in Figure 4.2a, is calculated from the applied voltage and measured current:

$$\sigma = \frac{It}{VA} = \frac{t}{AR} \tag{4.3}$$

where t is the film thickness, A is the area of electrode overlap (the presence of the 'stick' of the top lollipop electrode shown in Figure 4.2a needs to be accounted for), and R is the resistance. Some organic films (e.g. self-assembled monolayers (SAMs)) are of nano-metre dimension, so that a modest voltage can give rise to very high electric fields. For example, 100 mV applied across a 10 nm layer gives a field of 10^7 V m^{-1}. Although this is below the electrical breakdown strength of many solids (the DC breakdown strength for the polymer polytetrafluoroethylene (PTFE) is around 6×10^7 V m^{-1}, while that of SiO$_2$ can exceed 10^9 V m^{-1}), breakdown may occur because of an increase in field strength at the electrode edges or as a result of defects in the film. The field may also give rise to 'high-field' conduction processes (described in Section 6.6) for which the measured conductivity will not be independent of the applied voltage.

From an experimental point of view, the determination of the electrode dimensions for the out-of-plane configuration may not be entirely straightforward. For example, a high surface conductivity can effectively increase the electrode area A or reduce the value of t because of carrier screening. In the latter case, the creation of a virtual electrode beneath the real electrode may be considered.

To measure the in-plane conductivity, two metal electrodes are defined (e.g. by vacuum evaporation) onto an insulating (e.g. quartz) substrate and the organic film is deposited on top. The electrodes may also be deposited on top of the organic film. These two structures can possess different conductivities as the electrodes may not make good electrical contact with the entirety of the film thickness. This effect is discussed in more detail in Section 8.3, in which the different configurations of organic field effect transistors are compared. Figure 4.2b shows an arrangement of inter-digitated electrodes, which improves the sensitivity of the in-plane conductivity measurement. Neglecting contact resistance effects, the electrical conductivity can be evaluated from the current and voltage:

$$\sigma = \frac{Id}{VLt} \tag{4.4}$$

where d is the distance between the electrodes, and L is the total length of their overlap.

One problem with the in-plane measurements concerns the substrate. Clearly, when an electric field is applied in the plane of the organic film, a component of the current will flow through the substrate itself (the electrical resistance associated with the substrates will be in parallel with that due to the organic film). Therefore, it is essential to use only high resistance substrates and to establish a reference conductivity level by making a measurement on the uncoated substrate (no organic film). For both the architectures shown in Figure 4.2, it is possible to explore the effect of contact resistances – by varying the film thickness in the case of the out-of-plane structure or the electrode spacing for the in-plane interdigitated device. If contact resistances are negligible, then the measured current should scale, with t or d, as these parameters are changed.

A further issue concerning the interpretation of electrical data from organic thin films is related to the plethora of different DC and AC electrical conductivity process that may be operating. These are explored in detail in Chapters 6 and 7. Some of the physical mechanisms can exhibit similar current versus voltage behaviour and it is usually impossible to identify the physical process responsible for the electrical data from a measurement of the current versus voltage characteristic alone. This matter is discussed in Section 6.1.

4.3 Neutral, Ohmic, and Blocking Contacts

As noted in Section 2.6.4, when two materials are joined together, an important principle is that their Fermi levels must align in equilibrium (no applied electric field). A potential energy diagram

Figure 4.3 Potential energy diagram showing an idea metal/semiconductor contact, for which the metal work function Φ_m is equal to the semiconductor work function Φ_s. The Fermi levels in the metal and semiconductor will align, $E_{Fm} = E_{Fs}$. X_s = semiconductor electron affinity. CB = semiconductor conduction band. VB = semiconductor valence band.

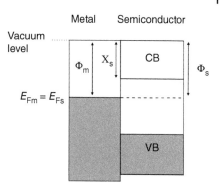

for the simplest metal/semiconductor contact is shown in Figure 4.3. The work function, Φ, of a material is the energy difference between its Fermi level and the vacuum level. (NB In the convention adopted in this book, Φ will have units of energy, J, or eV; in some texts Φ is measured in Volts.) The work function defines the energy at which the electron is free from that solid, and where the electron has zero kinetic energy. For the metal, the work function, Φ_m, is the minimum energy required to remove an electron from the solid. Typical values of Φ_m for very clean surfaces are 4.3 eV for Al and 5.1 eV for Au. These figures are changed if the surfaces become contaminated and differ for the same substance in different morphological forms (e.g. single crystal and polycrystalline). In the metal, there are electrons at the Fermi level, E_{Fm}, but in the semiconductor, there are usually none at E_{Fs} (i.e. the semiconductor Fermi level is probably located within the band gap). Nonetheless, the semiconductor work function, Φ_s, still represents the energy required to remove an electron from the semiconductor. In Figure 4.3, the energy difference between the lowest lying state in the conduction band and the vacuum level of the semiconductor is its electron affinity, X_s (Section 1.5.2; X_s also has units of energy in this text).

For the case under consideration in Figure 4.3, the work functions of the metal and semiconductor are identical. The energy bands in the semiconductor therefore remain flat to the interface with the metal. The contact is called neutral, in contrast to those contacts where the semiconductor bands bend up or down at the surface. The maximum current that can be drawn from such a contact is that due to thermionic emission (TE) from the metal over the potential step into the semiconductor. For an initial applied voltage, the metal can supply sufficient current to balance that flowing in the semiconductor and the conduction process is Ohmic. There is, however, a limit to the current that the metal electrode can supply, and this is the saturated thermionic (Richardson) current over the barrier. When this limit is reached, the conduction ceases to be Ohmic and becomes electrode limited (discussed in Chapter 6).

For an n-type semiconductor in which the electrons have an isotropic distribution of velocities, the number incident per second per unit area of boundary is given by simple kinetic theory as $\frac{n\bar{v}_t}{4}$, where n is the carrier concentration and $\bar{v}_t$ is the average thermal velocity. Hence, the maximum thermionic emission current density, $J_{te(max)}$, can be calculated:

$$J_{te(max)} = \frac{ne\bar{v}_t}{4} \tag{4.5}$$

The thermionic emission will be saturated at an electric field, $\mathcal{E}_{max}$, that results in a carrier drift velocity approximately equal to its thermal velocity:

$$J_{te\,(max)} = \frac{ne\bar{v}_t}{4} = \mathcal{E}_{max}\, ne\mu \tag{4.6}$$

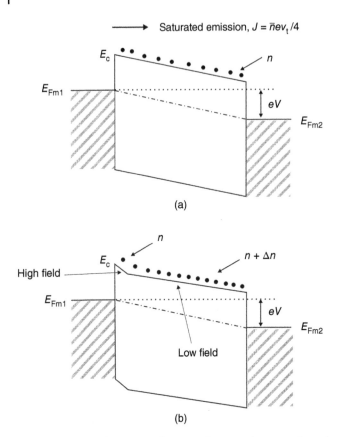

$$\longrightarrow \quad \text{Saturated emission, } J = \bar{n}ev_t/4$$

(a)

(b)

Figure 4.4 Potential energy diagrams of a neutral contact. A semiconductor is sandwiched between two metals 1 and 2 (Fermi levels, E_{Fm1} and E_{Fm2}) with a voltage V applied between them; Metal 2 is positive with respect to Metal 1. Metal 1 is able to supply carriers to the semiconductor up to the maximum value determined by thermionic emission. (a) Semiconductor in dark. (b) Semiconductor illuminated. The excess carriers Δn increase the electric field adjacent to the contact with metal 1 in order to supply the additional current.

where $\mathcal{E}_{max} = \frac{\bar{v}_t}{4\mu}$. Hence, a mobility of $100 \text{ cm}^2 \text{ V}^{-1} \text{ s}^{-1}$ and a thermal velocity of 10^7 cm s^{-1}, gives a saturating field of $2.5 \times 10^4 \text{ V cm}^{-1}$. At this field, departures from Ohm's law begin to occur because the carriers become 'hot' and their mobility decreases (Section 2.5).

For current densities less than the value given by Eq. (4.5), the current drawn through the semiconductor is less than that 'available' from the metal and the contact is Ohmic. In this case, an increase in the electric field in the semiconductor will provide a proportional increase in current, as indicated in Figure 4.4a, which shows the bands in a metal/semiconductor/metal structure (with different metals, 1 and 2, and corresponding Fermi energies E_{Fm1} and E_{Fm2}) with an applied voltage, V. With the polarity of the bias depicted, electrons pass from the left-hand contact with metal 1 into the semiconductor and out through metal 2. Metal 1 can also supply the additional current needed when the semiconductor is made more conductive by illumination of light (with energy $> E_g$). The effect of the excess carriers, Δn, is to increase the electric field at the cathode (negative) contact to supply the additional current, Figure 4.3b. The range of photocurrents for which the contact is Ohmic is, of course, limited to values less than the saturated thermionic emission. For light intensities higher than those needed to saturate the thermionic emission, the contact is blocking, i.e. it

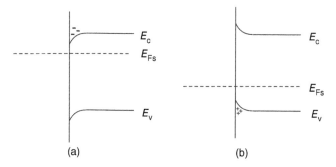

Figure 4.5 Potential energy diagrams showing ohmic contact formation on (a) n-type and (b) p-type semiconductors. For case (a) a metal with a small work function ($\Phi_m < \Phi_s$) is chosen. For (b) a high-work function metal ($\Phi_m > \Phi_s$) is used. In each case, the band bending at the semiconductor surface provides a region where there is a higher density of majority carriers.

supplies no additional carriers to replace those excited by the light in the volume of the semiconductor and drawn out at the anode. At these higher light intensities, the field becomes concentrated near the electrode where the carriers enter.

From the above discussion, it is evident than an Ohmic contact is a junction between a metal and a semiconductor that does not restrict the current flow. The contact, which acts as a reservoir of charge, is capable of supplying electrons to the semiconductor as required by the bias conditions. The conductivity is controlled by the semiconductor bulk and the current is essentially limited by the resistance of the semiconductor outside the contact region rather than by the thermal emission rate of carriers across a potential barrier at the contact. Very often such contacts are more difficult to fabricate than the Schottky barriers described in Section 4.4. Ideal metal/semiconductor contacts are Ohmic when majority carriers provide the charge induced in the semiconductor to align its Fermi level with that of the metal. This requires the use of low-work function metals (e.g. Al, $\Phi_m \approx 4.3\,\text{eV}$) for n-type semiconductors and high-work function metals (e.g. Au, $\Phi_m \approx 5.1\,\text{eV}$) for p-type semiconductors. An approach that is often used to make Ohmic contacts is to dope the semiconductor heavily in the contact region. If a barrier is then formed at the interface, the depletion width (discussed in Section 4.4.1) is sufficiently small to allow carriers to tunnel through the barrier (Section 6.3). Figure 4.5 shows Ohmic contact formation on n-type and p-type semiconductors. In both cases, the semiconductor energy bands bend at the surface to provide an increase in the majority carrier density.

4.4 Schottky Barrier

A good example of a blocking contact is a metal/semiconductor junction under reverse bias conditions. Such a device is referred to as a Schottky barrier following the work of Walter Schottky who assumed that the shape of the potential barrier in such a structure was determined by a uniform space-charge due to ionized impurities. Earlier work by Nevill Mott had suggested that the barrier region was devoid of charged impurities so that the electric field was constant [2].

4.4.1 Barrier Formation

The formation of a Schottky barrier between a metal and a semiconductor is frequently established by the thermal evaporation of a metal onto the surface of the semiconductor in a vacuum. Consider

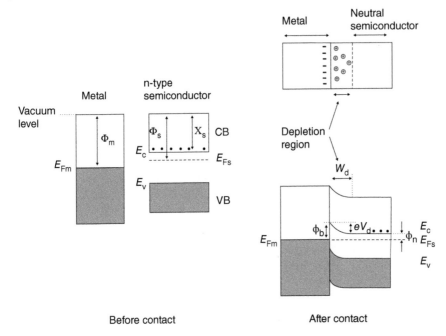

Figure 4.6 Formation of a Schottky barrier between a metal and an n-type semiconductor. The potential energy diagrams show the situations before and after contact has been established. CB and VB are the conduction and valence bands, respectively. E_c and E_v are the edges of the conduction and valence bands, respectively. E_{Fs} and E_{Fm} are the Fermi levels of the semiconductor and metal, respectively. Φ_s and Φ_m are the work functions of the semiconductor and metal, respectively. X_s is the electron affinity of the semiconductor. ϕ_b is the Schottky barrier height and V_d is the diffusion potential. W_d is the width of the depletion layer and ϕ_n is the distance of the Fermi level from the conduction band edge in the semiconductor bulk.

the case for an n-type semiconductor where the work function of the metal is greater than the work function of the semiconductor, $\Phi_m > \Phi_s$. The energy band diagrams before and after contact is made are shown in Figure 4.6. When the two solids come into contact, the more energetic electrons in the conduction band of the semiconductor can transfer to the metal, into lower empty energy levels (just above E_{Fm}). These transferred electrons accumulate at the surface of the metal and are contained within the Thomas–Fermi screening distance ($\sim$0.05 nm). The charge transfer leaves behind an electron-depleted region of width W_d in which there are uncompensated positively charged donor atoms, i.e. a net positive space charge. The positive charge within this so-called 'depletion' region matches the negative charge on the metal. A built-in or diffusion potential (sometimes also called the contact potential), V_d, is established between the metal and semiconductor. This prevents further electron diffusion from the semiconductor conduction band to the metal. In equilibrium, the value of V_d (measured in Volts) is

$$V_d = (\Phi_m - \Phi_s)/e \tag{4.7}$$

Associated with the diffusion potential is an internal electric field directed from the semiconductor to the metal surface (directed from positive to negative charge). This field is nonuniform and will have its maximum value at the metal/semiconductor interface (where there are a maximum number of field lines from positive to negative charge).

The Fermi level throughout the metal and semiconductor must be uniform in equilibrium (a Fermi level difference across the metal/semiconductor junction would result in an electric current

Figure 4.7 Variation of electrical parameters with distance, x, over the Schottky barrier region according to the depletion approximation. ρ = charge density; $|\mathcal{E}|$ = magnitude of electric field. W_d = depletion layer width.

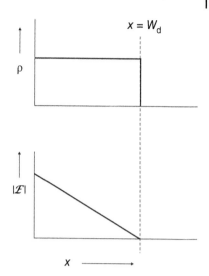

flowing in an external circuit). For this to occur, the energy bands in the semiconductor bend, as depicted in Figure 4.6. At the semiconductor edge of the depletion region, the semiconductor is still n-type. However, the carrier concentration in the conduction band n decreases towards the metal/semiconductor interface as $E_c - E_{Fs}$ increases (for intrinsic or undoped material, E_{Fs} is located close to the middle of the band gap, Section 2.6.1). The potential energy required for an electron to move from the semiconductor bulk to the metal is called the Schottky barrier height, ϕ_b:

$$\varphi_b = \Phi_m - \chi_s = eV_d + (E_c - E_{Fs}) = eV_d + \varphi_n \tag{4.8}$$

where ϕ_n is the energy difference between the semiconductor Fermi level and the bottom of the conduction band in the bulk semiconductor region (Figure 4.6). Note that the Schottky barrier height is greater than eV_d.

Figure 4.7 indicates how the charge density, ρ, and magnitude of the electric field, $|\mathcal{E}|$, vary throughout the depletion region of the semiconductor under the abrupt junction assumption that $\rho = eN_d$ for $x < W_d$ and $\mathcal{E} = 0$ for $x > W_d$. This premise is usually referred to as the depletion layer approximation. The electric field strength increases linearly with distance from the edge of the depletion layer to the interface with the metal. This is in accordance with Gauss' theorem. The depletion layer width and variation of the electric field may be obtained using Poisson's equation:

$$W_d = \sqrt{\frac{2\varepsilon_s\varepsilon_0}{eN_d}\left(V_d - V - \frac{k_B T}{e}\right)} \tag{4.9}$$

$$|\mathcal{E}(x)| = \frac{eN_d}{\varepsilon_s\varepsilon_0}(W_d - x) = \mathcal{E}_m - \frac{eN_d x}{\varepsilon_s\varepsilon_0} \tag{4.10}$$

The electrostatic potential within the depletion region may be evaluated by integrating the electric field [2] and will vary quadratically with distance x. At the surface ($x = 0$), the magnitude of this potential will be a maximum and equal to V_d:

$$V_d = \frac{eN_d W_d^2}{2\varepsilon_s\varepsilon_0} \tag{4.11}$$

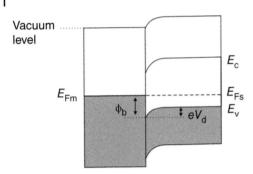

Figure 4.8 Formation of a Schottky barrier between a metal and an p-type semiconductor. E_c and E_v are the edges of the conduction and valence bands, respectively. E_{Fs} and E_{Fm} are the Fermi levels of the semiconductor and metal, respectively. ϕ_b is the Schottky barrier height, and V_d is the diffusion potential.

In the above equations, ε_s is the relative permittivity of the semiconductor, and $\mathcal{E}_m$ is the maximum electric field, which occurs at $x = 0$. This is

$$\mathcal{E}_m = \mathcal{E}(x = 0) = \sqrt{\frac{2eN_d}{\varepsilon_s\varepsilon_0}\left(V_d - V - \frac{k_BT}{e}\right)} = \frac{2(V_d - V - (k_BT/e))}{W_d} \tag{4.12}$$

The Schottky barrier can be modelled as a parallel plate capacitor as it is essentially two conductors (metal and semiconductor) separated by an insulator (depletion region). The space charge, Q_{sc}, per unit area of semiconductor and the depletion-layer capacitance, C_d, per unit area are

$$Q_{sc} = eN_dW_d = \sqrt{2e\varepsilon_s\varepsilon_0N_d\left(V_d - V - \frac{k_BT}{e}\right)} \tag{4.13}$$

$$C_d = \frac{\varepsilon_s\varepsilon_0}{W_d} = \sqrt{\frac{e\varepsilon_s\varepsilon_0N_d}{2(V_d - V - k_BT/e)}} \tag{4.14}$$

If N_d is constant throughout the depletion region, a straight line may be obtained by plotting $1/C_d^2$ versus applied voltage (Problem 4.1).

A Schottky barrier can also be made on a p-type semiconductor (most organic semiconductors exhibit p-type conductivity). In this case, the requirement is that the work function of the metal is less than that of the semiconductor, $\Phi_m < \Phi_s$. The band diagram is shown in Figure 4.8. At equilibrium, the alignment of the Fermi levels requires a positive charge on the metal and an equal negative charge distributed in the depletion region of the semiconductor. This is provided by ionized acceptors left uncompensated by holes.

4.4.2 Image Force

Before the electrical transport processes in Schottky barriers are considered, it is important to understand an important physical phenomenon that occurs in the vicinity of a metal surface – the effect of the image force.

Consider the interface between metal and vacuum. When an electron is at a distance x from the metal, a positive charge will be induced on the metal surface. The force of attraction between the electron and the induced positive charge is equivalent to the force which would exist between the electron and an equal positive charge, the image charge, located at $-x$. Coulomb's law gives the resulting attractive image force, A:

$$A = -\frac{e^2}{16\pi\varepsilon_0x^2} \tag{4.15}$$

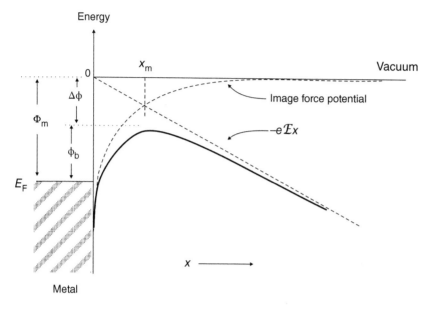

Figure 4.9 Potential energy versus distance x for a metal/vacuum interface. The metal work function is Φ_m. The effective barrier is lowered when an electric field is applied to the metal surface. This lowering is due to the combined effects of the field and the image force.

The work done by an electron in the course of its transfer from infinity to the point x defines the potential at x:

$$V(x) = \int_{\infty}^{x} A\,dx = -\frac{e^2}{16\pi\varepsilon_0 x} \tag{4.16}$$

This corresponds to the potential energy of an electron at a distance x from the metal surface, shown in Figure 4.9 and is measured downwards from the x-axis.

When an external electric field is applied (in the $-x$ direction), the total potential energy as a function of distance is given as

$$V_{\text{total}} = -\frac{e^2}{16\pi\varepsilon_0 x} - e|\mathcal{E}|x \tag{4.17}$$

The effect is shown in Figure 4.9. This lowering of the metal work function by an amount $\Delta\phi$ because of the image force and the applied electric field is sometimes called the Schottky effect (giving rise to some confusion between the Schottky *effect* and the Schottky *barrier*). The magnitude of the lowering and location of the potential maximum, x_m, are given by the condition $dV_{\text{total}}/dx = 0$:

$$x_m = \sqrt{\frac{e}{16\pi\varepsilon_0|\mathcal{E}|}} \tag{4.18}$$

$$\frac{\Delta\phi}{e} = \sqrt{\frac{e|\mathcal{E}|}{4\pi\varepsilon_0}} = 2|\mathcal{E}|x_m \tag{4.19}$$

For an electric field of 10^7 V m^{-1}, Eqs. (4.18) and (4.19) predict $x_m = 6$ nm and $\Delta\phi = 0.12$ eV, while for $\mathcal{E} = 10^9$ V m^{-1}, $x_m = 0.6$ nm, and $\Delta\phi = 1.2$ eV. Hence, at high fields, the effective metal work function for thermionic emission is significantly reduced.

The above results can be applied to a metal/semiconductor system. In this case, the electric field should be replaced by the appropriate field at the interface, $\mathcal{E}_m$, and the permittivity of free space, ε_0, should be replaced by the permittivity of the semiconductor, $\varepsilon_s\varepsilon_0$. The latter may be different from the static (low frequency) permittivity. This is because, during the emission process, if the electron transit time from the metal/semiconductor interface to the barrier maximum x_m is shorter than the dielectric relaxation time, the semiconductor medium will not have sufficient time become polarized, and a smaller permittivity than the static value (e.g. the high frequency value – see Section 7.3.2) may be expected. Equation (4.19) now becomes

$$\frac{\Delta\phi}{e} = \sqrt{\frac{e|\mathcal{E}_m|}{4\pi\varepsilon_s\varepsilon_0}} \tag{4.20}$$

Because of the larger values of permittivity in a metal/semiconductor system, the barrier lowering is less than that for the metal/vacuum system. For example, for a semiconductor permittivity $= 3\varepsilon_0$ (generally, organic semiconductors possess relative permittivity values in the range 2 to 4), $\Delta\phi$ is only 0.07 eV for $\mathcal{E} = 10^7$ V m^{-1} (Problem 4.2). The barrier lowering is small (and will depend on the voltage applied to the Schottky barrier) but can influence the carrier transport process [1, 2]. The relationship between $\Delta\phi$ and V can be shown to be approximately [2]:

$$\frac{\Delta\phi}{e} = \left(\frac{e^3 N_d}{8\pi^2\varepsilon_s^3\varepsilon_0^3}\left(\frac{\phi_b - \phi_n}{e} - V - \frac{k_B T}{e}\right)\right)^{1/4} \tag{4.21}$$

4.4.3 Current Versus Voltage Behaviour

Provided that no external connections are made to the Schottky barrier (open-circuit conditions) no net current will flow through the metal/semiconductor interface. The number of electrons thermally emitted over the potential energy barrier ϕ_b from the metal to the semiconductor is equal to the number of electrons thermally emitted over the barrier eV_d from the semiconductor to the metal. However, when the semiconductor side of the junction is connected to the negative terminal of an external DC power source (e.g. a battery) and the metal to the positive terminal, the effect will be to reduce the diffusion potential from V_d to $(V_d - V)$, where V is the magnitude of the external voltage. It is now easier for the electrons to overcome the potential energy barrier into the metal. This situation is called forward bias. When the polarity of the external power source is reversed, the Schottky barrier becomes reverse biased, and the diffusion potential is increased by an amount V, and it is more difficult for a current to flow. The Schottky contact will therefore exhibit rectifying, or diode, behaviour. Figure 4.10 shows the band diagrams for forward (Figure 4.10a) and reverse (Figure 4.10b) bias.

The current transport in metal/semiconductor contacts is due mainly to majority carriers (this is not the case in other electronic devices, such as p–n junctions). Figure 4.11 indicates four important transport processes that can occur under forward bias conditions for a Schottky diode based on a n-type semiconductor. These are emission of electrons from the semiconductor over the potential barrier into the metal (process a in the figure); quantum mechanical tunnelling of electrons through the barrier (process b); recombination in the space-charge region (process c); and diffusion of holes, injected from the metal, into the semiconductor (process d – equivalent to recombination in the neutral region). In addition, in a practical device there may be edge leakage currents due to high electric fields at the periphery of the metal contact and currents due to interface traps.

Before electrons can be emitted over the barrier into the metal, these must first be transported from the interior of the semiconductor to the interface. In traversing the depletion region of the semiconductor, their motion is governed by the usual mechanisms of diffusion and drift in the electric field of the barrier (Section 2.7.2). When the electrons arrive at the barrier, the number of

Figure 4.10 Schottky diode based on an n-type semiconductor. (a) Bands in forward bias. (b) Bands in reverse bias. V is the applied voltage and V_d is the diffusion potential. The shaded regions represent the presence of electrons.

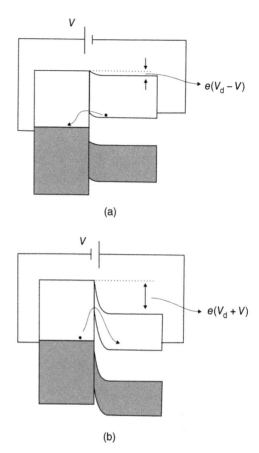

(a)

(b)

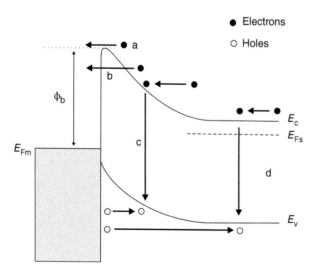

Figure 4.11 Possible carrier transport processes in a forward biased Schottky diode based on an n-type semiconductor: a emission of electrons from the semiconductor over the barrier; b tunnelling through the barrier; c recombination in the space charge region; and d recombination in the neutral region.

vacant energy states in the metal controls their emission. These two processes are in series, and the current is determined by whichever causes the largest impediment to the flow of electrons. Consequently, two distinct theoretical models have been developed to describe the current versus voltage behaviour of Schottky barriers, based on diffusion and thermionic emission.

4.4.3.1 Diffusion and Thermionic Emission Theories

According to diffusion theory, carrier drift and diffusion are the limiting factors for charge transport. Thermionic emission theory suggests that electron emission into the metal is more important [2]. A useful insight into the difference between these two approaches is provided by the behaviour of the electron quasi-Fermi level in the semiconductor (Section 2.7.1). The two possibilities for a Schottky diode based on an n-type semiconductor are depicted in Figure 4.12. In diffusion theory, the concentration of conduction electrons in the semiconductor immediately adjacent to the interface is not altered by the applied bias. This is equivalent to the assumption that, at the interface, the electron quasi-Fermi level, E_{Fqn}, coincides with the Fermi level in the metal. In this case, E_{Fqn} drops through the depletion region, as shown in Figure 4.12. For thermionic emission theory, E_{Fqn} remains constant up to the semiconductor/metal interface and falls in the metal as the emitted 'hot' electrons lose their energy. Since the gradient of the quasi-Fermi level provides the driving force for the flow of electrons, Figure 4.12 shows that, for diffusion theory, the main obstacle to current flow is drift and diffusion in the depletion region, whereas the process of emission of the electrons into the metal is the limiting factor for the thermionic emission model.

The predicted current versus voltage behaviours for the two theories are very similar. The starting point for the diffusion theory is the expression for the current density, J_d, due to drift and diffusion processes:

$$J_d = en\mu_n \mathcal{E} - eD_n \frac{dn}{dx} = eD_n \left(\frac{n}{k_B T} \frac{dE_c}{dx} - \frac{dn}{dx} \right) \tag{4.22}$$

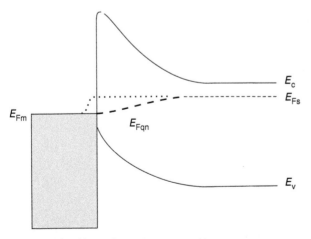

Figure 4.12 Electron quasi-Fermi level, E_{Fqn}, in a forward-biased Schottky diode based on a n-type semiconductor. Dotted lines according to thermionic emission theory. Dashed lines according to diffusion theory.

Using the abrupt junction approximation and neglecting the image force effect, it can be shown that the forward current through the Schottky junction is [1, 2]:

$$J_d \approx e\mu_n N_c \mathcal{E}_m \exp\left(-\frac{\phi_b}{k_B T}\right) \left(\exp\left(\frac{eV}{k_B T}\right) - 1\right) \tag{4.23}$$

i.e.

$$J_d = J_0 \left(\exp\left(\frac{eV}{k_B T}\right) - 1\right) \tag{4.24}$$

where J_0 is a constant for the Schottky device. When the voltage across the device is reversed, Eq. (4.24) predicts that the reverse current will saturate at $J_d = -J_0$. The diffusion theory predicts a current density versus voltage characteristic that is almost in the form of an ideal diode, i.e. $J = J_0(\exp(eV/k_B T) - 1)$, where J_0 is the reverse saturation current density. The difference arises because $\mathcal{E}_m$ is not independent of bias voltage but is proportional to $(V_d - V)^{1/2}$ [2]. According to diffusion theory, for large values of reverse bias, the current does not saturate but increases roughly as $|V|^{1/2}$ [2].

In the case of thermionic emission theory, the effects of drift and diffusion in the depletion region are assumed to be negligible, which is equivalent to assuming an infinite mobility [1, 2]. The current-limiting process is the transfer of electrons across the interface between the semiconductor and the metal and the actual shape of the barrier profile is unimportant. The electron concentration on the semiconductor side of the boundary is

$$n = N_c \exp\left(-\frac{(\phi_b - eV)}{k_B T}\right) \tag{4.25}$$

Hence, from simple kinetic theory (Eq. (4.5)), the current density from the semiconductor into the metal, $J_{s\to m}$, is

$$J_{s\to m} = \frac{eN_c \overline{v_t}}{4} \left(\exp -\frac{(\phi_b - eV)}{k_B T}\right) \tag{4.26}$$

There is also a flow of electrons from the metal into the semiconductor, $J_{m\to s}$, which is unaffected by the application of the bias because the barrier, ϕ_b, seen from the metal remains unchanged (neglecting any field dependence of the barrier height). For zero bias, the current from semiconductor to metal just balances this current:

$$J_{m\to s} = \frac{eN_c \overline{v_t}}{4} \exp\left(-\frac{\phi_b}{k_B T}\right) \tag{4.27}$$

The net thermionic emission current density, J_{te}, is therefore:

$$J_{te} = J_{s\to m} - J_{m\to s} = \frac{eN_c \overline{v_t}}{4} \exp\left(-\frac{\phi_b}{k_B T}\right) \left(\exp\left(\frac{eV}{k_B T}\right) - 1\right) \tag{4.28}$$

For a Maxwell–Boltzmann velocity distribution, kinetic theory gives $\overline{v_t} = \left(8k_B T/\pi m_e^*\right)^{1/2}$ while the conduction band effective density of states, N_c, has been previously defined in Eq. (2.34), $N_c = 2\left(2\pi m_e^* k_B T/h^2\right)^{3/2}$. Substitution into Eq. (4.28) yields:

$$J_{te} = A^* T^2 \exp\left(-\frac{\phi_b}{k_B T}\right) \left(\exp\left(\frac{eV}{k_B T}\right) - 1\right) \tag{4.29}$$

i.e.

$$J_{te} = J_0 \left(\exp\left(\frac{eV}{k_B T}\right) - 1\right) \tag{4.30}$$

where

$$J_0 = A^* T^2 \exp\left(-\frac{\phi_b}{k_B T}\right) \tag{4.31}$$

and

$$A^* = \frac{4\pi e m_e^* k_B^2}{h^3} \tag{4.32}$$

The constant A^* is the same as the Richardson constant for thermionic emission from a metal, with the substitution of the semiconductor effective mass for the free electron mass.

The current density expressions for the diffusion theory (Eq. (4.24)) and thermionic emission theory (Eq. (4.30)) are therefore very similar. However, as noted above, the saturation current density for diffusion theory is dependent on bias. It is also less sensitive to temperature compared to the saturation current density of the thermionic emission theory. Several authors have combined the thermionic emission and diffusion theories by considering the two processes to be in series [1, 2].

The effect of the image force on the current density versus voltage relationship can be calculated using thermionic emission theory. The effective barrier height ϕ_e may be written as

$$\phi_e = \phi_b - \Delta\phi \tag{4.33}$$

where the barrier lowering will be a function of applied voltage. If the change in the effective barrier height with voltage $(1/e)(d\phi_e/dV) = \beta$ is constant:

$$\phi_e = \phi_{b0} - \Delta\phi_0 + \beta eV \tag{4.34}$$

where ϕ_{b0} and $\Delta\phi_0$ refer to zero bias conditions. The coefficient β is positive because ϕ_e always increases with increasing forward bias. The current density now becomes

$$J_{te} = A^* T^2 \exp\left(-\frac{(\phi_{b0} - \Delta\phi_0 + \beta eV)}{k_B T}\right)\left(\exp\left(\frac{eV}{k_B T}\right) - 1\right)$$

$$= J_0 \exp\left(-\frac{\beta eV}{k_B T}\right)\left(\exp\left(\frac{eV}{k_B T}\right) - 1\right) \tag{4.35}$$

where

$$J_0 = A^* T^2 \exp\left(-\frac{(\phi_{b0} - \Delta\phi_0)}{k_B T}\right) \tag{4.36}$$

Equation (4.35) may be written in the form

$$J_{te} = J_0 \exp\left(\frac{eV}{n k_B T}\right)\left(1 - \exp\left(-\frac{eV}{k_B T}\right)\right) \tag{4.37}$$

where

$$\frac{1}{n} = 1 - \beta = 1 - \left(\frac{1}{e}\right)\left(\frac{d\phi_e}{dV}\right) \tag{4.38}$$

The parameter n is often called the ideality factor of the Schottky diode (Problem 4.3). If $(d\phi_e/dV)$ is a constant, then n is also a constant. For values of forward voltage greater than about $3k_B T$, Eq. (4.37) can be written as

$$J = J_0 \exp\left(\frac{eV}{n k_B T}\right) \tag{4.39}$$

Other effects, such as the presence of an interfacial layer (described in Section 4.4.4) can also give the current versus voltage relationship of Eq. (4.39).

4.4.3.2 Tunnelling Through the Barrier

Quantum mechanical tunnelling of electrons through the barrier (process b, in Figure 4.11) can be a significant conduction process for heavily doped (degenerate) semiconductors and/or at low temperatures. The origin of this mechanism is discussed in Section 6.3. Tunnelling at energies close to the semiconductor Fermi level, E_{Fs}, is known as field emission (FE). The tunnelling current from semiconductor to metal, $J_{s \to m}$, is proportional to the quantum transmission coefficient (tunnelling probability) multiplied by the occupation probability in the semiconductor and the unoccupied probability in the metal [1].

If the temperature is increased, electrons are excited to higher energies and the tunnelling probability increases rapidly as the electrons encounter a thinner barrier. This process is called thermionic-field emission (TFE). Of course, if the temperature is increased further, a point is reached at which all the electrons have sufficient energy to surmount the barrier – the process of thermionic emission described earlier. Figure 4.13 illustrates the processes of FE, TFE, and thermionic emission (TE) for a degenerate n-type semiconductor (note that E_{Fs} lies in the conduction band because of the heavy doping). To estimate the relative contributions of these components, it is useful to introduce a voltage term V_{00}, which is a parameter that plays an important role in tunnelling theory. For tunnelling through a Schottky barrier on a n-type semiconductor:

$$V_{00} = \frac{\hbar}{2} \sqrt{\frac{N_d}{m_e^* \varepsilon_s \varepsilon_0}} \tag{4.40}$$

V_{00} represents the diffusion potential of a Schottky barrier such that the transmission probability of an electron with energy coinciding with the bottom of the conduction band at the edge of the depletion region is equal to e^{-1}. Hence, when $k_B T \gg eV_{00}$, thermionic emission dominates; when $k_B T \ll eV_{00}$ tunnelling (field emission) is more important; and when $k_B T \approx eV_{00}$, thermionic-field emission is significant.

The current versus voltage behaviour for tunnelling are complex [3, 4]. Except for very low values of forward bias, the thermionic field-emission current may be described:

$$J = J_0 \exp\left(\frac{V}{V_0}\right) \tag{4.41}$$

Figure 4.13 Potential energy diagram showing tunnelling currents in a Schottky barrier based on a degenerate n-type semiconductor in forward bias. TE = thermionic emission. TFE = thermionic field emission. FE = field emission.

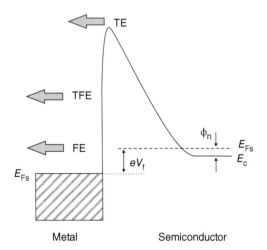

where the pre-exponential term J_0 depends weakly on bias and is a function of temperature, barrier height and semiconductor parameters, and V_0 is given as

$$V_0 = V_{00} \coth \left(\frac{eV_{00}}{k_B T} \right) \tag{4.42}$$

At low temperatures ($k_B T/eV_{00} \ll 1$), $V_0 \approx V_{00}$ and the slope of a graph of $\ln (J)$ versus V is independent of temperature. This is the case of field emission. At high temperatures ($k_B T/eV_{00} \gg 1$), eV_0 is slightly greater than $k_B T$ and the slope of a graph of $\ln (J)$ versus V can be written as $e/nk_B T$, where $n = eV_0/k_B T = (eV_{00}/k_B T)\coth(eV_{00}/k_B T)$.

4.4.3.3 Recombination in the Depletion Region

Recombination in the space-charge region (process c in Figure 4.11) usually takes place via localized centres [5]. As noted in Section 3.4, the most effective centres are those with energies lying close to mid-gap. The resulting recombination current density, J_r, versus voltage behaviour is the same as that for a p–n junction and is approximately given as

$$J_r = J_{r0} \left(\exp \left(\frac{eV}{2k_B T} \right) - 1 \right) \tag{4.43}$$

where J_{r0} is

$$J_{r0} = \frac{en_i W_d}{2\tau_r} \tag{4.44}$$

where n_i is the intrinsic carrier concentration, which is proportional to $\exp(-E_g/2k_B T)$, and τ_r is the lifetime in the depletion region. The total current density, J, will be the sum of the thermionic emission and recombination currents:

$$J = A^* T^2 \left(\exp \left(\frac{eV}{k_B T} \right) - 1 \right) \exp \left(-\frac{\phi_b}{k_B T} \right) + \frac{en_i W_d}{2\tau_r} \left(\exp \left(\frac{eV}{2k_B T} \right) - 1 \right) \tag{4.45}$$

For bias voltages greater than a few $k_B T$, the ratio of thermionic to recombination currents is

$$\frac{J_{te}}{J_r} \propto \tau_r \exp \left(\frac{(E_g + eV - 2\phi_b)}{2k_B T} \right) \tag{4.46}$$

Therefore, the recombination current is likely to be more important for a high barrier, low carrier lifetime, reduced temperature, and for small values of forward bias. When recombination dominates, the temperature variation of the forward current shows two activation energies. At high temperature, the activation energy (the slope obtained by plotting $\ln(J)$ versus T^{-1} for a fixed applied bias) is proportional to ($\phi_b - eV$), characteristic of the thermionic emission component, while at low temperature, it becomes proportional to ($E_g - eV$)/2, characteristic of the recombination process.

4.4.3.4 Minority Carrier Injection

The Schottky barrier is predominantly a majority carrier device. However, under certain circumstances, minority carriers can play an important role. For example, if the height of the Schottky barrier on n-type material is greater than half the band gap, the region of semiconductor adjacent to the metal becomes p-type and therefore contains a high density of holes. Some of these carriers will diffuse into the neutral region of the semiconductor under forward bias, thereby giving rise to hole injection – process d in Figure 4.11. In the case of thermionic emission theory, the minority-carrier injection ratio, γ, defined as the ratio of the minority current to the total current, is given by [1, 2]

$$\gamma = \frac{J_p}{J_p + J_n} \approx \frac{J_p}{J_n} = \left(\frac{\mu_p n_i^2 J_n}{\mu_n N_d^2 J_0} \right) + \left(\frac{eD_p n_i^2}{N_d d_{neut} J_0} \right) \tag{4.47}$$

where d_{neut} is the thickness of the neutral region (assumed to be less than the hole diffusion length), J_n, J_p are the electron and hole current densities, respectively, and J_0 is the electron (majority carrier) saturation current density for thermionic emission. The first term on the right-hand side of Eq. (4.47) is due to carrier drift and is bias (current) dependent. The second term is due to diffusion and is bias independent; this defines the injection ratio for small bias voltages. To reduce the minority carrier current, a Schottky barrier with a large N_d (corresponding to low resistivity material), a large J_0 (small barrier height), and a small n_i (large band gap) is required. Furthermore, high-bias operation should be avoided.

4.4.3.5 Reverse Bias Characteristics

According to thermionic emission theory, the reverse current density of an ideal Schottky diode should saturate at the value $J_0 = A^* T^2 \exp\left(-\frac{\phi_b}{k_B T} \right)$ (Eq. (4.31)). In contrast, diffusion theory predicts that for large values of reverse bias, the current increases roughly as $|V|^{1/2}$. There are other reasons why the reverse current might not saturate with bias. The image force effect is an obvious example. Since $\Delta\phi_b$ is proportional to $V^{1/4}$ for large values of reverse bias (Eq. (4.21)) a plot of ln (J) against $V^{1/4}$ should be a straight line. For inorganic semiconductors, the barrier lowering necessary to explain the lack of saturation of the reverse bias characteristics is often considerably less than that due to the image force. However, this might not be the case for organic materials, which generally possess smaller permittivity values.

Quantum mechanical tunnelling through the barrier becomes more significant at lower doping levels in reverse bias than in the forward direction because the bias voltages are much greater. The application of a moderately large reverse bias can cause the potential barrier to become thin enough for significant tunnelling of electrons from the metal to the semiconductor to take place. Tunnelling can again be described as either field or thermionic field assisted. In the thermionic-field emission regime, the current density versus reverse bias relationship has been evaluated [3]:

$$J = J_s \exp\left(\frac{V}{V'} \right) \tag{4.48}$$

where

$$V' = V_{00} \left(\left(\frac{eV_{00}}{k_B T} \right) - \tanh\left(\frac{eV_{00}}{k_B T} \right) \right)^{-1} \tag{4.49}$$

and

$$J_s = \frac{A^* T (\pi e V_{00})^{1/2}}{k_B} \left((eV - \phi_n) + \frac{\phi_b}{\cosh^2(eV_{00}/k_B T)} \right)^{1/2} \exp\left(-\frac{\phi_b}{eV_0} \right) \tag{4.50}$$

V_0 has been defined in Eq. (4.42). J_s is a slowly varying function of applied bias.

At higher doping concentrations, field emission may occur. Unlike the case of forward bias, field emission can take place in non-degenerate semiconductors because it involves tunnelling of electrons from the metal, which is always degenerate. At low temperatures, the predicted current density versus voltage behaviour simplifies [3]:

$$J = A^* \left(\frac{eV_{00}}{k_B} \right)^2 \left(\frac{\phi_b + eV - \phi_n}{\phi_b} \right) \exp\left(-\frac{2\phi_b^{3/2}}{3eV_{00}(\phi_b + eV - \phi_n)^{1/2}} \right) \tag{4.51}$$

An interesting consequence of field emission under reverse bias conditions is that the reverse current may exceed the forward current [2].

Tunnelling is often the cause of non-saturation of the reverse characteristics of Schottky barrier diodes. It is particularly important at the periphery of the metal contact, as the electric field will be enhanced (decreasing the width of the barrier and increasing the image force lowering). These edge effects can be reduced or even eliminated by the use of a guard ring (a concentric metal ring around the Schottky contact, at a separation comparable to the depletion width, and held at the same potential – Figure 7.16).

Once effects due to tunnelling and image force lowering have been reduced, a reverse current due to the generation of electron–hole pairs in the depletion region may become dominant. This is the inverse of process c in Figure 4.11. The resulting generation current density, J_g, will be given by Eq. (4.44):

$$J_g = \frac{e n_i W_d}{2 \tau_r} \tag{4.52}$$

where τ_r is the carrier (recombination) lifetime in the depletion region. This generation current will increase with reverse bias because the depletion layer width, W_d, is proportional to $(V_d + V)^{1/2}$. Like recombination, the generation current will be significant for high barriers and in low-lifetime semiconductors because it has a lower activation energy than the thermionic emission component.

4.4.4 Effect of an Interfacial Layer

The various analyses of the Schottky barrier outlined above assume that the metal is in intimate contact with a semiconductor surface. An inorganic semiconductor surface will contain incomplete covalent bonds and other defects. These are so-called 'surface states', which can act as traps for electrons or holes. Furthermore, the contact is seldom an atomically sharp discontinuity between the semiconductor and the metal. There is typically a thin interfacial layer, which is neither semiconductor nor metal. For example, silicon crystals are covered by a 1–2 nm oxide layer even after etching or cleaving in atmospheric conditions. Organic semiconductor surfaces are likely to possess fewer bond disruptions at their surfaces (Section 3.5.1). However, contamination layers and surface states will exist. For these reasons, it is often difficult to fabricate Schottky diodes with barrier heights predicted from work function differences (Eq. (4.8)).

The introduction of a thin insulating layer between a metal and a semiconductor can result in several effects:

(i) there will be a potential drop across the insulator and the zero-bias barrier height will be lower than it would be in an ideal diode;

(ii) when a bias voltage is applied, part of this will be dropped across the insulating layer so that the barrier height becomes a function of applied voltage;

(iii) the electrons will have to tunnel through the barrier presented to the insulator so that the current for a given bias will be reduced compared to the ideal case;

(iv) the presence of interface insulator can isolate states (traps) at the surface of the semiconductor so that these can be a property of the semiconductor with occupancy determined by the semiconductor Fermi level; and

(v) the insulator may incorporate fixed charge and/or electric dipoles, the effect of which is to change the work function of the metal.

Figure 4.14 depicts the energy band diagram for a metal/semiconductor contact based on an n-type semiconductor and incorporating a thin insulating layer. The interface traps are assumed to

Figure 4.14 Potential energy diagram for a junction between an n-type semiconductor and a metal, in the presence of a thin interfacial layer of thickness δ. Φ_m = metal work function. X_s = semiconductor electron affinity. ϕ_{b0} = Schottky barrier height (no image force lowering). V_i = voltage across interfacial layer. V_d = diffusion potential. ϕ_0 = neutral level (above valence band edge) of interface states. Q_m = surface charge density on metal. Q_{sc} = space-charge density (charge per unit area) in semiconductor. Q_{it} = interface trap charge density. E_{Fm}, E_{Fs} = Fermi levels in metal and semiconductor, respectively. E_c, E_v = edges of conduction and valence bands, respectively.

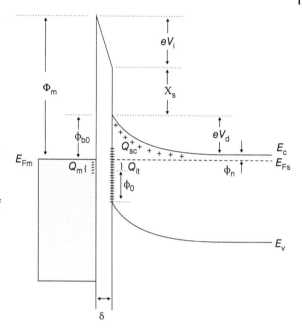

be as described in Section 3.5.1, i.e. the interface traps, of uniform density D_{it} states per unit area per unit energy, possess a neutral level, ϕ_0. Above ϕ_0 the traps are of an acceptor type (neutral when empty, negatively charged when full of electrons) and below ϕ_0 the traps are of donor-type (neutral when full of electrons, positively charged when empty).

From Figure 4.14, the interface trap density, Q_{it}, will be negative:

$$Q_{it} = -eD_{it}(E_g - \phi_0 - \phi_{b0}) \tag{4.53}$$

The quantity in parenthesis is simply the energy difference between the semiconductor Fermi level at the surface and the neutral level. The space charge that forms in the depletion layer at equilibrium is given by Eq. (4.13):

$$Q_{sc} = eN_dW_d = \sqrt{2\varepsilon_s\varepsilon_0 N_d \frac{(\phi_{b0} - \phi_n - k_BT)}{e}} \tag{4.54}$$

The total charge density on the semiconductor surface will be the sum of Eqs. (4.53) and (4.54). In the absence of any space charge in the interfacial layer, a balancing charge Q_m will develop on the metal surface:

$$Q_m = -(Q_{it} + Q_{sc}) \tag{4.55}$$

The potential across the interfacial layer, V_i, is obtained using Gauss' law:

$$V_i = \frac{Q_m\delta}{\varepsilon_i\varepsilon_0} \tag{4.56}$$

where δ and ε_i are the thickness and relative permittivity, respectively, of the interfacial layer. Also, by inspection of the band diagram in Figure 4.14:

$$eV_i = \Phi_m - (\chi_s + \varphi_{b0}) \tag{4.57}$$

Eliminating V_i from Eqs. (4.56) and (4.57):

$$\Phi_m - \chi_s - \varphi_{b0} = e \sqrt{\left(\frac{2\varepsilon_s\varepsilon_0 N_d \delta^2}{\varepsilon_i^2 \varepsilon_0^2}\frac{(\varphi_{b0} - \varphi_n - k_B T)}{e}\right) - \left(\frac{e^2 D_{it}\delta}{\varepsilon_i\varepsilon_0}(E_g - \varphi_0 - \varphi_{b0})\right)} \qquad (4.58)$$

The solution to Eq. (4.58) is complex. However, the square root term can be shown to be small [1]. If this is neglected (equivalent to assuming that the semiconductor space-charge density, Q_{sc}, is zero, which is accurate for conditions where the bands are flat in the semiconductor):

$$\Phi_m - \chi_s - \varphi_{b0} = \frac{e^2 D_{it}\delta}{\varepsilon_i\varepsilon_0}(\varphi_{b0} + \varphi_0 - E_g) \qquad (4.59)$$

Solving for ϕ_{b0} gives

$$\varphi_{b0} = \gamma(\Phi_m - \chi_s) + (1 - \gamma)(E_g - \varphi_0) \qquad (4.60)$$

where

$$\gamma = \frac{\varepsilon_i\varepsilon_0}{\varepsilon_i\varepsilon_0 + e^2 D_{it}\delta} \qquad (4.61)$$

Clearly, if D_{it} becomes small, the parameter γ tends to unity and Eq. (4.60) is identical to Eq. (4.8). Taking $\delta = 2$ nm as an upper limit to the thickness of the interfacial layer and using $\varepsilon_i = 3$, D_{it} must be less than about 10^{17} eV^{-1} m^{-2} for γ to be less than unity. In contrast, if $D_{it} = \infty$, Eq. (4.60) gives

$$\phi_{b0} = E_g - \phi_0 \qquad (4.62)$$

The Fermi level at the interface is *pinned* by the interface states to a value ϕ_0 above the valence band. The barrier height is now independent of the metal work function and is determined entirely by the surface properties of the semiconductor.

A Schottky barrier incorporating a thin interfacial layer is called a tunnel diode. It is also referred to as a metal/insulator/semiconductor or MIS structure. However, in this case, it is important to draw a distinction between an MIS structure in which the 'I' layer is of nanometre dimensions to those described in Chapter 5 in which the 'I' layer is sufficiently thick so that no charge transport occurs.

The current density versus voltage behaviour of a tunnel diode operating at a forward bias $>3k_B T/e$ can be shown to be [1, 6]:

$$J = A^* T^2 \exp(-B\delta\sqrt{\phi_t}) \exp\left(-\frac{\phi_b}{k_B T}\right)\left(\exp\left(\frac{eV}{nk_B T}\right) - 1\right) \qquad (4.63)$$

where ϕ_t is the effective height of the tunnelling barrier, and B is a constant. Equation (4.63) is like the standard thermionic emission equation (Eq. (4.30) with $V > 3k_B T/e$) with the incorporation of the ideality factor n (see Eq. (4.37)) and the inclusion the first exponential term which represents the tunnelling probability through the barrier presented by the thin interfacial layer. If ϕ_t is approximately 1 eV and $\delta > 5$ nm, this tunnelling probability is 10^{-22}, and so the current is negligible [1]. However, as δ and/or ϕ_t decrease, the current increases rapidly towards the thermionic-emission level.

The ideality factor in Eq. (4.63) accounts for the bias dependence of the barrier height. This parameter is related to the thickness and permittivity of the interfacial layer and to the density of interface traps. If it is assumed that the interface traps can be divided into two groups, such that those in the first group (density D_{itm}) are in equilibrium with the metal, while the traps in the other group (density D_{its}) are in equilibrium with the semiconductor, the ideality factor may be written [2] as

$$n = 1 + \left(\frac{\delta}{\varepsilon_i \varepsilon_0}\right) \frac{(\varepsilon_s \varepsilon_0 / W_d + e^2 D_{its})}{1 + (\delta/\varepsilon_i \varepsilon_0) e^2 D_{itm}} \tag{4.64}$$

States in equilibrium with the metal reduce n and therefore hold ϕ_b constant, while those in equilibrium with the semiconductor increase n and tend to hold V_d constant. Normally, when the interfacial layer is less than 3 nm, the interface traps are in equilibrium with the metal, while for thicker layers, these traps tend to be in equilibrium with the semiconductor.

An important effect of the interfacial layer is to increase the minority carrier-injection ratio under forward bias. For an n-type semiconductor with a very thin layer, the effect arises because the electron current is limited by thermionic emission and is therefore proportional to the probability of electrons tunnelling through the interfacial barrier. However, the hole current is controlled by diffusion in the neutral region of the semiconductor and will be relatively unaffected by the presence of the interfacial layer. In the case of thicker layers, there is a major realignment of the bands in the semiconductor with respect to the Fermi level in the metal, with the result that more holes can tunnel from the metal into the semiconductor. The effect can be exploited to improve the efficiency of electroluminescent structures and photovoltaic cells based on Schottky barriers [7–9].

In the reverse direction, the presence of an interfacial layer causes the effective barrier height to decrease with increasing bias so that the reverse current does not saturate. As a result, the reverse current of a Schottky barrier with a relatively thick interfacial layer may be greater than that of a device with a thin layer [8, 10].

If the departure of n from unity arises from image-force lowering or from interface effects, n should be independent of temperature, but if it is due to thermionic-field emission or to the effect of recombination in the depletion region, the ideality factor will depend on temperature. Many Schottky diodes exhibit n values that depend on temperature and show current versus voltage characteristics of the form:

$$J = A^* T^2 \exp\left(-\frac{\phi_b}{k_B(T + T_0)}\right) \left(\exp\left(\frac{eV}{k_B(T + T_0)}\right) - 1\right) \tag{4.65}$$

where T_0 is independent of voltage and temperature. This is equivalent to writing

$$n = 1 + \frac{T_0}{T} \tag{4.66}$$

There have been several attempts to explain this 'T_0' phenomenon, including tunnelling, particular distributions of interface states and a non-uniformly doped surface layer [2, 11].

4.4.5 Organic Schottky Diodes

A plethora of data can be found for inorganic metal/semiconductor devices, particularly those based on Si and GaAs. The experiments can readily be tested against the theories developed in the previous sections of this chapter, and general conclusions can be drawn. However, there are several obstacles faced by researchers endeavouring to establish the physical basis of electrical conductivity for organic Schottky diodes. The first is that reliable data are relatively scarce (although this situation continues to improve). Another is that there are many more organic semiconductors than there are inorganic semiconductors and, generally, the purity of the organic materials used in devices is inferior to that of their inorganic counterparts. Comparison between materials, prepared in different ways and with very different levels of impurities (intentional or unintentional), is an

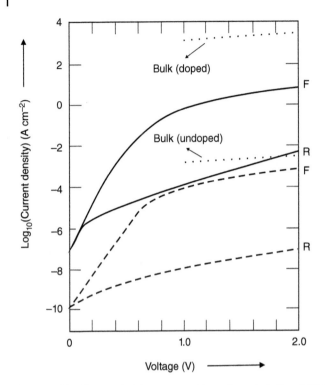

Figure 4.15 Current density versus voltage characteristics at 300 K for a Schottky diode based on a thiophene oligomer. Forward (F) and reverse (R) characteristic shown for doped (full lines) and undoped (dashed lines) materials. The horizontal dotted lines reveal the bulk conductivity values for the doped and undoped polymers. Source: Reprinted with permission from Lous et al. [12].

impossible task and it is therefore difficult to generalize about the physical basis of the electrical data. Different electrical processes may control the conductivity for different materials, and there is probably no *typical* current versus voltage behaviour. Nonetheless, the previous sections provide the concepts of drift and diffusion (together with the electrostatic framework) that can usefully be applied to provide insights into the electrical operation of metal/semiconductor diodes based on a wide range of organic solids. Notable exceptions are rectifying devices that function with single molecules, or small clusters of molecules. A separate section (Section 4.5) is devoted to these.

One example of the current versus voltage behaviour of organic diodes based on thiophene oligomers is shown in Figure 4.15 [12]. Both the forward (F) and reverse (R) characteristics measured at 300 K are shown. The top rectifying contacts were indium while gold was used to establish Ohmic back contacts. Two sets of characteristics are shown in Figure 4.15: the full lines show data for diodes that had been deliberately doped with 2,3-dichloro-5,6-dicyano-1,4-benzoquinone (DDQ), while the dashed lines are the forward and reverse currents for devices that were unintentionally doped. The bulk conductivity values for the polymer were 0.1 S cm^{-1} for the doped material and 10^{-7} S cm^{-1} for the unintentionally doped polymer (indicated by horizontal dotted lines in Figure 4.15). The current versus voltage behaviour for both types of diode exhibit four orders of magnitude rectification and both sets of data can be modelled using thermionic emission theory modified by the inclusion of an ideality factor, n, greater than unity (Eq. (4.35)). From the exponential slope between 0 and 0.5 V forward bias, n was found to be 2.0 and 1.8 for the unintentionally doped and doped devices, respectively. The increase in the saturation current

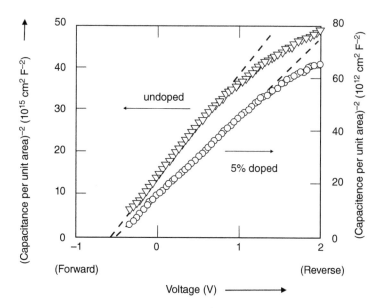

Figure 4.16 (Capacitance per unit area)$^{-2}$ versus voltage for undoped and doped thiophene oligomer Schottky diodes at 300 K, and measured at frequencies of 10 and 1000 Hz, respectively. Source: Reprinted with permission from Lous et al. [12].

(J_0 in Eq. (4.35)) by about three orders of magnitude suggests that the Schottky barrier height has decreased in the doped devices. At forward voltages greater than 1 V, the current density depends superlinearly on the forward voltage (approximately $J \propto V^{2.5}$), indicating that the charge transport is limited by space-charge effects (see Section 6.6). In this voltage regime, the forward current in the doped device is significantly lower than that anticipated based on the polymer bulk conductivity (indicated for both the doped and undoped devices in Figure 4.15).

Figure 4.16 shows capacitance versus voltage data – plotted in the form of C^{-2} versus V – for the doped and unintentionally doped thiophene oligomer diodes, measured at 1 kHz and 10 Hz, respectively (Problem 4.4) [12]. With increasing reverse bias, and up to reverse voltages of about 0.5 V, the capacitance versus voltage relationship is like that predicted by Eq. (4.14). However, at higher reverse voltages, the C^{-2} versus V curves deviate from linearity. This suggests that the doping density is not uniform throughout the depletion region. In the case of inorganic Schottky barrier diodes, there are many other reasons why a C^{-2} versus V plot may not be linear. These include series resistance effects, effective contact area variation with the depletion layer width and the presence of various distribution of interface traps [13, 14]. Nonlinear effects are often reported for Schottky diodes based on amorphous silicon [15]. These are thought to be related to the distribution of localized states. Interpreting that data in Figure 4.16 as resulting from a non-flat doping profile suggests the presence of a thin layer of lightly doped oligomer close to the metal-oligomer interface for both the doped and unintentionally doped devices [12]. The low acceptor density in this region makes this layer (in both diodes) highly resistive ($\sigma < 10^{-7}$ S cm^{-1}), limiting the forward current in both devices and leading to space-charge limited transport.

Other Schottky diodes show qualitatively similar behaviour to the results shown in Figures 4.15 and 4.16. Reported ideality factors range from about 1.2 to over 5 [16]. Figure 4.17 reveals how the ideality factors for Al/poly(3-methylthiophene) devices vary with temperature [16]. Data are shown for three devices (data points: O, x, +) and for one of the devices (data shown by '+' points) following a post-metallization annealing process at 90 °C for several minutes (*). Unannealed devices exhibit

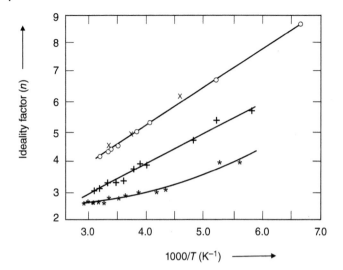

Figure 4.17 Ideality factor, *n*, as a function of reciprocal temperature for Schottky diodes based on a thiophene oligomer. Data are shown for three devices (data points: O, x, +) and also for one of the devices (data shown by '+' points) following a post-metallization anneal (*). Source: Taylor and Gomes [16].

the dependence of *n* on temperature suggested by Eq. (4.66). The annealing process results in an ideality factor that tends to become more temperature independent as the temperature is increased and approaches 300 K. A suggestion is that, as the temperature is raised, the limitation on the diode current imposed by the series resistance of the bulk polymer becomes less important [16].

As shown in Figure 4.15, the reverse bias characteristics of the thiophene oligomer diodes do not saturate, as predicted by thermionic emission theory, Eq. (4.30). This phenomenon is often observed in Schottky barriers based on organic semiconductors. The increase in the current with bias is also usually greater than that predicted by diffusion theory, Eq. (4.24). Figure 4.18 shows a plot of the $\ln J$ versus $(V_c - V_r)^{1/4}$, where $V_c = V_d - k_B T/e$ and V_r is the applied reverse bias, for the polythiophene diodes described above [16]. The results suggest that image force effects dominate the reverse current (Section 4.4.2).

Although the charge carrier mobilities in organic semiconductors are relatively low, organic diodes can operate to reasonably high frequencies. The frequency limit for any device based on a thin film will depend on the time for the charge carriers to cross the film. This transit time, t_t, will be related to the voltage applied to the film, V, and the film thickness t:

$$t_t = \frac{t^2}{\mu V} = \frac{1}{f} \tag{4.67}$$

where *f* is the operating frequency. In the case of a diode, the maximum operating frequency, f_{max}, is [17]:

$$f_{max} = \frac{\mu(V - V_{DC})}{t^2} \tag{4.68}$$

where *V* is the applied voltage to the diode and V_{DC} is the rectified DC voltage. A more stringent and realistic frequency limit can be calculated from the recharging time for the load capacitance. The maximum theoretical operating frequency is determined by the speed with which the charges consumed by the load resistor at a DC voltage V_{DC} during one frequency cycle can be recharged onto the load capacitance (by the current flowing through the organic diode during the fraction of the cycle in which the diode is in forward bias). This leads to the following expression for the

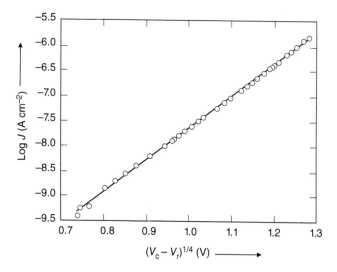

Figure 4.18 Log (current density) versus $(V_c - V_r)^{1/4}$ for polythiophene Schottky diode: V_r = applied reverse bias; $V_c = V_d - k_B T/e$, where V_d is the diffusion potential. Source: Taylor and Gomes [16].

maximum frequency [17]:

$$f_{max} = \frac{9\mu}{16t^2 \pi V_{DC}} \left((-3V_{DC} + V_f)\sqrt{V^2 - (V_{DC} + V_f)^2} + \left(V^2 + 2V_{DC}^2\right) \arccos\left(\frac{V_{DC} + V_f}{V}\right) \right)$$

(4.69)

where V_f is defined as the transition voltage of the diode – the voltage at which the conduction mechanism in an organic thin film undergoes the transition from Ohmic conduction to space-charge limited conduction (Section 6.6).

Figure 4.19 reveals the frequency dependence of the rectified DC voltage for an organic diode based on pentacene, using the following parameters: $\mu = 0.15\,\text{cm}^2\,\text{V}^{-1}\,\text{s}^{-1}$, $t = 160\,\text{nm}$, $V_f = 3.5\,\text{V}$,

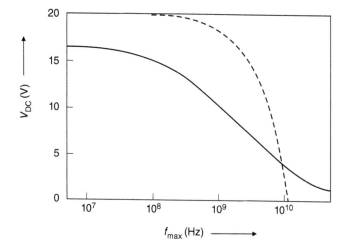

Figure 4.19 Frequency dependence of the rectified DC voltage, V_{DC}, for an organic diode. The solid line is calculated from Eq. (4.69) while the dashed line gives the frequency limit estimated from the transit time of the diode. For both calculations, the following parameters were used: $\mu = 0.15\,\text{cm}^2\,\text{V}^{-1}\,\text{s}^{-1}$, $t = 160\,\text{nm}$, $V_f = 3.5\,\text{V}$, and $V = 20\,\text{V}$. Source: Reprinted by permission from Steudel et al. [17].

and $V = 20\,V$ [17]. The solid line is calculated from Eq. (4.69), while the dashed line gives the frequency limit estimated from the transit time of the diode. Frequencies in the range 100 MHz to 1 GHz are seen to be within the operating range for organic diodes.

4.5 Molecular Devices

Many scenarios for future electronic devices envisage organic molecules as the 'switches' in computational architectures. Different approaches have been used to exploit the electrical properties of individual molecules or assemblies of molecules. Electrical contacts to the organic molecules can be made using solid or liquid electrodes, and both in-plane and out-of-plane measurements are possible using geometries like those depicted in Figure 4.2. A challenge for the out-of-plane measurements is that of depositing the top electrode without damaging the ultra-thin (e.g. molecular) layer. However, there are many other issues to be faced by the experimentalist using such deceptively simple device structures. These have, to some extent, been discussed earlier in Section 4.1.

Making electrical connections to molecules severely limits and hinders the reliable and reproducible study (and realization) of molecular devices. This is due mostly to the following difficulties [18]:

 (i) Finding ways to attach molecules reproducibly to the electrodes.
 (ii) Variability in the contacts. Computational analysis of some prototype molecules shows that small changes in the geometry of the molecule with respect to the gold electrode can significantly modify the conductance [19]. The conductance can change either by factors of 2–10 or by orders of magnitude, depending on the spectral density overlap between the molecular π-electron system and the gold electrode. Experiments have also revealed substantial changes in the measured electronic transport through conjugated 'molecular-wire' molecules by changing the type or geometry of their contacts to surfaces [18].
(iii) Understanding the interactions(s) between the electrodes and the different parts of the molecule. Such information is crucial for knowing how the details of contacting, in terms of materials and methods, affects the resulting molecule-based device characteristics.

Several methods that have been used in making electrical contacts to molecules are reviewed below.

4.5.1 Metal/Molecule Contacts

4.5.1.1 Chemical Bonding

Some contacting methods rely on the formation of chemical bonds between the contact and the molecules. Although this limits the choice of molecule/contact combinations, the approach offers some control of the reproducibility of the resulting electrical measurements. Much work exploits the Au—S bond, even though this is not particularly stable. Figure 4.20 shows a device configuration that utilizes such chemical bonding. The method is based on transfer printing (so-called 'soft lithography') and uses a solid or elastomeric stamp that is fabricated by casting and curing a liquid polymer, such as polydimethylsiloxane (PDMS), onto a silicon wafer (the 'master'), whose features have been defined by lithography. A thin film of metal is then evaporated onto the stamp and electrical contacts are established by bringing the metal-coated PDMS stamp into contact with the molecular layer. Following removal of the stamp, the pattern is transferred because the

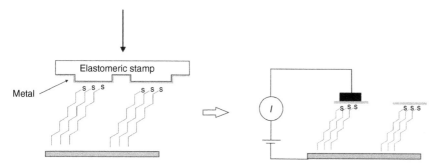

Figure 4.20 Left: a metal-coated solid (e.g. Si) or elastomeric (PDMS) stamp is brought into mechanical contact with a monolayer-coated substrate having a reactive terminal group (e.g. thiol). Right: when the stamp is removed from the substrate, the pattern is transferred because the metal adhesion to PDMS is significantly less than to the exposed thiol group.

metal adhesion to PDMS is significantly less than to the exposed thiol group (right-hand diagram in Figure 4.20). Although metal transfer is possible using a variety of metal-molecule combinations, the exploitation of chemical bonding improves the contact formation. For example, using the thiol-terminated molecules shown in Figure 4.20, the metal is chosen to be gold. The formation of a monolayer on a substrate, with unreacted thiol group oriented away from the substrate, is not a trivial process [18].

4.5.1.2 Nanoparticle-Coupled Scanning Probe

Another method of measuring the electrical properties of molecular structures uses the tip of a scanning probe microscope – a scanning tunnelling microscope (STM) or an atomic force microscope (AFM) to provide the top contact. In this approach, a molecule with a reactive functional group is inserted, via directed assembly, into a monolayer of other molecules that lack the reactive functional groups at their external terminals. The modified substrate is then exposed to a solution containing metallic nanoparticles that can react with the exposed functional groups, leading to the formation of chemical bonds between the nanoparticles and the incorporated molecules. Figure 4.21 shows an example of this [20]. A SAM of octanethiol on a gold surface is first obtained. Molecules of 1,8-octanedithiol are then implanted into the octanethiol monolayer using a replacement reaction whereby one of the two thiol groups becomes chemically bound to the gold surface. The octanethiol monolayer acts as a molecular insulator, isolating the dithiol molecules from one another. Incubating the monolayer with a suspension of gold nanoparticles then derivatizes the thiol groups at the top of the film. Finally, a gold-coated conducting AFM probe is used to locate and contact individual particles bonded to the monolayer. Measurements on over 4000 nanoparticles produced only five distinct families of curves [20]. Figure 4.22 shows representative curves from each family. The curves correspond to multiples of a fundamental curve, which is ascribed to a situation in which a single dithiol molecule links the gold nanoparticle to the underlying gold substrate. In the low-voltage region (between ±0.1 V) the single molecule was found to possess a resistance of 900 ± 50 MΩ.

A related strategy to establish electrical contacts to SAMs has used metallic nanoparticles conformally deposited, from solution, onto the SAM and the reinforcement of this contact by direct metal evaporation [21]. The intrinsic molecular properties are not affected by the nanoparticle layer and subsequent metallization. For certain molecules (those equipped with two anchor groups), this process provides a route to large-scale integration of molecular compounds into solid-state devices, which can be scaled down to the single-molecule level.

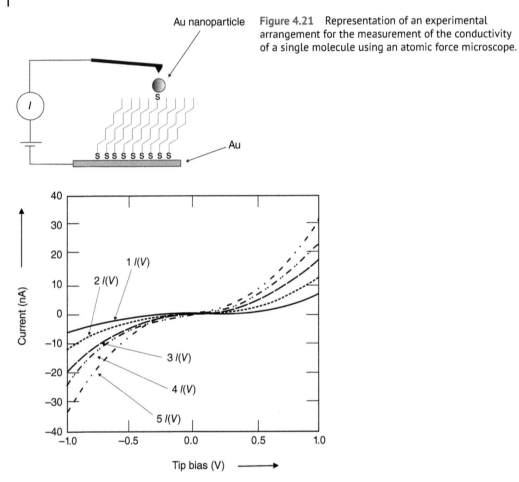

Au nanoparticle

Figure 4.21 Representation of an experimental arrangement for the measurement of the conductivity of a single molecule using an atomic force microscope.

Au

Figure 4.22 Current versus AFM tip bias, $I(V)$, curves measured with the equipment shown in Figure 4.21. The five curves shown are representative of distinct families, $N\,I(V)$ that are integer multiples of a fundamental curve $I(V)$ ($N = 1, 2, 3, 4,$ and 5). Source: Reproduced with permission from Cui et al. [20].

4.5.1.3 Liquid Metal

A liquid metal can serve to make mechanical contact to organic molecules. Alternatively, a monolayer can be formed on the liquid metal and the molecularly modified drop brought into contact with a second electrode [22]. Mercury (work function = 4.5 eV) is the most common choice for a liquid metal contact as it can be used in ambient conditions without too much concern about the formation of an oxide. The experimental arrangement is depicted in Figure 4.23. To form a contact, a monolayer is adsorbed on a solid substrate, after which mechanical contact is established. The advantages of contacts made by this method are that they are easy to make, stable and reproducible, and can be used on a variety of molecular systems. However, this contacting strategy is unlikely to find its way into practical molecular devices.

4.5.2 Break Junctions

Metal electrodes with very narrow (5 nm) spacings can be fabricated using lithographic methods, such as electron beam lithography. However, the resulting nano-gaps can still be large when

Figure 4.23 Schematic diagram showing the formation of a molecular junction using a liquid metal (e.g. Hg) contact.

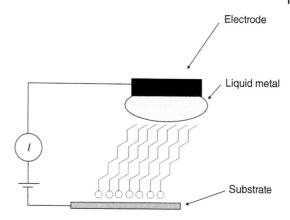

compared to the lengths of typical molecules (1–3 nm). The use of break junctions allows the conductance of single molecules to be measured [22, 23]. This makes use of two metal wires separated by a very thin gap, on the order of the inter-atomic spacing (less than a nano-meter). The gap can be obtained by physically pulling a single wire apart or through chemical etching or electromigration. As the wire breaks, the separation between the electrodes can be indirectly controlled by monitoring the electrical resistance of the junction. Figure 4.24 shows one example of the technology. A length of a metallic wire is fixed on a flexible substrate called a bending beam. Making a notch near the middle of the wire reduces its cross section. The bending substrate is normally fixed at both ends by counter supports. A vertical movement of a push rod, which can be precisely controlled by a piezoelectric actuator or motor, can exert a force on the bending beam. As the latter is bent, the wire starts to elongate, resulting in the reduction of the cross section at the notch and finally producing complete fracture of the metal wire. Two clean-facing nanoelectrodes are then generated; the bending or relaxing of the substrate controls their separation. After introducing organic molecules into the gap, these may bridge the two electrodes and the electronic properties of the molecules can be determined. Both two-terminal and three-terminal break junctions have been fabricated [23]. The break junction can be incorporated into a high vacuum system and integrated

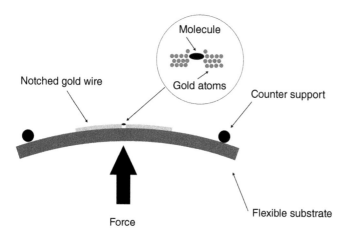

Figure 4.24 A mechanically controllable break junction to enable electrical measurements to be made on an organic molecule.

with measurement techniques, such as Raman spectroscopy, inelastic tunnelling spectroscopy or noise spectroscopy, to obtain fingerprint information of the molecules [23].

A significant disadvantage of mechanically fabricated break junctions is the uncontrollable nature of the breaking process. For instance, the local shape of the electrodes and the atomic configurations of the electrodes are unknown. Break junctions cannot be scaled in a manner like crossed bar architectures (Figure 9.6) to address a specific cross-section, which limits their potential applications.

A variation of the method above uses a gold STM tip. This is immersed in a solution of the molecules and driven into and pulled back from an Au substrate. As the tip and the substrate separate, a chain of Au atoms forms and, ultimately, breaks, allowing one or more molecules to be 'caught' in the freshly formed gap [24–26]. Figure 4.25 reveals the results of such an experiment using molecules of 4,4′ bipyridine [25]. This compound has two nitrogen atoms at its extremities that can bind strongly to the gold electrodes. During the initial stage of pulling the tip away from the substrate (Figure 4.25a), the conductance decreases in a stepwise fashion, with each step occurring preferentially at an integer multiple of the conductance quantum $G_0 = 2e^2/h$ (Eq. 3.23). A histogram constructed from approximately 1000 such conductance plots show pronounced peaks at $1G_0$, $2G_0$, and $3G_0$ (Figure 4.25b). This value is the well-known conductance quantization, which occurs when the size of a metallic contact is decreased to a chain of Au atoms. When the atomic chain is broken by pulling the tip away further, a new sequence of steps, in a lower conductance regime appears in the presence of the organic molecules (Figure 4.25c). The corresponding histogram shows pronounced peaks near $0.01G_0$, $0.02G_0$, and $0.03G_0$ (Figure 4.25d), which is two orders of

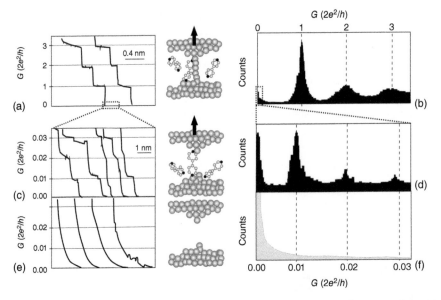

Figure 4.25 (a) Conductance of a gold contact formed between a gold STM tip and a gold substrate decreases in quantum steps as the tip is pulled away from the substrate. (b) A corresponding conductance histogram comprising 1000 curves as depicted in (a). (c) When the contact shown on (a) has been completely broken, a new series of conductance steps is evident, corresponding to the conductance through molecules of 4,4′ bipridine in solution. (d) A corresponding conductance histogram comprising 1000 curves as depicted in (a). (e) and (f) In the absence of molecules, no steps or peaks are observed. Source: Reproduced with permission from Xu and Tao [25]. Copyright (2003). Reproduced with permission from The American Association for the Advancement of Science.

magnitude lower than those that were seen through the conductance quantization. These were ascribed to the formation of stable molecular 'devices' incorporating one, two, and three molecules. In the absence of the organic molecules, no peaks below $1G_0$ are evident (Figure 4.25e,f).

Many challenges need to be overcome before single-molecule devices can be used as commercial products. Nonetheless, as a fundamental research technique, break junctions provide an important platform for investigations of single molecule behaviour.

4.5.3 Molecular Rectifying Diodes

The concept of molecular rectification, i.e. asymmetrical current versus voltage behaviour within a molecule, was predicted in 1974 by Aviram and Ratner [27]. The proposal was that an organic molecule containing a donor and an acceptor group separated by a short σ-bonded bridge, allowing quantum mechanical tunnelling, should exhibit diode characteristics. An example of such a molecule, together with its energy band structure, is given in Figure 4.26. The ground (HOMO) and excited (LUMO) states of the molecules are E_{D1}, E_{D2} (for the donor) and E_{A1}, E_{A2} (acceptor), respectively. The presence of the σ-bridge is essential to the device operation. If the electronic systems of the donor and acceptor are allowed to interact strongly with one another, a single donor level will exist on the timescale of the experiment. The donor and acceptor sites therefore need to be isolated from one another in order for the device to function.

If the molecule is sandwiched between two metallic electrodes, the passage of electrons through the device can be considered as a three-step process: metal cathode to acceptor, acceptor to donor, and finally from the donor to the anode. As the applied voltage is increased (the acceptor side of the

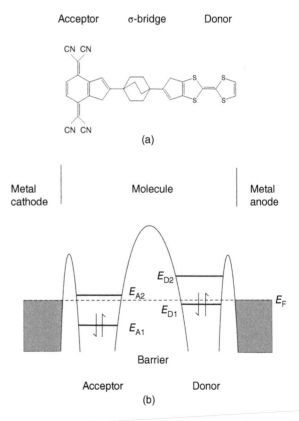

Figure 4.26 (a) An example of a molecular rectifier molecule based on acceptor/σ-bridge/donor architecture. (b) Energy band structure of molecular rectifier. Source: Reprinted from Aviram and Ratner [27].

molecule is biased negatively with respect to the donor side), electrons will be transferred from the cathode to the acceptor molecule (to energy state E_{A2}) and from the donor (from E_{D1}) to the anode. Electrons will move from the acceptor molecule LUMO level (E_{A2}) to the donor HOMO level (E_{D1}) by the process of quantum mechanical tunnelling (Section 6.3) if the σ-bridge is sufficiently short. When the direction of the applied bias is reversed, conduction can only take place at much higher voltages, i.e. the HOMO level of the donor would have to be higher than the LUMO level of the acceptor. This leads to predicted diode characteristics.

There have been many attempts to demonstrate this effect in the laboratory, particularly in organic thin films [e.g. [28–35]]. Asymmetric current versus voltage behaviour has certainly been recorded for many metal/insulator/metal structures, although these results are often open to several interpretations because of the experimental issues highlighted in the previous section. The asymmetry in the electrical characteristics may originate from work function differences or from Schottky barrier formation between the organic material and one of the electrodes. This effect may not be eliminated using the same metal for the electrodes as its method of preparation can affect its work function. For example, the work function of a metal film deposited onto a glass substrate may differ from that of the same metal deposited onto an organic layer [36]. A key test is to relate the observed rectification to a property of the molecule under test. Figure 4.27 shows the chemical structure of two similar self-assembling monolayers consisting of a bulky 4-methoxynaphthyl donor that is linked via a –CH—CH– σ bridge to a bulky quinolinium acceptor [34]. The steric hindrance provided by the donor and acceptor of the molecule depicted in Figure 4.27a enforces the required non-planarity, which, in turn provides an effective electron tunnelling barrier between the electroactive ends of the molecule. SAMs of this molecule exhibit rectification ratios of about 30 at ±1 V, as shown in Figure 4.28. These measurements were undertaken using the tip of an STM as the top electrode; the measured electrical data were indistinguishable for PtIr and Au tips. Significantly, the electrical asymmetry is suppressed in SAMs of a less bulky analogue, Figure 4.27b.

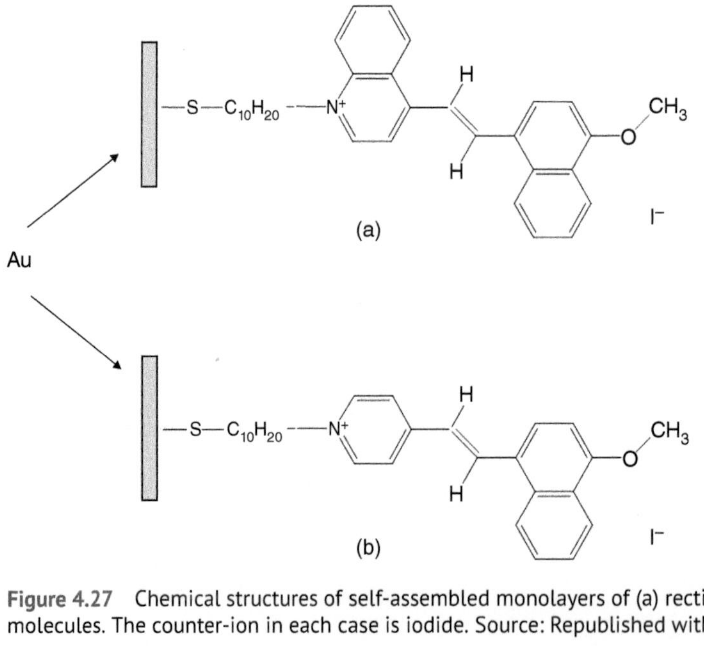

Figure 4.27 Chemical structures of self-assembled monolayers of (a) rectifying and (b) non-rectifying molecules. The counter-ion in each case is iodide. Source: Republished with permission Ashwell et al. [33].

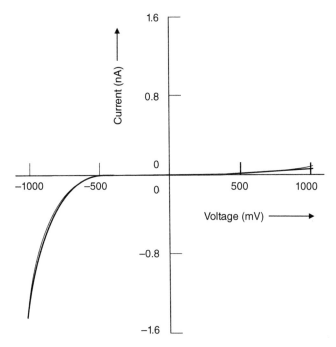

Figure 4.28 Current versus voltage characteristics of the iodide salt of the molecule depicted in Figure 4.27a. The polarity relates to the substrate electrode and the higher forward bias current corresponds to electron flow from the gold-coated substrate to the contacting tip of an STM. Source: Republished with permission of The Royal Society of Chemistry, from Ashwell et al. [33]. Copyright (2005); permission conveyed through Copyright Clearance Center, Inc.

It has been noted that mechanisms other than asymmetric couplings between the molecule and the electrodes may be the origin for the observed rectification in molecular junctions [19, 37, 38]. For example, a molecular bias drop may be responsible for the experimental data. First, some of the applied bias drops across the molecule, indicating that the electrode/molecule/electrode junction has a (permanent or induced) dipole (Section 7.2). In effect, the molecular channel energies change with the bias: a positive (negative) bias might bring a channel closer to resonance, thereby increasing the current, whereas a negative (positive) bias would push the channel away from resonance and decrease the current (see following section). Asymmetric electrode couplings may indirectly lead to rectification, even though these are not directly responsible for it. It is probable that the different linker groups used to produce asymmetric couplings may also lead to different induced dipoles with an applied bias. The change in induced dipole from one system to the next might then change the rectification ratio of the junction, giving the illusion that asymmetric couplings lead to rectification.

4.5.4 Molecular Resonant Tunnelling Diodes

Molecular resonant tunnelling diodes (RTDs) exploit the energy quantization of electron states in an organic molecule to control the current between two electrodes [39–42]. Unlike the rectifying device described in the previous section, current passes equally well in both directions in the RTD. Structurally and functionally, the diode is a molecular analogue of much larger solid state RTDs that may be fabricated using III–V inorganic semiconductors [1]. Figure 4.29 shows an example of a molecular RTD based on a polyphenylene-based molecule [39–41]. Two aliphatic

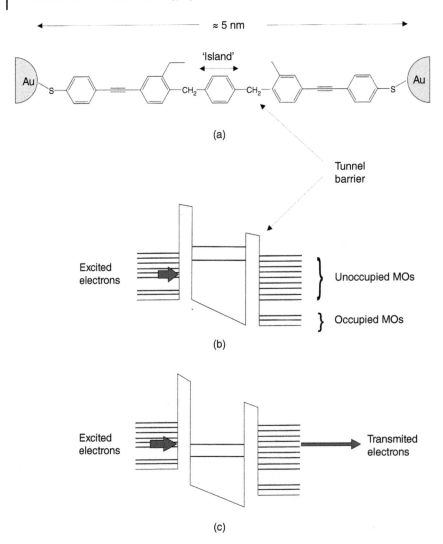

(a)

(b)

(c)

Figure 4.29 Molecular resonant tunnelling device. (a) Chemical structure of polyphenylene-based molecule with two potential energy tunnel barriers to electrons. (b) For a low applied voltage, there is no alignment of the electronic energy levels molecular orbitals (MOs) on the left and right sides of the molecule and the MOs in the 'island' region between the tunnel barriers; the current is that due to thermionic emission. (b) For a larger applied voltage, the MO levels align and there is a relatively large resonant tunnelling current through the molecule. Source: Reproduced by permission Ellenbogen and Love [41].

methylene groups, Figure 4.29a, provide potential barriers to carrier transport. Quantum mechanical confinement results in electronic energy levels that are relatively far apart on the 'island' between these barriers. In contrast, the unoccupied energy states are more densely spaced in the regions of the molecule to the left and right of the barriers. Electrons are injected from the gold electrode under a voltage bias into the lowest unoccupied energy levels on the left-hand side of the molecule. These electrons can then pass through the thin barrier regions to the right-hand side of the molecule by the process of quantum mechanical tunnelling. The probability that electrons can tunnel from the left-hand side to the molecule is dependent on the energies of the incoming

Figure 4.30 Expected current versus voltage behaviour of a resonant tunnelling device, as depicted in Figure 4.29. The large current peak is due to the resonant tunnelling current. This is followed by a region of negative differential resistance (NDR) until the baseline current due to thermionic emission is reached.

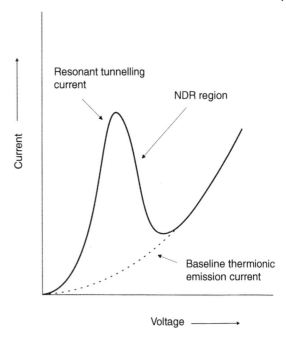

electrons compared to the widely spaced unoccupied energy levels on the island between the two barriers. As shown in Figure 4.29b, if the applied voltage bias across the molecule produces incoming electrons with kinetic energies that differ from the unoccupied energy levels available inside the potential well on the island, then current does not flow. The RTD is in its 'Off' state. However, if the bias is adjusted so that the energy of the incoming electrons aligns with one of the islands energy levels, the energy of the electrons outside the well are 'in resonance' with the allowed energy level inside the well. Electrons can then tunnel from the left-hand region of the molecule onto the island. If this process can occur and, simultaneously, an energy level inside the well is also in resonance with one of the many unoccupied energy levels in the region on the right-hand side of the molecule, current can flow through the molecule. The RTD is now in its 'On' state. Figure 4.30 shows the theoretical current versus voltage characteristic for the RTD. This reveals that the transmitted current through the device rises to a peak at the resonance voltage. (Of course, it might be possible to observed multiple peaks because of resonant tunnelling through different energy levels on the island.) The current then falls to a valley and a region of negative differential resistance (NDR) is observed. The *I–V* curve is quite like that shown in Figure 2.5, for the transferred electron effect, observed in certain inorganic semiconductors. Following the NDR region in Figure 4.30, the current rises, limited by thermionic emission for the direct metal/molecule barrier.

Generally, the current–voltage characteristics of RTD devices based on molecules such as that shown in Figure 4.29a would be expected to be symmetric about the zero-voltage axis. However, experimental problems such as variations in the composition or geometry of the electrical contacts can often lead to some asymmetry in the observed *I–V* characteristics [39]. A figure of merit for the operation of an RTD can be given by the measured On/Off peak-to-valley ratio in its current versus voltage characteristic. A figure of 1000 has been reported for a SAM of a molecule containing a nitroamine moiety [43].

Quite sophisticated molecules have been synthesized to be incorporated in RTDs. For example, DNA molecules have acted as a bridge between two carbon nanotube leads [42].

The forward current in the Aviram–Ratner molecular rectifier is a result of quantum mechanical tunnelling through the σ-bonded barrier, and the response of this current will be essentially instantaneous. However, rectifiers must not only conduct when they become forward biased but also cease to conduct when they are reverse biased. The device current will include a component flowing through the parasitic capacitance which is inherently present. This will not be a problem at low-switching frequencies but will limit the maximum frequency for which the device has useful characteristics. An estimate of the limiting frequency of the device proposed by Aviram and Ratner – about 1.5 kHz – can be obtained from the RC time constant of the device and by using some of the original parameters (Problem 4.5) [44]. The key factor is the current density through the device. Even if the parasitic capacitance is large, it can be charged and discharged faster if the device current is increased. Further increases lead to reductions in switching time, up to as point where transit delays due to charge carrier movement start to dominate. One strategy to both increase the molecular diode current density by many orders of magnitude while keeping the small capacitance (approximately 100 aF) is to base the rectifier on an array of molecules addressed by the tip of an interferometric STM. Such an approach, using approximately 150 molecules of ferrocenyl undecanethiol molecules, has been shown to produce a rectifying device operating up to 17 GHz, with an estimated cut-off frequency of 520 GHz [45]. The current in these diodes is thought to originate from a combination of direct tunnelling and resonant tunnelling through the organic molecules.

Problems

4.1 The capacitance per unit area, C, of an Schottky diode based on a p-type semiconductive polymer ($\varepsilon_s = 3.0$) varies with reverse applied voltage V_r as follows:

V_r (V)	C (nF cm^{-2})
0.2	35.4
0.4	30.9
0.4	27.7
0.8	25.4
1.0	23.6

Assuming ideal behaviour, calculate the diffusion potential and the doping concentration in the semiconductor.

4.2 A Schottky diode is formed between aluminium and p-type silicon, which has a doping concentration of 5×10^{16} donors cm^{-3}. If the Al work function $= 4.3$ eV and the Si electron affinity $= 4.0$ eV, calculate the values of barrier height and diffusion voltage at 300 K if the device is ideal. For an applied electric field of 10^5 V cm^{-1}, calculate the value of the barrier height lowering due to the image force effect. How does this figure compare to that for a p-type organic semiconductor/Al Schottky junction under the same conditions? Assume that the organic semiconductor has a relative permittivity $\varepsilon_s = 3$. (For Si at 300 K: $E_g = 1.1$ eV; $n_i = 1 \times 10^{10}$ cm^{-3}; $\varepsilon_s = 11.9$).

4.3 Using thermionic emission theory, calculate the ideal current density for a Schottky diode based on pentacene for a forward applied voltage of 0.2 V at 300 K. The barrier height is 0.7 eV. Take the hole effective mass to be the same as the free electron mass. If the image force effect causes an effective increase in barrier height in forward bias of 0.1 eV V^{-1}, calculate the ideality factor of the diode. How much will this factor alone reduce the forward current density at 0.2 V?

4.4 Using the experimental data shown in Figures 4.15 and 4.16, estimate the Schottky barrier heights and diffusion voltages for the doped and undoped thiophene diodes. Hence, estimate the position of the Fermi levels in the bulk p-type semiconductors. How might your estimates be improved?

4.5 The cut-off frequency f_c of a forward biased diode is given by

$$f_c = \frac{1}{2\pi RC}.$$

where R, C are the resistance and capacitance, respectively. Using data provided in the original paper by Aviram and Ratner (*Chem. Phys. Lett.* (1974) 29: 277–283) and reasonable values for other quantities, calculate the cut-off frequency for a molecular diode.

References

1 Sze, S.M. and Ng, K.K. (2007). *Physics of Semiconductor Devices*, 3e. Hoboken, NJ: Wiley-Interscience.

2 Rhoderick, E.H. (1978). *Metal-Semiconductor Contacts*. Oxford: Clarendon Press.

3 Padovani, F.A. and Stratton, R. (1966). Field and thermionic field-emission in Schottky barriers. *Solid State Electron.* 9: 695–707.

4 Crowell, C.R. and Rideout, V.L. (1969). Normalised thermionic-field (T-F) emission in metal-semiconductor (Schottky) barriers. *Solid State Electron.* 12: 89–105.

5 Yu, A.Y.C. and Snow, E.H. (1968). Surface effects in metal-silicon contacts. *J. Appl. Phys.* 39: 3008–3016.

6 Card, H.C. and Rhoderick, E.H. (1971). Studies of tunnel MOS diodes I. Interfacial effects in silicon Schottky diodes. *J. Phys. D: Appl. Phys.* 4: 1589–1601.

7 Pulfrey, D.L. (1978). MIS solar cells: a review. *IEEE Trans. Elec. Dev.* ED-25: 1308–1317.

8 Card, H.C. and Rhoderick, E.H. (1973). The effect of an interfacial layer on minority carrier injection in forward-biased silicon Schottky diodes. *Solid State Electron.* 16: 365–374.

9 Petty, M.C., Batey, J., and Roberts, G.G. (1985). A comparison of the photovoltaic and electroluminescent effects in GaP/Langmuir–Blodgett film diodes. *IEE Proceedings Part 1* 132: 133–139.

10 Card, H.C. and Rhoderick, E.H. (1971). Studies of tunnel MOS diodes II. Thermal equilibrium considerations. *J. Phys. D: Appl. Phys.* 4: 1602–1611.

11 Crowell, C.R. (1977). The physical significance of the T_0 anomalies in Schottky barriers. *Solid State Electron.* 20: 171–175.

12 Lous, E.J., Blom, P.W.M., Molenkamp, L.W., and de Leeuw, D.M. (1995). Schottky contacts on a highly doped organic semiconductor. *Phys. Rev. B* 51: 17251–17254.

13 Goodman, A.M. (1963). Metal-semiconductor barrier height measurement by the differential capacitance method – one carrier system. *J. Appl. Phys.* 34: 329–338.

14 Fonash, S.J. (1983). A reevaluation of the meaning of capacitance plots for Schottky-barrier-type diodes. *J. Appl. Phys.* 54: 1966–1975.

15 Snell, A.J., Mackenzie, K.D., Le Comber, P.G., and Spear, W.E. (1979). The interpretation of capacitance and conductance measurements on metal-amorphous silicon barriers. *Philos. Mag.* 40: 1–15.

16 Taylor, D.M. and Gomes, H.L. (1995). Electrical characterization of the rectifying contact between aluminium and electrodeposited poly(3-methylthiophene). *J. Phys. D: Appl. Phys.* 28: 2554–2568.

17 Steudel, S., Myny, K., Arkhipov, V. et al. (2005). 50 MHz rectifier based on an organic diode. *Nat. Mater.* 4: 597–600.

18 Haick, H. and Cahen, D. (2008). Making contact: connecting molecules electrically to the macroscopic world. *Prog. Surf. Sci.* 83: 217–261.

19 Basch, H., Cohen, R., and Ratner, M.A. (2005). Interface geometry and molecular junction conductance: geometric fluctuation and stochastic switching. *Nano Lett.* 5: 1668–1675.

20 Cui, X.D., Primak, A., Zarate, X. et al. (2001). Reproducible measurement of single-molecule conductivity. *Science* 294: 571–574.

21 Hellman, G.P., Venkatesan, K., Mayor, M., and Lörtscher, E. (2018). Metallic nanoparticle contacts for high-yield, ambient-stable molecular-monolayer devices. *Nature* 559: 232–235.

22 Xiang, D., Wang, X., Jia, C. et al. (2016). Molecule-scale electronics: from concept to function. *Chem. Rev.* 116: 4318–4440.

23 Xiang, D., Jeong, H., Lee, T., and Mayer, D. (2013). Mechanically controllable break junctions for molecular electronics. *Adv. Mater.* 25: 4845–4867.

24 Reed, M.A., Zhou, C., Muller, C.J. et al. (1997). Conductance of a molecular junction. *Science* 278: 252–254.

25 Xu, B. and Tao, N.J. (2003). Measurement of single-molecule resistance by repeated formation of molecular junctions. *Science* 301: 1221–1223.

26 He, J., Sankey, O., Lee, M. et al. (2006). Measuring single molecule conductance with break junctions. *Faraday Discuss.* 131: 145–154.

27 Aviram, A. and Ratner, M.A. (1974). Molecular rectifiers. *Chem. Phys. Lett.* 29: 277–283.

28 Zhang, G. and Ratner, M.A. (2015). Is molecular rectification caused by asymmetric electrode couplings or by a molecular bias drop? *J. Phys. Chem. C* 119: 6254–6260.

29 Ashwell, G.J. and Stokes, R.J. (2004). Do alkyl tunnelling barriers contribute to molecular rectification? *J. Mater. Chem.* 14: 1228–1230.

30 Metzger, R.M. (2002). Monolayer rectifiers. *J. Solid State Chem.* 168: 696–711.

31 Metzger, R.M. (2009). Unimolecular electronics and rectifiers. *Synth. Met.* 159: 2277–2281.

32 Honciuc, A., Jaiswal, A., Gong, A. et al. (2005). Current rectification in a Langmuir–Schaefer monolayer of fullerene-bis[4′diphenylamino-4″-(*N*-ethyl-*N*-2″-ethyl) amino-1,4-diphenyl-1,3-butadiene] malonate between Au electrodes. *J. Phys. Chem. B* 109: 857–871.

33 Ashwell, G.J., Mohib, A., and Miller, J.R. (2005). Induced rectification from self-assembled monolayers of sterically hindered π-bridged chromophores. *J. Mater. Chem.* 15: 1160–1166.

34 Ashwell, G.J., Mohib, A., Collins, C.J., and Aref, A. (2009). Molecular rectification: confirmation of its molecular origin by chemical suppression of the electrical asymmetry. *Synth. Met.* 159: 2282–2285.

35 Metzger, R.M. (2018). Quo vadis, unimolecular electronics? *Nanoscale* 10: 10316–10332.

36 Watkins, N.J., Yan, L., and Goa, Y. (2002). Electronic structure symmetry of interfaces between pentacene and metals. *Appl. Phys. Lett.* 80: 4384–4386.

37 Cuevas, J.C. and Scheer, E. (2017). *Molecular Electronics: An Introduction to Theory and Experiment*, 2e. Singapore: World Scientific.

38 Lamport, Z.A., Broadnax, A.D., Scharmann, B. et al. (2019). Molecular rectifiers on silicon: high performance by enhancing top-electrode/molecule coupling. *ACS Appl. Mat. Inter.* 11: 18564–18570.

39 Tour, J.M., Kozaki, M., and Seminario, J.M. (1998). Molecular scale electronics: a synthetic/computational approach to digital computing. *J. Am. Chem. Soc.* 120: 8486–8493.

40 Reed, M.A. (1999). Molecular-scale electronics. *Proc. IEEE* 87: 652–658.

41 Ellenbougen, J.C. and Love, J.C. (2000). Architectures for molecular electronic computers: 1. logic structures and an adder designed from molecular electronic diodes. *Proc. IEEE* 88: 386–426.

42 Mathew, P.T. and Fang, F. (2018). Advances in molecular electronics: a brief review. *Engineering* 4: 760–771.

43 Chen, J., Reed, M.A., Rawlett, A.M., and Tour, J.M. (1999). Large on-off ratios and negative differential resistance in a molecular electronic device. *Science* 286: 1550–1552.

44 Peterson, I.R. (1992). Langmuir–Blodgett films: a route to molecular electronics. In: *Nanostructures Based on Molecular Materials* (eds. W. Göpel and C. Ziegler), 195–208. Weinheim: VCH.

45 Trasobares, J., Vuillaume, D., Théron, D., and Clément, N. (2016). A 17 GHz molecular rectifier. *Nature Commun.* 7: 12850.

Further Reading

Blakemore, J.S. (1969). *Solid State Physics*. Philadelphia: Saunders.

Cuniberti, G., Fagas, G., and Richer, K. (eds.) (2005). *Introducing Molecular Electronics*. Berlin: Springer.

Kasap, S.O. (1997). *Principles of Electrical Engineering Materials and Devices*, 3e. Boston: McGraw-Hill.

Rose, A. (1963). *Concepts in Photoconductivity and Allied Problems*. New York: Robert E. Krieger Publishing Co.

Simmons, J.G. (1971). *DC Conduction in Thin Films*. London: Mills and Boon.

Streetman, B.G. and Banerjee, S. (2000). *Solid State Electronic Devices*, 5e. Hoboken, NJ: Prentice Hall.

Tour, J.M. (2003). *Molecular Electronics*. Singapore: World Scientific.

5

Metal/Insulator/Semiconductor Devices: The Field Effect

5.1 Introduction

The operation of many computational components, such as the metal/insulator/semiconductor (MIS) transistor or MISFET, is based on a phenomenon called the field effect (NB, the term 'MOSFET' refers to the specific case in which the insulator is an oxide, typically silicon dioxide). The physical principles can best be understood by first considering a simple MIS configuration. Transistors are discussed in Chapter 8.

5.2 Ideal MIS Device

The energy bands for an MIS architecture, based on a p-type semiconductor, are depicted in Figure 5.1. This device differs from the MIS structure noted in Section 4.4.4, only in the thickness of the insulating layer. For the configuration under consideration here, the insulator is assumed to be relatively thick so that charge transport between the metal and semiconductor does not occur. The structure is essentially that of a parallel plate capacitor in which one of the metallic plates has been replaced by a semiconductor. For the idealized case, Figure 5.1, it is assumed that the work function of the metal is equal to that of the semiconductor, $\Phi_m = \Phi_s$. With no external bias ($V = 0$ V), the Fermi levels of the metal and semiconductor align, and the various bands are flat throughout the MIS structure. This situation is referred to as the flat-band condition. The difference between the metal and semiconductor work functions, Φ_{ms}, can be written as

$$\Phi_{ms} = \Phi_m - \left(X_s + E_g - \varphi_p\right) = 0 \tag{5.1}$$

where X_s is the electron affinity of the semiconductor, ϕ_p is the energy difference between the semiconductor Fermi level, and the valence band edge (equivalent to the term ϕ_n for a n-type semiconductor). Indicated in Figure 5.1 are the quantities ϕ_i, the energy barrier between metal and conduction band of insulator, and X_i, the electron affinity of the insulator. A further parameter that is shown is ϕ_F, which is a measure of the energy difference between the semiconductor Fermi level E_{Fs} and the Fermi level for intrinsic material E_{Fi}; this becomes useful later in this chapter.

Figure 5.2 shows the effect of an external bias to the MIS structure. If a negative voltage is applied to the metal, or gate, contact (with respect to the semiconductor), a negative charge will be produced on the metal. Using the capacitor analogy, an equal and opposite charge must therefore be produced at the semiconductor surface. In the case under consideration, that of a p-type semiconductor, this occurs by hole (majority carrier) accumulation at the semiconductor/insulator interface, Figure 5.2a.

Electrical Processes in Organic Thin Film Devices: From Bulk Materials to Nanoscale Architectures, First Edition. Michael C. Petty.
© 2022 John Wiley & Sons Ltd. Published 2022 by John Wiley & Sons Ltd.
Companion Website: www.wiley.com/go/petty/organic_thin_film_devices

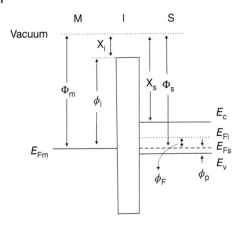

Figure 5.1 Energy band diagram for an ideal MIS structure based on a p-type semiconductor at equilibrium ($V = 0$). Φ_m, Φ_s = work functions of metal and semiconductor, respectively; X_s, X_i = electron affinities of semiconductor and insulator, respectively; E_c, E_v = conduction band and valence band edges, respectively; E_{Fs} = semiconductor Fermi energy; E_{Fi} = intrinsic Fermi energy; ϕ_p = energy difference between E_v and E_{Fi}; ϕ_F = energy difference between E_{Fs} and E_{Fi}. ϕ_i = energy barrier between metal and conduction band of insulator.

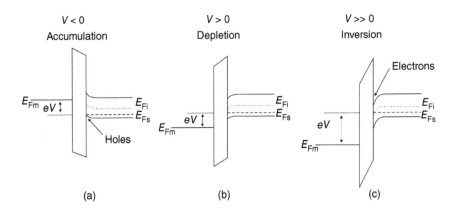

Figure 5.2 Energy band diagrams for an MIS structure based on a p-type semiconductor with different voltages applied to the gate (metal) electrode. (a) Negative voltage applied to gate, holes accumulate at semiconductor surface. (b) Positive voltage applied to gate, depletion layer forms at semiconductor surface. (c) Large positive voltage applied to gate, layer of negative charge (inversion layer) forms at semiconductor surface. V = applied voltage. E_{Fm} = metal Fermi level; E_{Fs} = semiconductor Fermi level; E_{Fi} = intrinsic Fermi level.

Since the applied negative potential reduces the electrostatic potential of the metal relative to the semiconductor, the electron energies are raised in the metal relative to the semiconductor. As a result, the metal Fermi level E_{Fm} lies above its equilibrium position by an amount eV, where V is the applied voltage. As no current flows through the ideal MIS structure, there can be no variation in the semiconductor Fermi level, E_{Fs}, within the semiconductor. Moving E_{Fm} higher in energy with respect to E_{Fs} results in an electric field in the insulator and hence a 'tilt' of its energy bands.

To accommodate the accumulation of holes at the semiconductor surface, the conduction and valence bands in the semiconductor are displaced upwards with respect to E_{Fs}. The carrier density in the valence band depends exponentially on the energy difference between the valence band edge E_v and E_{Fs} (Section 2.6).

If a positive voltage is now applied to the metal, a negative fixed charge appears in the semiconductor. In the p-type material, this arises from the depletion of holes from the region near the surface, leaving behind uncompensated ionized acceptors. The situation is called depletion and is like that for the Schottky diode discussed in Section 4.4. In the depleted region, shown in Figure 5.2b, the hole concentration decreases, bending down the bands at the semiconductor/insulator interface.

If the positive voltage on the gate is increased further, the semiconductor bands bend down more strongly. At some point, E_{Fi}, the Fermi-level position for intrinsic material becomes lower than E_{Fs} at the surface. The semiconductor region adjacent to the surface now has the electrical properties of n-type material. This situation is referred to as inversion and is illustrated in Figure 5.2c. The inversion layer, isolated from the underlying p-type material by a depletion region, is the key to the operation of many MIS transistor devices.

The modulation of the surface conductivity of a semiconductor by an electric field applied orthogonally to its surface, as described in the above sections, is called the field effect.

A more detailed energy band diagram of the surface of the p-type semiconductor is shown in Figure 5.3. The potential $\psi_p(x)$ is defined as the potential with respect to the bulk, or neutral region, of the semiconductor:

$$\psi_p(x) = -\frac{\left(E_{Fi}(x) - E_{Fi}(\infty)\right)}{e} \tag{5.2}$$

In the semiconductor bulk $\psi_p(\infty)$ is zero. At the semiconductor surface, $\psi_p(0)$ is denoted ψ_s, which is called the surface potential (units of voltage). The electron and hole concentrations can be written as functions of ψ_p as follows:

$$n_p(x) = n_{pb} \exp\left(\frac{e\psi_p}{k_B T}\right) \tag{5.3}$$

$$p_p(x) = p_{pb} \exp\left(\frac{-e\psi_p}{k_B T}\right) \tag{5.4}$$

where n_{pb} and p_{pb} are the equilibrium densities of electrons and holes, respectively, in the bulk of the semiconductor. As depicted in Figure 5.3, ψ_p is positive when the bands are bent downwards.

Figure 5.3 Energy band diagram at the surface of a p-type semiconductor. The potential energy $e\psi_p$ is measured with respect to the intrinsic Fermi energy E_{Fi} in the bulk. The surface potential ψ_s is positive as shown. ϕ_F = energy difference between E_{Fs} and E_{Fi} Accumulation occurs when $e\psi_s < 0$. Depletion occurs when $\phi_F > e\psi_s > 0$. Inversion occurs when $e\psi_s > \phi_F$. E_c, E_v = conduction and valence band edges, respectively. E_g = band gap. E_{Fs} = semiconductor Fermi energy. W_d = depletion layer width.

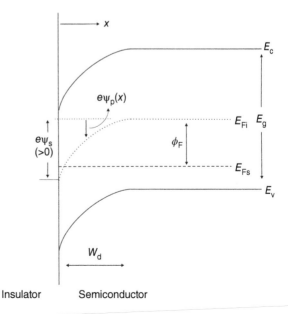

At the surface, the carrier concentrations, n_{p0} and p_{p0}, are the following:

$$n_{p0} = n_{pb} \exp\left(\frac{e\psi_s}{k_B T}\right) \tag{5.5}$$

$$p_{p0} = p_{pb} \exp\left(\frac{-e\psi_s}{k_B T}\right) \tag{5.6}$$

The following regions of surface potentials can therefore be distinguished:

$\psi_s < 0$	Accumulation of holes (bands bent upwards)
$\psi_s = 0$	Flat-band conditions
$2\phi_F/e > \psi_s > 0$	Depletion of holes (bands bent downwards)
$\psi_s = \phi_F/e$	Fermi level at mid-gap, $E_{Fs} = E_{Fi}(0)$, $n_{p0} = p_{p0} = n_i$
$2\phi_F/e > \psi_s > \phi_F/e$	Weak inversion (electron enhancement, $n_{p0} > p_{p0}$)
$\psi_s > 2\phi_F/e$	Strong inversion ($n_{p0} > p_{p0}$ or N_a)

where ϕ_F, introduced earlier, is the energy difference between the semiconductor Fermi level E_{Fs} and the Fermi level for intrinsic material E_{Fi}. (A reminder – in the convention used in this book, ϕ_F represents energy while ψ_s is a voltage).

The charge distribution, electric field, and electrostatic potential for an MIS structure based on a p-type semiconductor in strong inversion are shown in Figure 5.4; the energy band structure is given in Figure 5.4a. For simplicity, the depletion approximation (Section 4.4.1) is used, which assumes complete depletion for $0 < x < W_d$ and neutral material for $x > W_d$. In equilibrium, the positive charge per unit area on the metal surface, Q_m, is balanced by the negative charge in the semiconductor, Q_s, which is due to both the ionized acceptors in the depletion region $Q_d (=eN_a W_d)$ and the charge in the inversion region Q_n:

$$Q_m = -(Q_n + eN_a W_d) = -Q_s \tag{5.7}$$

This is shown in Figure 5.4b.

The distribution of charge carriers changes continuously from the insulator/semiconductor interface to the semiconductor bulk. This distribution can be estimated using Poisson's equation. To a first approximation, the thickness of the inversion and accumulation layers is $\pi L_D/\sqrt{2}$, where L_D is the Debye length (Eq. (3.18)), which is of the order of a few tens of nm (10 nm). However, most of the charges are located very close (a few nm) to the semiconductor surface.

The profiles of the electric field, $\mathcal{E}(x)$, Figure 5.4c, and potential, $V(x)$, Figure 5.4d, can be obtained by the first and second integrations of the Poisson equation. In the absence of any work function differences between the metal and semiconductor, the voltage that is applied across the MIS device will be dropped partly across the insulator and partly across the semiconductor:

$$V = V_i + \psi_s \tag{5.8}$$

The voltage across the insulator, V_i, may be written as

$$V_i = -\frac{Q_s d}{\varepsilon_i \varepsilon_0} = -\frac{Q_s}{C_i} \tag{5.9}$$

where C_i is the capacitance per unit area of the insulator. Using the depletion approximation, the width of the depletion region may be estimated in a similar fashion to the calculation for the Schottky barrier:

$$W_d = \sqrt{\frac{2\varepsilon_s \varepsilon_0 \psi_s}{eN_a}} \tag{5.10}$$

Figure 5.4 (a) Energy band diagram of an ideal MIS structure of a p-type semiconductor under strong inversion conditions. (b) Charge distribution $\rho(x)$. (c) Electric field distribution $\mathcal{E}(x)$. (d) Potential distribution $V(x)$ relative to the semiconductor bulk. V = applied voltage. E_{Fm} = metal Fermi level; E_{Fs} = semiconductor Fermi level; E_{Fi} = intrinsic Fermi level. d, W_d = insulator and depletion layer thicknesses, respectively. Q_m, Q_n, Q_d are charges per unit area in metal, inversion layer and depletion layer, respectively. V_i = voltage across insulator; ψ_s = surface potential.

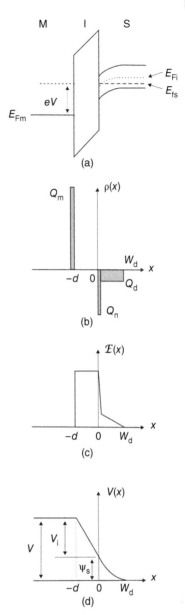

In the depletion region, and using the above expression, it can be shown that the doping density in the semiconductor in an MIS device can be obtained using the following relationship [1]:

$$\frac{\partial}{\partial V}\left(\frac{1}{C^2}\right) = \frac{2}{e\varepsilon_s\varepsilon_0 N_a} \tag{5.11}$$

where C is the measured capacitance of the MIS device per unit area. This is like the result for an ideal Schottky barrier (Eq. (4.14)). The doping density is therefore obtained from the slope of a plot and the reciprocal of the measured capacitance per unit area squared versus the applied gate bias. If the doping density is not uniform, then its profile may be obtained by evaluating the doping density at each value of gate bias. The corresponding value of W_d can then be calculated:

$$W_d = \frac{\varepsilon_s \varepsilon_0}{C_d} = \varepsilon_s \varepsilon_0 \left(\frac{1}{C} - \frac{1}{C_i} \right) \tag{5.12}$$

The depletion layer grows with increased voltage across the MIS structure until strong inversion is reached. At this point, the depletion layer width reaches a maximum. When the bands are bent down far enough that $\psi_s = 2\phi_F/e$, the semiconductor is effectively shielded from further penetration of the electric field by the inversion layer. Therefore, the maximum depletion region under steady-state conditions, W_{dmax}, is

$$W_{dmax} = \sqrt{\frac{2\varepsilon_s \varepsilon_0 \psi_s(\text{strong inversion})}{eN_a}} \tag{5.13}$$

Note that the criterion for strong inversion given above ($\psi_s > 2\phi_F/e$) is that the surface of the semiconductor should be as strongly n-type as the semiconductor bulk is p-type, i.e. E_{Fi} should lie as far below E_{Fs} at the surface as it is above E_{Fs} in the neutral region:

$$\psi_s(\text{strong inversion}) = 2\frac{\phi_F}{e} = 2\frac{k_B T}{e} \ln \frac{N_a}{n_i} \tag{5.14}$$

Substituting Eq. (5.14) into Eq. (5.13):

$$W_{dmax} = \sqrt{\frac{4\varepsilon_s \varepsilon_0 k_B T \ln (N_a/n_i)}{e^2 N_a}} \tag{5.15}$$

An important parameter in MIS or MOS device physics is the turn-on or threshold voltage, V_t, at which strong inversion occurs. This is discussed further in Section 8.2, on transistor devices. Using Eq. (5.8) and substituting from Eqs. (5.9) to (5.14):

$$V_t = \frac{|Q_s|}{C_i} + 2\frac{\phi_F}{e}$$

$$= \frac{\sqrt{2\varepsilon_0 \varepsilon_s eN_a (2\phi_F/e)}}{C_i} + 2\phi_F/e \tag{5.16}$$

The threshold voltage must be sufficiently large to create the maximum depletion region plus the required surface potential for strong inversion (Problem 5.1). The phenomenon of a maximum depletion layer width is unique to the MIS structure. It does not occur in Schottky barriers (Section 4.4) or in p–n junction devices. However, under conditions where the bias voltage is varied rapidly from accumulation towards inversion, the inversion layer will not have time to generate, and it is possible for the depletion region to exceed the maximum value given by Eq. (5.15). This situation is known as deep depletion.

It is important to note that in the case of MIS devices in which the semiconducting layer is very thin (as may be the case for organic device architectures), the maximum theoretical depletion width given by Eq. (5.15) may exceed the semiconductor thickness. The measured capacitance will then saturate with increasing applied voltage before inversion conditions are attained.

The semiconductor capacitance, C_s, that is associated with the depletion region, C_d, will be a nonlinear function of applied voltage:

$$C_s = \frac{dQ}{dV} = \frac{dQ_d}{d\psi_s} \tag{5.17}$$

This is properly called the differential capacitance (to distinguish this from the capacitance calculated using the absolute values of Q and V). The overall capacitance of the MIS capacitor is a

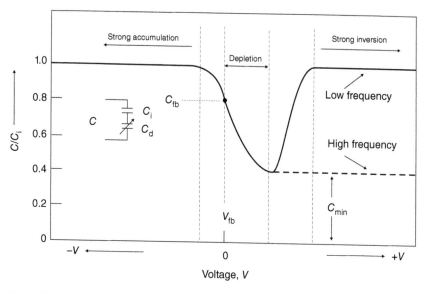

Figure 5.5 Normalized capacitance versus voltage characteristics for an MIS structure based on a p-type semiconductor. The dashed curve for positive applied voltages is expected at high measurement frequencies. V_{fb} = flat band voltage, C_{fb} = flat band capacitance; C_i, C_d = insulator and depletion layer capacitances, respectively. C_{min} = minimum measured capacitance in inversion region. All capacitances are per unit area.

series combination of a fixed insulator capacitance and the voltage-dependent semiconductor contribution, C_d. The capacitance versus voltage curve for an ideal MIS structure, based on a p-type semiconductor, is shown in Figure 5.5. The inset provides the equivalent circuit. The flat-band voltage of this ideal structure, V_{fb}, is assumed to be zero (i.e. the work functions of the metal and semiconductor are equal). For non-ideal devices, V_{fb} is the applied voltage required to produce flat bands throughput the MIS device ($\psi_s = 0$) and C_{fb} is the corresponding value of capacitance.

In accumulation, C_d will be very high (the accumulation charge changes rapidly with surface potential). The result is that the total capacitance is simply that of the insulator, C_i. In this regime, the MIS device behaves as a voltage-independent parallel plate capacitor:

$$C_i = \frac{\varepsilon_i \varepsilon_0}{d} \tag{5.18}$$

(NB, as with the other capacitance values in this section, C_i is the insulator capacitance per unit area).

As the voltage applied to the metal electrode increases (becomes more positive), the semiconductor surface is depleted and the overall capacitance is that of a depletion layer capacitance, C_d, in series with the insulator capacitance. The former is given by

$$C_d = \frac{\varepsilon_s \varepsilon_0}{W_d} \tag{5.19}$$

The total capacitance is therefore

$$C = \frac{C_i C_d}{C_i + C_d} = \frac{\varepsilon_i \varepsilon_s \varepsilon_0}{\varepsilon_s d + \varepsilon_i W_d} \tag{5.20}$$

The semiconductor flat-band capacitance depends on the Debye screening length, L_D. This can be shown to be [1]

$$C_d(\text{flat-band}) = \frac{\varepsilon_s \varepsilon_0}{L_D} \qquad (5.21)$$

The total flat-band capacitance, C_{fb}, is therefore given by Eqs. (5.20) and (5.21):

$$C_{fb}\left(\psi_s = 0\right) = \frac{\varepsilon_i \varepsilon_s \varepsilon_0}{\varepsilon_s d + \varepsilon_i L_D} = \frac{\varepsilon_i \varepsilon_s \varepsilon_0}{\varepsilon_s d + \varepsilon_i \sqrt{\varepsilon_s \varepsilon_0 k_B T / e^2 N_a}} \qquad (5.22)$$

where the expression given by Eq. (3.18), has been used for the Debye length.

After inversion is reached, the differential capacitance depends on whether the measurements are made at high (typically 1 MHz for inorganic semiconductors such as Si) or low ($\sim$1–100 Hz) frequency. In this context, 'high' or 'low' are with respect to the generation-recombination rate of the minority carriers in the inversion layer. If the AC signal that is applied to the metal electrode (to measure the differential capacitance) is varied too rapidly, the charge in the inversion layer cannot respond and will therefore not contribute to the small AC signal capacitance. As noted earlier, in the inversion regime, the presence of the inversion charge effectively prevents further penetration of the electric field into the semiconductor. The result is that the capacitance saturates with increasing voltage at a minimum value, C_{min}, shown by the dashed line in Figure 5.5. This minimum value can be calculated by substitution the value for W_{dmax} from Eq. (5.15) into Eq. (5.20):

$$C_{min} = \frac{\varepsilon_i \varepsilon_s \varepsilon_0}{\varepsilon_s d + \varepsilon_i W_{dmax}} \qquad (5.23)$$

However, if the voltage applied to the metal contact is changed very slowly, there will be time for the minority carriers to be generated in the neutral region of the semiconductor and drift across the depletion region to the inversion layer. The semiconductor capacitance becomes very large, and the overall capacitance approaches the insulator value again, as shown by the full line in Figure 5.5. Carrier response is not so important in the accumulation region as only the majority carriers are required to respond. These can react on a timescale of the semiconductor dielectric relaxation time ($\varepsilon_s \varepsilon_0 \rho$), which can be around 10^{-13} seconds for an inorganic semiconductor. In contrast, minority carriers respond on the time scale of generation-recombination times, 100s of μs in the case inorganic semiconductors. However, these times will be different (generally longer) for organic semiconductors.

If the maximum theoretical depletion layer thickness exceeds the semiconductor thickness, the minimum measured capacitance will correspond to a series combination of the insulator capacitance and the geometrical capacitance associated with the semiconductor, i.e. W_{dmax} in Eq. (5.23) will be replaced by the semiconductor layer thickness (Problems 5.2 and 5.3).

Although the high-frequency capacitance of MIS structures is low in the inversion region, if the capacitor structure is incorporated into a transistor, i.e. as a part of a MISFET (Chapter 8), the inversion charges can flow rapidly into the semiconductor from the source and drain electrodes rather than be generated in the bulk.

5.3 Departures from Ideality

Several physical effects can influence the characteristics of the ideal MIS device described in the previous section. These include work function differences between the metal and the semi-conductor, fixed and mobile charge in the insulator, and the presence of traps in the vicinity of the metal/semiconductor interface. The electrical properties of such defects have already been introduced in Sections 3.4 and 3.5, and their influence on Schottky barrier devices discussed in Section 4.4.4. Non-ideal insulators may also exhibit a finite conductivity, which will invalidate the assumption that no current passes through the insulator in the MIS device. The various conductivity mechanisms that might occur are described, in detail, in the following chapter. Below, some of the other departures from ideality are examined.

5.3.1 Insulator Charge and Work Function Differences

Electrical charges which might be found in the insulator and have been described in Section 3.3 (see Figure 3.8). These include fixed insulator charge, Q_{fix}, which is immobile under an applied electric field as well as mobile ionic charge, Q_{mob}, such as metal ions, which will drift when a field is applied. Fixed charges result in a simple shift in the flat-band voltage and a parallel shift in the theoretical $C–V$ curve. The flat-band voltage shift, ΔV_{fb}, resulting from fixed insulator charge may be calculated from Gauss' law:

$$\Delta V_{\text{fb}} = -\frac{1}{C_{\text{i}}}\left(\frac{1}{d}\int_0^d x\rho(x)\mathrm{d}x\right)$$

(5.24)

where $\rho(x)$ is the charge density per unit volume in the insulator. Eq. (5.24) reveals that the effect of fixed insulator charge on the flat-band voltage will depend precisely on where the charge is located in the insulator. For instance, charge positioned very close to the insulator/semiconductor interface will have more effect on the flat-band voltage than the same charge located close to the metal. The effect of the sign of the fixed charge on the high-frequency capacitance versus voltage curve for an MIS structure based on a p-type semiconductor is shown in Figure 5.6. Addition of negative charge is equivalent to an increase in the negative gate bias for the semiconductor, so a more positive gate bias is needed to achieve the same band bending, Figure 5.6a. The new flat-band voltage is shifted from zero, for the ideal device, to a positive value. The opposite is true for positive fixed insulator charge, Figure 5.6b. Similar effects will occur for an n-type semiconductor: positive charge will cause the $C–V$ curve to shift to a more negative values of gate bias, while negative charge will cause a shift to more positive gate biases. The presence of fixed charge results in a simple displacement of the capacitance versus voltage curve along the voltage axis; the shape of the $C–V$ curve is unaltered.

The shift in the flat-band voltage provides insufficient information to calculate the charge distribution. However, in the case of MIS devices based on the silicon/silicon dioxide system, the fixed

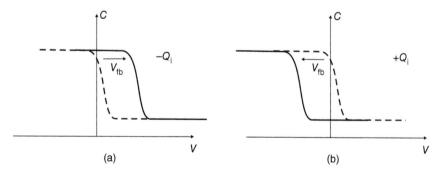

Figure 5.6 Effect of insulator charge on the high frequency capacitance versus voltage characteristics of an MIS structure based on a p-type semiconductor. Curves without any insulator charge are shown as dashed. (a) Negative insulator charge, $-Q_{\text{i}}$, shifts $C–V$ curve to more positive voltages. (b) Positive insulator charge, $+Q_{\text{i}}$, shifts $C–V$ curve to more negative voltages. V_{fb} = flat band voltage.

charge is known to be located, in a sheet, very close to the Si/SiO$_2$ interface [1]. This charge is generally positive. Its density is not greatly affected by the oxide thickness or by the type or concentration of impurities in the silicon, but it depends on oxidation and annealing conditions, and on silicon surface orientation. For this case,

$$\Delta V_{fb} = -\frac{Q_{fix}}{C_i} \tag{5.25}$$

A further influence on the flat-band voltage is from the presence of mobile ionic charge. In the Si/SiO$_2$ system, this is most commonly caused by the presence of ionized alkali metals, such as sodium or potassium. This type of charge is located either at the metal/oxide interface, where it originally entered the oxide layer, or at the Si/SiO$_2$ interface, where it has drifted under an applied field. Drift can occur because such ions are mobile in SiO$_2$ at relatively low temperatures. Fixed and mobile insulator charges can usually be distinguished by a bias-temperature ageing experiment [1]. The gate bias is applied for a given length of time while the sample is held at a moderately high temperature. The density of fixed oxide charge may change during this experiment, but the centres responsible for this charge do not move. In contrast, mobile ionic charge can be cycled back and forth between the metal/insulator and the insulator/semiconductor interfaces, and a resulting ionic current can be detected. Organic MIS architectures will also be affected by both fixed and mobile charges as a result of their processing – frequently undertaken in 'wet' environments.

In the previous section, it was assumed that the work function difference between the metal and the semiconductor (as defined by Eq. (5.1)) is zero. If this is not the case, then the experimental C–V curve will be shifted from the theoretical curve by a simple displacement along the voltage axis, as for the case of fixed insulator charge. Figure 5.7 illustrates the effect of reducing the work function of the gate electrode. The work function difference, Φ_{ms}, defined by Eq. (5.1), now becomes negative. With no bias applied, the metal becomes positively charged and the semiconductor surface becomes negatively charged to accommodate this work function difference, Figure 5.7b. Note that if Φ_{ms} is sufficiently negative, the semiconductor surface can become inverted with no applied bias. To obtain flat-band conditions, a negative voltage (Φ_{ms}/e) must be applied to the metal, Figure 5.7c.

A work function difference between the metal and semiconductor can clearly be obtained by using a different metal. However, this can also be achieved by keeping the metal the same but changing its environment. For example, if a Pd top contact in an MIS device is exposed to hydrogen gas, gas molecules may penetrate to the metal/insulator interface where these become adsorbed and give rise to a dipole layer at the interface. This changes the work function difference between the Pd and the semiconductor [2]; the phenomenon can form the basis of a simple hydrogen gas sensor.

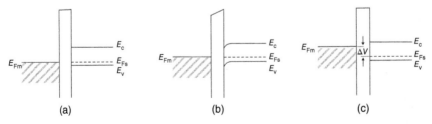

Figure 5.7 Energy band diagram for an ideal MIS structure based on a p-type semiconductor. (a) Flat-band conditions with identical metal and semiconductor work functions. (b) Lower gate work function at zero applied bias. (c) Negative voltage ΔV applied to the metal electrode restores flat-band conditions. E_c, E_v = conduction band and valence band edges, respectively; E_{Fs}, E_{Fm} = semiconductor and metal Fermi energies, respectively.

The effect of the work function difference between the metal and the semiconductor is in addition to that resulting from the fixed and mobile insulator charges (Problem 5.4). Taking all these into consideration, and assuming that the fixed and mobile charges are located at the semiconductor/insulator interface, the flat-band voltage becomes

$$V_{fb} = \frac{\Phi_{ms}}{e} - \frac{(Q_{fix} - Q_{mob})}{C_i} \tag{5.26}$$

5.3.2 Interface Traps

The existence of traps located at the semiconductor/insulator interface has been introduced in Section 3.5; their effects on the properties of a Schottky barrier were discussed in Section 4.4.4. When a voltage is applied to the MIS device, the semiconductor Fermi level moves up or down with respect to the interface trap distribution, and a change in the charge associated with these traps occurs.

The C–V characteristics of MIS structure can be affected by interface traps in three ways: (i) the traps provide an additional capacitance; (ii) these introduce a frequency dependence of the curve; and (iii) these affect the dependence of the surface potential, ψ_s, on the applied voltage. Figure 5.8 shows the qualitative features of the effect on the interface traps. Two C–V curves are contrasted, one for an ideal MIS structure based on a p-type semiconductor (full curve) and the other (dashed curve) for a device that includes interface traps. A very noticeable effect is that the introduction of traps has stretched out the C–V curve along the voltage axis. This is simply due to extra charge being needed to fill the traps; hence, a greater applied voltage is required to achieve the same surface potential (or band-bending). This effect occurs for MIS devices based on either p-type or n-type semiconductors.

Interface traps invariably give rise to hysteresis in the capacitance versus voltage (and conductance versus voltage) curves, i.e. the C–V plots are not superimposed when the voltage bias is scanned from negative to positive (or vice versa) and then the direction of scan is reversed. The degree of hysteresis, usually measured by the shift in the flat-band voltage, increases with the rate of the voltage scan. However, interface traps are not the only phenomenon that can give rise to such hysteresis. Figure 5.9 contrasts common hysteresis effects on the C–V curves for MIS structures based on a p-type semiconductor. Ion drift and insulator polarization (see Section 7.2) both result in a clockwise hysteresis while interface traps give rise to hysteresis in the counter-clockwise direction.

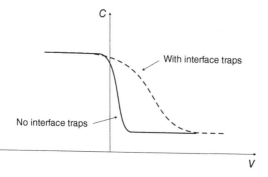

Figure 5.8 Influence of interface traps on the high-frequency capacitance versus voltage curve for an MIS structure based on a p-type semiconductor. The C–V curve becomes more stretched out in the presence of the interface traps (dashed line).

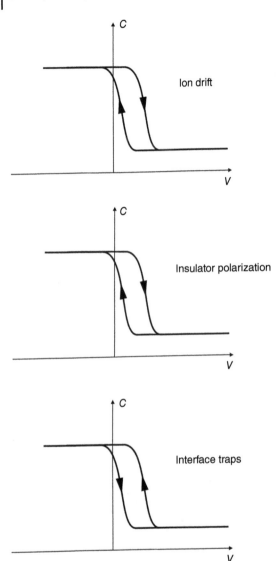

Figure 5.9 Hysteresis effects on the high-frequency capacitance versus voltage curve for an MIS structure based on a p-type semiconductor. Ion drift and insulator polarization (top and middle diagrams) produce clockwise hysteresis, while the presence of interface traps results in anti-clockwise hysteresis.

These directions are preserved for n-type semiconductors, i.e. counter-clockwise for interface traps and clockwise for ion drift and insulator polarization.

The electrical effect associated with the charging and discharging of the traps can be obtained by modelling the traps using Shockley–Read–Hall statistics (Section 3.4). The basic equivalent circuit for an MIS device incorporating interface traps is shown in Figure 5.10a [3, 4]. In the figure, C_{it} and R_{it} are the capacitance and resistance associated with unit areas of interface traps. These parameters will be functions of energy (i.e. of the location of the traps within the semiconductor band gap). The resistance R_{it} is associated with an energy loss as the traps charge and discharge with the changing applied bias; the product $R_{it} C_{it}$ defines the interface trap lifetime, τ_{it}, which will determine their frequency behaviour. The parallel branch of the equivalent circuit shown in Figure 5.10a can be converted into a frequency-dependent capacitance, C_p, in parallel with a frequency-dependent

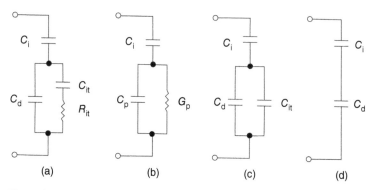

Figure 5.10 (a) to (b) Equivalent circuits of an MIS structure including interface traps, which can be modelling using the series resistance-capacitance circuit, R_{it} and C_{it}. (c) Low-frequency equivalent circuit. (d) High-frequency equivalent circuit. C_i, C_d = insulator and depletion layer capacitances, respectively; G_p, C_p are the parallel conductance and capacitance, respectively. All electrical parameters are for unit device area. Source: Adapted from Nicollian and Goetzberger [3, 4].

conductance, G_p, as shown in Figure 5.10b:

$$C_p = C_d + \frac{C_{it}}{1 + \omega^2 \tau_{it}^2} \tag{5.27}$$

$$\frac{G_p}{\omega} = \frac{C_{it}\omega\tau_{it}}{1 + \omega^2 \tau_{it}^2} \tag{5.28}$$

It is interesting to note that Eqs. (5.27) and (5.28) take a similar form to the Debye equations for a dielectric (Section 7.3.2) in which the polarization can relax exponentially with a characteristic time following a change in the applied electric field.

The low-frequency and high-frequency equivalent circuits shown in Figure 5.10a are given in Figure 5.10c,d, respectively. In the low-frequency limit, R_{it} is set to zero (i.e. there is no energy loss associated with the charging and discharging of the traps as these processes will be able to follow the relatively slow variations in the applied voltage), leaving C_d in parallel with C_{it}. At high frequencies, the C_{it}–R_{it} branch of the equivalent circuit is ignored as the charging and discharging of the traps will not be fast enough to follow the changes in the high-frequency signal used for the admittance measurements. However, the traps will be able to respond to the much slower changes in the DC gate bias as the MIS device is scanned from accumulation to inversion (or vice versa). This will affect the relationship between the applied bias and the surface potential. The equivalent capacitances for these two limiting cases, the low-frequency capacitance, C_{lf}, and the high-frequency capacitance, C_{hf}, are given by

$$C_{lf} = \frac{C_i \left(C_d + C_{it}\right)}{C_i + C_d + C_{it}} \tag{5.29}$$

$$C_{hf} = \frac{C_i C_d}{C_i + C_d} \tag{5.30}$$

A few methods have been devised to extract interface trapping information. These have been developed for Si/SiO$_2$ systems but have yet to be widely applied to MIS structures based on organic semiconductors and insulators. Either capacitance or conductance data can be used as these parameters both contain information about the interface traps. The conductance method is generally the more sensitive (certainly for Si/SiO$_2$ MIS structures). However, the capacitance data can provide a rapid measurement of the total interface trapped charge. Three capacitance methods have been developed [5]: (i) low-frequency capacitance – comparing the low-frequency C–V curve to the theoretical curve; (ii) high-frequency capacitance – comparing the high-frequency curve to the theoretical curve; and (iii) high-low-frequency capacitance – comparing the measured low-frequency C–V curve to that measured at high frequencies.

The advantage of the high-low-frequency method is that no theoretical calculation is needed for comparison [6]. Using Eqs. (5.29) and (5.30), the capacitance due to the interface traps can be written as

$$C_{it} = \left(\frac{1}{C_{lf}} - \frac{1}{C_i}\right)^{-1} - C_d = \left(\frac{1}{C_{lf}} - \frac{1}{C_i}\right)^{-1} - \left(\frac{1}{C_{hf}} - \frac{1}{C_i}\right)^{-1} \tag{5.31}$$

Noting that $dQ_{it} = eD_{it}dE$ and $dE = ed\psi_s$:

$$C_{it} = \frac{dQ_{it}}{d\psi_s} = \frac{eD_{it}dE}{dE/e} = e^2 D_{it} \tag{5.32}$$

If the capacitance difference between the high-frequency and low-frequency curves is defined $\Delta C = C_{lf} - C_{hf}$, and using the above relationship for D_{it}, the trap density (Eq. (3.17)) is

$$D_{it} = \frac{C_i}{e^2}\left(\left(\frac{1}{\Delta C/C_i + C_{hf}/C_i} - 1\right)^{-1} - \left(\frac{1}{C_{hf}/C_i} - 1\right)^{-1}\right)$$

$$= \frac{\Delta C}{e^2}\left(1 - \frac{C_{hf} + \Delta C}{C_i}\right)^{-1}\left(1 - \frac{C_{hf}}{C_i}\right)^{-1} \tag{5.33}$$

By evaluating Eq. (5.33) at each bias point, the variation of D_{it} with applied voltage can be obtained. To a first approximation (order of magnitude estimate), the trap density is proportional to ΔC.

To calculate the density of traps versus position in the semiconductor band gap, the variation of the surface potential with applied voltage must be known. One technique is to compare the high-frequency curve with a theoretical curve [7]. However, an approach based on the integration of the low-frequency capacitance curve is generally preferred as it is more accurate [8]. The equivalent circuit at low frequencies, Figure 5.10c, shows that part of the applied voltage, V, will be dropped across the insulator and the rest will appear across the semiconductor (Eq. (5.8)). The voltage across the semiconductor is the surface potential, ψ_s. The relationship between ψ_s and V is then determined by the potential divider network formed by the various capacitances:

$$\frac{d\psi_s}{dV} = \frac{C_i}{C_i + C_d + C_{it}} = 1 - \frac{C_d + C_{it}}{C_i + C_d + C_{it}} = 1 - \frac{C_{lf}}{C_i} \tag{5.34}$$

Integrating Eq. (5.34) over two applied voltages gives

$$\psi_s(V_2) - \psi_s(V_1) = \int_{V_1}^{V_2}\left(1 - \frac{C_{lf}}{C_i}\right)dV + \text{constant} \tag{5.35}$$

Hence, the surface potential at any value of applied voltage can be determined by integrating $(1 - C_{lf}/C_i)$. The constant of integration can be the starting point in accumulation or strong inversion where the surface potential depends weakly on the applied voltage.

One problem with capacitance methods for the evaluation of the interface trap density is that errors occur because the interface trap capacitance must first be extracted from the measured capacitance, which consists of the insulator capacitance, depletion capacitance, and surface state capacitance. In contrast, the measured conductance arises directly from the interface traps [3, 4]. The principle of the conductance technique is best understood by reference to the equivalent circuits shown in Figure 5.10a,b. The admittance of the MIS structure is first measured; a measurement in the region of strong accumulation will provide a measurement of the insulator capacitance. The MIS admittance is then converted into an impedance, the reactance of the insulator capacitance subtracted, and the resulting impedance converted back to an admittance. This leaves C_d in parallel with the series R_{it}–C_{it} network. The capacitance and equivalent parallel conductance are given by Eqs. (5.27) and (5.28). The equivalent parallel conductance expression only contains information on the interface states (unlike Eq. (5.27), Eq. (5.28) does not include C_d). The expression to convert the measured admittance, C_m in parallel with G_m, is

$$\frac{G_p}{\omega} = \frac{\omega C_i^2 G_m}{G_m^2 + \omega^2 \left(C_i - C_m\right)^2} \tag{5.36}$$

At a given bias, G_p/ω can be measured at a function of frequency. Eq. (5.28) indicates that a plot of G_p/ω versus $\omega\tau_{it}$ will go through a maximum when $\omega\tau_{it} = 1$. This gives τ_{it} directly. The value of G_p/ω at the maximum is $C_{it}/2$. Once C_{it} is known, the interface trap density can be calculated using Eq. (5.32), i.e. $D_{it} = C_{it}/e^2$. These equations will only be approximately true as they have been derived from a simple equivalent circuit. A full analysis will consider a continuum of traps over the band gap of the semiconductor and dispersion in the interface trap time constant resulting from surface-charge fluctuations [1, 4, 5].

5.4 Organic MIS Devices

MIS devices based on Si/SiO$_2$ have been explored extensively over the last 50 years and much is known about this system [1]. Generally, MIS structures incorporating organic insulators or organic semiconductors (or both) exhibit (qualitatively) similar characteristics to those outlined in the previous sections. Significant differences result from the longer carrier response times in organic semiconductors. The organic semiconductor layers in the MIS devices are also relatively thin, which means that the minimum capacitance observed in their C–V response is usually attributable to the full depletion of the semiconductor layer rather than to the generation of inversion charges. Relatively large hysteresis effects are often observed in the C–V and G–V curves for organic MIS devices. However, it is not clear that these are always associated with chemical bonding disruptions between the insulator and semiconductor. It is quite likely that such effects are related to the lower purity of the organic compounds and/or to the device processing environments, which have not been developed to the same degree of cleanliness as those for inorganic semiconductors. The following sections review some of the MIS data that are now available for structures incorporating organic materials as the insulator, the semiconductor, or both.

5.4.1 Inorganic Semiconductor/Organic Insulator Structures

In many instances, organic insulators have been used in conjunction with inorganic semiconductors to probe the electronic behaviour of the semiconductor/insulator interface and as the basis of certain electronic devices, e.g. transistors. The Langmuir–Blodgett (LB) deposition technique provides a simple way of building-up, one monomolecular layer at a time, an insulating film of a desired thickness. Figure 5.11 shows the measured C–V and G–V curves of a p-type CdTe/LB film MIS structure, at a frequency of 100 kHz [9]. The organic film comprised 31 individual layers of cadmium stearate, giving a total insulator thickness of approximately 80 nm. The capacitance curve reveals accumulation, depletion, and inversion behaviour characteristic of ideal MIS structures, i.e. Figure 5.5 (Problem 5.5). No hysteresis was evident on reversing the voltage scan (at a

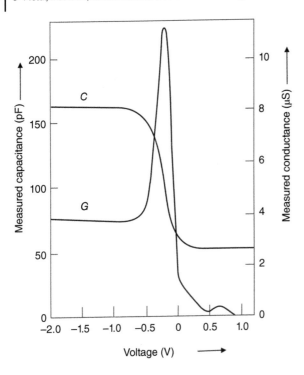

rate of $10 \, \text{mV s}^{-1}$), and the flat-band voltage was about $-0.3 \, \text{V}$ [9]. The expected increase in capacitance due to minority carrier response in the inversion region was not observed, even at the lowest frequency used (8 Hz). The conductance curve shows a peak in the depletion region, consistent with losses associated with interface traps. As the measuring frequency is reduced, the height of the G_m/ω peak decreases and its position shifts towards the weak-inversion region. The constant value of the conductance in the accumulation region is related to the series resistance of the CdTe substrate (NB, this factor is ignored in the equivalent circuits depicted in Figure 5.10). Assuming that the equivalent circuit in accumulation can be represented by a series combination of the insulator capacitance, C_i, and substrate resistance, R_s, then these values may be calculated using the equations:

$$R_s = \frac{G_m}{\left(G_m^2 + \omega^2 C_m^2\right)} \tag{5.37}$$

$$C_i = C_m \left(1 + \frac{G_m^2}{\omega^2 C_m^2}\right) \tag{5.38}$$

Figure 5.12 shows the parallel conductance (G_p/ω) versus frequency data for the CdTe/LB film device at a gate bias corresponding to flat-band conditions [9]. The curve is broader than predicted from a single time constant model described above (dashed curve in Figure 5.12). Dispersion of the time constant can arise from fluctuations of the surface potential or from tunnelling of carriers from the semiconductor surface to traps located in the insulator. Complex data processing and curve fitting are required to extract accurate trapping parameters [1]. However, using the single time constant approach to provide an order of magnitude estimate for the density of surface stats produced a value of $D_{it} \approx 10^{12} \, \text{cm}^{-2} \, \text{eV}^{-1}$ at a position approximately $0.2 \, \text{eV}$ above the CdTe valence band edge. For the CdTe/LB MIS structure, the device preparation was thought to result in the formation of a very thin passivating 'oxide' layer on the semiconductor surface. There will therefore

Figure 5.12 Parallel conductance (G_p/ω) versus frequency data at a gate bias (0.225 V) corresponding to flat-band conditions for a p-type CdTe/cadmium stearate LB film MIS structure. The dashed curve is that expected for a single time constant model. Source: With permission from Petty and Roberts [9].

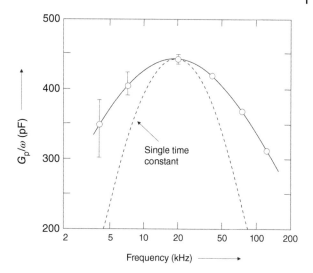

be two interfaces where the measured traps could be located. This situation will almost certainly occur for other inorganic semiconductor/organic insulator combinations.

The LB method also provides a means to explore the properties on MIS devices as a function of the insulating layer thickness and to study the effects of charge incorporation. For example, Figure 5.13 reveals how the flat-band voltage of n-type Si/SiO$_2$/LB film MIS devices changes with the number of deposited LB layers [10]. The organic film in this instance was 22-triconsenoic acid, which is a long chain fatty acid with a terminal group containing a double bond. Monolayers of this compound, each with a thickness of approximately 3 nm, were built-up on an n-type silicon substrate, onto which had been grown a 32 nm layer of oxide. As shown in Figure 5.13, a 'reference' device with no organic layer, possessed a V_{fb} of about −0.9 V (single data point on the ordinate corresponding to zero monolayers). The deposition of a single monolayer produced a shift in V_{fb} to a more negative value; the addition of subsequent monolayers then resulted in a gradual change in V_{fb} to more positive values as further monolayers were built-up. Three curves are shown in Figure 5.13: curve a – for the as-deposited MIS device; curve b – following three days storage in a desiccator; and curve c – following re-exposure to air under laboratory conditions for five days. The curves reveal the same basic shape, but there is a very noticeable difference in the data for the devices measured in air and following desiccation, presumably associated with the removal and re-absorption of water molecules in the organic film. Again, this will be an issue for many organic dielectrics, which, unless encapsulated, will exhibit a propensity for absorbing water vapour from the atmosphere.

The negative shift in V_{fb} when one monolayer is deposited suggests positive charge incorporation in the organic layer (cf. Figure 5.6b). This was thought to be associated with the organic film itself or to impurities trapped at the SiO$_2$ surface during the LB deposition. The latter is perhaps more likely as changes in the deposition conditions (aqueous subphase pH and/or addition of metallic cations into the subphase) did not have an appreciable effect on V_{fb} following the deposition of the single LB monolayer [10]. The marked changes in V_{fb} following desiccation suggest that water molecules are directly responsible for this effect. Although, the molecules will be uncharged, these can orient within the LB array in the vicinity of the polar fatty acid head group regions, producing a dipole layer and a consequent shift in the flat-band voltage, as noted above in Section 5.3.1.

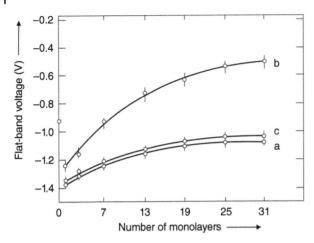

Figure 5.13 Flat-band voltage as a function of number of monolayers in MIS structures based on n-type Si and fatty acid Langmuir–Blodgett films. Curve a – before desiccation; curve b – after desiccation; curve c – following exposure to air for five days. The single data point on the ordinate corresponding to 0 monolayers represents the flat-band voltage for a reference device. Source: Reprinted with permission from Evans et al. [10].

5.4.2 Organic Semiconductor Structures

Much of the early work on MIS structures incorporating organic semiconductors exploited silicon diode as the dielectric layer. This is a highly resistive insulator that can be conveniently grown on highly doped silicon, which forms the top electrode. Figure 5.14 shows the capacitance at a measurement frequency of 111 Hz and a temperature of 270 K as a function of bias for an Au/silicon dioxide/poly(3-hexylthiophene) (P3HT) MIS structure [11]; the thicknesses of the layers was 150 nm for the P3HT (formed by spin-coating) and 50 nm in the case of the SiO_2. The accumulation and depletion regions are clear. At applied voltages greater than about 2 V, the capacitance levels out, indicative of the high-frequency inversion characteristics shown in Figure 5.5. However, it is more likely that the sample becomes fully depleted at large positive gate voltages, leading to a minimum capacitance that is the series sum of the insulator and semiconductor capacitances. As noted above, this will always be an issue when using a very thin (<100 nm) film of an organic semiconductor as the basis for an MIS device. The capacitance versus voltage scan shown in Figure 5.14 showed little dependence on the direction of the voltage sweep and the flat-band voltage is close to 0 V, suggesting minimal trapped charge in the insulator of interface traps at the SiO_2/P3HT interface.

Capacitance versus voltage data for a spin-coated P3HT-based MIS device incorporating an organic insulator, in this case polysilsesquioxane (PSQ), are shown in Figure 5.15; the layer thicknesses were 150 nm for the P3HT and 400–450 nm in the case of the PSQ [12]. The device configuration was Au/P3HT/PSQ, with a bottom electrode of indium tin oxide (ITO). A monolayer of hexamethyldisilazane was deposited on top of the PSQ (before the semiconductive polymer was deposited) to render the surface hydrophobic. Similar treatments have been used between other organic insulator/semiconductor combinations and generally improve the electronic properties of the semiconductor/insulator interface. After fabrication, the devices were annealed, in stages, under vacuum at elevated temperatures to remove adventitious dopants from the semiconductor. The anneal sequence consisted of three successive heating cycles at 90 °C for one hour, followed by a fourth cycle at 100 °C for one hour and finally a fifth heating cycle at 90 °C for eight hours.

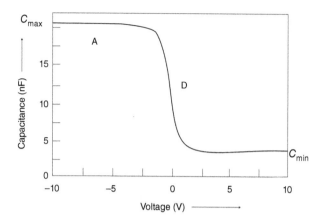

Figure 5.14 Capacitance versus voltage behaviour for an MIS structure comprising Si/SiO$_2$/polythiophene/Au measured at 111 Hz and at 270 K. The accumulation A and depletion D regions are indicated. Source: Reprinted with permission from Grecu et al. [11].

Figure 5.15 Capacitance versus voltage behaviour for an MIS structure comprising a polysilsesquioxane insulator and a poly(3-hexylthiophene) semiconductor, measured at room temperature and 1 kHz. The curves are shown after taking the device through various heating cycles: First (data not shown), second, and third (data not shown) annealing all at 90 °C for one hour; fourth cycle (data not shown) annealing at 100 °C for one hour; fifth cycle at 90 °C for eight hour. Source: Reprinted with permission from Torres and Taylor [12].

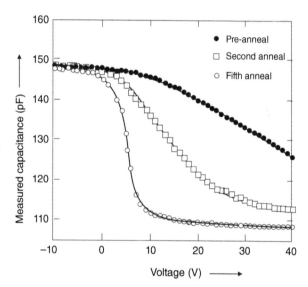

The device characteristics were measured at room temperature after each heating cycle without breaking the vacuum (~10^{-5} mbar).

The data in Figure 5.15 reveal a sharpening up of the C–V characteristics, with the flat-band voltage moving towards 0 V with progressive annealing. This is consistent with the removal of interface traps (Figure 5.8) and/or fixed charge from the insulator (Figure 5.6). Again, the standard accumulation and depletion characteristics are evident in the C–V curve. The constant capacitance at high positive voltages is attributed to the depletion region extending across the entire thickness of the P3HT layer [12]. A large positive voltage therefore fully depletes the P3HT layer rather than generates an inversion layer at the P3HT/PSQ interface. Figure 5.16 shows the parallel conductance (G_p/ω) versus voltage data at different measurement frequencies for a P3HT/PSQ MIS device following the fifth annealing stage [12]. Two loss peaks are evident, at approximately +6 V and at 0 V for a measurement frequency of 1 kHz. Both conductance peaks move to more negative voltages, i.e. less band bending, as the measurement frequency increases. This suggests that these arise from

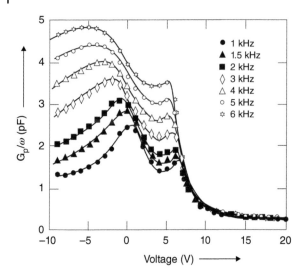

Figure 5.16 Parallel conductance, G_p/ω, versus voltage behaviour for an MIS structure comprising a polysilsesquioxane insulator and a poly(3-hexylthiophene) semiconductor, measured at room temperature and at different frequencies. Data from same devices as Figure 5.15, after the fifth anneal. Source: Reprinted with permission from Torres and Taylor [12].

two distinct interface traps distributions (for a single interface trap, the loss should peak at the same voltage for all frequencies but vary in magnitude). The peak at around 0 V is attributed to shallow hole traps located close to the equilibrium Fermi level, while the +6 V peak is thought to originate from deep hole traps. Interface traps densities, obtained by fitting the response of theoretical models to the experimentally measured admittance of the device, indicated that the traps densities after the final annealing stage were around 3×10^{12} cm^{-2} eV^{-1} for the shallow traps and approximately 1×10^{10} cm^{-2} eV^{-1} in the case of the deep traps, which were thought to be centred approximately 0.5 eV above the equilibrium Fermi level.

An experiment to influence the electrical nature of the semiconductor/insulator interface by using nanoparticles is revealed by the data in Figure 5.17. This shows C–V curves for a different organic semiconductor/organic semiconductor combination – with poly(methyl methacrylate) (PMMA) as the insulator and pentacene as the semiconductive layer [13]. The device configuration was Au/PMMA/pentacene/Al/glass. The thicknesses of the pentacene and PMMA layers were both approximately 40 nm; the PMMA was formed by spin-coating, while the pentacene semiconductor was thermally evaporated. For some devices, gold nanoparticles were deposited between the PMMA and pentacene layers using solution adsorption. A schematic diagram for the MIS architecture is shown in Figure 5.17a. The capacitance versus voltage curve, measured at 100 kHz, for a control device, with no nanoparticles, is shown in Figure 5.17b. This shows the expected accumulation and depletion regions evident for other organic semiconductor/insulator combinations in Figures 5.13 and 5.14. No hysteresis is evident on reversing the direction of the voltage scan (voltage scan rate = 100 mV s^{-1}). However, the addition of the layer of Au nanoparticles produces distinct changes in the C–V curve, as shown in Figure 5.17c. On reversing the direction of the voltage scan, significant hysteresis is evident in the measured curve. This is attributed to the presence of the nanoparticle layer becoming charged and discharged during the voltage cycle. The counter-clockwise direction of the hysteresis indicates that the Au nanoparticles become charged from the semiconductor surface (rather than by charge transport through the insulating layer). It is suggested that, in accumulation, when a negative bias is applied to the metal gate electrode, holes are injected from the pentacene surface into the nanoparticle layer, charging up the latter. A smaller positive voltage needs to be applied to the Au electrode to produce depletion in the pentacene surface (i.e. the C–V curve is displaced to more negative voltages). The decreased value for

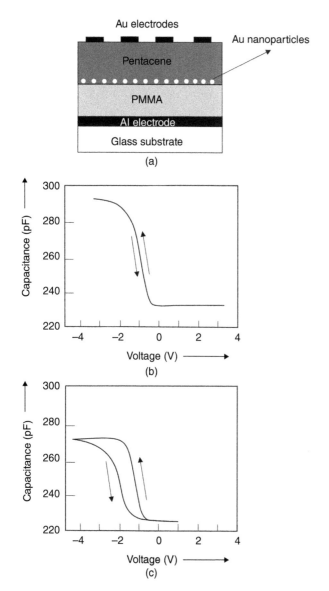

Figure 5.17 (a) Schematic diagram for an MIS structure based on Al/PMMA/gold nanoparticles/
pentacene/Au. (b) Capacitance versus voltage curve measured at 100 kHz for a control device, with no
nanoparticles; no hysteresis is evident. (c) Addition of layer of nanoparticles results in anticlockwise
hysteresis in the $C-V$ curve. Source: With permission from Mabrook et al. [13].

the accumulation capacitance for the nanoparticle-containing MIS device results from the additional insulator thickness resulting from a capping layer surrounding the nanoparticles and an addition surfactant layer used to aid their deposition. The hysteresis effect shown in Figure 5.17 increases with increasing voltage sweep and the resulting voltage 'window' can be exploited in memory devices, as discussed in Section 9.2.

The sections above show that both organic insulators and semiconductors may be exploited in MIS devices. The field effect is evident using organic semiconductors. Accumulation and depletion of the semiconductor surface is possible using a voltage applied to an external gate electrode.

However, the minimum capacitance observed in *C–V* scans is usually the result of a fully depleted semiconductor layer rather than to the generation of an inversion layer. The exploitation of MIS architectures based on organic semiconductors is described in Chapters 8 and 9.

Problems

5.1 An MOS structure is based on p-type silicon, with $N_a = 2 \times 10^{16}$ cm^{-3}. If $\Phi_{ms} = -0.95$ eV and there is a fixed positive interface charge of 2×10^{-8} C cm^{-2}, calculate the threshold voltage for strong inversion at room temperature. For Si, $\varepsilon_s = 11.9$; $n_i = 1.5 \times 10^{10}$ cm^{-3}. Assume silicon oxide has a relative permittivity = 3.9 and thickness = 10 nm.

5.2 Calculate the maximum width of the depletion region at room temperature (293 K) for an ideal MIS capacitor using SiO$_2$ as the insulator and based on
(i) A 10 μm thin film of p-type Si ($\varepsilon_s = 11.9$; $n_i = 1.5 \times 10^{10}$ cm^{-3}; $N_a = 10^{16}$ cm^{-3}).
(ii) A 100 nm thin film of p-type pentacene ($\varepsilon_s = 3.6$; $n_i = 10$ cm^{-3}; $N_a = 10^{16}$ cm^3).
Comment on your answers.

5.3 Using the data from Question 1 and a 10 nm SiO$_2$ layer as the insulator, calculate the following values of capacitance per cm^2 for both MIS structures:
(i) Accumulation capacitance.
(ii) Maximum depletion layer capacitance.
(iii) Minimum capacitance.
The relative permittivity of SiO$_2$ is 3.9.

5.4 An MIS structure based on a p-type semiconductive polymer has a metal to semiconductor work function difference of −1.05 eV (i.e. $\Phi_s > \Phi_m$) at room temperature. What sign and value of fixed charge density, located at the insulator/semiconductor interface, will compensate for this work function difference? The insulator in the MIS device has a thickness of 100 nm and a relative permittivity of 3.0.

5.5 Using the capacitance data in Figure 5.11, estimate the doping density in the CdTe semiconductor ($\varepsilon_s = 10.2$). Assume that the top electrodes are circular with a diameter of 1 mm. Is it feasible to obtain a value of doping density using the maximum and minimum capacitance values from the *CV* curve?

References

1 Nicollian, E.H. and Brews, J.R. (1982). *MOS (Metal Oxide Semiconductor) Physics and Technology*. New York: Wiley-Interscience.

2 Lundström, I., Shivaraman, M.S., Svensson, C., and Lundkvist, L. (1975). A hydrogen sensitive MOS field effect transistor. *Appl. Phys. Lett.* 26: 55–57.

3 Nicollian, E.H. and Goetzberger, A. (1965). MOS conductance technique for measuring surface state parameters. *Appl. Phys. Lett.* 7: 216–219.

4 Nicollian, E.H. and Goetzberger, A. (1967). The Si-SiO$_2$ interface - electrical properties as determined by the MIS conductance technique. *Bell Syst. Tech. J.* 46: 1055–1130.

5 Sze, S.M. and Ng, K.K. (2007). *Physics of Semiconductor Devices*, 3e. Hoboken, NJ: Wiley-Interscience.

6 Castagné, R. and Vapaille, A. (1971). Description of the SiO$_2$-Si interface properties by means of very low frequency MOS capacitance measurements. *Surf. Sci.* 28: 157–193.

7 Terman, L.M. (1962). An investigation of surface states at a silicon/silicon dioxide interface employing metal-oxide-silicon diodes. *Solid State Electron.* 5: 285–299.

8 Berglund, C.N. (1966). Surface states at steam-grown silicon-silicon dioxide interface. *IEEE Trans. Electron Devices* ED-13: 701–705.

9 Petty, M.C. and Roberts, G.G. (1980). Analysis of p-type CdTe-Langmuir film interface. *Inst. Phys. Conf. Ser.* 50: 186–192.

10 Evans, N.J., Petty, M.C., and Roberts, G.G. (1988). Charge incorporation in ω-tricosenoic acid Langmuir-Blodgett multilayers. *Thin Solid Films* 160: 177–185.

11 Grecu, S., Bronner, M., Opitz, A., and Brütting, W. (2004). Characterisation of polymeric metal-insulator-semiconductor diodes. *Synth. Met.* 146: 359–363.

12 Torres, I. and Taylor, D.M. (2005). Interface states in polymer metal-insulator-semiconductor devices. *J. Appl. Phys.* 98: 073710.

13 Mabrook, M.F., Yun, Y., Pearson, C. et al. (2009). Charge storage in pentacene/polymethylmethacrylate memory devices. *IEEE Electron Device Lett.* 30: 632–634.

Further Reading

Petty, M.C. (1996). *Langmuir-Blodgett Films.* Cambridge: Cambridge University Press.

Streetman, B.G. and Banerjee, S. (2000). *Solid State Electronic Devices*, 5e. Hoboken, NJ: Prentice Hall.

6

DC Conductivity

6.1 Introduction

At a fundamental level, the presence of a DC voltage will produce a drift in free charges within a solid. The movement of the charges constitutes an electric current. If, at the same time, excess carriers are generated, e.g. by optical means, the process of diffusion will also contribute to the current (Section 2.7). For those materials with no or very few of these charges, the main effect of the applied field is to polarize the material. The result is a separation of the centres of positive and negative charge and the creation of electric dipoles (Section 7.2). The application of a step DC voltage across a low conductivity sample will produce an initial displacement current (Section 7.3.1) that can exceed the small ionic and electronic contributions over a long period. In extreme cases, a steady current reading will never be obtained, and a direct measurement of conductivity will not be possible. It can therefore be more appropriate to use AC methods to study electrical phenomena; this is discussed in Chapter 7.

Some of the important DC, or low frequency, electrical conductivity processes observed in organic solids are outlined in Sections 6.2–6.9. These can occur simultaneously, and each may dominate at different values of the applied electric field and/or over different temperature ranges. It can therefore be a challenge to identify the physical mechanism(s) at work. Moreover, the equations governing the individual conductivity processes may contain many unknowns, each of which needs to be determined before definite conclusions can be drawn. John von Neumann famously said, 'With four parameters I can fit an elephant, and with five I can make him wiggle his trunk'. By this, he meant that one should not be impressed when a complex model fits a data set well. For the experimentalist, measurements over a wide temperature range and using different electrode materials (i.e. different work functions) and separations may help to elucidate the underlying physics. Organic chemistry also provides the means to synthesize subtle and systematic variations to the molecular structure of organic semiconductors (e.g. the inclusion of different chemical groups, different aliphatic chain lengths), which may provide insights into their electrical conductivity.

6.2 Electronic Versus Ionic Conductivity

Although most of the electrical processes described in this book focus on electrons as the transporters of electric charges in signal processing devices, ions can contribute to the electrical conductivity. Nature favours ions as charge carriers. There are many different types of ions. These possess size, shape as well as charge, and different ions can be exploited effectively in natural processes concerned with the transfer and storage of data. Ion channels in biological membranes can

Electrical Processes in Organic Thin Film Devices: From Bulk Materials to Nanoscale Architectures, First Edition. Michael C. Petty.
© 2022 John Wiley & Sons Ltd. Published 2022 by John Wiley & Sons Ltd.
Companion Website: www.wiley.com/go/petty/organic_thin_film_devices

be highly specific filters for ions of a very similar character, e.g. Na$^+$ and K$^+$, providing immense scope for information processing (Section 12.3.4). On the other hand, present-day technology (e.g. the microelectronics industry) is based on solid-state devices and prefers electrons as charge carriers as these can respond rapidly to applied electrical fields. A moot point is whether this will continue to be the case as scientific and technological progress is made in the twenty-first century.

The preparation methods of many organic solids involve the use of aqueous solvents and other substances containing ions. Ionic conduction can also be a significant factor in polymeric materials, particularly those that contain counterions to balance the charges on the polymer backbone. An ionic contribution to the measured conductivity in organic semiconductors is therefore to be expected. It is important to be able to separate these ionic processes from electronic conduction.

The most definitive evidence for ionic conduction is the detection of electrolysis products formed on discharge of the ions as they arrive at the electrodes. However, the very low level of conductivity in most organic materials usually precludes this. Even at a conductivity of 10^{-9} S m^{-1} (and many organic compounds possess lower ionic contributions to their conductivity), 100 V applied across a sample 100 mm^2 in area and 1 mm thick will produce only about 10^{-11} m^3 of gas at standard temperature and pressure (STP) per hour [1].

Ions travelling through a sample under the influence of an applied electric field will accumulate at defects (e.g. grain boundaries) or at one of the solid electrodes. The resulting polarization will reduce the ionic current to zero over a period. Ionic conductivity is expected therefore to give rise to time-dependent currents. Furthermore, ions do not respond as readily as electrons to high frequency electric fields and the ionic contribution to conductivity should therefore be diminished at high frequencies. For organic materials based on charge-transfer complexes (Section 1.5.4), the presence of a charge-transfer band in the infrared part of the electromagnetic spectrum can be used to identify electronic conductivity.

A strong correlation between the measured conductivity of a sample and its permittivity indicates the presence of ionic conductivity [1]. This can be explained by the reduction of the Coulomb forces between ions in a high permittivity medium. For example, the absorption of water, which has a relatively high permittivity, generally enhances the conductivity, σ, of a polymer, and polymer-water systems often conform to the equation:

$$\log \sigma = -\frac{X}{\varepsilon_{\text{lf}}} + Y \tag{6.1}$$

where X, Y are constants, and ε_{lf} is the low frequency value of the real part of the relative permittivity (Section 7.3.2).

Further evidence of ionic conduction may be obtained from detailed studies of the dependence of current on electric field. A simple model gives [1]:

$$J \propto \sinh\left(\frac{ea\mathcal{E}}{2k_{\text{B}}T}\right) \tag{6.2}$$

where a is the distance between potential wells associated with the ionic movement. (Of course, this equation provides a linear relationship between J and $\mathcal{E}$ for small values of the argument of the hyperbolic function – like Ohmic electronic conductivity.)

6.3 Quantum Mechanical Tunnelling

If the energy of an electron is less than the interfacial potential barrier at a metal/insulator interface upon which it is incident, classical physics predicts reflection of the electron at the interface. The

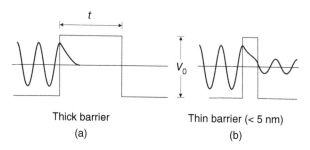

Electrons

Thick barrier

(a)

Thin barrier (< 5 nm)

(b)

Figure 6.1 Schematic representation of the quantum mechanical tunnelling of an electron wave through a finite rectangular potential energy barrier: (a) thick barrier; (b) thin barrier. V_0 = barrier potential energy; t = barrier width.

electron cannot penetrate the barrier and its passage from one electrode to the other is inhibited. Quantum mechanics contradicts this view. The wave nature of the electron allows penetration into the forbidden region of the barrier. This is the origin of quantum mechanical tunnelling. Figure 6.1 depicts an electron incident on a rectangular barrier of potential energy V_0 and width t. At the interface, part of the electron wave will be reflected while the remainder will be transmitted into the barrier region. The wave function associated with the transmitted electron decays rapidly with depth of penetration from the electrode/insulator interface and, for barriers of macroscopic thickness, is essentially zero at the opposite interface, as shown in Figure 6.1a. This indicates a zero probability of finding the electron here. However, if the barrier is very thin ($t < 5$ nm), the wave function has a non-zero value at the opposite interface. For this case, there is a finite probability that the electron can pass from one electrode to the other by penetrating the barrier, shown in Figure 6.1b. Solutions to the Schrödinger wave equation (Section 1.2.1) may be obtained for the three spatial regions shown in Figure 6.1. To the left and right of the barrier, the electron wave function is described by a travelling wave. Within the barrier region, the mathematical solution is that of a wave with an amplitude that is exponentially decaying with distance x (rather than an oscillatory function). The electron wave will have a reduced probability of penetrating the barrier, which will depend on the barrier height and thickness. An important point is that the electron does not loose energy during the idealized (elastic) process of quantum mechanical tunnelling (Problem 6.1). The relative probability that an electron will tunnel through the barrier is given by the transmission probability P. Analytical expressions for P are complex, but in the case of a wide or high barrier, it may be shown that [2]:

$$P = P_0 \exp(-2\alpha t) \tag{6.3}$$

where

$$P_0 = \frac{16E\left(V_0 - E\right)}{V_0^2} \text{ and } \alpha = \sqrt{\frac{2m_e\left(V_0 - E\right)}{\hbar^2}}$$

and t is the barrier thickness or tunnelling distance. V_0 is the potential barrier height, as shown in Figure 6.1, and E is the electron kinetic energy. The constant α (units m^{-1}) in Eq. (6.3) is related to the rate of decay of the electron wave function. Equation (6.3) is valid for $\alpha t \gg 1$ (Problem 6.2).

Figure 6.2 shows a more detailed potential energy diagram for two metal electrodes, cathode and anode, separated by an insulating tunnel barrier of arbitrary shape, with the image force (Section 4.4.2) taken into account. The barrier height $\bar{\phi}_t$ represents the average value above the Fermi level

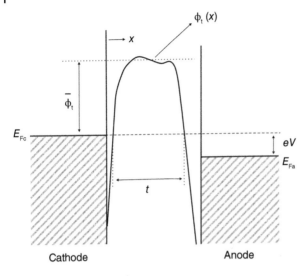

Figure 6.2 Potential energy band diagram of an arbitrary tunnelling barrier, $\phi_t(x)$, between metal electrodes, including the effects of the image force. E_{Fc} and E_{Fa} are the Fermi levels in the cathode and anode electrodes, respectively. t is the width of the tunnel barrier and $\overline{\phi_t}$ is equal to the mean barrier height above E_{Fc}.

of the cathode electrode (i.e. in contrast to the definition of V_0 in Figure 6.1). The current density due to electron tunnelling through the forbidden energy gap of the insulator is given by [3–5]:

$$J = \frac{4\pi e m_e}{h^3} \int_0^\infty dE \left(f_c(E) - f_a(E) \right) \int_0^E P(E_x)\, dE_x \tag{6.4}$$

where e, m_e are the electronic charge and free electron mass, respectively, $f_c(E)$ and $f_a(E)$ are the Fermi–Dirac distribution functions in the cathode and anode electrodes, respectively, and $P(E_x)$ is the electron transmission probability, in the x-direction, through the insulator. Equation (6.4) assumes the classical parabolic relationship between the electron energy and momentum (Eq. (1.13)).

The simplest generalized equation relating the tunnel current density $J(V, T)$ at temperature T K to that at 0 K, $J(V, 0)$ is given by [6, 7]:

$$\frac{J(V, T)}{J(V, 0)} = \frac{\pi A k_B T / 2 \left(\overline{\phi_t}\right)^{1/2}}{\sin\left(\pi A k_B T / 2 \left(\overline{\phi_t}\right)^{1/2}\right)} \tag{6.5}$$

where

$$J(V, 0) = \left(\frac{eht^2}{2\pi}\right) \left(\overline{\phi_t} \exp\left(-A\left(\overline{\phi_t}\right)^{1/2}\right) - \left(\overline{\phi_t} + eV\right) \exp\left(-A\left(\overline{\phi_t} + eV\right)^{1/2}\right) \right) \tag{6.6}$$

and

$$A = \frac{4\pi t \left(2m_e\right)^{1/2}}{h} \tag{6.7}$$

From Eq. (6.5) and using typical values of $\overline{\phi_t} = 2\,\text{eV}$ and $t = 2\,\text{nm}$:

$$J(V, T) = J(V, 0) \left(1 + 6 \times 10^{-7} T^2\right) \tag{6.8}$$

Therefore, for a constant voltage, and for moderate temperatures, the current exhibits very little dependence on temperature.

The complete current versus voltage relationships for tunnelling depend on the magnitude of the applied voltage and whether the tunnel barrier is symmetric or asymmetric (i.e. whether the two electrodes are the same or different metals) [3]. For very low applied voltages (much less than the energy barrier height divided by the electronic charge), the tunnelling current density is proportional to the applied voltage (Ohmic conductivity) and the tunnelling probability varies exponentially with the barrier thickness. The tunnelling conductivity, σ_t, may be given by an equation [6]:

$$\sigma_t = B\exp(-Ct) \tag{6.9}$$

where B and C are constants. However, at higher voltages, the conductivity data for insulating films have been shown to deviate from linearity. For example, one theoretically predicted current versus voltage dependence for a symmetrical rectangular barrier takes the form [4]:

$$I = 2I_0\left(\frac{\pi Gk_B T}{\sin\left(\pi Gk_B T\right)}\right)\exp\left(-HV^2\right)\sinh\left(\frac{GV}{2}\right) \tag{6.10}$$

where I_0 is a constant, and G and H are coefficients related to the tunnelling barrier.

The ability to form thin organic films with precisely defined thickness, such as Langmuir–Blodgett films, should offer a simple means to study quantum mechanical tunnelling. Even so, experimental data must be treated with caution because of the presence of oxide and other layers on the metallic electrodes that will have comparable thicknesses to the organic film under investigation. With careful attention to experimental detail, some of the theoretical predictions for tunnelling can be observed. Figure 6.3 shows the results of experiments using monolayer films of different hydrocarbon chain length fatty acids (C_{14}–C_{20}) sandwiched between various metallic electrodes (Al–C_n–Mg, Al–C_n–Au, and Al–C_n–Al) [8]. The tunnelling conductivity (measured at an applied voltage of 10 mV) dependence upon film thickness is clearly of the form of Eq. (6.9). Other published data for monolayer organic films show a similar dependence on the tunnel insulator thickness, although the absolute values of conductivity vary by several orders

Figure 6.3 Logarithm of conductivity versus monolayer thickness for different metal/monolayer/metal structures. C_n is the number of carbon atoms in the monolayer compound. Source: Polymeropoulos [8].

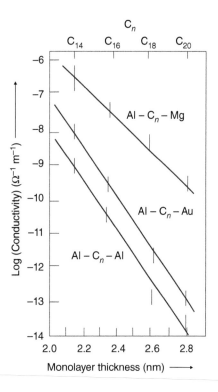

of magnitude. It is possible that this is related to the ambient in which the measurements are undertaken [9].

As the applied voltage is increased, the current versus voltage behaviour of organic monolayer films can follow the equations predicted for quantum mechanical tunnelling, e.g. Eq. (6.10) [9].

One useful experiment that can be used to support the view that quantum mechanical tunnelling occurs in monolayer organic films is inelastic tunnelling spectroscopy [10]. In addition to the normal elastic tunnelling of electrons (i.e. tunnelling through a potential barrier to a state at the same energy level), it is possible for a small proportion, typically <1%, of the electrons to tunnel inelastically, thereby losing energy to excite molecular vibrations in the barrier [11]. Thus, the observation of this small number of relatively rare inelastic tunnelling events would seem to be strong evidence for tunnelling through organic monolayer films. The inelastic tunnelling spectra can be obtained by electronically measuring d^2V/dI^2 as a function of bias. The onset of inelastic events associated with the excitation of molecular oscillations is revealed at a particular bias as a local peak in d^2V/dI^2. Data reported for barium stearate monolayers show a good correlation with infrared and Raman spectra for stearic acid [10].

The above equations for quantum mechanical tunnelling assume that the major contribution to the current is from tunnelling of carriers close to the Fermi energy. If, in the electrode, an electron acquires additional thermal energy, then some modifications to the tunnelling equations are required. This process is termed thermally assisted tunnelling [12]. The resulting current can vary rapidly with temperature, in contrast to the variation suggested by Eq. (6.8), and for large applied electric fields an exponential dependence on applied voltage is predicted. This can make the electrical behaviour for thermally assisted tunnelling similar to that for a Schottky barrier (Section 4.4.3).

6.4 Variable Range Hopping

As noted in Chapter 1, the electronic band structure of an organic semiconductor may resemble that of an amorphous material. States in the band tails are localized and there exists a mobility edge that separates the region of the extended states in the interior of the band from the region of the localized states. Whether the organic semiconductor behaves as a metal or an insulator is determined by the relative position of the Fermi level with respect to the mobility edge.

At high temperature, the conductivity will be determined by electrons (or holes) thermally activated to extended states above (below) the mobility edge (E_c for electrons and E_v for holes). The conductivity can be expressed in the form of Eq. (2.21), i.e. as for crystalline case. However, for disordered materials, the carrier mobility μ cannot be calculated assuming the electrons are free (with their paths interrupted only by occasional scattering events) because the mean free path is of the order of the interatomic distance. It is generally assumed that the transport is diffusive (Section 2.7) and the Einstein relation determines the carrier mobility (Eq. (2.69)):

$$\mu = \frac{eD}{k_B T} \tag{6.11}$$

where D, the diffusion coefficient, is $v_{el}a^2/6$. Here, v_{el} is an electronic frequency which, from quantum theory, is approximately $\hbar/m_e a^2$ where a is taken to be an interatomic spacing or as the coherence length of an electron wave. Therefore, at room temperature:

$$\mu \approx \frac{e\hbar}{6m_e k_B T} \approx 6\,\mathrm{cm^2 V^{-1} s^{-1}} \tag{6.12}$$

This is similar to the lower limit for μ calculated from simple kinetic theory considerations in Eq. (2.25). Carrier mobilities of around $1\,\text{cm}^2\,\text{V}^{-1}\,\text{s}^{-1}$ are measured in some organic molecular materials; the values for conductive polymers are usually much less, Table 2.1.

If the localized energy levels in a disordered material are unoccupied, these provide a possible alternative conduction path for carriers, in parallel with the band (extended state) conduction discussed above. Clearly, the electron motion in localized states is much smaller than in extended states but at low temperatures the number of electrons activated to the conduction band can be sufficiently small to allow localized state conduction to become the predominant conduction process.

Electrons can move between localized states in three different ways. First, they can be thermally activated over the potential barrier separating the two states. However, the barrier height is usually of the same order as the energy separating the localized states from the extended states. If the temperature is sufficiently high to facilitate jumping, extended state conduction will be the more favoured process. The emission at lower temperatures can be aided by the electric field. Secondly, electrons may tunnel through the potential barrier and thirdly they can move by a combination of activation and tunnelling. Since in the disordered solid, the localized states will not be degenerate, some measure of activation will generally be required to allow the occupied state and the empty state to resonate and hence allow tunnelling. This thermally (phonon) assisted quantum mechanical tunnelling is called hopping and is an important feature of conduction in non-crystalline materials.

There have been many theoretical treatments of this hopping process [13, 14]. It is generally assumed that the states are distributed randomly in energy and space and that the Fermi level lies within these localized states, as depicted in Figure 6.4. There are localized states, with an average separation R within the gap, above and below the Fermi energy, which is located around mid-gap. Electrons will hop (tunnel) from occupied to empty states. Most of these hops will have to be to states at higher energies, i.e. upward hops. At high temperatures, there will be sufficient thermal energy (phonons) available to assist such electron transitions. However, as the temperature is reduced, the electrons will need to look further afield for an accessible energy state. The average hopping distance will decrease as the temperature decreases – hence the term 'variable range hopping'. Since the tunnelling probability decreases with increasing distance, the conductivity also decreases. This decrease is smoother than for the case for a semiconductor with a well-defined energy gap because there is a continuous distribution of activation energies. This results in a 'soft' exponential equation for the conductivity [15]:

$$\sigma = \sigma_0 \exp\left(-\left(\frac{T_0}{T}\right)^\gamma\right) \tag{6.13}$$

where T_0 and γ are constants. The latter depends on the dimensionality d of the hopping process:

$$\gamma = \frac{1}{1+d} \tag{6.14}$$

For three-dimensional hopping ($d = 3$), $\gamma = 1/4$. This leads to the variable range hopping model of Mott [13]:

$$\sigma = \sigma_0 \exp\left(-\left(\frac{T_0}{T}\right)^{1/4}\right) \tag{6.15}$$

σ_0 is a constant defined:

$$\sigma_0 = e^2 S\left(E_F\right) R^2 v_{ph} \tag{6.16}$$

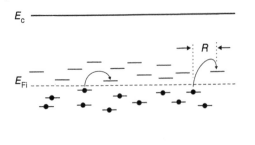

E_c

E_{Fi}

E_v

R

E_g

Figure 6.4 Schematic diagram to illustrate hopping conductivity between localized states. E_c, E_v are conduction and valence band edges, respectively. E_{Fi} is the intrinsic Fermi level. E_g is the semiconductor band gap. R represents the local distance between states or hopping distance. Filled states and empty states are depicted by closed and open circles, respectively.

where R is the average hopping distance:

$$R = \frac{3}{4}\left(\frac{3}{(2\pi\alpha S(E_F)k_B T)}\right)^{1/4} \tag{6.17}$$

and

$$T_0 = \frac{24\alpha^3}{\pi k_B S(E_F)} \tag{6.18}$$

In the above equations, $S(E_F)$ is the density of states at the Fermi energy, α is the inverse localization length (spatial extension of the localized electron wave function) and v_{ph} represents a typical phonon frequency. The last parameter may be considered as an attempt to hop rate.

The variable range hopping model can also be applied to the conductivity of a material resulting from the application of an alternating electric field. In this case, the conductivity is expected to be higher than the case for a DC field as backward hops will be relatively easy. This situation is discussed in the Chapter 7, dealing with AC processes.

Equation (6.13) is consistent with the observed temperature dependence of the DC conductivity of conductive polymers, such as polyacetylene doped with halogens (see, for example, [15, 16]). However, in practice, it can be difficult to distinguish between one-, two-, and three-dimensional hopping (i.e. any value of γ between 1/2 and 1/4 can be used to provide acceptable fits to the experimental data) [15]. Figure 6.5 shows experimental data for samples of polyacetylene doped with bromine, $(CHBr_y)_x$, in the form of log (normalized conductivity) versus $T^{-1/4}$ [16]. Acceptable straight lines are evident, suggesting that variable range hopping is the physical explanation for the DC conductivity. However, the temperature range of the data in Figure 6.5 is limited. Moreover, electron microscopy of the polymer samples suggests it is the contacts between the individual polymer fibres that dominate the conductivity [16]. An improved model to account for the DC conductivity of such materials is described in Section 6.5.

6.5 Fluctuation-Induced Tunnelling

The model of variable range hopping assumes a random distribution of localized states. There are categories of disordered materials, such as some conductor-insulator composites, disordered semiconductors, and doped organic semiconductors, in which most of the conduction electrons are delocalized and free to move over distances that are very large compared to atomic dimensions. For these random systems, the electrical conduction is dominated by electron transfer between conductive segments rather than by hopping between localized states. If the distribution is not

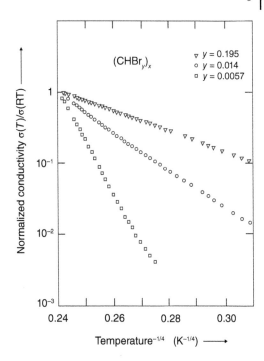

Figure 6.5 Log of normalized DC conductivity ($\sigma(T)/\sigma$(RT)) for bromine doped polyacetylene, with different concentrations of Br, versus temperature$^{-1/4}$. RT = room temperature. Source: Reprinted with permission of Chiang et al. [16].

random and the defects tend to cluster, the theory of fluctuation-induced tunnelling conduction, introduced by Sheng [17], provides an appropriate description of the electrical conductivity. This model assumes metallic islands in an insulating matrix, for example the coexistence of doped and undoped regions in a conjugated polymer. The essential physics of this mechanism is contained in the observation that, since the electrons tend to tunnel between conducting regions at points of their closest approach, the relevant tunnel junctions are usually small in size and are therefore subject to large (thermally activated) voltage fluctuations across the junction. By modulating the potential barrier, the voltage fluctuations directly influence the tunnelling probability and introduce a characteristic temperature variation to the normally temperature-independent tunnelling conductivity. The resulting temperature dependence of the conductivity takes the form:

$$\sigma = \sigma_0 \exp\left(-\frac{T_1}{T + T_0}\right) \tag{6.19}$$

where σ_0, T_0, and T_1 are constants depending on the geometry of the tunnel barrier and the size of the conducting particles. Since T_0 appears as an additive constant to T, it can be viewed as the temperature above which the fluctuation effects become significant. For $T \ll T_0$, Eq. (6.19) predicts a constant conductivity. Experimental data for heavily (iodine) doped polyacetylene (in both the *trans* and *cis* forms) are shown in Figure 6.6 [18]. The full lines show theoretical fits to Eq. (6.19). The fluctuation-induced tunnelling model seems to provide a better explanation for the DC conductivity of halogen-doped polyacetylene than variable range hopping. The reader is directed to Problem 6.3, which illustrates the difficulty in fitting experimental conductivity data to the available theoretical models.

Attempts to fit the fluctuation-induced tunnelling model to the conductivity of other polymers and to carbon nanotubes have also been made. In such cases, a term is added to Eq. (6.19) to account for the metallic conductivity that can be observed at high temperatures [19]. The dependence of

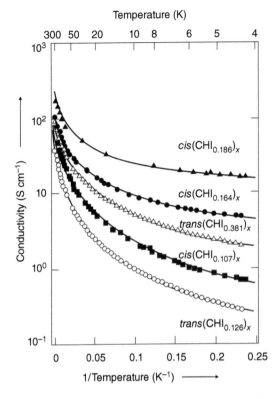

Figure 6.6 Temperature dependency of the DC conductivity of heavily (iodine) doped polyacetylene (experimental data points) and the fits (solid lines) to the model of fluctuation induced tunnelling. Source: Ehinger and Roth [18].

resistivity, ρ, on temperature, is then written [20]:

$$\rho = \alpha T + \rho_t \exp\left(\frac{T_1}{T_0 + T}\right) \tag{6.20}$$

where α is the temperature coefficient arising from the metallic resistivity and ρ_t is a constant. Figure 6.7 shows that the above equation provides a good fit to data for ropes and mats of carbon nanotubes over a wide temperature range [20] and demonstrates the similarity to the conductivity data for conductive polymers. The increase in the non-metallic terms in the carbon nanotube mats relative to the single rope indicates that inter-rope contacts make a significant contribution to the resistance in the mats, and the large tunnelling term for the rope with tangled regions suggests a contribution from disordered semiconductor-like conduction in the tangled regions.

6.6 Space Charge Injection

If the electrical contacts to an insulating or semiconducting sample are Ohmic, these allow electron transfer between the electrodes and the sample, and the resulting current is proportional to the applied voltage (Section 4.3). Under certain conditions, however, the contacts can become 'super-Ohmic' and the current is only limited by the space charge between the electrodes. This conductivity regime is called space-charge-limited and arises from a space charge of excess carrier injected from one (or both) of the electrodes. The resulting current is referred to as the space-charge-limited current (SCLC).

An initial case to consider is that of the injection of carriers from an electrode into a vacuum (the simplest insulator). Figure 6.8 shows a voltage applied to a vacuum diode (two electrodes sealed in

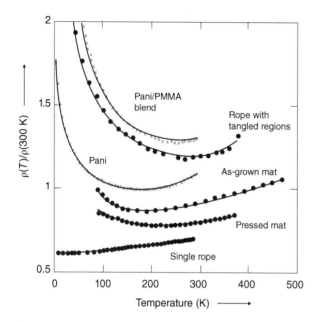

Figure 6.7 Fits of Eq. (6.20) to normalized resistivity data for single rope, rope with tangled regions, as-grown mat and pressed mat carbon nanotube samples and fits to typical polymer data (Pani [polyaniline] and Pani/PMMA blend). Different data sets are displaced vertically for clarity. Source: Kaiser et al. [20].

Figure 6.8 Cathode and anode electrodes with intervening vacuum. The cathode is heated and electrons (solid circles) injected into the vacuum. All the lines of electric field terminate on charges. L is the cathode to anode spacing.

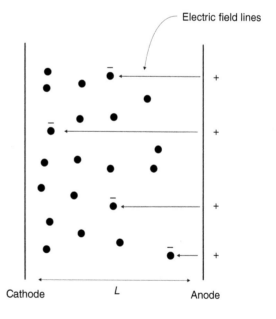

a vacuum) that has a heated cathode. This electrode is assumed to supply a reservoir of free carriers for the adjacent vacuum. The applied field draws a part of this reservoir into the space between the cathode and the anode, of separation L. All the field lines end on charges (Gauss' Law) in space. Hence, the field at the cathode (for zero emission velocity) is zero.

From the viewpoint of electrostatics, current injection is very similar to the physics of a parallel plate capacitor. A given voltage V between cathode and anode can support only a limited total charge Q per unit area injected into the insulator from the reservoir contact; in this case, $Q = CV$. If the injected charge was uniformly distributed between cathode and anode, its average distance from the anode would be $L/2$ and so the capacitance C per unit area would be twice the geometric capacitance, C_0, given by:

$$C_0 = \frac{\varepsilon_0}{L} \tag{6.21}$$

Since the charge is all injected at the cathode, its distribution is expected to be nonuniform – with a greater concentration close to the cathode. Therefore, the average distance of the injected charge from the anode will be slightly greater than $L/2$, but clearly is smaller than L, the limiting distance corresponding to no injection. On this physical argument, $C_0 < C < 2C_0$, a relationship that was established by Lampert and Mark [21]. An approximation (better than a factor of two) that can be used is:

$$Q \approx C_0 V = \left(\frac{\varepsilon_0}{L}\right) V \tag{6.22}$$

The current density, J, that flows between the anode and the cathode is:

$$J = \frac{Q}{t} = \frac{\varepsilon_0 V}{Lt} = \frac{\varepsilon_0 V \overline{v_d}}{L^2} \tag{6.23}$$

where t is the electron transit time from cathode to anode and $\overline{v_d}$ is the mean electron drift velocity. The acceleration, a, of an electron between the cathode and the anode is given by Newton's second law:

$$a = \frac{eV}{m_e L} \tag{6.24}$$

Assuming that the mean electron velocity is one half of the final electron velocity:

$$\overline{v_d} = \left(\frac{1}{2}\right)\left(\frac{2eV}{m_e}\right)^{1/2} \tag{6.25}$$

Substitution of Eq. (6.25) into Eq. (6.23) then gives:

$$J = \frac{\varepsilon_0}{2}\left(\frac{2e}{m_e}\right)^{1/2}\frac{V^{3/2}}{L^2} \tag{6.26}$$

A more complete analysis uses the Poisson equation [21]. A similar solution to that given by Eq. (6.26) is obtained, except that the first term becomes $4\varepsilon_0/9$. This relationship, which predicts that the current density will vary as the three halves power of the applied voltage, is known as Child's Law.

There are several significant modifications needed to the above analysis if a solid insulator replaces the vacuum, in which there is no viscous force on the electrons, and their behaviour is governed by simple particle dynamics. In the solid state, the electrons give up energy to the lattice via scattering processes and there will be a unique (linear) relationship between the applied field and the drift velocity of the electrons ($v_d = \mu V/L$), which is assumed to be valid in the range of electric fields under consideration. Another difference between the solid insulator and the vacuum is that in the former there may be some carriers present in the absence of net charge. In most theories of space-charge-limited conductivity in solids, diffusion is not taken into consideration (in the case of the vacuum, there is no diffusion term because of the absence of scattering). This is an adequate model in most situations of practical importance [21]. Diffusion currents are sizable only in the immediate vicinity of the contacts and neglect of such currents is consistent with ignoring the role of the contacts.

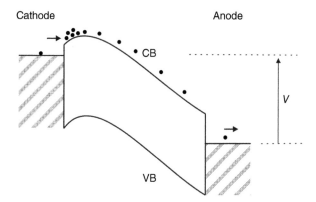

Figure 6.9 Band diagram to show space charge injection into a solid. CB and VB are conduction band and valence band, respectively. V = applied voltage between the cathode and the anode electrodes.

For carrier injection into a semiconductor or insulator, a similar approach to that above may be taken, with the permittivity of free space in Eq. (6.23) replaced by the permittivity of the solid and noting the relationship between transit time and carrier mobility given by:

$$t = \frac{L^2}{V\mu} \tag{6.27}$$

The equivalent form of Child's law, using the complete analysis, then becomes:

$$J = \left(\frac{9}{8}\right)\frac{\varepsilon_r\varepsilon_0\mu V^2}{L^3} \tag{6.28}$$

where ε_r is the relative permittivity of the semiconductor or insulator (more appropriately, the low-frequency permittivity – see Section 7.3.2).

Figure 6.9 depicts an energy diagram for the solid-state equivalent of the vacuum diode; this figure is essentially that of Chapter 4, with a bias applied. A low work function cathode is assumed to supply a reservoir of electrons to the conduction band of the solid, adjacent to the cathode. This reservoir is partly drawn into the space between the two electrodes by the applied field. The SCLC in a solid is considerably smaller than that in vacuum for the same electrode spacing and applied voltage because the electron velocity is much smaller than in vacuum.

6.6.1 Effect of Traps

In a real solid, the existence of traps must be considered. The presence of electron traps will generally result in a greatly reduced current at lower injection ratios, since those traps initially empty will capture, and thereby immobilize, some (maybe most) of the injected carriers. On the other hand, the amount of excess charge that can be supported in the insulator at an applied voltage V, namely $Q = CV$, is the same whether the excess charges are free or trapped. (The amount of charge that can be stored on the plates of a parallel-plate capacitor, at a given voltage, is independent of whether the charge is free or immobilized in surface states.) An injected electron will only spend part of its time in the conduction band and thus an effective mobility, μ_{eff}, must be used in Eq. (6.28). This will be dependent on both the actual mobility, μ, (i.e. in the absence of trapping) and the density of free and trapped charge carriers – n and n_t, respectively:

$$\mu_{\text{eff}} = \frac{\mu n}{n + n_t} \tag{6.29}$$

Three cases may be considered.

6.6.1.1 Shallow Traps

Shallow traps are characterized by a fixed ratio of free to trapped carriers. If the ratio of free to trapped electrons, θ, is small, Eq. (6.28) becomes:

$$J = \left(\frac{9}{8}\right) \frac{\varepsilon_r \varepsilon_0 \theta \mu V^2}{L^3} \tag{6.30}$$

where

$$\theta = \frac{n}{n_t} \tag{6.31}$$

6.6.1.2 Single Deep Trapping Level

If the insulator or semiconductor contains deep traps located at a single energy level within the band gap, then the injected charge will fill these as the applied voltage increases. Once sufficient charge has been injected, the traps will become saturated. The voltage at which this occurs corresponds to the trap-filled limit, V_{TFL}, which is given by [21]:

$$V_{TFL} \approx \frac{e N_t L^2}{\varepsilon_r \varepsilon_0} \tag{6.32}$$

where N_t is the trap density. Figure 6.10 shows the expected current versus voltage behaviour under these conditions (Problem 6.4). The lowest voltage region of the curve corresponds to the situation in which the injection of excess carriers in negligible. At these voltages, the volume conductivity (i.e. Ohm's law) dominates. Only when the injected carrier density exceeds the volume generated carrier density, will space-charge effects be seen and the quadratic current versus voltage dependence become apparent. The trap-filled limit is characterized by a rapid rise in the current. Beyond V_{TFL}, the material behaves as if it was trap free, and the quadratic current versus voltage relationship given in Eq. (6.28) is again observed.

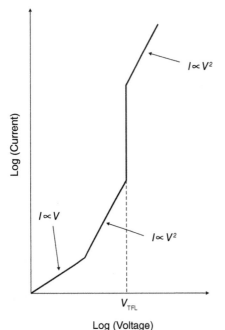

Figure 6.10 Predicted single-carrier space-charge-limited current versus voltage characteristics for a semiconductor (or insulator) containing a single deep trapping level. V_{TFL} is the voltage corresponding to the trap-filled limit.

6.6.1.3 Distributions of Traps

In the case of deep traps (i.e. located more than a few $k_B T$ from a band or mobility edge), the form of the trap distribution in the energy gap must first be known to calculate the injected current density. A common assumption is that the number of traps per unit volume per unit energy, $N_t (E)$, can be described by an exponential relationship:

$$N_t(E) = A \exp\left(-\frac{|E, E_c|}{k_B T_c}\right)$$

(6.33)

where A is a constant. The notation $|E, E_c|$ is used to represent the absolute value of the energy interval ΔE between E and E_c (i.e. $\Delta E = |E, E_c| = |E - E_c|$). This scheme follows that of Rose [22] and avoids ambiguity either about the sign of the exponential or the location of the zero-reference level of energy.

In the above equation, T_c is a characteristic temperature used to approximate the rate at which the trap density changes with energy; $T_c \to \infty$ means a uniform distribution in energy. For T_c less than the ambient temperature, T, the problem reduces to that of shallow traps given by Eq. (6.30). For, $T_c \geq T$ the density of traps is assumed to be considerably larger than the density of free carriers.

A first approximation to the solution for SCLC in the presence of a continuum of traps is obtained by assuming that the injected charge is uniformly distributed between cathode and anode. Furthermore, as the trap density is large compared with the free-carrier density, the injected charge will, to a high degree of approximation, lie in the trapping states. Hence, the Fermi level may be considered as having been raised from its initial value E_F to E_{Fn} by filling the trapping states with the injected charge. Figure 6.11 illustrates such a trap distribution, located below the conduction band edge, with filled (full circles) and empty (open circles) trapping states.

The number of states in the interval $|E_F, E_{Fn}| \gg k_B T_c$ is given by:

$$\int_{E_F}^{E_{Fn}} N_t(E)dE = \int_{E_F}^{E_{Fn}} A \exp\left(-\frac{|E, E_c|}{k_B T_c}\right) dE \approx k_B T_c N_t\left(E_{Fn}\right)$$

$$= k_B T_c A \exp\left(-\frac{|E_{Fn}, E_c|}{k_B T_c}\right)$$

(6.34)

The charge required to fill these states is:

$$eLk_B T_c A \exp\left(-\frac{|E_{Fn}, E_c|}{k_B T_c}\right) = VC$$

(6.35)

Figure 6.11 Exponential trap distribution within the energy gap of a semiconductor. E_c, E_v are the conduction and valence band edges, respectively. E_F is the Fermi level position before charge is injected into the conduction band and E_{Fn} represents the new Fermi level after the injected charge has filled some of the trapping states: full circles represent filled states while open circles represent empty states.

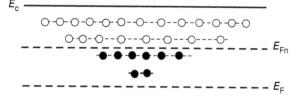

where C is the capacitance per unit area and V is the applied voltage. To calculate the carrier density, the above equation can be rewritten:

$$\left(\exp\left(-\frac{|E_{\mathrm{Fn}}, E_{\mathrm{c}}|}{k_{\mathrm{B}} T} \right) \right)^{T/T_{\mathrm{c}}} = \frac{VC}{eLk_{\mathrm{B}} T_{\mathrm{c}} A} \tag{6.36}$$

or

$$\exp\left(-\frac{|E_{\mathrm{Fn}}, E_{\mathrm{c}}|}{k_{\mathrm{B}} T} \right) = \left(\frac{VC}{eLk_{\mathrm{B}} T_{\mathrm{c}} A} \right)^{T_{\mathrm{c}}/T} \tag{6.37}$$

The density of free carriers in the conduction band of the insulator will be given by:

$$n = N_{\mathrm{c}} \exp\left(= \frac{|E_{\mathrm{Fc}}, E_{\mathrm{c}}|}{k_{\mathrm{B}} T} \right) \tag{6.38}$$

Substitution of Eq. (6.37) into Eq. (6.38) provides an expression for the free carrier density:

$$n = BV^{T_{\mathrm{c}}/T} \tag{6.39}$$

where

$$B = N_{\mathrm{c}} \left(\frac{C}{eLk_{\mathrm{B}} T_{\mathrm{c}} A} \right)^{T_{\mathrm{c}}/T} = N_{\mathrm{c}} \left(\frac{\varepsilon_{\mathrm{r}} \varepsilon_0}{eL^2 k_{\mathrm{B}} T_{\mathrm{c}} A} \right)^{T_{\mathrm{c}}/T} \tag{6.40}$$

The SCLC then follows:

$$I = \frac{V}{L} ne\mu = N_{\mathrm{c}} e\mu \left(\frac{\varepsilon_{\mathrm{r}} \varepsilon_0}{ek_{\mathrm{B}} T_{\mathrm{c}} A} \right)^{T_{\mathrm{c}}/T} L^{-(2(T_{\mathrm{c}}/T)+1)} V^{(T_{\mathrm{c}}/T)+1} \tag{6.41}$$

Equation (6.41) reveals that the current increases as the $(T_{\mathrm{c}} + T)/T$ power of the voltage and even more rapidly, as the $(2(T_{\mathrm{c}}/T) + 1)$ power of the reciprocal electrode spacing. This is often written:

$$I \propto L \left(\frac{V^m}{L^{2m+1}} \right) \tag{6.42}$$

where m is an integer.

While the high-power dependency of current on voltage is often reported for organic semiconductors, data regarding the power dependence on electrode spacing are not as commonplace. Figure 6.12 shows current versus voltage data, in the form of a $\log(I)$ versus $\log(V)$ plot, for a thin film sample of copper phthalocyanine, a molecular crystalline compound (Section 1.5.2), sandwiched between gold electrodes [23]. The transitions from Ohmic conductivity, where the exponent $m = 1$ to regions for which $m > 1$ are evident. The values of m in the high voltage regions are shown on the figure. The dependence of m on the temperature is a further indication that the organic solid contains an exponential distribution of trapping centres within the energy band gap.

Space-charge-limited conductivity that is dominated by exponential traps distributions seems a common conductivity process in different types of organic semiconductive materials. Figure 6.13 reveals the current density versus applied voltage, presented in the form of a log–log plot, for a 94 nm thick film of a conductive polymer – poly(p-phenylene vinylene) (PPV) – sandwiched between indium⁻tin-oxide (ITO) and Al electrodes [24]; measurements were taken at different temperatures. The power law relationship of Eq. (6.42) is clear, with values of m ranging from about 7 at 290 K to about 18 at 11 K. As with the data shown in Figure 6.12 for the phthalocyanine film, m decreases with increasing temperature. Measurements were also taken on the PPV films for two different film thicknesses. The data are in good agreement with the thickness dependence of current suggested by Eq. (6.42).

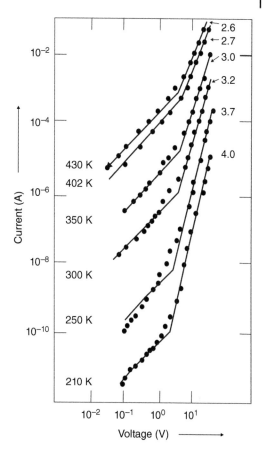

Figure 6.12 Current versus voltage characteristics of a Au/copper phthalocyanine/Au sandwich at different temperatures. The linear dependence of current on voltage at low biases is the result of Ohmic conductivity. The super-Ohmic regions at higher biases indicate space-charge-limited conductivity. The value of m from the equation $I \propto V^m$ is indicated for each curve. Source: Gould [23].

6.6.2 Two Carrier Injection

A further consideration in the development of SCLC theory is that of double, or two carrier, injection. A single carrier injection current, consisting exclusively of injected electrons or injected holes, is necessarily a space-charge current since each injected carrier contributes one excess electronic charge, negative for electrons and positive for holes. By making one contact to an insulator or semiconductor electron injecting and the other hole injecting and applying a voltage of appropriate polarity, it is possible to obtain double injection, that is, simultaneous injection of electrons and holes, as depicted by the band diagram in Figure 6.14. Because the injected electrons and holes can largely neutralize each other, a two-carrier injection current will be larger than either one-carrier current in the same material. In contrast, there is a new limitation on current flow – a loss of carriers through recombination. The injected electrons and holes can mutually recombine before they complete their respective transits between cathode and anode. Normally, this recombination is a two-step process (electron capture followed by capture of a hole, or vice versa) that takes place through localized recombination centres, labelled N_r in Figure 6.14. Also included in this figure are localized electron traps, labelled N_{tn} and localized hole traps, denoted N_{tp}. In the steady state, the net rates of electron capture and hole capture by each set of recombination centres must, of course, be equal. Because of this requirement of kinetic equilibrium, the occupancies of recombination centres are not normally tied to either the conduction band or the valence band, i.e. a quasi-Fermi level does not determine the occupancies. In addition to the normally dominant two-step recombination process, there can also occur a one-step recombination directly across the band gap.

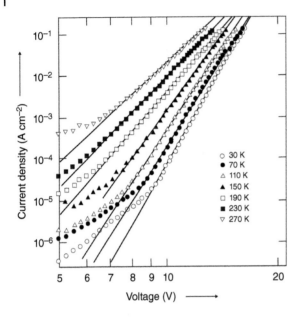

Figure 6.13 Variation of current with voltage at different temperatures for a 94 nm thick poly(*p*-phenylene vinylene) film sandwiched between indium tin oxide and aluminium electrodes. The full lines are fits to Eq. (6.42). Source: Campbell et al. [24].

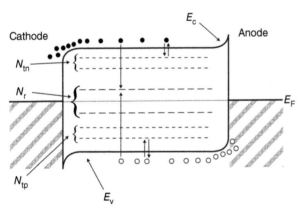

Figure 6.14 Potential energy diagram illustrating the process of two carrier space-charge injection. Electrons (full circles) enter the conduction band of an insulator or semiconductor from the cathode electrode while holes (open circles) enter the valence band from the anode. The injected carriers can fall into shallow trapping centres (densities, N_{tn} and N_{tp}) or recombine at recombination centres (density N_r). E_c and E_v are conduction and valence band edges, respectively. E_F is the Fermi level.

The double injection problems that are simplest to analyse are the 'injected plasma' situations. These are cases in which the injected electrons and holes remain free and approximately neutralize each other. It is assumed that significant trapping does not occur and, furthermore, that the only role of the recombination centres is to provide a conduit through which recombination may take place, i.e. changes in occupancy of these centres are not important. An injected plasma will always be the limiting regime reached in any double injection problem at voltages sufficiently high that the injected carriers outnumber the defect states. Under these conditions, the current density versus voltage relationship can be approximated by [21]:

$$J \approx 8\varepsilon_r \varepsilon_0 \bar{\tau} \mu_n \mu_p \frac{V^3}{L^5} \tag{6.43}$$

where $\bar{\tau}$ is the common average lifetime for the injected electrons and holes. The above derivation assumes that the average carrier lifetime is independent of the average injection level. In the case where there is such dependence, then $\bar{\tau}$ must be eliminated from Eq. (6.43) [21].

6.7 Schottky, Fowler–Nordheim, and Poole–Frenkel Effects

Other conductivity regimes dominate for large applied electric fields ($>10^7$ V m^{-1}). Three important physical processes are the Schottky, Fowler–Nordheim, and Poole–Frenkel effects. The essential features of these are contrasted in the potential energy diagrams shown in Figure 6.15.

When the potential barrier at an electrode/insulator interface is too thick to permit quantum mechanical tunnelling, or at sufficiently high temperature, the current flowing through the insulator is limited principally by the rate at which electrons are thermally excited over the interfacial potential barrier, ϕ_b, into the insulator conduction band. The top diagram in Figure 6.15 illustrates this process. The emission current is given by the Richardson equation, which forms the basis for the thermionic emission theory for the conductivity of a Schottky barrier, discussed in Section 4.4.3. The current density may be written:

$$J = A^* T^2 \exp\left(-\frac{\phi_b}{k_B T}\right)$$

(6.44)

Figure 6.15 Potential energy diagrams to illustrate Schottky emission (top), Fowler–Nordheim tunnelling (centre) and the Poole–Frenkel effect (bottom). In each case, ϕ_b represents the appropriate energy barrier. E_F is the metal Fermi level.

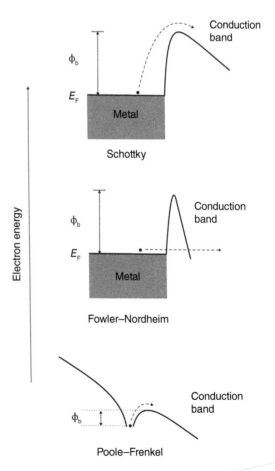

where the constant A^*, the Richardson constant, given by Eq. (4.32), has the value of $120\,\mathrm{A\,cm^{-2}\,K^{-2}}$.

The application of an applied field will reduce the interfacial barrier by the image force effect (Section 4.4.2). The barrier lowering $\Delta\phi$ will be given by Eq. (4.20) and therefore ϕ_b in Eq. (6.44) may be written:

$$\phi_b = \phi_{b0} - \Delta\phi = \phi_{b0} - \beta_S \mathcal{E}^{1/2} \tag{6.45}$$

where ϕ_{b0} refers to zero bias conditions and β_S is a constant, called the Schottky constant:

$$\beta_S = \left(\frac{e^3}{4\pi\varepsilon_r\varepsilon_0}\right)^{1/2} \tag{6.46}$$

The resulting expression for the current density is therefore:

$$J = AT^2 \exp\left(-\frac{\phi_{b0}}{k_B T}\right) \exp\left(\frac{\beta_S \mathcal{E}^{1/2}}{k_B T}\right) \tag{6.47}$$

This thermionic emission mechanism due to image force lowering at a neutral contact is often called the Schottky effect, which as noted in Chapter 4 causes confusion with the Schottky barrier. The Schottky effect forms only part of the explanation of the operation of a Schottky barrier (or metal/semiconductor) device. Image force lowering in the latter leads to a $\ln(J) \propto V_r^{1/4}$ dependence for reverse bias (an example of a blocking contact, Section 4.4.3).

If the applied electric field is sufficiently large, the barrier to electrons at the interface becomes very thin and electrons can tunnel directly from the metal to the conduction band of the insulator. This conductivity mechanism is termed Fowler–Nordheim field emission and is illustrated by the middle diagram in Figure 6.15. The resulting current density versus voltage behaviour is:

$$J \propto \mathcal{E}^2 \exp\left(-\frac{\gamma}{\mathcal{E}}\right) \tag{6.48}$$

where γ is the Fowler–Nordheim constant. The process can also be thermally assisted, the electrons being excited to an energy state, at which these are able to tunnel through the barrier. As noted earlier, in Section 6.3, this thermally assisted tunnelling mechanism can result in a current that varies rapidly with temperature [12].

A similar process to Schottky emission can take place at impurity centres in the bulk of the material. This is the Poole–Frenkel effect and is illustrated by the bottom potential energy diagram in Figure 6.15. The application of a high electric field will result in the lowering of the potential barrier around the centre, allowing a carrier to escape into the conduction band of the insulator. The current versus voltage behaviour for Poole–Frenkel conduction has the form:

$$J \propto \mathcal{E} \exp\left(\frac{\beta_{PF} \mathcal{E}^{1/2}}{k_B T}\right) \tag{6.49}$$

where $\beta_{PF} = 2\beta_S$. This equation is similar to Eq. (6.47) for Schottky emission, although the Schottky and Poole–Frenkel coefficients differ. It should be noted that the Poole–Frenkel model (Eq. (6.49)) includes a pre-exponential electric field term; therefore, to be strictly correct, log(conductivity) should be plotted against (electric field)$^{1/2}$ to obtain a linear relationship. Particular distributions of traps can reduce the Poole–Frenkel constant by a factor of two [14]. Such complications, together with the uncertainty in estimating the permittivity of the solid, make it impossible using a single measurement to distinguish experimentally between the Schottky and Poole–Frenkel effects. However, Poole–Frenkel conduction is essentially a bulk phenomenon and should show little dependence upon electrodes or upon the polarity of the field.

Poole–Frenkel conductivity is often observed for inorganic thin films (see, for example, [14]) and has been reported for thin films of organic–inorganic composites [25]. For organic thin films and devices (e.g. field effect transistors – see Section 8.4), the carrier mobility sometimes follows a Poole–Frenkel type field dependence $\ln \mu \propto \sqrt{\mathcal{E}}$ [26]. Poole–Frenkel conductivity has also been noted for large area composite films of single wall carbon nanotubes [27]. While Ohmic conduction was observed in films having a high nanotube concentration, the Poole–Frenkel mechanism was thought to be responsible for the in-plane conduction in composites with a lower nanotube loading. The temperature (293 to 77 K) dependence of the conductance provided a further confirmation of this. However, a larger theoretical than experimental value of β_{PF} suggested that the effective electrode separation was somewhat less than the measured value [28]. This implies a varying degree of homogeneity of electrical field (probably resulting from the connectivity between the conductive nanotube paths in the composite films) (Section 4.2). Two levels of ionized traps were thought to contribute to the conduction.

6.8 Electrical Breakdown

If the voltage across a solid is increased, there will come a point at which electrical breakdown will occur. The maximum electric field that a homogeneous solid can sustain indefinitely, sometimes referred to as the intrinsic dielectric strength, is usually greater than $100 \, \text{MV m}^{-1}$. The breakdown field for diamond is around $2000 \, \text{MV m}^{-1}$, while insulating organic polymers may have dielectric strengths as low as $20 \, \text{MV m}^{-1}$. Electrical breakdown in solids is not a scientific topic that currently attracts a lot of attention [29]. However, there are two good reasons that this might change. The first is that, on the nanoscale, the breakdown field of a solid can be achieved by the application of quite a modest voltage, e.g. 1 V applied across a 10 nm thin film. This clearly has technological relevance as the thickness of the gate insulator in a current (2021) metal/oxide/semiconductor field effect transistor is approximately 1 nm. A further reason to study electrical breakdown is that it may be possible to develop electronic devices that exploit the phenomenon. For example, electrical breakdown in solid can lead to filament formation, which might be exploited in memory (and other) devices (see, for example, [30]).

Although electrical breakdown in a solid would appear to be a fundamental physical property, it is difficult to measure reliably. This is a consequence of the irreproducibility of the breakdown process itself, which is stochastic in nature, and the fact that electrical breakdown may often occur because of non-intrinsic processes, for example, material purity, sample history, or environmental effects. The metallic electrodes can also play a significant role. Enhanced electric fields can exist in the vicinity of electrode edges and point contacts. A brief review of the various electrical breakdown processes that may be at work is provided in the sections below [1]. Each has characteristic time associated with the mechanism: intrinsic (electronic) breakdown occurs very quickly after the application of an applied field, while erosion and electrochemical effects can persist over a long time (perhaps, many years). The variation of the breakdown field with time (on a log scale) following the application of an electric field to a solid is depicted in Figure 6.16. The various regimes indicated are discussed below.

6.8.1 Intrinsic Breakdown

Intrinsic breakdown is expected to occur only in very pure materials. The effect occurs in times of the order of 10^{-8} s and therefore is assumed to be electronic in nature. When an electric field is

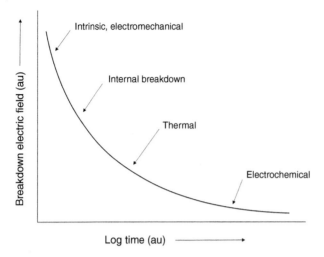

Figure 6.16 Variation of breakdown electric field with time (log scale) following application of an electric field to a solid. The various breakdown regimes are indicated.

applied, electrons gain energy from the field and cross the forbidden energy gap from the valence band to the conduction band. In an early model [28, 29], the charge carriers were assumed to be scattered by phonons (rather than by, say, electron–electron interactions). Continual thermalization by this process sets a limit to the kinetic energy that an electron can acquire through acceleration in the electric field. A consequence is that a temperature increase, which will lead to increased carrier scattering and therefore to an increase in the rate at which electrons lose the energy gained from the applied field, will increase the dielectric strength. This is found experimentally for certain highly pure ionic materials. However, if the solid contains a significant number of charge carriers (e.g. as a result of impurities), then electron–electron scattering may predominate. Hence, the breakdown field strength can decrease as the temperature rises. A noteworthy experimental result is that of an increase in breakdown field with increasing temperature for pure crystalline quartz, in contrast to the opposite effect for fused (amorphous) quartz [28]. Experimental data suggest that insulating polymers, such as polyethylene, exhibit a reduction in dielectric strength with increasing temperature [31].

6.8.2 Electromechanical Breakdown

Electrodes attached to the surfaces of a specimen during a dielectric breakdown experiment will exert a compressive force on the material by Coulombic attraction between the electrodes. If this is sufficient to cause appreciable deformation at fields below the intrinsic breakdown value, the dielectric strength will be reduced. A simple calculation reveals the following relationship between the breakdown electric field, $\mathcal{E}_b$, and the Young's modulus, Y, of the solid [28]:

$$\mathcal{E}_b \approx 0.6 \left(\frac{Y}{\varepsilon_0 \varepsilon_r} \right)^{1/2} \tag{6.50}$$

The values of $\mathcal{E}_b$ predicted by Eq. (6.50) can be quite large. For example, for polyethylene ($Y \approx 0.7$ GPa; $\varepsilon_r \approx 2.4$), $\mathcal{E}_b \approx 3.4$ GV m^{-1}. However, the derivation of Eq. (6.50) ignores any departure from linear elastic behaviour at large strains.

6.8.3 Thermal Runaway

If there are sufficient free charge carriers in a material, the application of an electrical field will generate Joule (I^2R) heating. In a semiconductor or insulator, this heating will increase the conductivity further and thermal breakdown, or thermal runaway, may take place. This process will occur for both DC and AC applied voltages. However, in the latter case, there may be additional heat generation via one of more of the relaxation mechanisms described in Chapter 7.

The onset on thermal runaway will depend on the balance between the heat generated and the ability of the solid to conduct this to the surroundings:

> Electrical power dissipated per unit volume = rate of increase in heat content
>
> +rate at which heat is conducted away

i.e.

$$\sigma \mathcal{E}^2 = C\rho \frac{dT}{dt} - div(K \text{ grad } T) \tag{6.51}$$

where C is the specific heat capacity of the solid (units J kg^{-1} K^{-1}), ρ is its density, and K is the material's thermal conductivity.

Analytic solutions to Eq. (6.51) are not possible as the parameters σ, C, and K are all functions of temperature. However, approximate numerical solutions exist [1]. The main feature is the prediction of a critical voltage above which the temperature increases indefinitely. This will either result in the melting or chemical decomposition of the solid, or a reduction in the intrinsic dielectric strength until breakdown occurs. Computer simulations are available for the Joule heating of nanowires grown on different substrates [32].

An estimate of an upper limit for the heating rate in a sample can be obtained by ignoring the thermal conductivity term in Eq. (6.51). Assuming that both the current density and electrical conductivity of the sample are uniform, and that C is temperature independent (Problem 6.5):

$$\frac{dT}{dt} = \frac{J^2}{C\rho\sigma} \tag{6.52}$$

6.8.4 Contact Instability

Occasionally, failure of the conductive electrodes can be responsible for the electrical breakdown of a material under test or an electronic device. For example, ITO is an optoelectronic material that is applied widely in both research and industry. The compound combines high electrical conductivity with optical transparency. ITO can therefore find use in many applications, such as flat-panel displays, smart windows, polymer-based electronics and thin film solar cells. However, this electrode material can decompose at elevated temperatures, in high electric fields and under high current densities. This can result in diffusion of the ITO constituents (particularly indium) into material beneath the electrode or disruption of its planar nature. Figure 6.17 shows a dramatic example of this – an atomic force microscope image of an electroluminescent polymer device (Chapter 10) built up on an ITO-coated glass substrate [33]. The image, 100 µm × 100 µm, has been taken following operation of the device and after the top electrode and organic layers have been removed. A step region between the ITO and glass, running from the top left to the bottom right of the image, is evident. The surface roughness of the ITO is considerable (compared to the glass) with several large spikes and a 'crater' present. X-ray chemical analysis of the spikes has confirmed that these contain both In and Sn, suggesting that the ITO electrode is able to reform under electrical stress. The inclusion of a buffer layer, such as the conductive polymer poly(3,4-ethylenedioxythiophene)

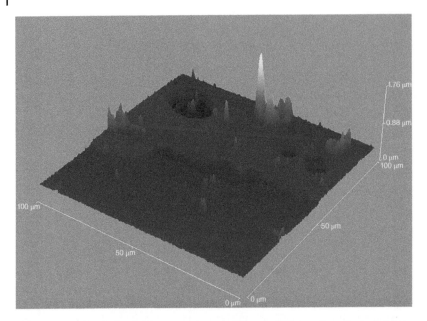

Figure 6.17 Atomic force microscope image, 100 μm × 100 μm, of an electroluminescent organic polymer device built on an ITO electrode, following device operation and removal of the organic layers. The step region between the ITO and glass substrate is evident. Source: Widdowson [33].

(PEDOT), is frequently used to smooth the surface of ITO and minimize effects such as those shown in Figure 6.17.

Material migration under the influence of an applied voltage from electrodes has also been observed for organic thin film devices using aluminium electrodes. For example, it has been suggested at the Al pillars can grow on an electrode under the influence of an applied voltage, forming a metallic bridge across an adjacent organic thin film. The presence of these filaments was thought to be responsible for switching and memory effects in a metal/organic film/metal device architecture (Section 9.3) [34].

6.8.5 Other Effects

The dielectric strength of a gas is very much less than that of a solid material, around $3\,\text{MV}\,\text{m}^{-1}$. As a consequence, during the application of a high voltage to a solid specimen, discharges are likely to occur at an early stage in any gas that is at the edges of the electrodes or that may be contained within a void in the solid. Such discharges are likely to damage the sample or lead to dielectric failure; this is sometimes referred to as internal breakdown. Organic materials (e.g. polymers) are often affected by this type of failure as bombardment of the molecules by ions in the discharge can break chemical bonds. The physics of the gas discharge process is governed by Paschen's law, which states that the minimum voltage necessary to produce an electrical discharge across a gap depends on the product of the gap length and the density of the gas. Dielectric breakdown resulting from these internal discharges often leads to the formation of channels, generally progressing in the direction of the field. The channels follow pathways of weakness (e.g. interfacial boundaries in non-homogeneous materials). There is a tendency for branching and electrical discharge networks called electric trees are formed [1]. Internal discharges are a common feature of bulk polymeric insulators as voids are easily incorporated into their structure during processing (e.g. injection moulding), but probably less significant in thin films.

A further cause of breakdown in solids can occur over a long period and is electrochemical in origin. Chemical changes, such as oxidation or hydrolysis, may take place under electrical stress. Even in the absence of an applied electric field, there may be chemical deterioration of the material, leading to its ultimate electrical failure. Moisture content in the solid can exacerbate these effects, and organic solids are particularly prone to such problems. For this reason, electronic devices based on organic thin films are invariably encapsulated.

6.9 Electromigration

Standard electrical conductors (e.g. as found in the home) possess a current carrying capacity of around 10^4 A cm^{-2}, limited by Joule heating. However, the conductors used in microelectronic circuits are in contact with heat sinks and their current carrying ability can be significantly higher, around 10^6 A cm^{-2}. The failure mechanism in such instances is not due to Joule heating but results from an effect called electromigration. This is the electric field-induced transport of matter [35, 36]. The phenomenon is quite general and occurs in all materials. It should not be confused with electrophoresis, which is the direct response of charge species to an applied electric field. To observe significant electromigration effects, a high electron density and large scattering of the carriers are required. For this reason, electromigration occurs predominately in metals but may also be observed in heavily doped semiconductors. Figure 6.18 illustrates the process. Two forces act on the metal ions, which make up the lattice of the interconnect material. The first, F_{field} in Figure 6.18, is due to the action of the applied electric field on the metal ions, which causes the positively charged metal ions to be attracted to the cathode. This effect is relatively small due to screening from negative charges. Electromigration is the result of the dominant force, the momentum transfer from the electrons, which move in the applied electric field. This so-called 'electron wind', F_{wind}, depends on the local diffusion coefficient. In Figure 6.18, the metal ions move in the same direction as the negative charge carriers, the same direction as the current flow. However, in a heavily doped p-type semiconductor the ion metal ion movement will be opposite to that of the conventional current flow.

Electromigration is a thermally activated diffusion process, so the rate increases with increasing temperature. A simple model predicts that the particle flux is proportional to the current density [36, 37]. From a practical (microelectronic device) point of view, it has been found that the mean

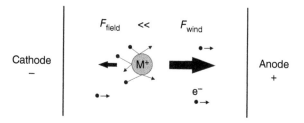

Figure 6.18 Schematic diagram showing the phenomenon of electromigration. Two forces act on the metal ions M (which make up the lattice of a solid). Electromigration is the result of the dominant force, i.e. the momentum transfer from the electrons that move in the applied electric field. F_{field} = force due to electrostatic interaction with the applied electric field; F_{wind} = force due to momentum transfer from the charge carriers (electrons, depicted by small solid circles).

time taken for a wire to fail for a current density J is given by the following relationship, known as Black's equation [38]:

$$\text{MTF} = CJ^{-n} \exp\left(\frac{E_a}{k_B T}\right) \tag{6.53}$$

where MTF is the mean time to failure, C is a constant dependent on the material properties (e.g. geometry, diffusion constant), J is the current density, E_a is the electromigration activation energy, and n is an integer, which in most cases is two.

The reader might wonder why this section, concerning the limitations of high current-carrying conductors, is included in a book that concerns organic thin films. The relevance is that carbon nanotubes may provide an alternative material to conventional (inorganic metals) conductors for use as high current carrying interconnects in microelectronic circuits [39]. It has been suggested that the limitation to the current density in carbon nanotubes imposed by electromigration effects may be close to 10^9 A cm^{-2}, 3 orders of magnitude greater than the figure given above for conventional interconnects [39]. There are several technological hurdles to be overcome before carbon nanotubes can be exploited as high current carrying conductors. Means are required to manufacture the materials with reproducible properties and methods need to be found to organize these nonconductors on solid substrates. Nonetheless, such one-dimensional conductors may become the electronic device interconnects of the future.

6.10 Measurement of Trapping Parameters

DC conductivity measurements provide a useful way of obtaining the characteristics of trapping levels (e.g. density, capture cross section) in organic semiconductors [40]. Methods based on two-terminal and three-terminal (transistor) devices are available. However, rarely do these experiments provide information on the nature, i.e. chemical origins, of the defect levels. For this, the electrical tests must be conducted in conjunction with analytical chemistry. In the sections below, two popular electrical experiments, extensively applied to inorganic semiconductors, are outlined, and some data for organic semiconductors are provided. The first uses a DC current technique, whilst the other is based on a capacitance method. In both cases, a variation in temperature is used to reveal a spectrum representing the traps.

6.10.1 Thermally Stimulated Conductivity

The method of thermally stimulated conductivity (TSC) requires that the trapping levels in the semiconductor are first populated at a low temperature. This can be achieved electrically, by driving a large current through the sample, or optically via irradiation by band gap light. The experiment can be conducted on a simple thin film sample, if it is sufficiently insulating, or using a device with a depletion layer, such as a Schottky barrier. On heating the sample at a steady rate, charges are released from the traps. Under the application of an external bias (or internal field in the case of a Schottky barrier) these charges can be monitored as an increase in conductivity. Generally, a series of current peaks is observed during the experiment (as the temperature increases) with each peak related to a particular trapping level. There is a related experiment – thermoluminescence (TL) in which the light output of the sample is monitored as trapped charge is released and recombines at a recombination centre. Over the years, numerous techniques have been used to extract trapping parameters from TSC data. There now exist several reviews on this topic (see, for example [41]).

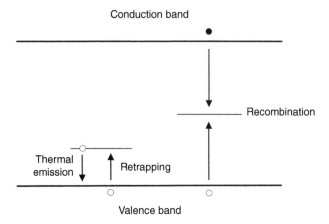

Figure 6.19 Schematic diagram showing carrier emission from a single hole trap in a semiconductor. The hole can be thermally emitted from the trap to the valence band increasing the free hole concentration in the semiconductor. Subsequently, the hole can be captured (retrapped) by the same trap or can recombine with an electron at a recombination centre.

A simple model is that of a single deep level hole trap of density N_t that can exchange charge with the valence band, as shown in Figure 6.19. Once in the valence band, the charge can be either retrapped or removed by recombination at a recombination centre. The thermal emission rate e_p (units s^{-1}) of a hole trapped in such a defect with activation energy ΔE (equal to the energy difference between the trapping level and edge of the valence band) may be written [40]:

$$e_p = N_v \overline{v_t} \sigma_p \exp\left(-\frac{\Delta E}{k_B T}\right) = v_0 \exp\left(-\frac{\Delta E}{k_B T}\right)$$

(6.54)

where N_v is the effective density of states of the valence band, $\overline{v_t}$ is the average carrier thermal velocity and σ_p is the capture cross section of the hole trap. The product $\overline{v_t}\sigma_p$ is equal to the capture coefficient of the trap β defined in Section 3.4.2. Finally, v_0 is called the attempt to escape frequency and defines the maximum rate of detrapping cycles.

Following the initial filling of the trapping centre, the densities of free carriers $p(t)$ and trapped carriers $p_t(t)$ at any time are related by:

$$\frac{dp}{dt} = -\frac{dp_t}{dt} - \frac{p}{\tau}$$

(6.55)

where τ is the average lifetime of a free hole before recombination. The first term on the right-hand side of Eq. (6.55) represents the rate of change of the trapped hole charge while the second term is the rate of recombination. The variation of the trapped hole density can be expressed:

$$\frac{dp_t}{dt} = -p_t e_p + p\left(N_t - p_t\right)\sigma_p \overline{v_t}$$

(6.56)

Analytical solutions to the above differential equations can only be obtained if some simplifications are made [41]; for example, that detrapped electrons recombine immediately (slow retrapping). This leads to different approaches to extract the trapping parameters from TSC curves. A very simple technique, valid if little thermal emission from the trap has taken place, uses the initial current increase of a TSC peak. Other methods rely on the position of current maximum, the peak shape, and the variation of the peak temperature with heating rate. Alternatively, curve-fitting procedures can be used.

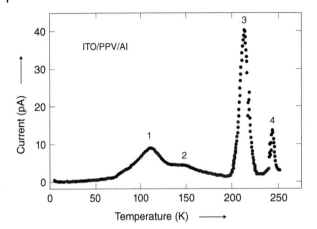

Figure 6.20 Thermally stimulated current spectrum for an ITO/PPV/Al device. Four trapping centres (labelled 1 to 4) are identified. These are discussed in the text. Source: Meier et al. [43].

Thermally stimulated current experiments have been undertaken on a variety of small organic molecule and polymer semiconductors (see, for example [42, 43]). Figure 6.20 shows a TSC spectrum obtained for a sample of PPV [43]. The current peaks labelled 1 and 2, between 100 and 150 K, were thought to result from the chemical reaction of the polymer with the ITO substrate. The associated trap depths were estimated as 0.03–0.06 eV and 0.13–0.18 eV (the trapping parameters were extracted by curve fitting). The relatively narrow peaks at higher temperatures, labelled 3 and 4, possessed trap energies of 0.6–0.9 eV (peak 3) and between 0.9 and 1.0 eV for peak 4. It was suggested that these were associated with the oxidizing effect of air on the PPV.

Currents can also be observed in insulating, polar organic films as these are heated. This is evident in alternate-layer (acid/amine) Langmuir–Blodgett architectures, in which permanent electric dipoles have been built into the ordered multilayer array during film deposition; such thin films exhibit the pyroelectric effect (Section 7.2.2). As these are heated, a pyroelectric current is first measured (because of a change in sample polarization with temperature) followed by a current resulting from the disordering of the dipoles [44].

6.10.2 Capacitance Spectroscopy

There are several analytical electrical techniques based on the measurement of capacitance [40]. The temperature evolution of capacitance, rather than current, is a variation of the TSC approach described above. A popular transient capacitance method is that of deep level transient spectroscopy (DLTS) [45]. This has emerged as a powerful technique to study defects in inorganic semiconductors. The experiment uses a two-terminal device such as a Schottky diode or MIS structure. The device is reverse biased, and a forward bias voltage pulse momentarily reduces the depletion layer width, thereby filling bulk (or surface) traps. At the quiescent voltage, the capacitance of the device is monitored as a function of time and temperature. Figure 6.21 shows a series of capacitance transients obtained immediately after applying a voltage pulse at different temperatures. The DLTS experiment involves measuring the capacitance difference at two times, t_1 and t_2, after the application of the initial voltage pulse. In the original experimental set-up, this was accomplished using a dedicated analogue integrator (boxcar detector), but DLTS systems are now invariably based on computerized data acquisition systems. A normalized DLTS signal $S(T)$

Figure 6.21 Evolution of a deep level transient spectroscopy (DLTS) spectrum. The left-hand side of the figure shows capacitance transients at various temperatures. The capacitance values at two times (t_1 and t_2) following the application of a forward bias voltage pulse are measured. The difference in these, ($C(t_1) - C(t_2)$), constitutes the DLTS signal, shown as a function of temperature on the right of the figure. Source: Lang [45].

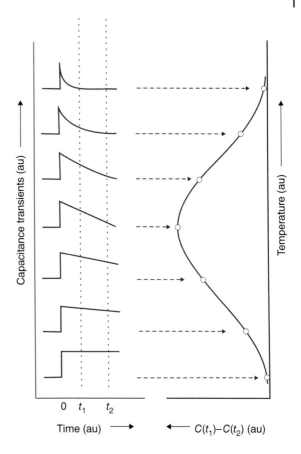

may be defined:

$$S(T) = \frac{\left(C\left(t_1\right) - C\left(t_2\right)\right)}{\Delta C(0)} \tag{6.57}$$

where $\Delta C(0)$ is the capacitance change due to the pulse at $t = 0$. As can be seen in Figure 6.21, $S(T)$ will have a small value at low and high temperatures, corresponding to slow and rapid capacitance decay, respectively. At an intermediate temperature $S(T)$ will go through a maximum. Therefore, the presence of each trap is indicated by a positive or negative peak on a flat base line as a function of temperature. The heights of these peaks are proportional to their respective trap concentrations, the sign of each peak indicates whether it is due to a majority- or a minority-carrier trap, and the positions of the peaks are determined by the thermal emission properties of the traps (Eq. (6.54)). In this respect, there is a significant advantage over TSC, which is unable to distinguish between electron and hole traps as these both result in peaks of the same current polarity.

For a thermal emission rate given by Eq. (6.54):

$$S(T) = \left(\exp\left(-\frac{t_1}{\tau}\right) - \exp\left(-\frac{t_2}{\tau}\right)\right) \tag{6.58}$$

where $\tau = 1/e_p$. This equation may be rewritten:

$$S(T) = \exp\left(-\frac{t_1}{\tau}\right)\left(1 - \exp\left(-\frac{\Delta t}{\tau}\right)\right) \tag{6.59}$$

where $\Delta t = t_2 - t_1$. If τ_{max} represents the value of τ at the maximum DLTS versus temperature signal, then the relationship between τ_{max} and t_1, t_2 can be obtained by differentiating $S(T)$ with respect to τ [40]:

$$\tau_{max} = (t_1 - t_2)\left(\ln\left(\frac{t_1}{t_2}\right)\right)^{-1} \tag{6.60}$$

The temperature at which the maximum DLTS signal occurs does not reveal the activation energy of the trap ΔE directly unless the pre-exponential factor in Eq. (6.54) is known. If not, measurements can be made with different time windows (varying t_1, t_2). The trap concentration may be evaluated from the magnitude of capacitance change corresponding to completely filling the trap [45].

A hybrid TSC/DLTS technique is Q-DLTS. Instead of measuring the capacitance transient, the charge ΔQ emitted from the device under test is analysed. An advantage to this method is that as no capacitance is measured, no depletion layer is needed. Q-DLTS is therefore not restricted to Schottky barrier and MIS devices. However, electron and hole traps are not distinguished.

DLTS techniques have been applied to study the defect states in a variety of organic semiconductors (see, for example [46–49]). Figure 6.22 shows a Q-DLTS spectrum for polyfluorene (PF) compound in the light-emitting device configuration ITO/PEDOT/PF/Ca/Al [48]. For clarity, a selection of the published data is presented, for two temperatures. The Q-DLTS spectrum is in the form ΔQ versus $\log(\tau)$ where τ is the time-window parameter:

$$\tau = (t_1 - t_2)\left(\ln\left(\frac{t_1}{t_2}\right)\right)^{-1} \tag{6.61}$$

Similar to the conventional DLTS spectrum based on the capacitance signal, this has a maximum ΔQ_{max} at $\tau_{max} = 1/e_p$. Five traps, corresponding to the five charge peaks (labelled A to E) are identified in Figure 6.22, with trap depths varying from 0.13 to 0.58 eV. It was suggested that the traps labelled B ($\Delta E = 0.22$ eV) and E ($\Delta E = 0.58$ eV) could be associated with defects along the PF polymer chain. No additional peaks were noted on degraded devices, implying that no additional defect levels were formed because of the ageing process.

The value of all the above current and capacitance spectroscopic methods is not only to extract trapping parameters but also to reveal the presence of certain impurities through their fingerprint spectra. This provides a useful platform from which to study the effects of material or device processing.

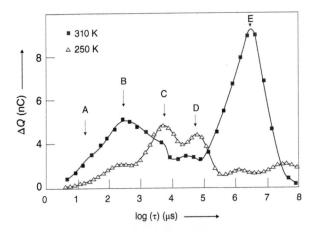

Figure 6.22 Q-DLTS spectrum of an ITO/PEDOT/PF/Ca/Al architecture at two different temperatures. The peaks correspond to charge released from individual, marked traps A to E. Charging time = 1 s, charging voltage = +6 V. Source: Nguyen et al. [48].

Problems

6.1 Do electrons loose energy in the barrier region during the process of quantum mechanical tunnelling? Explain your reasoning.

6.2 A monolayer organic film is sandwiched between two metal electrodes. If the energy barrier to electrons is 8 eV, calculate the probability of conduction electron in the metal electron with a kinetic energy of 4.5 eV tunnelling through the barrier if the monolayer is (i) 3 nm thick and (ii) 1.5 nm thick. Comment on your answers.

6.3 The conductivity of a sample of doped polyacetylene exhibits the following variation of conductivity with temperature.

Temperature (K)	Electrical conductivity ($S\,cm^{-1}$)
290	6.11×10^{-3}
250	5.67×10^{-3}
200	4.96×10^{-3}
150	4.03×10^{-3}
120	3.26×10^{-3}
100	2.68×10^{-3}
80	2.03×10^{-3}
60	1.32×10^{-3}
40	6.22×10^{-4}
20	1.21×10^{-4}

What conclusions can be drawn regarding the physical process of conductivity?

6.4 The doping density of a conductive polymer thin film is 2×10^{16} cm^{-3}; the film thickness is 100 nm and the relative permittivity of the polymer is 3.0. As the voltage applied across the thin film is increased, holes are injected into the organic layer (single carrier space-charge injection). Calculate the voltage at which $I \propto V^2$ will be observed. If a rapid rise in the current is observed at $V = 1.8$ V (trap-filled limit) estimate the deep trap density, assuming that a single level is dominant.

6.5 A free-standing wire, diameter 200 nm, is composed of a bundle of single wall carbon nanotubes. If a current of 1 μA is passed through the wire, calculate its maximum rate of temperature rise. Research appropriate material parameters for the nanotubes.

References

1 Blythe, T. and Bloor, D. (2005). *Electrical Properties of Polymers*. Cambridge: Cambridge University Press.

2 Kasap, S.O. (2008). *Principles of Electrical Engineering Materials and Devices*, 3e. Boston: McGraw-Hill.

3 Simmons, J.G. (1971). *DC Conduction in Thin Films*. London: Mills and Boon.

4 Price, P.J. and Radcliffe, J.M. (1959). Esaki tunneling. *IBM J. Res. Dev.* 3: 364–371.

5 Stratton, R. (1962). Volt-current characteristics for tunnelling through insulating films. *J. Phys. Chem. Solids* 23: 1177–1190.

6 Simmons, J.G. (1963). Generalised formula for the electric tunnel effect between similar electrodes separated by a thin insulating film. *J. Appl. Phys.* 34: 1793–1803.

7 Simmons, J.G. (1964). Potential barriers and emission-limited current flow between closely spaced parallel metal electrodes. *J. Appl. Phys.* 35: 2472–2481.

8 Polymeropoulos, E.E. (1977). Electron tunnelling through fatty acid monolayers. *J. Appl. Phys.* 48: 2404–2407.

9 Petty, M.C. (1992). Characterisation and properties. In: *Langmuir-Blodgett Films* (ed. G.G. Roberts), 133–221. New York: Plenum Press.

10 Ginnai, T.M., Oxley, D.P., and Pritchard, R.G. (1980). Elastic and inelastic tunnelling in single-layer Langmuir films. *Thin Solid Films* 68: 63–68.

11 Hansma, P.K. (1977). Inelastic electron tunnelling. *Phys. Rep.* 30: 145–206.

12 Roberts, G.G. and Polanco, J.I. (1970). Thermally assisted tunnelling in dielectric films. *Phys. Status Solidi A* 1: 409–420.

13 Mott, N.F. and Davis, E.A. (1979). *Electronic Properties in Non-crystalline Materials*, 2e. Oxford: Clarendon Press.

14 Roberts, G.G., Apsley, N., and Munn, R.W. (1980). Temperature dependent electronic conduction in polymers. *Phys. Rep.* 60: 59–150.

15 Roth, S. (1995). *One-Dimensional Metals*. Weinheim: VCH.

16 Chiang, C.K., Park, Y.W., Heeger, A.J. et al. (1978). Conducting polymers: halogen doped polyacetylene. *J. Chem. Phys.* 69: 5098–5104.

17 Sheng, P. (1980). Fluctuation-induced tunnelling conduction in disordered materials. *Phys. Rev. B* 21: 2180–2195.

18 Ehinger, E. and Roth, S. (1986). Non-solitonic conductivity in polyacetylene. *Philos. Mag. B* 53: 301–320.

19 Fischer, J.E., Dai, H., Thess, A. et al. (1997). Metallic resistivity in crystalline ropes of single-wall nanotubes. *Phys. Rev. B* 55: R4921–R4924.

20 Kaiser, A.B., Park, Y.W., Kim, G.T. et al. (1999). Electronic transport in carbon nanotube ropes and mats. *Synth. Met.* 103: 2547–2550.

21 Lampert, M.A. and Mark, P. (1970). *Current Injection in Solids*. New York: Academic Press.

22 Rose, A. (1978). *Concepts in Photoconductivity and Allied Problems*. New York: Robert E. Krieger Publishing.

23 Gould, R.D. (1966). Structure and electrical conduction properties of phthalocyanine thin films. *Coord. Chem. Rev.* 156: 237–274.

24 Campbell, A.J., Bradley, D.D.C., and Lidzey, D.G. (1997). Space-charge limited conduction with traps in poly(phenylene vinylene) light emitting diodes. *J. Appl. Phys.* 82: 6326–6342.

25 Senthilarasu, S., Sathyamoorthy, R., Lalitha, S., and Subbarayan, A. (2006). Electrical conduction properties of $ZnPc/TiO_2$ thin films. *Sol. Energy Mater. Sol. Cells* 90: 783–797.

26 Köhler, A. and Bässler, H. (2015). *Electronic Processes in Organic Semiconductors*. Weinheim: Wiley-VCH.

27 Jombert, A.S., Koleman, K.S., Wood, D. et al. (2008). Poole–Frenkel conduction in single wall carbon composite films built up by electrostatic layer-by-layer deposition. *J. Appl. Phys.* 104: 094503.

28 Mareš, J.J., Krištofik, J., and Šmíd, V. (1988). On space-charge-limited conduction in semi-insulating GaAs. *Solid-State Electron.* 31: 1309–1313.

29 O'Dwyer, J.J. (1973). *The Theory of Electrical Conduction and Breakdown in Dielectrics*. Oxford: Oxford University Press.

30 Petty, M.C. (2018). Organic electronic memory devices. In: *Handbook of Organic Materials for Optical and Optoelectronic Devices*, 2e (ed. O. Ostroverkhova), 843–874. Oxford: Elsevier.

31 Lawson, W.G. (1966). Effects of temperature and techniques of measurement on the intrinsic electric strength of polyethylene. *Proc. Inst. Electr. Eng.* 113: 197–202.

32 Fangohr, H., Chernyshenko, D.S., Franchin, M. et al. (2011). Joule heating in nanowires. *Phys. Rev. B* 84: 054437.

33 Widdowson, N.E. (2006). Electroluminescent devices based on blended polymeric films. PhD thesis. Durham: University of Durham.

34 Pearson, C., Bowen, L., Lee, M.-W. et al. (2013). Focused ion beam and field-emission microscopy of metallic filaments in memory devices based on thin films of an ambipolar organic compound consisting of oxadiazole, carbazole, and fluorene units. *Appl. Phys. Lett.* 102: 213301.

35 Durkan, C. (2017). *Current at the Nanoscale: An Introduction to Nanoelectronics*. London: Imperial College Press.

36 Hoffmann-Vogel, R. (2017). Electromigration and the structure of metallic contacts. *App. Phys. Rev.* 4: 031302.

37 Mokry, G., Pozuelo, J., Vilatela, J.J. et al. (2019). High ampacity carbon nanotube materials. *Nanomaterials* 9: 383.

38 Black, J.R. (1969). Electromigration – a brief survey and some recent results. *IEEE Trans. Elect. Dev.* ED-16: 338–347.

39 Collins, P.G., Hersam, M., Arnold, M. et al. (2001). Current saturation and electrical breakdown in multiwalled carbon nanotubes. *Phys. Rev. Lett.* 86: 3128–3131.

40 Stallinga, P. (2010). *Electrical Characterization of Organic Electronic Materials and Devices*. Chichester: Wiley.

41 Kivits, P. and Hagebeuk, H.J.L. (1977). Evaluation of the model for thermally stimulated luminescence and conductivity; reliability of trap depth determinations. *J. Lumin.* 15: 1–27.

42 Karg, S., Steiger, J., and von Seggern, H. (2000). Determination of trap energies in Alq_3 and TPD. *Synth. Met.* 111–112: 277–280.

43 Meier, M., Karg, S., Zuleeg, K. et al. (1998). Determination of trapping parameters in poly(*p*-phenylenevinylene) light-emitting devices using thermally stimulated currents. *J. Appl. Phys.* 84: 87–92.

44 Jones, C.A., Petty, M.C., Davies, G., and Yarwood, J. (1988). Thermally stimulated discharge of alternate-layer Langmuir–Blodgett film structures. *J. Phys. D: Appl. Phys.* 21: 95–100.

45 Lang, D.V. (1974). Deep-level transient spectroscopy: a new method to characterize traps in semiconductors. *J. Appl. Phys.* 45: 3023–3032.

46 Jones, G.W., Taylor, D.M., and Gomes, H.L. (1997). DLTS investigation of acceptor states in P3MeT Schottky barrier diodes. *Synth. Met.* 85: 1341–1342.

47 Yang, Y.S., Kim, S.H., Lee, J.-I. et al. (2002). Deep-level defect characteristics in pentacene organic thin films. *Appl. Phys. Lett.* 80: 1595–1597.

48 Nguyen, T.P., Renaud, C., Huang, C.H. et al. (2008). Effect of electrical operation on the defect states in organic semiconductors. *J. Mater. Sci: Mater. Electron.* 19: S92–S95.

49 Neugebauer, S., Rauh, J., Deibel, C., and Dyakonov, V. (2012). Investigation of electronic trap states in organic photovoltaic materials by current-based deep level transient spectroscopy. *Appl. Phys. Lett.* 100: 263304.

Further Reading

Barford, W. (2005). *Electronic and Optical Properties of Conjugated Polymers*. Oxford: Oxford University Press.

Cuniberti, G., Fagas, G., and Richer, K. (eds.) (2005). *Introducing Molecular Electronics*. Berlin: Springer.

Farchioni, R. and Grosso, G. (eds.) (2001). *Organic Electronic Materials*. Berlin: Springer.

Groves, C. (2017). Simulating charge transport in organic semiconductors and devices: a review. *Rep. Prog. Phys.* 80: 026502.

Hummel, R.E. (1992). *Electronic Properties of Materials*, 2e. Berlin: Springer-Verlag.

Kaiser, A.B. (2001). Electronic transport properties of conducting polymers and carbon nanotubes. *Rep. Prog. Phys.* 64: 1–49.

Nabook, A. (2005). *Organic and Inorganic Nanostructures*. Boston: Artech House.

Petty, M.C. (2019). *Organic and Molecular Electronics: From Principles to Practice*, 2e. Chichester: Wiley.

Simmons, J.G. (1971). *DC Conduction in Thin Films*. London: Mills and Boon.

Stubb, H., Punkka, E., and Paloheimo, J. (1993). Electronic and optical properties of conducting polymer thin films. *Mat. Sci. Eng.* 10: 85–140.

Tredgold, R.H. (1966). *Space Charge Conduction in Solids*. Amsterdam: Elsevier.

7

Polarization and AC Conductivity

7.1 Introduction

In the previous chapter, the application a DC voltage to a sample was assumed to result only in the drift of charges (electrons, holes, or ions) through the material. Two further important matters are now introduced. The first is the process of polarization, significant if there is no free or trapped charge within the material. The other consideration is the physical processes that may occur when the DC applied field is replaced by an alternating AC field. Both these topics are frequently grouped together in the literature under the terminology 'dielectrics' [1].

7.2 Polarization

Dipoles can either be present in a material by the nature of its chemical bonding or can be induced by an applied electric field. These will create their own electric field, which can then interact with any external field. The permittivity, or dielectric constant, of the material is an important physical quantity that describes how an electric field influences and is affected by a dielectric medium or a material's ability to transmit (or 'permit') an electric field. The frequency dependence of this parameter is key to understanding electrical phenomena occurring at high frequencies.

7.2.1 Dipole Creation

A static electric field can induce an electric dipole in a solid material. This is the process of polarization and is illustrated in Figure 7.1, which shows the application of an electric field to a neutral atom (or molecule). The diagram depicts electronic polarization, which is one of the processes that may occur in materials, as discussed later in Section 7.3.2. Before the electric field is applied, the centres of positive and negative charge in the atom coincide (Figure 7.1a). The field then produces a very small shift (much exaggerated in Figure 7.1b) in the electron cloud with respect to the nucleus. This results in the formation of an induced dipole. The induced dipole can be modelled as two equal and opposite charges ($\pm\Delta q$) at a distance apart of a. The dipole moment $\mathbf{p}$ is a vector quantity and can be defined:

$$\mathbf{p} = \mathbf{a}\Delta q \qquad (7.1)$$

The distance a (a scalar quantity) is replaced by the vector $\mathbf{a}$ directed from negative to positive charge in Eq. (7.1). The SI units of dipole moment are (C m); however, many workers prefer to use

Electrical Processes in Organic Thin Film Devices: From Bulk Materials to Nanoscale Architectures, First Edition. Michael C. Petty.
© 2022 John Wiley & Sons Ltd. Published 2022 by John Wiley & Sons Ltd.
Companion Website: www.wiley.com/go/petty/organic_thin_film_devices

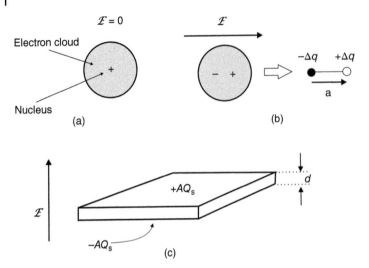

Figure 7.1 Formation of a dipole moment. (a) The centre of the negative electron cloud in a neutral atom coincides with the nucleus. (b) Application of an electric field, $\mathcal{E}$, results in a shift in the centre of positive and negative charge and creates an electric dipole, which can be represented by an equal positive and negative charge Δq at a distance a apart: magnitude of dipole moment = $a\Delta q$. (c) The effect of polarization on a thin film, thickness d, area A, with electric field applied perpendicular to the film plane is to induce equal and opposite charges per unit area Q_s on the upper and lower surfaces: magnitude of dipole moment = dAQ_s.

the CGS unit of Debye ($1\,D = 3.37 \times 10^{-30}\,C\,m$). The ability of a material to acquire an electric dipole in the presence of an applied electric field is referred to as its polarizability.

The individual electric dipoles will align with the applied field in a solid, for example, in the form of a thin film. If the field is orthogonal to the plane of the film of surface area A, a bound charge will appear on its upper and lower surfaces. This is called the polarization **P** (units $C\,m^{-2}$) and is defined as the dipole moment per unit volume. Therefore, if there are N dipoles (originating from individual atoms or molecules) per unit volume and each has a dipole moment **p**:

$$\mathbf{P} = N\mathbf{p} \tag{7.2}$$

The polarized thin film can also be represented in terms of the bound charges per unit area $+Q_s$ and $-Q_s$, which appear on its upper and lower surfaces, as shown in Figure 7.1c. If the film thickness is d, then this arrangement can be thought of as one large dipole moment, $\mathbf{p}_{total}$. Focusing on magnitudes (rather than vector quantities):

$$P = \frac{p_{total}}{\text{Volume}} = \frac{Q_s A d}{A d} = Q_s \tag{7.3}$$

In this case, the polarization may also be defined as the surface charge per unit area.

Polarization is also an important material parameter at high frequencies. For example, the AC electric field of an electromagnetic, EM, wave produces oscillating dipoles in any matter with which it can interact. The electron cloud in an atom is relatively light with respect to the nucleus and can respond to EM fields up to very high frequencies (visible to ultraviolet light). The fluctuating dipoles, in turn, produce their own oscillating electric and magnetic fields, which radiate EM waves – the dipoles acting as 'molecular aerials'. These combine with the incident wave to produce an electromagnetic wave that travels through the material with a phase velocity slower than the velocity of light in vacuum. If v is the phase velocity of light in the material and c is the velocity of light in vacuum:

$$n = \frac{c}{v}$$

(7.4)

where n is the refractive index. Typical values are about 1.5 for glass and can be in the range 1.5–3.0 for many organic semiconductors. The refractive index may also be linked to the high frequency values of relative permittivity, ε_r, and relative permeability, μ_r, of the medium:

$$n = \sqrt{\varepsilon_r \mu_r}$$

(7.5)

For nonmagnetic materials, $\mu_r = 1$ and $n = \sqrt{\varepsilon_r}$. Permittivity and its frequency dependence are explored later, in Section 7.3.2.

7.2.2 Permanent Polarization

Many organic compounds contain dipoles in the absence of an applied electric field by virtue of their chemical bonding. These ground-state dipoles result from the electronegativities of the constituent atoms. A simple example is the carbonyl group (C=O), which has a small positive charge on the carbon atom and a small negative charge on the oxygen (the latter is more electronegative than the former). Table 7.1 lists some dipole moments (in Debye units) of this and other common chemical groups found in typical organic compounds.

Most organic thin films that are made up from polar molecules possess a polycrystalline or amorphous morphology. Therefore, the individual dipoles will be randomly oriented, and the thin film will possess no overall polarization. However, the application of an external electric field will cause the dipoles to attempt to orient in the field direction (opposed by thermal energy).

In the late nineteenth century, materials with permanent overall polarization were made by first melting a suitable dielectric material, such as a polymer or wax containing polar molecules, and then allowing it to re-solidify in a high electric field. Oliver Heaviside called such materials electrets, to suggest the electrical equivalent of a magnet [1]. Modern electrets are usually made by embedding excess charges into a highly insulating dielectric, e.g. by means of an electron beam, a corona discharge, injection from an electron gun or electric breakdown across a gap or a dielectric barrier. Charge densities as high as $1 \, mC \, cm^{-2}$ can be obtained with a time constant (the dielectric relaxation time – see Section 7.3.2) for decay more than 20 years. Polymer electrets, which have been metallized on one side, have been extensively exploited as components of commercial

Table 7.1 Dipole moments of organic compounds containing particular chemical groups.

Chemical group	Example compound	Dipole moment of compound (D)
CO	Acetone (C_2H_6CO)	2.91
OOH	Propanoic acid (C_2H_5COOH)	1.75
OH	Propanol (C_3H_7OH)	1.68
NO_2	Nitropropane ($C_3H_7NO_2$)	3.66
NH_2	Propylamine ($C_3H_5NH_2$)	1.17
CN	Cyanoethane (C_2H_5CN)	4.02
Benzene ring	Benzene (C_6H_6)	0

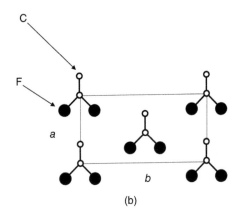

Figure 7.2 Structural forms of poly(vinylidene difluoride), PVDF. (a) Section of $(CH_2CF_2)_n$ chain. (b) β-phase. The projections of C atoms (small open circles) and F atoms (large full circles) onto the *ab* planes of the unit cell are shown. The H atoms have been omitted for clarity.

microphones, as these provide an excellent frequency response without the requirement of an external bias.

Other materials with permanent polarization exploit particular molecular and crystallographic configurations. An important example of a polymer with applications in the field of molecular electronics is poly(vinylidene difluoride) (PVDF) [2, 3]. The chemical structure of this material is shown in Figure 7.2. A section of the polymer chain is depicted in Figure 7.2a; an individual carbon-fluorine bond will possess a strong dipole moment. PVDF is a semicrystalline polymer with a monomer, or mer, unit (CH_2CF_2). The dipole moment is 7.0×10^{-30} C m (approximately 2.1 D) perpendicular to the chain direction. In the bulk solid, PVDF has several crystalline phases. The α-phase is the most common structure, while other forms can be obtained from this parent phase by applications of mechanical stress, heat, and electric field. The polymer chain in the α-phase results in the dipole moments associated with the carbon-fluorine bonds arranged in opposite directions so that there is no net polarization within the crystal.

When the α-phase of PVDF is mechanically deformed by stretching or rolling at temperatures below 100 °C, the β-phase is formed. The atomic arrangements in this phase are shown in Figure 7.2b, which gives the projection of the carbon (small, open circles) and fluorine (large, filled circles) onto one of the planes (the *ab* plane) of the unit cell [2]. The unit cell of the β-phase has a net dipole moment (normal to the chain direction). If all the monomer dipoles were aligned along the chain direction, a maximum microscopic polarization of about 100 mC m^{-2} could be achieved. However, semi-crystalline PVDF is approximately 50% crystalline and the observed polarization is around one half of this maximum value [3, 4].

If the β-phase of PVDF is in the form of a thin film, there will be no net polarization because of the random orientation of the crystallites. The application of a strong electric field, a process called poling (like the approach of Heaviside), is needed to confer the PVDF with an overall dipole moment. The poled PVDF can then be exploited in piezoelectric and pyroelectric applications.

An alternative method of conferring an overall polarization to an organic thin film is to incorporate this at the molecular level as the film is formed, for instance using self-assembly

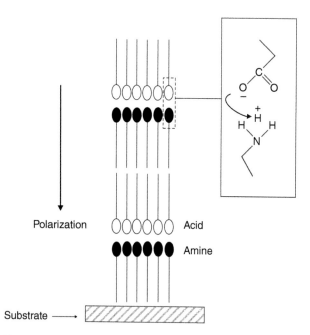

Figure 7.3 Schematic diagram of an organic superlattice formed by the alternate-layer Langmuir–Blodgett deposition of a long-chain fatty acid and a long-chain fatty amine. The orientations of the polar head groups are shown in the inset.

or the Langmuir–Blodgett (LB) technique. Figure 7.3 shows an example – an organic 'superlattice' – that can be realized in the laboratory by LB deposition [5]. The polar heads of the acid- and amine-terminated LB molecules are each associated with an electric dipole moment. In an alternating ABABABA…film, the dipoles of the two different molecules A and B will not cancel, and the multilayer will exhibit an overall polarization in the direction normal to the substrate plane. The figure also indicates that the deposition can result in a proton transfer from the acid to the amine head group, thereby creating a large dipole moment.

7.2.3 Piezoelectricity, Pyroelectricity, and Ferroelectricity

Polar materials, either in their bulk or thin film forms, may exhibit the properties of piezoelectricity, pyroelectricity, and ferroelectricity. Piezoelectricity is the ability of certain crystals (a well-known example is quartz) to generate a voltage in response to applied mechanical stress [2]. The piezoelectric effect is reversible; piezoelectric crystals subject to an externally applied voltage can change shape by a small amount. (There is a related physical phenomenon – electrostriction – by which all materials expand in an electric field; this is a small effect and not reversible.) Piezoelectricity can find many applications in electronics such as the generation and detection of sound at high (MHz) frequencies, and ultrafine manipulation of optical assemblies.

On a microscopic scale, piezoelectricity results from a nonuniform charge distribution within the unit cell of a crystal. When the crystal is mechanically deformed, the positive and negative charge centres are displaced by differing amounts. Although the overall crystal remains electrically neutral, the relative displacements of the centres of charge results in an electric polarization. Only certain classes of single crystals – those without a centre of symmetry – can be piezoelectric [2, 3]. In the form of a thin film, organic materials are either polycrystalline or amorphous (or a mixture of the two morphologies). These structures will possess a centre of symmetry and therefore will not be piezoelectric. However, the process of poling, described above, can break the symmetry in the thin film architecture, resulting in piezoelectric behaviour. The relationship between applied mechanical stress T (units $N\,m^{-2}$) and polarization can be written:

$$P = gT \tag{7.6}$$

where g is the piezoelectric coefficient (units $C\,N^{-1}$). To be strictly correct, Eq. (7.6) is a tensor equation, as the polarization is a vector quantity and stress is itself a second-rank tensor [2]; consequently, the piezoelectric third-rank tensor will form an array of $3 \times 3 \times 3$ coefficients. As a result, an applied stress in one direction in a piezoelectric crystal can give rise to a polarization in other crystal directions. For a thin film, there are two main situations to consider: the film can be mechanically deformed by applying a compressive or tensile stress orthogonal to the film plane; or the film can be subjected to bending. In the converse piezoelectric effect, the application of an electric field will produce a strain of the crystal S (dimensionless units):

$$S = g\mathcal{E} \tag{7.7}$$

The g coefficients in Eqs. (7.6) and (7.7) are the same (the units of electric field can be expressed as of $V\,m^{-1}$ or $N\,C^{-1}$).

A subset of piezoelectric crystals exhibits the property of pyroelectricity. If a pyroelectric material is heated, a change in the material polarization is produced. This can be measured as a voltage developed across the crystal. The pyroelectric coefficient p (units $C\,m^{-2}\,K^{-1}$) is given by the rate of change of polarization with temperature:

$$p = \frac{dP}{dT} \tag{7.8}$$

Very small temperature changes ($\sim 10^{-6}\,°C$) can give rise to voltages that are easily measurable (Problem 7.1). Because pyroelectric materials are also piezoelectric, they may develop an additional polarization when heated due to temperature-induced stresses. This will depend on if and how the material is mechanically constrained, for example in the form a thin film of material on a substrate with a different thermal expansion coefficient. The phenomenon is referred to as a secondary pyroelectric effect.

The polar multilayer structure depicted in Figure 7.3 will exhibit pyroelectric behaviour [5]. If the temperature of the thin film is varied, then the polarization across the film will change. If metallic contacts are provided above and below this alternate-layer LB film, the magnitude of the current I measured perpendicular to the substrate plane is:

$$I = pA\frac{dT}{dt} \tag{7.9}$$

where A is the sample area and dT/dt is the rate of temperature change. This equation can be derived easily from the definition of p given by Eq. (7.8). As shown in Figure 7.4, a current will only be produced while the temperature is varied. A positive current is measured on heating; if the multilayer film is cooled, the direction of current is reversed. The pyroelectric coefficients of such LB films are of the order 1–$10\,\mu C\,m^{-2}\,K^{-1}$. These are considerably less than observed in inorganic single crystals and ceramics, and slightly less than found for organic materials produced in other forms. For example, PVDF possesses a pyroelectric coefficient of $40\,\mu C\,m^{-2}\,K^{-1}$. However, the LB film structure has the advantage that it is relatively easy to fabricate – PVDF must be poled to produce a noncentrosymmetric structure.

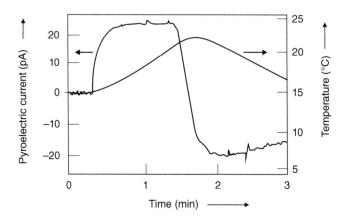

Figure 7.4 Time variation of temperature and pyroelectric current for a 99-layer acid/amine LB film. The arrows show which axis should be consulted for each curve. Source: Jones et al. [5].

Table 7.2 Piezoelectric and pyroelectric properties of polymers.

Polymer	Mer unit	T_g (°C)	T_m (°C)	ε'_r	g (pC N^{-1})	p (μC m^{-2} K^{-1})
PVC	(structure: —C—C— with H,H / H,Cl)	83	212	3.5	0.7	1
PVDF	(structure: —C—C— with F,H / F,H)	−35	175	12	28	40
Nylon-11	(structure: —N—C—(CH$_2$)$_{10}$ with H, O)	68	195	3.7	0.3	5

g = piezoelectric coefficient; p = pyroelectric coefficient; ε'_r = real part of relative permittivity; T_g = glass transition temperature; T_m = melting point [3, 4, 6]. The mer (repeat) groups of the polymers are given.

Table 7.2 compares the piezoelectric and pyroelectric properties of some polymeric compounds (poly(vinyl chloride) (PVC), PVDF, and nylon-11) along with data on their glass transition temperatures T_g (the temperature at which segments of the polymer chains begin to move) and melting points T_m [3, 4, 6]. The values of the g and p coefficients of organic materials are generally inferior to those of their inorganic counterparts. However, organic (particularly polymeric) sensors and actuators are lightweight, tough, readily manufactured into large areas, and can be cut and formed into complex shapes.

Ferroelectric materials have the additional property that the polarization can be reversed by changing the direction of the applied field. This effect disappears above a certain temperature, the Curie temperature, T_C (NB, for many polymeric materials T_C is greater than the melting point). PVDF is an example of an organic ferroelectric polymer. The material can be taken through a

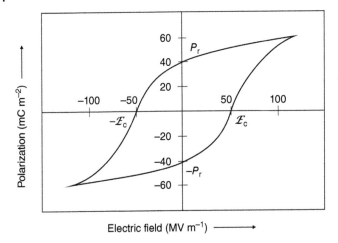

Figure 7.5 Typical polarization versus electric field hysteresis loop for PVDF. P_r = remanent polarization; $\mathcal{E}_c$ = coercive field.

hysteresis loop, as shown in Figure 7.5, which is in the form of a plot of the polarization as a function of the applied electric field [4]. The applied electric field required to reduce the polarization to zero, $\mathcal{E}_c$ is called the coercive field while the polarization retained by the material with no applied field, P_r, is the remanent or residual polarization.

Copolymers of vinylidene difluoride (VDF) with trifluoroethylene (TrFE) and tetrafluoroethylene (TFE) have also been shown to exhibit strong piezoelectric, pyroelectric, and ferroelectric effects. An attractive morphological feature of the comonomers is that they force the polymer into an all-*trans* molecular conformation that has a polar crystalline phase. This eliminates the need for mechanical stretching to yield a polar phase. Poly(VDF-TrFE) crystallizes to a much greater extent than PVDF (up to 90% crystalline) yielding a higher remanent polarization, lower coercive field, and much sharper hysteresis loops. TrFE also extends the operational temperature by about 20° C, to close to 100 °C. Conversely, copolymers with TFE have been shown to exhibit a lower degree of crystallinity and a suppressed melting temperature as compared to the PVDF homopolymer.

Piezoelectricity and pyroelectricity are well known throughout the natural world. For example, piezoelectric effects were found in keratin (a fibrous protein found in hair, skin and nails) in 1941 [6]. Subsequently, the phenomenon has been observed in a wide range of other biopolymers including collagen, polypeptides such as polymethylglutamate and polybenzyl-L-glutamate, oriented films of DNA, poly-lactic acid, and chitin. The piezoelectric constants of biopolymers are small relative to synthetic polymers (Table 7.2) ranging in value from 0.01 pC N^{-1} for DNA to 2.5 pC N^{-1} for collagen. The physiological significance of piezoelectricity in many biopolymers is not well understood, but it is believed that such electromechanical phenomena may have a distinct role in biochemical processes. For example, it is known that electric polarization in bone influences bone growth. Most biological membranes are electrically polarized with the negative inside the cell (Section 12.3.4). Experiments with alternating asymmetric LB layers of phospholipids, one of the main constituents of cell membranes, reveal modest pyroelectric activity [7], but, again, it is unclear if this is physiologically important.

The bistable polarization of ferroelectrics makes these candidates for binary memory applications in a similar way that the bistable magnetization of ferromagnetics is exploited. The memory is non-volatile and does not require a holding voltage. This is discussed further in Chapter 9. It should be noted that many organic ferroelectric materials possess high second-order nonlinear optical properties and can be used for second-harmonic generation and electro-optic switching. Ferroelectricity can also be exploited in molecules of liquid crystals (LCs) to increase the switching times in liquid crystal displays (LCDs) [2].

7.3 Conductivity at High Frequencies

Direct current and high-frequency phenomena are essentially separate and, for the most part, independent quantities. However, at low frequencies, both can contribute to a measured current. The frequency-dependent conductivity $\sigma(\omega)$ of a sample may be expressed in the form:

$$\sigma(\omega) = \sigma_{DC} + \sigma_{AC} \tag{7.10}$$

where σ_{DC} is the DC conductivity, discussed in Chapter 6, and σ_{AC} is the AC component. For highly insulating materials, the AC conductivity is largely determined by their permittivity. This and other phenomena are explored in the following sections. But first, an important electrical feature of all dielectric media is reviewed.

7.3.1 Displacement Current

A conduction current flowing in a wire is associated with a magnetic field. According to Ampère's law, the line integral of the magnetic field around a loop surrounding the wire is equal to the current enclosed. This situation holds for both DC and AC currents. However, if a parallel plate capacitor is inserted in series with the wire, the situation changes. Assuming that there is a vacuum between the plates of the capacitor, then no conventional current can flow between its plates. If a step voltage is applied across the ends of the wire, a conduction current will flow into one plate of the capacitor and out of the other plate, while the capacitor charges (Problem 7.2). To account for the continuity of the magnetic effect (a magnetic field will be present both around the wire and inside the capacitor plates), Maxwell introduced the concept of a different type of current between the capacitor plates. He called this the displacement current, which is proportional to the rate of change of the electric field. The size of this displacement current between the plates of a capacitor being charged and discharged is equal to the magnitude of the conduction current in the wires leading to and from the capacitor. Maxwell's fourth equation, which is a modified form of Ampère's law, includes both types of current and, in vector form, is:

$$\nabla \times \mathbf{B} = \text{curl } \mathbf{B} = \mu_0 \mathbf{J} + \mu_0 \varepsilon_0 \frac{\partial \mathbf{\mathscr{E}}}{\partial t} \tag{7.11}$$

where $\mathbf{B}$ is the magnetic field vector (units T), $\mathbf{J}$ is the conduction current density in the wire. The second term on the right-hand side of Eq. (7.11) is the displacement current.

   If a dielectric material is now placed between the plates of the capacitor, the material will become polarized as it charges. Equation (7.11) becomes:

$$\text{curl } \mathbf{B} = \mu_0 \mathbf{J} + \mu_0 \varepsilon_0 \frac{\partial}{\partial t} \left(\frac{\mathbf{P}}{\varepsilon_0} + \mathbf{\mathscr{E}} \right) = \mu_0 \left(\mathbf{J} + \frac{\partial \mathbf{D}}{\partial t} \right) \tag{7.12}$$

where the vector $\mathbf{D}$ ($=\varepsilon_0 \mathbf{\mathscr{E}} + \mathbf{P}$) is known as the electric displacement (same units as polarization, $C\,m^{-2}$).

7.3.2 Frequency-Dependent Permittivity

The presence of a dielectric material between the plates of a parallel plate capacitor increases its capacity for storing charge. This is illustrated in Figure 7.6, showing a capacitor connected to a

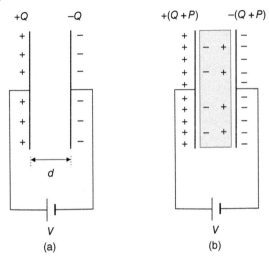

+Q   −Q   +(Q + P)   −(Q + P)

Figure 7.6   Charges on the plates of a parallel-plate capacitor with (a) a vacuum between the plates and (b) a dielectric between the plates. V = voltage applied by battery. Q = charge per unit area to maintain electric field between plates. P = polarization due to dielectric. d = distance between plates.

V   V
(a)   (b)

battery. With a vacuum between the plates (Figure 7.6a), the capacitor will become charged with a charge magnitude Q per unit area stored on its plates:

$$Q = \varepsilon_0 \mathcal{E} \tag{7.13}$$

If a dielectric is now inserted between the plates (Figure 7.6b), this will become polarized and additional charge must flow from the battery to compensate for the bound surface charge, and to maintain the same electric field between the plates ($\mathcal{E} = V/d$). The ratio of the new charge stored to that for the vacuum capacitor defines the relative permittivity, ε_r:

$$\varepsilon_r = \frac{Q + P}{Q} \tag{7.14}$$

Substituting from Eq. (7.13):

$$\varepsilon_r = \frac{(\varepsilon_0 \mathcal{E} + P)}{\varepsilon_0 \mathcal{E}} = 1 + \frac{P}{\varepsilon_0 \mathcal{E}} = 1 + \chi \tag{7.15}$$

where χ is the electric susceptibility of the material. It is also evident that the electric displacement and electric field are related:

$$D = \varepsilon_r \varepsilon_0 \mathcal{E} \tag{7.16}$$

The flux of electric displacement begins and ends on free charges and is otherwise continuous, even at the interface between two media. In contrast, electric field is discontinuous at an interface between two different materials because of the different degrees of polarization (Problem 7.3). In the foregoing, Eqs. (7.13)–(7.16) should formally be written in vector format. Therefore, the permittivity and susceptibility will be tensor quantities (second rank tensors with 3 × 3 coefficients), which is important if the material is anisotropic (e.g. in the form of a single crystal).

Once charged, the surface charge on the capacitor plates, $Q(t)$, will decay exponentially with time:

$$Q(t) = Q(0) \exp(-t/\tau) \tag{7.17}$$

where $Q(0)$ is the initial charge and τ is a characteristic time constant. If the capacitor is connected to an external circuit, τ will be equal the RC (resistance × capacitance) time constant of the network. If the charged capacitor is isolated from any external connections, then the time constant will depend on the dielectric's permittivity and resistivity:

$$\tau = \varepsilon_r \varepsilon_0 \rho \tag{7.18}$$

In this case, τ is called the dielectric relaxation time and will determine the decay of excess charge within, as well as on the surface, of a material.

The relative permittivity ε_r of a material is a complex quantity:

$$\varepsilon_r = \varepsilon_r' - j\varepsilon_r'' \tag{7.19}$$

where ε_r' and ε_r'' are the real and imaginary parts, respectively; both parameters are frequency dependent. The real part of the permittivity is a measure of the ability of an electric field to polarize the medium while the imaginary part represents losses in the material as an AC field attempts to orient the electric dipoles within the material. For engineering applications of dielectrics (e.g. capacitors), it is required to minimize ε_r'' for a given ε_r'. In this respect, a useful dimensionless parameter is the loss tangent, $\tan \delta$:

$$\tan \delta = \frac{\varepsilon_r''}{\varepsilon_r'} \tag{7.20}$$

which is also frequency dependent. The loss tangent is a measure of the energy dissipated per cycle divided by the energy stored per cycle when an AC voltage is applied to a capacitor.

A material will exhibit several polarization mechanisms, dependent on its constitution. Each process will have a characteristic frequency dependence. Electronic polarization was introduced in Section 7.2.1. The orientation of permanent dipoles in an electric field is called dipolar (or orientational) polarization. Other processes are ionic polarization, due to the displacement of positive and negative ions, and interfacial polarization, resulting from the accumulation of charges at interfaces (surfaces, grain boundaries) within the material. These effects will produce a change in ε_r' and a peak in ε_r'' over a particular frequency range. In ionic solids, this corresponds to the infrared part of the electromagnetic spectrum where the energy stored in the induced dipoles is the same as the lattice vibrational frequency. At higher frequencies, the positive and negative ions in the solid will not be able to respond sufficiently fast to the applied AC electric field and the ionic contribution to the material's permittivity is lost.

Figure 7.7 is a generic diagram indicating the frequency dependence of the real and imaginary parts of the permittivity. Although the figure shows distinct peaks in ε_r'' and transitions in ε_r', in real materials, these features can be broader and overlap. For single crystals, the polarization effects also depend on the orientation of the electric field with respect to the crystallographic axes (as permittivity is a tensor quantity). At low frequencies, the interfacial (or space charge) features are very broad because there can be a number of different conduction mechanisms contributing to the charge accumulation. At high frequencies – usually microwave and beyond – the polarization processes that take place are undamped and are invariably observed as resonances.

The theory of dielectric behaviour, pioneered by Debye, is based on an exponential approach of the polarization to equilibrium, following the removal of an applied electric field, with a characteristic relaxation time τ [1]. This leads to the following expression for the complex permittivity:

$$\varepsilon_r = \varepsilon_r' - j\varepsilon_r'' = \varepsilon_{hf} + \frac{\varepsilon_{lf} - \varepsilon_{hf}}{1 - j\omega\tau} \tag{7.21}$$

where and ε_{lf} and ε_{hf} represent the low- and high-frequency values of the real parts of the relative permittivity, respectively. Separating Eq. (7.21) into real and imaginary parts:

$$\varepsilon_r' = \varepsilon_{lf} + \frac{\varepsilon_{lf} - \varepsilon_{hf}}{1 + \omega^2\tau^2} \tag{7.22}$$

$$\varepsilon_r'' = \frac{\left(\varepsilon_{lf} - \varepsilon_{hf}\right)\omega\tau}{1 + \omega^2\tau^2} \tag{7.23}$$

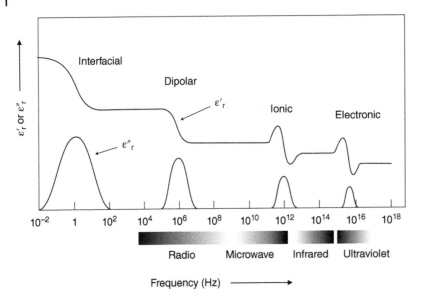

Figure 7.7 The frequency dependence of the real ε'_r and imaginary ε''_r parts of the relative permittivity in the presence of interfacial, dipolar, ionic, and electronic polarization processes.

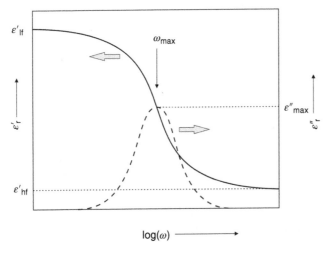

Figure 7.8 Debye dispersion curves. Predicted real ε'_r (left ordinate) and imaginary ε''_r (right ordinate) parts of the relative permittivity versus frequency ω. ε'_{lf} = low frequency real part of permittivity; ε'_{hf} = high frequency real part of permittivity. ω_{max} = frequency of maximum loss.

Equations (7.22) and (7.23) are called the Debye equations. These are plotted in Figure 7.8. The dielectric loss peak has a half width of 1.14 decades and its maximum value occurs at $\omega_{max} = 1/\tau$.

There is clearly some correspondence between Debye's predictions and the data shown in Figure 7.7 (certainly for the lower frequency polarization processes). Experimental results for many polar liquids show excellent agreement with the theoretical Debye curves, with relaxation times of the order of 10^{-11} seconds. However, relaxations in solids, including polymers, exhibit broader dispersion curves and lower loss maxima than those precited by the Debye model. This can be explained by a range of relaxation times, symmetrically distributed about τ, which may be present in solids. The complex permittivity is often better described by a modification to Eq. (7.21):

$$\varepsilon_r = \varepsilon_{hf} + \frac{\varepsilon_{lf} - \varepsilon_{hf}}{1 - (j\omega\tau)^\alpha} \tag{7.24}$$

where α is a parameter, between zero and unity, related to the distribution of relaxation times. Departures from the assumed exponential form of the approach to equilibrium may also be responsible for the breadth of the relaxation process in some materials [1]. Dielectric resonances observed for ionic and electronic polarization mechanisms (Figure 7.7) result from much higher frequencies of the electromagnetic radiation. The interaction of radiation and the electronic systems of atoms and molecules is usually described in terms of the refractive index of the medium rather than its relative permittivity [1, 3].

The temperature dependences of ε_r' and $\tan \delta$ for a material (and hence the stability of associated capacitors) will depend on the dominant polarization mechansism. For example, polar polymers have permanent dipole groups attached to the polymer backbone chains, as in poly(ethylene terephathate) (PET). At room temperature, these groups cannot respond easily to an AC field and the dipolar contribution to ε_r' is small. As the glass transition (softening) temperature of the polymer is approached, the dipolar side groups and polarized chains are able to respond. Consequently, both ε_r' and $\tan \delta$ increase with temperature. In contrast, for nonpolar polymers, such as polypropylene, the polarization is mainly due to electronic polarization, and the materials do not reveal marked changes in their dielectric behaviour with temperature. Capacitors manufactured from these polymers are relatively stable (a disadvantage, however, is that such materials possess lower permittivity values).

7.3.3 AC Conductivity

The AC conductivity of a dielectric is related to its permittivity. The real part of permittivity is the value that is used when calculating the capacitance of a parallel plate capacitor:

$$C = \frac{\varepsilon_r' \varepsilon_0 A}{d} \tag{7.25}$$

Application of an AC voltage of RMS (root-mean-square) amplitude V to this capacitor will result in an AC current, I, given by:

$$I = YV = \frac{V}{Z} = j\omega CV \tag{7.26}$$

where Y is called the electrical admittance (units S or Ω^{-1}) and Z is the electrical impedance. (units Ω). The presence of j in the above equation indicates a 90° phase shift between the applied voltage and the current (the current leads the voltage by 90° in a capacitor – only after charge has accumulated on the capacitor plates is the voltage established). Substituting Eq. (7.25) into Eq. (7.26) and recognizing that permittivity is the complex quantity of Eq. (7.19) gives:

$$I = V(j\omega C + G) \tag{7.27}$$

where

$$C = \frac{A\varepsilon_r' \varepsilon_0}{d} \quad \text{and} \quad G = \frac{\omega A\varepsilon_r'' \varepsilon_0}{d} \tag{7.28}$$

Therefore, the admittance of the dielectric medium is a parallel combination of an ideal, lossless capacitor C, with a relative permittivity ε_r' and a conductance G (or resistance $R = 1/G$) proportional to ε_r''. This is illustrated in Figure 7.9 (Problem 7.4).

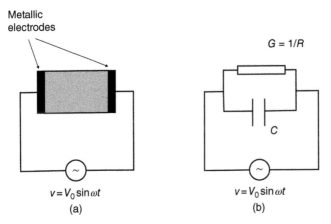

Metallic
electrodes

$G = 1/R$

$v = V_0 \sin \omega t$
(a)

$v = V_0 \sin \omega t$
(b)

Figure 7.9 (a) A dielectric simulated by a sinusoidal AC voltage of frequency ω. (b) Equivalent electrical circuit: C = capacitance, R = resistance, G = conductance.

From the above, it is expected that the measured AC conductivity, σ_{AC}, will increase with frequency:

$$\sigma_{AC} = \omega \varepsilon_r'' \varepsilon_0 \qquad (7.29)$$

If the sample is lossless over the frequency range under consideration (e.g. between the loss peaks in Figure 7.7), then ε_r'' will be constant and the AC conductivity will increase linearly with frequency.

Many inorganic and organic solids exhibit a simple power-law relationship of the form:

$$\sigma_{AC} \propto \omega^n \qquad (7.30)$$

where n is less than unity (usually in the range $0.7 < n < 1$). This relationship is often referred to as the Universal law for the response of dielectrics [7].

Figure 7.10 shows AC conductivity data for a thin film of an anthracene derivative [8]. The low frequency behaviour $G \propto \omega^{0.85}$ agrees with that expected from the discussion above. However, at frequencies above 10^4 Hz, the power dependence changes to approximately 1.8. The explanation is that at these high frequencies, contact resistances associated with the metallic electrodes become important (Problem 7.5). A full analysis of the equivalent circuit of the sample including these additional resistances reveals the $G \propto \omega^{2n}$ relationship. In the case $n = 1$, a square law dependence of conductance on frequency is seen at high frequencies [9].

Equation (7.10) recognizes that the DC mechanisms for charge transport in organic thin films may also contribute to the measured conductivity at high frequencies. These can include transport by carriers excited into the extended states near the conduction and valence band edges and hopping transport by carriers with energies near the Fermi level. The first process is expected to be associated with very short relaxation times, and a frequency dependence of the conductivity associated with free carriers is not observed (up to 10^7 Hz) [10]. Consequently, if the valence (and/or) conduction bands of a material contain many free carriers, the measured conductivity will be constant with frequency.

There have been several theories of hopping transport by electrons with energies near the Fermi level. These predict an increase in conductivity with frequency and a small dependence of AC conductivity with temperature. One model, developed by Mott and co-worker, gives the following

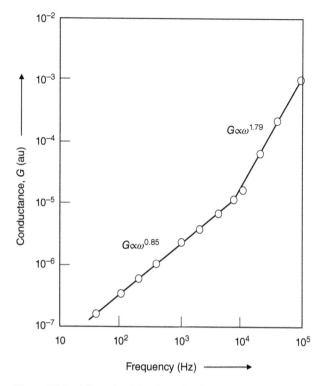

Figure 7.10 AC conductivity data showing the variation in conductance, G, with frequency, ω, for a thin film of an anthracene derivative. Source: Roberts et al. [8].

equation for the frequency dependence of AC conductivity [10]:

$$\sigma(\omega) = \frac{1}{3}\pi e^2 k_B T \left(S\left(E_F\right)\right)^2 \alpha^{-5} \omega \left(\ln\left(\frac{v_{ph}}{\omega}\right)\right)^4 \tag{7.31}$$

where $S(E_F)$ is the density of states at the Fermi energy, α is the wavefunction decay constant (inverse localization length) and v_{ph} is a characteristic phonon frequency (parameters encountered previously in Eqs. (6.16) and (6.17)). The frequency dependence of conductivity predicted by Eq. (7.31) can be written $\sigma(\omega) \propto \omega^n$, where n is a weak function of frequency if $\omega \ll v_{ph}$. The predicted hopping conductivity is therefore consistent with the empirical relationship given by Eq. (7.30).

Figure 7.11 shows conductivity versus frequency data for 150 nm thick film of copper(II) 2,9,16,23-tetra-*tert*-butyl-29H,31H-phthalocyanine (CuTTBPc) [11]. Plots are presented for different temperatures. At low frequency, the conductivity exhibits only a slight frequency dependence. This region corresponds to the DC conductivity of the sample. As the frequency increases, the conductivity conforms to the frequency dependence given by Eq. (7.30), with a value of n that varies from approximately unity at 303 K to 0.7 at 393 K.

As noted in Chapter 6, electrical experiments with assemblies of molecular (e.g. LB) films allow the thickness dependence of conductivity to be explored. Care needs to be taken to establish electrical contacts to the relatively fragile organic architectures and allowance must be made for interfacial layer(s) that might be present on the metallic electrodes. Figure 7.12a shows the AC conductivity for a series of seven-layer cadmium palmitate data set (A), cadmium stearate (B), and cadmium

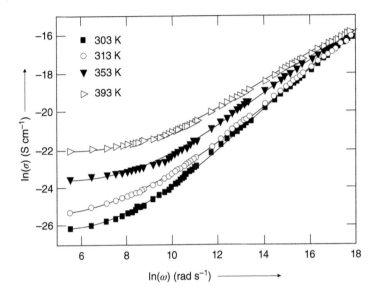

Figure 7.11 AC conductivity versus frequency for CuTTBPc films at different temperatures. Source: Darwish et al. [11].

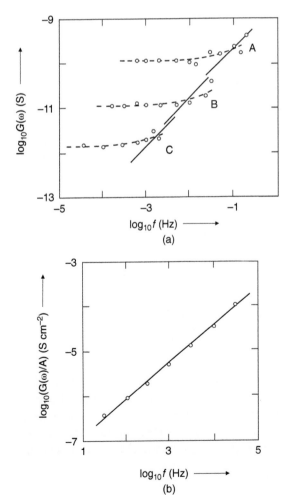

Figure 7.12 (a) Conductance, $G(\omega)$, versus frequency at 77 K for seven-layer LB films of C_{16} (A), C_{18} (B), and C_{20}(C) fatty acid assemblies. (b) An example of the almost linear variation of $G(\omega)/A$ at room temperature for a 7-layer C_{18} LB film. Source: Sugi et al. [12].

arachidate (C) LB films [12]. The number of carbon atoms in the hydrocarbon chain in these compounds is C_{16} (palmitate), C_{18} (stearate), and C_{20} (arachidate). As for the case of the CuTTBPc thin film discussed above, the conductance becomes independent of frequency as the measurement frequency is reduced. The following expression has been derived for the low frequency (and DC) conductance of the multilayer films [13]:

$$\sigma = e^2 S\left(E_F\right)(2\alpha)^{3/2} R^{5/2} \tau_0^{-1} \exp\left[-2\alpha R - \left(\frac{4\alpha}{\pi N\left(E_F\right) R k_B T}\right)^{1/2}\right] \tag{7.32}$$

where R is the hopping distance and τ_0 is a time constant. Eq. (7.32) was developed assuming the dominance of one-layer hops and starting from a particular mathematical expression (Miller-Abrahams) for the hopping relaxation time [10, 14, 15]. Thus, an electron at the Fermi level hops into one of the localized states distributed with a density $S(E_F)$ on the neighbouring interface. The dominant thickness dependence in Eq. (7.32) is due to the term $\exp(-2\alpha R)$, while the second term in the exponent is a minor correction. Experimental data (Figure 7.12a) reveal a good agreement with Eq. (7.32). The exponential dependence of the conductance upon the multilayer thickness is seen at both room temperature and 77 K. Furthermore, the predicted $T^{-1/2}$ law has also been verified [12].

Above 10 Hz, the data shown in Figure 7.12b reveal the conductivity versus frequency relationship given by Eq. (7.30), with an exponent n of approximately unity.

7.4 Impedance Spectroscopy

If the product $\omega\tau$ is eliminated between Eqs. (7.22) and (7.23), the following relationship between ε_r' and ε_r'' is obtained:

$$\left(\varepsilon_r' - \frac{\varepsilon_{lf} + \varepsilon_{hf}}{2}\right)^2 + \left(\varepsilon_r''\right)^2 = \left(\frac{\varepsilon_{lf} - \varepsilon_{hf}}{2}\right)^2 \tag{7.33}$$

This is the equation of a circle, centre $(\varepsilon_{lf} + \varepsilon_{hf})/2$, 0 radius $(\varepsilon_{lf} - \varepsilon_{hf})/2$. Therefore, a plot of ε_r' against ε_r'' should give a semicircle, as depicted in Figure 7.13. This diagram is known as a Cole-Cole plot. Again, a good agreement is obtained with data for polar liquids. The broader relaxations observed for many solids give rise to a semicircle that has its centre depressed below the absicca. This corresponds to a superposition of Debye-like relaxation processes, as noted earlier.

The measurement of the electrical impedance of a sample as a function of frequency is often referred to as impedance spectroscopy (or admittance spectroscopy in the case of a measurement of Y with frequency). To investigate the equivalent electrical circuit of the material under investigation, the measured data are plotted in the form of the real part of the impedance Z' versus the imaginary part Z''. Cole-Cole semicircles are revealed in the complex impedance plane.

Figure 7.14a,b show two simple resistor–capacitor circuits and their corresponding impedance spectra, Figure 7.14c,d. The arrows on the figures indicate the direction of increasing frequency. The circuit with two resistor–capacitor combinations (Figure 7.14b) will possess two time constants (or relaxation times), $R_1 C_1$ and $R_2 C_2$, corresponding to distinct semicircles in the impedance plot (Figure 7.14d), so long as the time constants are well-separated, i.e. $R_1 C_1 \gg R_2 C_2$. The high and low frequency limits can be used to determine some of the circuit components.

Experimental data are shown in Figure 7.15 for a layer of the conductive polymer poly (p-phenylene vinylene) (PPV) between aluminium and indium tin oxide (ITO) electrodes [16].

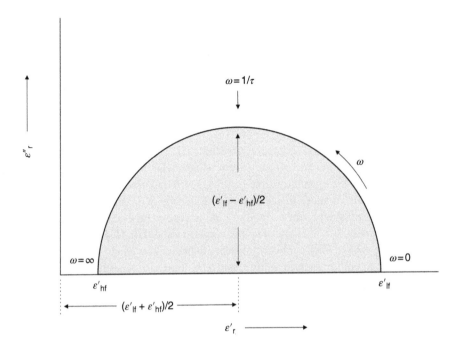

Figure 7.13 Cole-Cole plot. The imaginary part of the relative permittivity, ε_r'', plotted against the real part, ε_r'. Arrow on the semcircle indicates the direction of increasing frequency. See text for details.

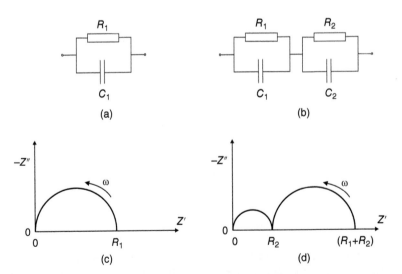

Figure 7.14 Diagrams (a) and (b) represent two simple resistor-capacitor circuits. Figures (c) and (d) are plots in the complex impedance plane of the imaginary part of the impedance, Z'', versus the real part, Z', for circuits (a) and (b), respectively. Arrows on the semcircles indicate the direction of increasing frequency.

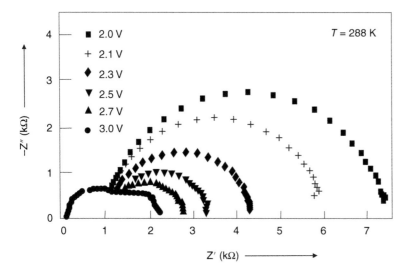

Figure 7.15 Imaginary Z'' versus real Z' parts of the complex impedance for an ITO/PPV/Al sandwich structure for various values of the forward bias. The frequency increases around the semicircles from right to left. The left semicircles are assigned to the polymer bulk and the right semicircles to a Schottky barrier. Source: Scherbel et al. [16].

This structure forms an organic light emitting device (Section 10.7) and will emit visible light when a negative voltage, or forward bias, is applied to the Al contact. The impedance spectra are shown for different values of forward bias voltages. Two semicircles are evident, suggesting that the equivalent circuit of the Al/PPV/ITO device is more complex than that of a lossy parallel plate capacitor, i.e. a parallel combination of a resistor and capacitor as depicted in Figure 7.14a. The small semicircle close to the origin in Figure 7.15 (corresponding to the high frequency region) is independent of voltage, whereas the larger right-hand semicircle (low frequency) shrinks as the applied bias is increased. The interpretation is that the device comprises of two electrical networks, a voltage-independent bulk region accounting for the left semicircle in the figure and a Schottky barrier (Section 4.4) at the Al/PPV interface. The resistance of the latter will depend on the applied voltage, accounting for the behaviour of the right semicircle in Figure 7.15. The equivalent circuit of the Al/PPV/ITO sandwich structure can therefore be modelled by the circuit shown in Figure 7.14b.

Conductivity measurements over a wide frequency range provide a very useful insight into the electrical networks associated with these and other devices. However, the interpretations are not all straightforward. In some cases, impedance spectroscopy reveals arcs of semicircles rather than full semicircles. As discussed above, this can be attributed to a distribution of relaxation times for the material under study.

7.5 AC Electrical Measurements

The method of determining the AC conductivity of a sample depends on the measurement frequency. The techniques generally fall into two groups: lumped circuit and distributed circuit methods [1]. The former is appropriate at lower frequencies (up to 10^8 Hz) and involves determining the electrical equivalent circuit of the sample at a given frequency. For example, the previous section

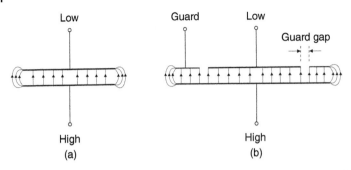

Figure 7.16 Pattern of electric field in (a) standard parallel plate capacitor (side view) and (b) a capacitor with a guard ring.

has shown how the components of the complex permittivity may be expressed as an equivalent parallel circuit of a resistor and a capacitor (Figure 7.9). At higher frequencies (infrared and above), the wavelength of the electromagnetic radiation becomes comparable to the sample dimensions and the lumped-circuit methods must be abandoned in favour of distributed-circuit methods in which the sample becomes the medium for propagation of electromagnetic waves. The following section will focus on lumped circuit methods, which are more appropriate to the subject of this book.

As with DC conductivity, meticulous attention to experimental detail is required for accurate measurements. For example, if the material under investigation is in the form of a parallel plate capacitor, then electric field effects at the periphery of the device (fringing fields) must be considered. Correction factors may be employed, but the difficulty is avoided using a guard ring electrode arrangement, as depicted in Figure 7.16. The guard is held at the same potential as the guarded electrode but is not connected to it. Provided that the sample and guard electrode extend beyond the guarded electrode by at least twice the sample thickness, and so long as the guard gap is small in comparison with the sample thickness, the field distribution in the guarded area is uniform between the electrodes. The guard electrode approach is also useful for accurate DC conductivity measurements. Furthermore, it may serve to eliminate the effect of surface conduction around the sample edges and prove advantageous in certain device configurations (e.g. transistors discussed in Section 8.3).

Stray capacitance (from leads and connectors) must also be avoided or compensated. This is particularly important when sample or device capacitances are very small (pFs). For this reason, the guard electrode is usually connected to zero potential and a completely shielded three-terminal cell used [1]. As the measurement frequency is increased, the inductances of the connecting leads can also introduce errors.

Common low frequency ($10-10^6$ Hz) measurement methods have traditionally relied on bridge circuits (e.g. a Wheatstone bridge) where the sample is compared to standard components [1, 17]. Such techniques have high sensitivity due to the null detection method (zero current) when the bridge is balanced. At higher frequencies (up to 10^8 Hz), resonance techniques may be used. For low frequencies (down to 10^{-4} Hz), step-response methods are employed – a voltage step is applied, and the subsequent current monitored over time.

7.5.1 Lock-in Amplifier

A popular system to measure capacitance and conductance simultaneously can be assembled from a two-channel lock-in amplifier [17]. This can readily be integrated into a customized digital data acquisition system. Lock-in amplifiers are used to detect and measure very small AC signals (a

Figure 7.17   Schematic diagram of a single-channel lock-in amplifier. The reference signal is generated internally by the lock-in equipment.

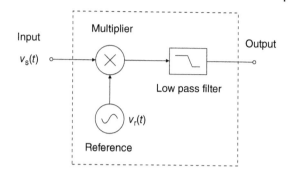

few nanovolts) and exploit a technique known as phase-sensitive detection (PSD) to extract the component of the signal at a specific reference frequency and phase. Noise signals, at frequencies other than the reference frequency, are rejected and do not affect the measurement.

A schematic diagram showing the principal features of a single-channel lock-in amplifier is shown in Figure 7.17. The arrangement is essentially that of a multiplier followed by a low-pass filter (an electrical circuit that removes high frequency signals). The signal from the sample is first multiplied by a reference derived internally from the lock-in amplifier. If the sample signal $v_s(t)$ and the reference signal $v_r(t)$ are simple sinusoids (square wave reference signals are often used in commercial equipment) with frequencies ω_1 and ω_2, respectively:

$$v_s(t) = V_{s0} \sin\left(\omega_1 t + \phi\right)$$
$$v_r(t) = V_{r0} \sin\left(\omega_2 t\right) \tag{7.34}$$

where ϕ is a phase angle. The output from the multiplier is then:

$$V_{s0}V_{r0} \sin\left(\omega_1 t + \phi\right) \sin\left(\omega_2 t\right) = \frac{V_{s0}V_{r0}}{2} \left(\cos\left(\left(\omega_1 - \omega_2\right) t + \phi\right) - \cos\left(\left(\omega_1 + \omega_2\right) t + \phi\right)\right) \tag{7.35}$$

The output product will consist of two AC signals, one at the difference frequency $(\omega_1 - \omega_2)$ and the other at the sum frequency $(\omega_1 + \omega_2)$. However, when the sample and reference frequencies are equal, the output from the multiplier becomes:

$$v_s(t)v_r(t) = \frac{V_{s0}V_{r0}}{2} \left(\cos(\phi) - \cos\left(\left(2\omega_1 t\right) + \phi\right)\right) \tag{7.36}$$

This takes the form of a sinusoidal output signal offset by a DC component. The former can be removed by the low-pass filter. The final output signal from the phase-sensitive detector is a DC voltage V_{psd} given by:

$$V_{psd} = \frac{V_{s0}V_{r0}}{2} \cos(\phi) \tag{7.37}$$

which is proportional to the amplitude of the original measurement signal.

For a two-channel lock-in amplifier, one channel multiplies the input signal with the reference signal, while the other performs the multiplication with a 90° phase-shifted reference signal. Following filtering, the two outputs from the system are directly proportional to the sample's capacitance and conductance (to C and G/ω) [17]. An advantage of using the lock-in amplifier is that the measurement system is relatively immune to noise and high accuracy can be obtained. This is useful, as it is important that only a small AC voltage signal is applied to the sample (of the order of tens of mV) to avoid nonlinear effects.

7.5.2 Scanning Microscopy

For nanomaterials and molecular device architectures, scanning microscopy techniques can be adopted to study dielectric properties [18]. Figure 7.18 shows the basic atomic force microscope (previously encountered in Section 4.5.1 to probe the DC electrical behaviour of molecular junctions). The equipment may be used to study a local electrical property – such as capacitance, electrical conductivity, or electrostatic potential – with a probe or tip placed very close to the sample. The small probe-sample separation (on the order of the instrument's resolution) makes it possible to take measurements over a small area. To acquire an image, the microscope raster-scans the probe over the sample while measuring the local property in question.

Scanning tunnelling microscopy (STM) measures the current flowing between the tip and the surface of the sample. Most commercial instruments use piezoelectric ceramic tubes to position the tip. The method may provide lateral and vertical resolutions of less than 0.3 and 0.02 nm, respectively. Furthermore, the electron energies are usually less than 3 eV, thus avoiding the degradation of an organic thin film, which can be a problem with other imaging methods. Whereas STM records the overlap of the local electron density of states between a tip and a surface (or its modulation by adsorbate molecules in the gap), atomic force microscopy (AFM) measures the interatomic forces between a cantilevered spring tip and the sample surface. The image contrast in AFM is achieved by probing the elastic response of the molecules to the force exerted by the scanning tip. The microscope, which essentially consists of a tip attached to a cantilever, can operate in three different modes: contact, noncontact, and tapping. At a relatively large distance from the surface, the force of attraction between the tip and the surface dominates, while the force of repulsion is the most significant at very small distances.

The instrument can generally measure the vertical deflection of the cantilever with picometre resolution. To achieve this, most AFMs use an optical lever, a device that achieves resolution comparable to an interferometer while remaining inexpensive and easy to use. It works by reflecting a laser beam off the cantilever. Angular deflection of the cantilever causes a twofold increase in the angular deflection of the laser beam. The reflected light then strikes a position-sensitive photodetector, consisting of two side-by-side photodiodes. The difference between the two photodiode signals indicates the position of the laser spot on the detector and thus the angular deflection of the cantilever. The optical lever arrangement greatly magnifies movement of the tip.

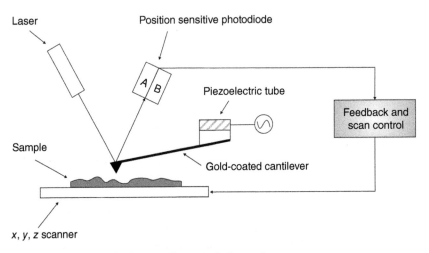

Figure 7.18 Schematic diagram of an atomic force microscope.

When a voltage V_0 is applied between a conducting AFM tip and a surface, there is an attractive force, F:

$$F = \frac{V_0^2}{2} \frac{dC}{dz} \tag{7.38}$$

where C and z are the capacitance and vertical distance between the tip and sample, respectively. The exact value of C will depend on the tip's geometry and the sample's conductivity. This is a long-range force, which may be used to detect local variations in surface potential in contact mode, or even in noncontact mode, where the tip is scanned at a fixed height above the surface with the feedback loop disabled. Any electric field due to potential variations in the surface will cause the cantilever to deflect as it scans over the surface. This technique is limited in resolution to approximately 100 nm due to the long-range nature of the electrostatic force. To overcome this problem, scanning Kelvin-probe microscopy (SKPM), which is also known as electrostatic force microscopy has been developed [18–20]. The methodology is to add an AC voltage to the tip. The AFM system operates in noncontact mode, and a conductive tip is oscillated at the first resonant frequency of the cantilever over the sample surface as it is scanned laterally. The topographic data are recorded by controlling the atomic force between the tip and sample. The long-range electrostatic force is detected with the AC tip voltage and using a lock-in amplifier. The AC voltage frequency is set at either the second resonant frequency of the cantilever or a low frequency (~20 kHz) that is far off the first resonant frequency, to avoid the interaction between the topographic and electrical signals. The electrostatic force is zero when the contact potential difference is completely compensated by a DC voltage applied to the tip. In this case, the contact potential difference is equal to the applied DC voltage. With the use of lock-in amplifier techniques, simultaneous topographic and potential maps of sample surfaces may be obtained.

A related AFM-based technique is scanning capacitance microscopy (SCM) [21]. This consists of a conductive metal probe tip and a highly sensitive capacitance sensor in addition to the normal AFM components. The probe electrode is positioned in contact or in close proximity to a sample's surface and scanned. SCM characterizes the surface of the sample using information obtained from the change in electrostatic capacitance between the surface and the probe. SCM has been widely used to obtain doping profiles of semiconductor surfaces.

7.6 Electrical Noise

Noise is a major limitation to the performance of any electronic device. This will become more important as devices continue to become smaller and more sophisticated. However, noise can also provide useful information concerning fundamental physical processes in electronic materials and devices [22, 23]. Noise is the fluctuation of voltage across or current through an electronic component, which is a result of random processes occurring during device operation. It is characterized by its power spectral density, $S(\omega)$, which is the frequency component of noise power per unit frequency band, or bandwidth. The noise in electronic devices is usually measured by biasing the current or voltage of the device. If a voltage is applied, the current noise can be determined, and vice versa.

Noise spectra are classified into several different types, according to their origins and frequency spectra. The main categories are thermal noise (also known as Johnson-Nyquist noise), shot noise, random telegraph (or burst) noise and $1/f$ (or flicker) noise. Thermal noise originates from the random fluctuations of charge carriers because of their thermal motion. This type of noise is often referred to as white noise as it has a constant power density spectrum. The current noise power spectral density S_I (units $A^2\,Hz^{-1}$) for thermal noise is frequency independent and for a resistance R can be expressed [23]:

$$S_I = 4k_B T/R \tag{7.39}$$

A similar expression is obtained for the voltage noise power spectral density S_V ($V^2\,Hz^{-1}$):

$$S_V = 4k_B TR \tag{7.40}$$

Thus, thermal noise depends only on the temperature and the resistance of the electrical component in which it is generated. To evaluate the total (RMS) noise voltage or current, Eqs. (7.39) or (7.40) are multiplied by the bandwidth. Noise in any electrical system can be reduced by limiting its bandwidth. This is one reason that the lock-in amplifier (Section 7.5.1) is immune to noise, as the bandwidth of this instrumentation system can be limited to a fraction of a Hz.

Shot noise is caused by the discrete nature of charge carriers. A current flow is required to observe shot noise. Consequently, it is observed in organic electronic devices such as transistors, photovoltaic cells, and organic light emitting structures. Shot noise also occurs in single-molecule junctions [23]. The current noise power spectrum density is:

$$S_I = 2eF|\bar{I}| \tag{7.41}$$

where $\bar{I}$ is the average current and F is called the Fano factor, which is determined by the interactions between the charge carriers. In the classical limit, $F = 1$. The shot noise spectral power is frequency independent, i.e. the noise is white.

Random telegraph noise (RTN) indicates an electronic noise in which discrete random transitions appear between two or more voltage/current levels (as if someone is randomly pressing the key of a telegraph machine). Trapping and detrapping events can provide a source for RTN. As with admittance spectroscopy, a discrete trap will manifest itself in the noise spectrum by a peak at a frequency related to the trap thermalization time τ. The noise density spectrum takes on the following Lorentzian form:

$$S_I(\omega) \propto \frac{1}{1 + (\omega\tau)^2} \tag{7.42}$$

In organic electronic devices, RTN has been observed in organic resistive memory devices, self-assembled monolayer junctions and single-molecule junctions due to contributions from the traps or conformation changes in the molecules [23].

Any electronic noise that exhibits a power spectrum density with a $1/f^n$ frequency dependence ($0.9 < n < 1.1$) is called $1/f$ noise – sometimes referred to as pink noise. This type of noise is dominant at low frequencies (typically <1 kHz) in most inorganic semiconductor devices, such as MOSFETs, light emitting diodes and solid-state memories. Because $1/f$ noise appears ubiquitously in electronic systems from the bulk to molecular size and from crystals to disordered systems, a well-defined process or mechanism explaining all the $1/f$ noises in different systems does not exist. Models based on the superposition of relaxation processes and distributed traps (a trap with a single relaxation time is expected to exhibit a noise power with the Lorentzian frequency dependence $1/f^2$, as in Eq. (7.42)) can provide some insight into the phenomenon in specific cases.

$1/f$ noise is usually the result of conductivity fluctuations (variations in the numbers of charge carriers and/or their mobility) within a material and so can reveal information about dynamic processes such as trapping and detrapping [22, 23]. Unlike, admittance spectroscopy where measurements are undertaken in the frequency domain, noise measurements are usually performed in the time domain; a Fourier transform links the two. Figure 7.19 shows the $1/f$ noise spectra

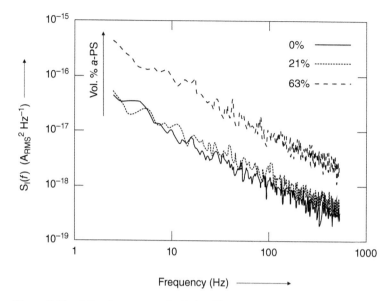

Figure 7.19 $1/f$ noise spectra of pristine P3HT and a-PS:P3HT blends. Different curves refer to volume percent a-PS. Source: Williams et al. [24]. Licenced under CC BY-3.0.

measured for a current of 3 mA for blends of the semiconductive polymer P3HT and the insulating polymer amorphous polystyrene (a-PS) [24]. The $1/f$ noise for the pristine P3HT and the sample containing 79% (by volume) of P3HT are of a similar magnitude. However, the noise level for the sample with 37% P3HT is significantly enhanced. This is attributed to the conductive polymer concentration approaching the percolation threshold. An increase in the n parameter as the polystyrene content in the blend was increased (approaching unity for pristine a-PS) was attributed to the sparseness of the charge conduction networks. This, in turn, suggested that a smaller proportion of the density of states was sampled by the charge carriers.

In organic electronics, $1/f$ noise has been found in almost all devices, and this depends on the gate voltage, luminescence, voltage bias regime, resistance, and morphology [23–26]. Although $1/f$ noise in disordered organic materials is hard to explain, analysing its dependence on various factors has provided meaningful insights into the charge transport mechanisms and microscopic structures of thin films.

Figure 7.20 shows noise data measured on a single-molecule junction [27]. In this case, the voltage noise power spectral density, S_V, is shown as a function of frequency. The measurements were recorded on self-assembled molecules of 1,4-benzenedithiol (BDT) in the gap created in a break junction created in a gold wire (Section 4.5.2). The noise spectrum was obtained by averaging 100 data sets using a spectrum analyser. In the case of a molecule-free gap junction (a reference structure), only $1/f$ and thermal components of noise were observed. However, with the molecule bridging the junction, a $1/f^2$ noise component was identified as the frequency was increased. The overall noise spectrum could therefore be described:

$$S_V(f) = \frac{A}{f} + \frac{B}{1 + (f/f_0)^2} + 4k_B TR \qquad (7.43)$$

where A, B, and f_0 are constants. The fit of this equation to the experimental data is as shown in Figure 7.20, together with guidelines indicating the $1/f$ and $1/f^2$ responses. The high frequency ($>10^5$ Hz) noise magnitude corresponds to the thermal noise. For the BDT molecule, f_0 was found

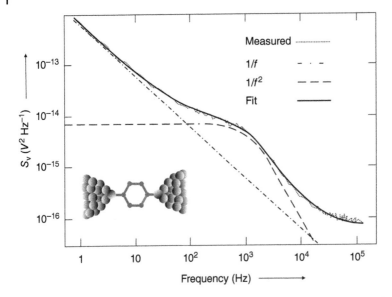

Figure 7.20 Voltage noise power spectral density of a single BDT molecule bridging two electrodes (inset cartoon). Source: Reprinted with permission from Xiang et al. [27].

to be 1.2 kHz. This parameter varied (0.07 kHz to 1.4 kHz) for different molecules. No correlation was found with the bonding strength at the contact/molecule interface (bond types Au-COOH, Au-S and Au-NH$_2$ were studied). However, a relationship between f_0 and the molecular weight, M, of the bridging molecule was evident ($f_0 \propto M^{-1/2}$).

The above work suggests that noise measurements can prove important in understanding the relationship between electrical conductivity and structure in organic materials, in both bulk and single-molecule forms. This area is likely to expand as organic electronics matures.

Problems

7.1   A pyroelectric heat sensor uses a thin film of the polymer PVDF. The radiation falling onto the detector is 'chopped' so that the radiation, intensity 100 µW cm^{-2}, is absorbed by the polymer for a time Δt seconds every τ seconds, where $\Delta t \ll \tau$. If $\Delta t = 0.1$ seconds, calculate the voltage developed across the sensor. Assume that all the incident radiation is absorbed by the thin film and that heat losses are negligible during Δt. For PVDF: density = 1.78 g cm^{-3}; specific heat capacity = 1.2 kJ kg^{-1} K^{-1}; relative permittivity = 12; and pyroelectric coefficient = 40 μC m^{-2} K^{-1}.

7.2   A parallel plate capacitor is fabricated with an ideal organic thin film insulator of relative permittivity 3.0. The capacitor dimensions are plate area = 0.5 cm^2 and distance between the plates = 1 μm. If the capacitor is connected to a 10 V battery in series with a 100 MΩ resistor, what is the initial displacement current that flows through the capacitor? What will be the value of this displacement current after 100 seconds? How are these values changed if the distance between the capacitor plates is reduced to 1 nm?

7.3 A parallel plate capacitor is made from two immiscible polymers, 1 and 2, with $\varepsilon_{1r} = 1.93$ and $\varepsilon_{2r} = 3.25$, respectively. The figure below shows two extreme arrangements, (a) and (b) in which the materials could segregate. For (b), assume that the two polymers take up equal areas of the capacitor plates. What is the electric field in each polymer for the two architectures for a voltage of 1 V applied between the capacitor plates? Calculate the overall capacitance for the two different structures. Plate area = 1 cm²; distance between metal plates = 2 μm.

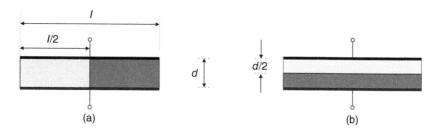

(a)   (b)

7.4 Values for the real part of permittivity and tan δ at 1 kHz for the polymer PET at 25 °C and 120 °C are given below.

Temperature	ε'_r	tan δ
25 °C	2.6	0.002
120 °C	2.8	0.01

Calculate the equivalent circuits at both temperatures for a 2.7 nF parallel capacitor (at 25 °C) that has a polymer thickness of 500 nm.

7.5 The AC conductivity of an organic thin film is being measured. Assume that the material behaves as an ideal dielectric. If there is a series contact resistance R_s with the film show that, under certain conditions, the measured conductance G_m will be given by:

$$G_m = \omega^2 C_f^2 R_s$$

where C_f is the capacitance of the thin film.

References

1 Blythe, T. and Bloor, D. (2005). *Electrical Properties of Polymers*. Cambridge: Cambridge University Press.

2 Petty, M.C. (2019). *Organic and Molecular Electronics: From Principles to Practice*, 2e. Chichester: Wiley.

3 Kasap, S.O. (2008). *Principles of Electrical Engineering Materials and Devices*, 3e. Boston, MA: McGraw-Hill.

4 Das-Gupta, D.K. (1995). Piezoelectric and pyroelectric materials. In: *An Introduction to Molecular Electronics* (eds. M.C. Petty, M.R. Bryce and D. Bloor), 47–71. London: Edward Arnold.

5 Jones, C.A., Petty, M.C., and Roberts, G.G. (1988). Langmuir-Blodgett films: a new class of pyro-electric materials. *IEEE Trans. Ultrason. Ferroelectr. Freq. Control* 35: 736–740.

6 Harrison, J.S. and Ounaies, Z. (2001). *Piezoelectric Polymers*. NASA ICASE Report No. 2001-43. Washington DC: NASA.

7 Jonscher, A.K. (1983). *Dielectric Relaxation in Solids*. London: Chelsea Dielectric Press.

8 Roberts, G.G., McGinnity, T.M., Barlow, W.A., and Vincett, P.S. (1980). AC and DC conduction in lightly substituted Langmuir films. *Thin Solid Films* 68: 223–232.

9 Street, R.A., Davies, G., and Yoffe, A.D. (1971). The square law dependence with frequency of the electrical conductivity of As_2Se_3. *J. Non-Cryst. Solids* 5: 276–278.

10 Mott, N.F. and Davis, E.A. (1979). *Electronic Processes in Non-crystalline Materials*, 2e. Oxford: Oxford University Press.

11 Darwish, A.A.A., Alharbi, S.R., Hawamdeh, M.M. et al. (2020). Dielectric properties and AC conductivity of organic films of copper (II) 2,9,16,23-tetra-*tert*-butyl-29*H*,31*H*-phthalocyanine. *J. Electron. Mater.* 49: 1787–1793.

12 Sugi, M., Fukui, T., and Iizima, S. (1979). Structure-dependent feature of electron transport in Langmuir multilayer assemblies. *Mol. Cryst. Liq. Cryst.* 50: 183–200.

13 Iizima, S. and Sugi, M. (1976). Electrical conduction in mixed Langmuir films. *Appl. Phys. Lett.* 28: 548–549.

14 Miller, A. and Abrahams, E. (1960). Impurity conduction at low concentrations. *Phys. Rev.* 120: 745–755.

15 Groves, C. (2017). Simulating charge transport in organic semiconductors and devices: a review. *Rep. Prog. Phys.* 80: 026502.

16 Scherbel, J., Nguyen, P.H., Paasch, G. et al. (1998). Temperature dependent broadband impedance spectroscopy on poly-(*p*-phenylene-vinylene) light-emitting diodes. *J. Appl. Phys.* 83: 5045–5055.

17 Stallinga, P. (2009). *Electrical Characterization of Organic Electronic Materials and Devices*. Chichester: Wiley.

18 Durkan, C. (2007). *Current at the Nanoscale*. London: Imperial College Press.

19 Sadewasser, S. and Glatzel, T. (eds.) (2018). *Kelvin Probe Force Microscopy: From Single Charge Detection to Device Characterization*. Cham: Springer.

20 Palermo, V., Palma, M., and Samori, P. (2006). Electronic characterization of organic thin films by Kelvin probe force microscopy. *Adv. Mater.* 18: 145–164.

21 Park, S.-E., Nguyen, N.V., Kopanski, J.J. et al. (2006). Comparison of scanning capacitance microscopy and scanning Kelvin probe microscopy in determining two-dimensional doping profiles of Si homostructures. *J. Vac. Sci. Technol., B* 24: 404–407.

22 Cuevas, J.C. and Scheer, E. (2010). *Molecular Electronics: An Introduction to Theory and Experiment*. Singapore: World Scientific.

23 Song, Y. and Lee, T. (2017). Electronic noise analyses on organic electronic devices. *J. Mater. Chem. C* 5: 7123–7141.

24 Williams, A.T., Farrar, P., Gallant, A.J. et al. (2014). Characterisation of charge conduction networks in poly(3-hexylthiophene)/polystyrene blends using noise spectroscopy. *J. Mater. Chem. C* 2: 1742–1748.

25 Kaku, K., Williams, A.T., Bendis, B.G., and Groves, C. (2015). Examining charge transport networks in organic bulk heterojunction photovoltaic diodes using 1/*f* noise spectroscopy. *J. Mater. Chem. C* 3: 6077–6085.

26 Djidjou, T.K., Bevans, D.A., Li, S., and Rogachev, A. (2014). Observation of shot noise in phosphorescent organic light-emitting diodes. *IEEE Trans. Electron Devices* 61: 3252–3257.

27 Xiang, D., Sydoruk, V., Vitusevich, S. et al. (2015). Noise characterization of metal-single molecule contacts. *Appl. Phys. Lett.* 106: 063702.

Further Reading

Anderson, J.C. (1964). *Dielectrics*. London: Chapman and Hall.

Barford, W. (2005). *Electronic and Optical Properties of Conjugated Polymers*. Oxford: Oxford University Press.

Bhushan, B. (ed.) (2013). *Scanning Probe Microscopy in Nanoscience and Nanotechnology*, NanoScience and Technology, vol. 3. Cham: Springer.

Harrop, P.J. (1972). *Dielectrics*. London: Butterworth.

Macdonald, J.R. (ed.) (1988). *Impedance Spectroscopy: Emphasising Solid Materials and Systems*. New York: Wiley.

Roberts, G.G., Apsley, N., and Munn, R.W. (1980). Temperature dependent electronic conduction in polymers. *Phys. Rep.* 60: 59–150.

Tessler, N., Preezant, Y., Rappaport, N., and Roichman, Y. (2009). Charge transport in disordered organic materials and its relevance to thin-film devices: a tutorial review. *Adv. Mater.* 21: 2741–2761.

8

Organic Field Effect Transistors

8.1 Introduction

This chapter reviews the physical principles of organic transistors with both micrometre and nanometre dimensions. The workings of all these devices share some similarities. However, a thin film transistor based on a molecular semiconducting material, such as pentacene, can become governed by different physical principles as the length of its channel is shortened to hundreds and even to tens of nanometres. Similarly, a transistor might operate with a single carbon nanotube or with a nanotube bundle. In the former case, the semiconductor can be treated as a ballistic conductor, while nanotube bundles, or ropes, might possess electrical characteristics comparable to conductive polymers. The chapter begins by looking at the more traditional organic thin film transistor (OTFT), the operation of which can be derived from the physics of inorganic semiconductor devices.

Metal/insulator/semiconductor field effect transistors (MISFETs) are three-terminal structures: a voltage applied to a metallic gate affects an electric current flowing between source and drain electrodes. The field effect has already been introduced in Chapter 5. Figure 8.1 shows how the MISFET (Figure 8.1a) is evolved from the metal/insulator/semiconductor (MIS) structure (Figure 8.1b). Application of a voltage to the gate electrode V_g alters the charge density at the semiconductor surface. The mode of operation of organic field effect transistors (FETs) is invariably in the accumulation region. Hence, the majority carrier current I_d flowing between the source electrode, held at a potential of $0\,V$ (where carriers enter the conductive channel provided by the accumulation charge) and the drain electrode, at $V = V_d$ (where carriers leave the channel), is changed by the gate voltage. This control feature allows the amplification of small AC signals or the device to be switched from an On state to an Off state, and back again. These two important operations, amplification and switching, are the basis of many electronic functions.

8.2 Physics of Operation

MISFETs using organic semiconductors are often referred to by the acronym OFET (organic field effect transistor). The following analysis of the electrical behaviour of the OFET is based on equations for conventional insulating gate field effect transistors, e.g. [1]. The latter operate in the region of strong inversion (Section 5.2). As noted above, OFETs usually operate in the accumulation regime, i.e. there is no depletion region that separates the thin surface accumulation charge (and electrically isolates it) from the semiconductor bulk (see Section 8.4.4). This important feature must therefore be taken into consideration in developing the theory for OFETs. A schematic

Electrical Processes in Organic Thin Film Devices: From Bulk Materials to Nanoscale Architectures, First Edition. Michael C. Petty.
© 2022 John Wiley & Sons Ltd. Published 2022 by John Wiley & Sons Ltd.
Companion Website: www.wiley.com/go/petty/organic_thin_film_devices

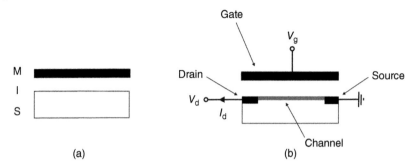

Figure 8.1 Contrast between (a) metal/insulator/semiconductor (MIS) structure and (b) metal/insulator/semiconductor field effect transistor (MISFET). V_g and V_d are the gate and drain voltages, respectively. The source electrode is at a potential of 0 V. I_d = drain current.

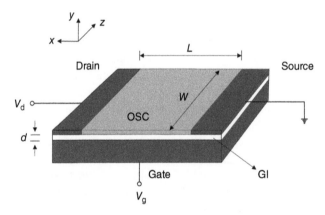

Figure 8.2 Schematic diagram of OFET device. Co-ordinate axes and parameters used in the text: OSC = organic semiconductor, GI = gate insulator, L = channel length, W = channel width, d = gate insulator thickness, V_g, V_d are gate and drain voltages, respectively.

diagram of a simple OFET, including its device parameters, is provided in Figure 8.2. Alternatively, vertical (rather than planar) architectures are considered in Section 8.3. The derivation below is based on the planar configuration and presumes that the transverse field (y-direction) is greater than the longitudinal field (x-direction). This is the gradual-channel approximation [1–5].

The total charge per unit area induced at the semiconductor/insulator interface on the application of a gate voltage V_g at a distance x from the source is given by:

$$Q_s(x) = C_i \left(\psi_s(x) - V_g \right) \tag{8.1}$$

where C_i is the gate capacitance per unit area ($= \varepsilon_i \varepsilon_0 / d$, where d is the insulator thickness) and $\psi_s(x)$ is the surface potential (Section 5.2) at distance x from the source electrode:

$$\psi_s(x) = V(x) + V_0 \tag{8.2}$$

where $V(x)$ is the voltage at a distance x from the source electrode (assumed to be at zero potential) and V_0 is a constant related to the device under consideration.

If it is assumed that the carrier mobility is constant in the accumulation layer, at any point the current $I(x)$ will be given by the product of the local free charge density $Q_s(x)$ (assuming there is no trapping), the local electric field $\mathcal{E}(x)$ ($\mathcal{E}_x = -dV(x)/dx$), the carrier mobility, and the device width W:

$$I_x(x) = -Q_s(x)\mu W \frac{dV(x)}{dx} \tag{8.3}$$

Substituting Eqs. (8.1) and (8.2) into Eq. (8.3) and integrating from the source ($x = 0$; $V = 0$) to the drain ($x = L$; $V = V_d$) gives:

$$\int_0^L I_x dx = W\mu C_i \int_0^{V_d} \left(V_g - V_0 - V(x) \right) \, dV_x \tag{8.4}$$

Hence:

$$I_d = \frac{W\mu C_i}{L} \left(\left(V_g - V_0 \right) V_d - \frac{V_d^2}{2} \right) \tag{8.5}$$

For MISFET devices operating in the inversion regime, the constant V_0 is the threshold voltage V_t – the voltage required for strong inversion to occur (Section 5.2). In the ideal accumulation OFET, there is no such requirement, and V_0 should be zero. However, for practical devices, a finite V_0 can arise from work function differences between the gate and the semiconductor, fixed charges in the insulator or surface traps located at the semiconductor/insulator interface. Despite the differences in operation of inversion and accumulation FETs, the parameter V_0 is frequently referred to as the threshold, or turn-on, voltage.

For low drain voltage, $V_d \ll V_g$, Eq. (8.5) reveals the channel acts as a resistance and the drain current increases linearly with V_d. A useful device parameter, called the transconductance of the OFET, g_m, can be defined:

$$g_m = \left(\frac{\partial I_d}{\partial V_g} \right)_{V_d=\text{constant}} = \frac{WC_i}{L} \mu V_d \tag{8.6}$$

This allows for the evaluation of the carrier mobility from the first derivative of the I_d versus V_g curve, known as the transistor transfer characteristic. This mobility term is often denoted μ_{FET} to distinguish it from the microscopic mobility defined by Eq. (2.23).

As the magnitude of V_d increases, I_d tends to saturate. This is the result of pinch-off of the accumulation layer. The effect is illustrated in Figure 8.3 for increasing values of drain voltage ($V_{d3} > V_{d2} > V_{d1}$). The combination of the voltages applied to the gate and drain will produce a wedge-shaped accumulation layer (Figure 8.3b). At pinch-off, the effective cross-sectional area of the conduction channel becomes zero at the drain (Figure 8.3c). The saturation region will occur when $Q_s(L) = 0$. From Eqs. (8.1) and (8.2), this is:

$$V(L) = V_d = V_g - V_0 \tag{8.7}$$

Inserting this into Eq. (8.5):

$$I_{d,sat} = \frac{WC_i}{2L} \mu \left(V_g - V_0 \right)^2 \tag{8.8}$$

The variation of the saturation drain current, $I_{d,sat}$, with V_g can also be used to calculate the semiconductor charge carrier mobility. However, in practical devices, the carrier mobility can depend on the gate bias (see Section 8.4.2 below). The mobility value obtained in this way may therefore differ from that calculated from the transfer characteristic (which is generally the preferred method to evaluate μ). Figures 8.4 and 8.5 show the current versus voltage characteristics for an ideal OFET

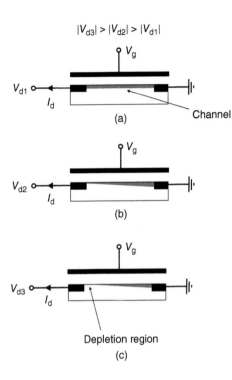

$|V_{d3}| > |V_{d2}| > |V_{d1}|$

(a)

Channel

(b)

(c)

Depletion region

Figure 8.3 Diagrams depicting the effect of applying an increasing drain voltage V_d ($|V_{d3}| > |V_{d2}| > |V_{d1}|$) to an OFET (at constant gate bias). (a) Linear regime – channel extends between the source and drain electrodes. (b) Pinch-off – cross-sectional area of channel becomes zero at the drain electrode. (c) Saturation region – carriers must cross the depletion layer to reach the drain electrode.

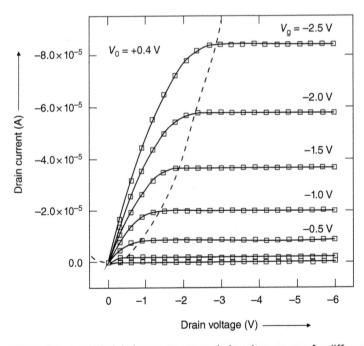

Figure 8.4 Modelled drain current versus drain voltage curves for different applied gate biases (output characteristics) for p-type OFET. The dashed line indicates where the channel of the transistor becomes pinched off as a result of increasing V_d. Source: Sandberg [6].

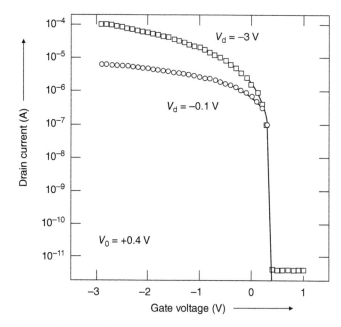

Figure 8.5 Modelled drain current versus gate voltage curves for two different drain voltages (transfer characteristics) for p-type OFET. V_d, V_0 are the drain and threshold/turn-on voltages, respectively. Source: Sandberg [6].

based on a p-type channel [6]. Figure 8.4 reveals the drain current versus drain voltage behaviour for different values of gate voltage. These curves are called the transistor output characteristics. For a particular gate bias, the current saturates with increasing drain voltage, as discussed above. The transition between the saturated and the linear region appears approximately where $V_d = V_g - V_0$ and is indicated in Figure 8.4 with a dashed line. Figure 8.5 shows the transfer characteristics of the OFET – the drain current as a function of the gate voltage – for fixed values of the drain voltage. The curves in Figures 8.4 and 8.5 were generated using SPICE (Simulation Program with Integrated Circuits Emphasis), a general-purpose analogue circuit simulator, which was developed in 1975 at the Electronics Research Laboratory of the University of California, Berkeley.

On inspection of the expression for the drain current in the saturation regime given by Eq. (8.8), it appears that I_d becomes zero when V_g is reduced to V_0. There is still some drain current conduction below threshold; this is known as the subthreshold conduction. In this regime, the drain current is due to carriers that have sufficient thermal energy to overcome the gate-voltage-controlled energy barrier near the source contact and diffuse, rather than drift, through the semiconductor to the drain contact. Above threshold, the densities depend linearly on the potential and drift currents exceed the diffusion currents.

The diffusion current (Section 2.7.2) will be proportional to the carrier density gradient in the channel. The latter will depend on the surface potential at the semiconductor surface at the source and drain electrodes. For large values of V_d, there will be a higher carrier concentration at the source; this will therefore control the channel diffusion current. For a p-type semiconductor, the surface hole concentration will depend exponentially on the surface potential ψ_s according to Eq. (5.6) (with ψ_s as a negative quantity in accumulation). The drain current will therefore be given by:

$$I_d = I_0 \exp\left(\frac{e\psi_p}{k_B T}\right) \tag{8.9}$$

where I_0 is a constant. The relationship between the gate voltage and surface potential will be determined by the potential divider network formed by the various capacitances in the device (insulator, depletion layer, interface traps) as indicated in Eq. (5.34). This can be written as follows:

$$\frac{d\psi_s}{dV_g} = \frac{1}{n} \tag{8.10}$$

where n is a constant, which is sometimes referred to as an ideality factor [4]. Hence:

$$I_d = I_0 \exp\left(\frac{eV_g}{nk_B T}\right) \tag{8.11}$$

If the OFET possesses a non-zero value of threshold voltage V_0, then V_g in the above equation is replaced by $(V_g - V_0)$. An important parameter that quantifies how sharply the transistor is turned off by the gate voltage is called the subthreshold swing S. This is defined as change in V_g required to change I_d by one order of magnitude; the units are Volts per decade. For an ideal device:

$$S = \frac{dV_g}{d\left(\log_{10} I_d\right)} = \ln 10 \frac{dV_g}{d\left(I_d\right)} = 2.3n \frac{k_B T}{e} \tag{8.12}$$

which has a minimum value ($n = 1$) of about 58 mV per decade at room temperature. For state-of-the-art MOSFETs, S is typically 70 mV per decade. This means that an input voltage variation of 70 mV will change the output I_d by an order of magnitude. Clearly, the smaller the value of this subthreshold slope, the better the transistor as a switch.

A further important property of a transistor is its On/Off ratio, which determines both the switching efficiency and leakage in the Off state. A simple definition of the On/Off ratio is the drain current in the saturation regime at a normal (maximum) gate voltage divided by the current at $V_g = 0$ V.

In the accumulation layer OFET, the channel is not isolated from the bulk (neutral) region by a depletion layer (as for inversion layer Si MOSFETs), and there will be an Ohmic current I_Ω that will always flow in parallel with the channel current I_d:

$$I_\Omega = GV_d \tag{8.13}$$

where G is the bulk conductance of the organic semiconductor, given by:

$$G = \frac{pe\mu Wt}{L} \tag{8.14}$$

where p is the density of free carriers (assumed to be holes) and t is the thickness of the semiconductor. The total measured source to drain current is:

$$I_{total} = I_d + I_\Omega \tag{8.15}$$

Optimizing an OFET will consist of first increasing the ratio I_d/I_Ω. At saturation, $V_d = V_g - V_0$. From Eqs. (8.13)–(8.15):

$$\frac{I_{On}}{I_{Off}} = \frac{I_d}{I_\Omega} = \frac{C_i V_d}{2pet} \tag{8.16}$$

To maximize the On/Off ratio, it is therefore important to decrease both the free carrier density and the semiconductor thickness (Problem 8.1). Of course, in practical devices, there may be sources of Off current in addition to I_Ω.

In Section 5.2, it was noted that, to a first approximation, the thickness of the inversion and accumulation layers in an MIS structure is $\pi L_D/\sqrt{2}$, where L_D is the Debye, or screening length (Eq. (3.18)). Typical values of this 'effective' thickness range from 0.1 to 1 nm. For this reason, it is often stated that, for practical purposes, all the charge in the channel resides within the layer of molecules adjacent to the semiconductor–insulator interface. This allows approximations to be made in OFET calculations.

The frequency response of an OFET is important in many device applications that require a high switching speed. A key limitation is the transit time t_t of a carrier from the source to the drain. This will depend on the channel length and the carrier drift velocity:

$$t_t = \frac{L}{v_d} = \frac{L^2}{\mu V_d} \tag{8.17}$$

The reciprocal of the transit time will indicate the maximum transistor operating frequency. From an AC perspective, an alternating gate voltage will charge and discharge the gate capacitance. This flow of charge will take the form of a displacement current (Section 7.3.1):

$$I_g = j\omega C_g V_g \tag{8.18}$$

where I_g is the AC gate current (90° out-of-phase with V_g) and C_g is the absolute value of gate capacitance (rather than the capacitance per unit area). The AC gate current therefore increases linearly with frequency. In contrast, the drain current I_d is frequency independent. This is related to the gate voltage via the OFET transconductance (Eq. (8.6)). The ratio (I_d/I_g) represents the current gain. When this falls below unity, the transistor can no longer be operated in a useful manner. The frequency at which this occurs is defined as the cut-off frequency f_T (Problem 8.2):

$$\frac{|I_d|}{|I_g|} = \frac{g_m V_g}{\omega C_g V_g}$$

$$f_T = \frac{g_m}{2\pi C_g} \tag{8.19}$$

The transconductance and gate capacitance are important quantities in determining the dynamic operation of the OFET. In real devices, the total gate capacitance will include the gate-to-source and gate-to-drain capacitances as well as the gate dielectric capacitance (due to the Miller effect – an effective amplification of the capacitance [4]). Moreover, additional parasitic capacitances may be associated with particular OFET architectures and connecting leads. The cut-off frequency is therefore likely to be reduced from that given by Eq. (8.19). The measured value of this parameter is often referred to as the extrinsic cut-off frequency (in contrast to the intrinsic – or theoretical maximum – value).

8.3 Transistor Fabrication

For efficient transistor operation, charge must be injected easily from the source electrode into the organic semiconductor and the carrier mobility should be high enough to allow a useful source-drain current to flow. The organic semiconductor and other materials with which it is in contact must also withstand the operating conditions without thermal, electrochemical, or photochemical degradation.

There are four main types of planar OTFT architecture, depending on the relative positions of the three electrodes to the semiconductor and insulating layers. These are contrasted in Figure 8.6.

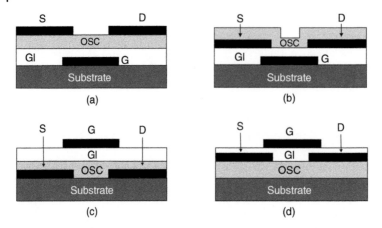

Figure 8.6 Schematic diagrams for different OFET architectures. (a) Bottom gate, top contact; (b) bottom gate, bottom contact; (c) top gate, bottom contact; and (d) top gate, top contact. S = source, D = drain, G = gate, OSC = organic semiconductor, GI = gate insulator.

In the bottom gate structures, depicted in (a) and (b), the semiconductor layer is formed after the metal gate and gate insulator. This has the advantage of the organic thin film being deposited in the final stage of device fabrication; it will therefore not be exposed to subsequent thermal or chemical processing steps. However, exposure of the top surface to the environment can lead to a short device lifetime (this can be exploited for chemical sensing, Section 8.8). The top gate structures, shown in (c) and (d) can reduce or eliminate this problem. It is important to note that the electronic nature of the metal/organic semiconductor junction can depend on whether the semiconductor is deposited onto the metal or vice-versa [7]. Top source and drain contact transistors generally exhibit the lowest contact resistances. This probably results from the increased metal/semiconductor contact area, as the metal (formed by an energetic process such as thermal evaporation) can penetrate into the underlying film with a resulting intermixing of materials. However, in the case of single-crystal devices, evaporation of the gate metal onto the semiconductor may affect the crystalline order of the active surface and generate defects. A major contribution to contact resistance in the top contact architectures is the access resistance. This is a consequence of the requirement that charge carriers must pass between the source contact to the accumulation layer (channel) and then back to the drain contact to be extracted. To minimize access resistance, the thickness of the organic semiconductor should not be too large.

Several OFET architectures have been proposed to reduce the process complexity, in particular the patterning of the semiconductor. One approach is to use a Corbino gate design [8]. This transistor configuration exploits a nested ring for the source and drain. The outer electrode, the source, acts as a guard ring (Figure 7.16) and is at the same potential as other sources in the vicinity, eliminating leakage through the unpatterned semiconductor layer.

Alternative OFET structures use vertically spaced source and drain electrodes, instead of the common in-plane arrangement [8–12]. These so-called vertical organic transistors (VOFETs) can operate at relatively high powers and high frequencies because of the short distance between the electrodes. Early VOFETs were step-edge devices – the architecture is composed of vertically displaced source and drain electrodes with an insulator as a spacer. Two examples are shown in Figure 8.7: (a) is a top gate configuration [9] while (b) depicts a bottom gate arrangement [10]. In both cases, the semiconductor/gate insulator interface defines the channel of transport, which

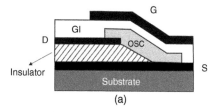 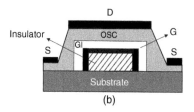

Figure 8.7 Designs for vertical OFETs (VOFETs). (a) Top gate configuration [9]. (b) Bottom gate configuration [10]. S = source, D = drain, G = gate, OSC = organic semiconductor, GI = gate insulator. An additional insulator is incorporated as part of the two designs. Note that the second source electrode in (b) is associated with an adjacent VOFET in an array. Source: (a) Modified from Stutzmann et al. [9], (b) Modified from Uno et al. [10].

is formed in the first few molecular layers of the organic semiconductor adjacent to the gate dielectric interface. Inspection of Figure 8.7 reveals that both architectures possess only a partly vertical channel. These devices should therefore be considered as pseudo-vertical FETs. The charge transport in these devices comprises of both lateral and vertical components. The relative contributions will be determined by the lateral overlap of the electrodes and the anisotropy of the charge carrier mobility in the organic semiconductor [12]. Contact resistance effects can be significant, and the device current can be controlled by injection from the source electrode (Section 8.4.1).

The organic permeable base transistor (OPBT) and organic static induction transistor (OSIT, also known the space-charge-limited transistor) have also been developed [8, 12]. These offer almost pure vertical transport and are based on different physical principles to the planar OFET and the step-edge VOFETs described above. The OPBT resembles a vacuum (valve) triode where a thin (permeable to charge carriers) electrode, known as the base, is sandwiched between a charge-injecting electrode, the emitter, and a charge-extraction electrode, the collector. The base electrode controls the current flowing between the emitter and collector, creating the transistor operation. In the OSIT, insulating layers are located above and below the permeable base electrode to isolate it from the other electrodes.

The semiconducting and insulating layers in organic transistors can be produced using a variety of different thin film techniques, including atomic layer deposition, spin-coating, thermal evaporation, and Langmuir–Blodgett assembly, together with processing methods such as nanoimprint lithography [13]. The use of additive printing techniques is particularly appealing [14]. Figure 8.8 illustrates a series of processing steps that can be used to fabricate an OFET using the method of inkjet printing. In the first stage, a substrate (either rigid or flexible) is patterned. A conductive compound is inkjet printed to form the transistor source and drain electrodes. A layer of organic semiconductor is then printed over the channel defined by the source and drain, and a layer of dielectric is deposited by spin-coating. Finally, a conducting gate electrode is printed. Gravure printing, where organic semiconducting inks are dispensed onto large area substrates in a roll-to-roll process, is a further method for the large-scale production of OFETs.

8.4 Practical Device Behaviour

The operating characteristics of organic transistors have advanced significantly since they were first demonstrated in the 1980s [5]. This has been brought about by both improvements in material synthesis and purity, and with the realization that organic compounds can be degraded by water and/or oxygen in the atmosphere. Consequently, the processing of devices and their subsequent

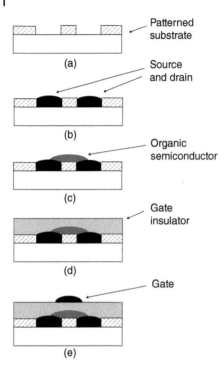

(a) — Patterned substrate

(b) — Source and drain

(c) — Organic semiconductor

(d) — Gate insulator

(e) — Gate

Figure 8.8 Processing steps to form an OFET using the technique of inkjet printing. (a) Patterned substrate. (b) Organic source and drain electrodes printed. (c) Organic semiconducting layer printed. (d) Gate insulator deposited by spin-coating. (e) Organic gate electrode printed.

storage usually take place in an inert ambient. The lifetimes of organic devices can be considerably extended by encapsulation. Even so, OFETs do not always exhibit the idealized output and transfer curves depicted in Figures 8.4 and 8.5.

Figure 8.9 shows the output characteristics for an OFET based on thermally evaporated pentacene [15]. This structure (channel length = 50 μm, width = 500 μm) was fabricated using a cross-linked polymer gate insulator (33 nm thickness) deposited by spin-coating. The electrical characteristics show features that may typically be observed for organic transistors produced in the laboratory. First, some hysteresis is evident between the forward and reverse voltage scans of some of the curves. This instability reflects dispersive transport and charge trapping in the organic semiconductor or at the interface (discussion in Section 8.4.2) [5]. The slight curvature of the output curves near the origin ('crowding' of the current versus voltage plots) is indicative of a nonlinear contact resistance (see following section). Figure 8.10 shows the transfer characteristic and a plot of $I_d^{0.5}$ versus V_g for the same device; the measurements were made in the saturation regime, with $V_d = -8\,$V. The square root of drain current plot is reasonably linear for high values of V_g and provides an average mobility figure of approximately 1.1 cm^2 V^{-1} s^{-1}. However, close examination of this curve reveals a slight decrease in its slope as the magnitude of the gate bias increases. A plausible explanation is that at lower gate voltages, the accumulation region is not as tightly confined to the interface and extends further into the bulk than at high gate voltages [5]. Alternatively, this phenomenon may be attributed to contact resistance effects. The pentacene FET has a subthreshold swing of 219 mV per decade and an On/Off current ratio of about 10^6.

These and other nonidealities that may occur in OFETs are examined in the following sections.

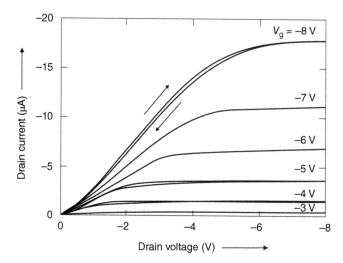

Figure 8.9 Drain current versus drain voltage characteristics of an OFET based on pentacene and using cross-linked PMMA as the gate insulator. Forward and reverse scans are indicated. Source: Yun et al. [15].

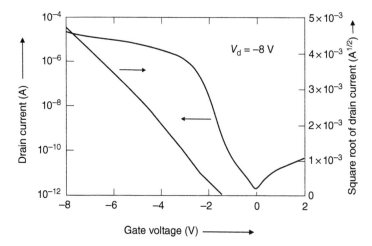

Figure 8.10 Drain current and square root of drain current versus gate voltage characteristics of an OFET based on pentacene and using cross-linked PMMA as the gate insulator; the I_d versus V_d characteristics of this device have been given in Figure 8.9. Data measured at $V_d = -8$ V. Reverse scan is shown. Source: Yun et al. [15].

8.4.1 Contact Resistance

In the previous section, it has been assumed that the gate dielectric is a perfect insulator and that there is no contact resistance between the source and drain electrodes and the organic semiconductor. Figure 8.11 shows how the electrical equivalent circuits of an OFET can include such effects.

Leakage from the gate electrode can be divided into different types depending on the precise geometry of the source, drain, and gate electrodes [3]. For example, current may flow directly between the source/drain and gate electrodes or between the gate and the channel. For devices with large gate areas, a current may also flow between the extended gate and the semiconductor beyond the source and drain contacts. Leakage between the source/drain and gate electrodes

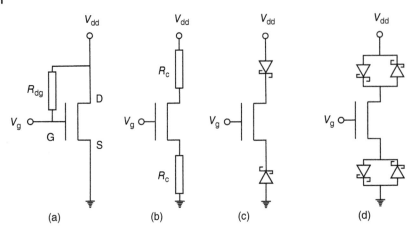

Figure 8.11 Equivalent circuits for OFET devices. (a) Incorporating leakage resistance R_{dg} between the drain (D) and gate (G). (b) With Ohmic contact resistance R_c at the source (S) and drain electrodes. (c) Schottky barrier contacts at the source and drain electrodes. (d) Back-to-back Schottky contacts at the source and drain electrodes. V_{dd} = power supply voltage.

is always present and can be modelled by a shunt resistance R_{dg} between the drain and gate, as shown in Figure 8.11a (Problem 8.3). The effect manifests itself by a crossing over of the I_d versus V_d curves in the output characteristics at a voltage $V_d = L/(C_i\mu W R_{dg})$ [3]. Leakage between the gate and the semiconductor region beyond the source and drain electrodes (and thereby to other OFETs manufactured on the same substrate) is avoidable and can be eliminated by patterning the gate so that it does not extend beyond the channel region.

Contact resistance between the source or drain electrodes can take many forms [2]. This resistance strongly depends on the nature of the electrode, e.g. its work function, and can be dependent on the gate bias (decreasing when the gate bias is increased). The total resistance at the source and drain electrodes may also be a series sum of the access resistance (Section 8.3) and the actual contact resistance. Figure 8.11 depicts the OFET equivalent circuits with contacts in the form of (b) an Ohmic resistance, R_c, (c) a Schottky barrier, and (d) a double Schottky barrier. In the case of non-Ohmic contacts that can be modelled by a Schottky barrier, it should be noted that if one of the contact barriers (e.g. at the drain electrode) is forward biased, then the other will be reverse biased. Double Schottky barriers have been used successfully to model experimental data, e.g. [16], although the physical interpretation is unclear [3]. Contact resistance is often described as a sum of source and drain resistances of equal magnitude. However, it is reasonable to expect that the injecting contact has the major influence, and the contact voltage drop will mostly occur at the source electrode. A straightforward method to determine the total contact resistance of an OFET is to measure the total resistance between source and drain (which will include the series combination of the channel and contact resistances) for a series of transistors with different channel lengths (but with identical channel widths). Extrapolating to zero channel length should then yield the contact resistance. These measurements are usually made at a fixed gate voltage in the linear regime. However, this experiment can take a long time and relies on an accurate knowledge of the channel lengths [2, 3]. Alternative methods that can separate the individual contributions of the source and drain contacts are based on fixed or scanning probe techniques, such as the scanning Kelvin probe microscopy, described in Section 7.5.2 [2].

For series resistors, the modification to OFET theory involves rewriting Eq. (8.5) (linear regime and neglecting the $V_d^2/2$ term) by introducing an additional voltage drop $R_c I_d$. This is accomplished by replacing V_d by $V_d - R_c I_d$. The following expression is obtained for the drain current:

$$I_d = V_d \left(\frac{1}{(W/L)C_i \mu \left(V_g - V_t \right)} + R_c \right)^{-1} \tag{8.20}$$

An Ohmic contact resistance results in 'crowding' of the output characteristic I_d versus V_d curves, evident at low values of drain voltage [3]. In the case of contact resistances in the form of Schottky barriers, the output characteristic curves can become both crowded and nonlinear near the origin (as in Figure 8.9).

A more general approach, appropriate if the contact is non-Ohmic, is to divide the channel into a main channel and small contact region where there is a voltage drop V_c [17]. The channel length is reduced to $L - d$, where d is the length of the contact region, and the voltage drop across the entire channel to $V_c - V_d$. The drain current then becomes:

$$I_d = C_i \mu \frac{W}{L-d} \left(\left(V_g - V_0 \right) V_d - \frac{V_d^2}{2} - \left(\left(V_g - V_0 \right) V_c - \frac{V_c^2}{2} \right) \right) \tag{8.21}$$

For an Ohmic contact resistance ($V_c = R_c I_d$) and $d \ll L$, neglecting $V_d^2/2$ term (i.e. linear regime), Eq. (8.21) reduces to Eq. (8.20).

In the presence of either Ohmic or non-Ohmic contact resistances, the mobility values calculated from the output or transfer curves of the OFET will be inaccurate. For example, if the (Ohmic) contact resistance becomes larger than the channel resistance (this usually occurs if the channel length of the device is reduced to about 1 µm), then the effective mobility of the transistor μ_{eff} may drop significantly below the intrinsic mobility μ of the organic semiconductor. The relationship between these two parameters is given by [18]:

$$\mu_{\mathrm{eff}} \approx \mu \left[1 - \left(\frac{\mu C_i W R_c \left(V_g - V_0 \right)}{L + \mu C_i W R_c \left(V_g - V_0 \right)} \right)^2 \right] \tag{8.22}$$

This equation accounts for the common observation that measured (effective) mobility of short channel organic transistors is often less than $1\,\mathrm{cm}^2\,\mathrm{V}^{-1}\,\mathrm{s}^{-1}$, even though the semiconductor is known to possess a much larger intrinsic mobility. This is also the reason that the highest operating frequencies of some devices are limited to a few MHz. Judicious attention to device design is needed to increase the operating frequency, which may be increased to 10–30 MHz [19].

8.4.2 Material Morphology and Traps

Although Eqs. (8.5) and (8.8) are widely used to describe the current versus voltage behaviour of OFETs, it has become clear that the carrier mobility of most devices is dependent on the gate voltage, the mobility increasing as the gate voltage increases [2, 3, 5, 20]. As an increase in V_g results in an increase in carrier concentration, it seems reasonable to attribute gate bias dependence of mobility with the carrier density.

The organic semiconductor used in an OFET will usually be in the form of a polycrystalline layer, or a mixed polycrystalline-amorphous film, and is expected to contain defects such as those described in Chapter 3. The idea of grain boundary potential barriers, as shown in Figure 3.12, was originally formulated to explain electrical data for inorganic thin film transistors. The charge states at a grain boundary present an electrostatic barrier to carrier transport, but charge within each grain screens the potential, reduces the barrier and increases the effective mobility. For such systems, Levinson calculated a linear dependence of I_d on V_d but a nonlinear dependence on V_g [21]. For a hole channel (negative V_g), the drain current in the linear regime is:

$$I_d = -\mu C_i \frac{W}{L} V_d V_g \exp\left(a/V_g\right) \tag{8.23}$$

Therefore, a plot of $\ln(I_d/V_g)$ versus $1/V_g$ should give a straight line. This is called a Levinson plot. The parameter a in the above equation (the slope of the Levinson plot) is dependent on temperature and provides an estimate of the grain boundary charge [21]. In this model, the mobility is thermally activated:

$$\mu = \mu_0 \exp\left(-E_b/k_B T\right) = \mu_0 \exp\left(-a/V_g\right) \tag{8.24}$$

where E_b is the energy barrier and μ_0 is the mobility within the grains.

Many researchers have found that this grain boundary model does not work adequately with OFETs, even for those in which the semiconductor morphology is better described as polycrystalline rather than amorphous [22, 23]. Alternative theories to explain the experimental output and transfer curves of OFETs that are based on carrier trapping have been formulated [3].

A simple approach assumes a flat band edge with localized states that trap carriers rather than create potential barriers [22]. The barrier and trap models are contrasted in Figure 8.12; E_t is the trap depth. The presence of traps will affect both the OFET output and transfer curves [3]. The Poole-Frenkel effect (Section 6.7) can produce a mobility dependence of both temperature and electric field in the direction of current (in-plane):

$$\mu = \mu_0 \left(-\left((E_t - E_v) - e\sqrt{e|\mathcal{E}_x|/\pi\varepsilon_s\varepsilon_0}\right)/k_B T\right) \tag{8.25}$$

Trapping accounts for the nonlinear output characteristics of an OFET, particularly at small values of V_d and at low temperatures [3]. Under these conditions, a larger proportion of the charge is trapped. It is also predicted that the nonlinearities become more pronounced as the channel length is reduced.

For an ideal OFET, the field effect mobility μ_{FET} is defined via the derivative of the transfer curve, Eq. (8.6):

$$\mu_{FET} = \frac{L}{WC_i V_d} \frac{\partial I_d}{\partial V_g} \tag{8.26}$$

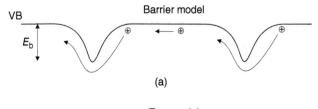

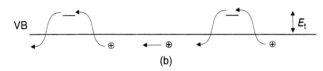

Figure 8.12 Comparison of potential barrier model (a) and trap model (b) to explain mobility dependence on gate voltage for a p-type OFET. VB = valence band. E_b = barrier height, E_t = trap depth. Source: Modified from Street and Salleo [17].

At low drain bias (linear regime), the charge density and electric field can be considered to be uniform in the channel. In this case, I_d is proportional to the free surface charge density Q_s – holes in the case of a p-channel OFET. The mobility in the linear regime is therefore proportional to the derivative of the charge density as a function of the gate bias. From Eq. (8.3):

$$\mu_{\text{FET}} = \frac{\mu_0}{C_i} \frac{\partial |Q_s (V_g)|}{\partial V_g} \tag{8.27}$$

For trap-free OFETs, this relationship is linear, and the measured mobility is equal to μ_0 (since $C_i = Q_s/V_g$). When the organic semiconductor is full of traps, changes in the Fermi level caused by changes in the gate bias will influence the occupation of the traps and the field effect mobility will be dependent on the gate bias. The exact relationship will depend on the mathematical form of the trap distribution. In the case that the traps have an exponential distribution in energy, and of a similar form to that use to model space-charge-limited currents (Eq. (6.34)), the number of traps per unit area per unit energy $N_t(E)$ for a p-type semiconductor can be written:

$$N_t(E) = \frac{N_{t0}}{k_B T_c} \exp \left(\frac{E_v - E}{k_B T_c} \right) \tag{8.28}$$

where N_{t0} is the surface density of traps at the valence band edge and T_c is a characteristic temperature, which describes the trap distribution. This leads to the following relationship between mobility and gate voltage [20]:

$$\mu_{\text{FET}} = \mu_0 \frac{N_v}{N_{t0}} \left(\frac{C_i V_g}{e N_{t0}} \right)^{\frac{T_c}{T} - 1} \tag{8.29}$$

where N_v is the density of states per unit area in the valence band. This equation justifies the empirical power law relationships that are often used to describe the dependence of carrier mobility on gate voltage $\mu \propto V_g^n$ or $\mu \propto (V_g - V_0)^n$ [20].

Note that the above expressions assume that the charge associated with the traps is located at the semiconductor/insulator interface. A similar expression to Eq. (8.29) is derived in the book by Stallinga [3]. Moreover, this text contains discussion of other modifications to the simple theory, for example an exponential distribution for the valence band states (tail states) as well as for the trapping states, and some helpful examples from the modelling. A mobility enhancement factor, to encompass the mobility dependence on gate bias, is incorporated in a generic analytical simulation for OFETs by Marinov et al. [24]. This model also includes the effect of contact resistances. Several other computer models for OFETs have been developed. These have been reviewed by Kim et al. [25] and Groves [26]. The interrelationship between the field effect mobility with the temperature and bias of has been observed to take the form of the Meyer–Neldel rule, Section 2.6.5 [3, 27].

The mobility dependence on gate voltage in Eq. (8.29) disappears when $T = T_c$ and the subsequent expression for the mobility may be contrasted to that used in certain expressions for space-charge limited conductivity in the presence of traps (Eq. (6.30)).

The presence of traps can also account for finite values of the threshold voltage V_0, for example if the organic semiconductor contains a positive trapped interface charge density N_t^+. In the conduction band, these are compensated by electrons. To remove the charges, a negative bias must be applied to the gate. Only an applied gate voltage beyond this threshold will start attracting holes to the valence band. Hence:

$$V_0 = -\frac{e N_t^+}{C_i} \tag{8.30}$$

Charge incorporated in the semiconductor (or the insulator) in an OFET can result in the transistor already being in the On state (i.e. surface accumulation charge) before any gate bias is applied. In this respect, the finite value of threshold voltage is similar to the change in the flat-band voltage of an MIS structure when additional charge is incorporated at the semiconductor/insulator interface (Eq. (5.25)).

The presence of trapped charge can also explain the time dependence of the measured output and transfer curves for OFETs as there will be a characteristic time associated with the trapping and detrapping processes [3]. The origin of this effect is like that responsible for hysteresis in the *C–V* curves of MIS structures (Section 5.3.2). Hysteresis is a common feature in inorganic (e.g. amorphous silicon) thin film transistors and OFETs (for example, Figure 8.8). The phenomenon of a continuing shift in the threshold voltage with prolonged application of a voltage or current – known as 'stressing' – is often encountered [3, 4, 28]. Traps can be a feature of the purity of the organic semiconducting layer and, in this respect, the performance of OFETs is expected to improve with better material purity and device processing conditions. However, the electrical behaviour of OFETs is also influenced by environmental factors (e.g. presence of water) indicating that encapsulation is essential for reliable device operation over an extended period [3, 29]. In the case of flexible structures (e.g. assembled on polymer supports), organic semiconductor devices are often fabricated with an additional buffer layer between the active device and the plastic base to prevent the ingress of air and water molecules.

The additional capacitance associated with interface traps, C_{it} ($=e^2 D_{it}$), will affect the subthreshold swing of the OFET. Assuming that the applied gate voltage is divided between C_i and C_{it}, the relationship between the semiconductor surface potential and the gate voltage (Eq. (8.11)) becomes:

$$\frac{dV_g}{d\psi_s} = \frac{C_i + C_{it}}{C_i} = n \tag{8.31}$$

The ideality factor n now is greater than unity, which will increase S. It is therefore possible to estimate the interface trap density from the subthreshold conduction characteristic [4, 30].

8.4.3 Short Channel Effects

Reducing the channel length of a field effect transistor will improve speed and reduce operating voltages (Problem 8.4). The derivation of the OFET output and transfer curves has assumed that the gate insulator thickness is considerably less than the channel length. However, as device scaling produces ever smaller transistors, deviations from this theory are expected. For example, the potential distribution in the channel will now depend on both the transverse field $\mathcal{E}_x$ (controlled by the gate) and the longitudinal field $\mathcal{E}_y$ (controlled by the drain bias). This distribution becomes two-dimensional, and the gradual channel approximation ($\mathcal{E}_x \gg \mathcal{E}_y$) is no longer valid. Other effects include the carrier velocity becoming independent of field for high $\mathcal{E}_x$ (see below) and the injection of hot carriers into the gate insulator, leading to insulator charging and a subsequent threshold voltage and transconductance degradation. These issues and methods to avoid them in the design of Si MOSFETs have been discussed extensively by Sze and Ng [1] and Streetman and Banerjee [31]. However, it should be noted that not all the short channel effects observed in silicon devices are applicable to OFETs. This is because the former usually operate in inversion mode and the source and drain contacts have associated depletion regions.

Although the dimensions of planar OFETs are relatively large compared to the current generation of Si MOSFET devices (channel lengths of μm rather than nm), short channel effects can become important, and several different (non-planar) architectures have been proposed and demonstrated (Section 8.3). These include vertical OFETs in which a short channel length is not defined by a lateral patterning step, but by the thickness of a film [12]. The short channel effects that are observed in OFETs are like those noted for amorphous silicon FETs [32]. Non-saturation of the output characteristics and increased influence of the source and drain contact resistances are often reported. The calculated mobilities from the linear and saturation regions of the output characteristics are significantly less than the actual mobilities, as discussed above in Section 8.4.1 [33]. Practical approaches to the reduction of contact resistance affects rely on the choice of appropriate source and drain electrode materials, modification of the semiconductor morphology with the use of interfacial layers and electrode doping in the contact regions [34].

The effects of the non-linear dependence of the carrier mobility upon the electric field are explored in the book by Sze and Ng [1]. For low values of applied field, the carrier drift velocity increases linearly with the field. However, in the case of short channels, the drift velocity can reach its saturation value, v_{sat}. Figure 2.4 reveals that v_{sat} is achieved for a field of about 5×10^6 V m^{-1} in Si at 300 K. Such a field may be obtained by the application of a few volts across a transistor channel that is a few hundred nanometres in length. In this case, Q_s in Eq. (8.3) becomes fixed for current continuity and is approximated by $(V_g - V_0)C_i$. The expression for the saturated drain current is [1]:

$$I_{d,sat} = W \left(V_g - V_0 \right) C_i v_{sat} \tag{8.32}$$

The transconductance is therefore independent of gate bias. Furthermore, the saturation current no longer depends on the channel length.

Saturation of the drift velocity is an equilibrium phenomenon at high field, with many scattering events occurring. However, in the case where the device dimensions are less than the carrier mean free path (e.g. in a very short channel device or for a material such as a SWNT), the carriers will cross the semiconductor from source to drain with no scattering. The carriers gain energy from the electric field but do not lose it to the lattice. The resulting ballistic transport (Section 3.8.2) can give rise to drift velocities higher than v_{sat}. Consequently, both the channel current and the transconductance can be higher than values corresponding to the saturation velocity. Computer simulations show that the electric field and the carrier velocity are very non-uniform in the channel [1]. In the limit of ballistic transport:

$$I_{d,sat} \propto \left(V_g - V_0 \right)^{3/2} \tag{8.33}$$

8.4.4 Organic Semiconductors

State-of-the-art OFETs possess characteristics similar (or even superior) to devices prepared from hydrogenated amorphous silicon, with carrier mobilities around 1 cm^2 V^{-1} s^{-1} and On/Off ratios greater than 10^6 [35]. Operating frequencies of tens of MHz are possible, and devices can be fabricated on flexible (plastic or paper) backing supports [36]. Many of the highest performing p-type transistors using small molecules are based on acenes (e.g. pentacene, rubrene) and fused heteroacene materials (Section 1.5.2) [37, 38]. As noted earlier, OTFTs normally operate in the accumulation regime. However, there is some evidence that inversion transistors can be realized, for example by doping a thin layer of pentacene at the interface to the gate insulator [39]. A challenge is to make OFET materials (both semiconductors and gate dielectrics) solution processable to aid device manufacture. Much progress has been achieved by exploiting the excellent solubility characteristics of the triisopropylsilylethynyl (TIPS) group with substitution on the central pentacene core [5, 10, 37, 40]. The molecular structure of one derivative is depicted in Figure 8.13a.

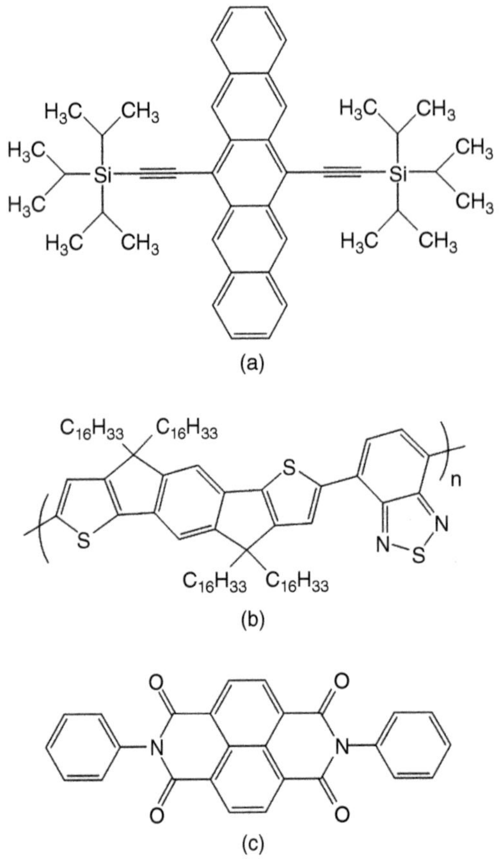

Figure 8.13 Examples of organic semiconductors used in transistor applications. (a) TIPS pentacene (p-type) [38]. (b) indacenodithiophene copolymer (p-type) [41]. (c) napthalene diimide (n-type) [42]. Source: (a) Modified from Braga et al. [38], (b) Zhang et al. [41], (c) Modified from Zhao et al. [42].

Promising polymer semiconductors include donor–acceptor copolymers, such as indacenodithiophene (IDT)-based compounds [5, 41], the molecular structure of one such compound is provided in Figure 8.13b. While these materials generally exhibit lower mobilities than their small molecule counterparts, they can be more readily processed from solution. A further approach uses blends of small molecules and insulating polymers [36, 43]. The motivation here is to combine the high mobility of the molecular semiconductor with the ability of the polymer to confer desirable thin film morphologies. Mixtures of molecular semiconductors and polymeric semiconductors may also be exploited [35]. Carbon nanotube OFETs with relatively high mobilities (above $20\,cm^2\,V^{-1}\,s^{-1}$) and operating frequencies (GHz range) can be conveniently fabricated using techniques such as inkjet printing [44]. The exploitation of both carbon nanotubes and graphene in transistor devices is covered separately in Section 8.6.

Although a number of p-channel organic MISFETs are available with performances comparable to amorphous silicon, n-channel devices have not received the same attention [42, 45]. Materials with high electron affinities (Section 1.5.2) are needed for n-type behaviour. Many such compounds exhibit instability (oxidation) in air. To some extent, this can be circumvented by developing molecules that favour dense packing. The introduction of electron-withdrawing groups, such as cyano or carbonyl, is a common strategy for lowering the LUMO energy level. Aromatic diimides are an example of useful n-type organic semiconductors, exhibiting high electron affinities, high mobilities, and excellent stabilities [35, 43]. The molecular structure of one example is shown in Figure 8.13c. Other n-type materials can be based on perylene, TCNQ, and fullerenes [45].

8.4.5 Gate Dielectric

The insulating gate dielectric forms an integral part of the OFET architecture, although its importance is often overlooked. This thin layer must, by definition, isolate the gate electrode from the rest of the transistor and therefore possess a high resistivity. A dielectric with a large permittivity (a so-called high-k dielectric) will also provide a high drain current (Eq. (8.5)) and allow for low-voltage operation. Additionally, the gate insulator should not incorporate fixed or mobile charge and must provide an electrically clean interface with the organic semiconductor. Otherwise, hysteresis will be prevalent in the transfer and output curves, as discussed for the MIS structure, Section 5.3.

Silicon dioxide ($\varepsilon_i = 3.9$) was favoured as the gate insulator in early OFETs. The ease by which this highly insulating oxide could be grown on single crystal silicon allowed the latter (often in a highly doped form) to form the gate electrode. Organic insulators are now used as gate dielectric layer as their formation can be compatible with OFET processing (e.g. inkjet printing), leading to all-organic devices [10, 45]. Poly(vinyl alcohol) (PVA, $\varepsilon_i \approx 10$) or poly(4-vinyl phenol) (PVP, $\varepsilon_i \approx 5$) have relatively large dielectric constants and have been incorporated in low-voltage OFETs [10]. However, these materials contain hydroxyl groups, which attract water molecules and cause operational reliability issues in ambient conditions. Other organic gate insulators that have been used include poly(methyl methacrylate) (PMMA, $\varepsilon_i \approx 3.5$) and polyimide polymers. Surface treatments, for example a hydrophobic self-assembled monolayer such as n-octadecyltrichlorosilane (OTS) or hexamethyldisilazane (HMDS), can modify and stabilize the organic semiconductor/insulator interface surface. High-k poly(vinylidene difluoride) (PVDF, $\varepsilon_i \approx 12$) copolymers without hydroxyl groups are popular. Increased dielectric constants can be achieved by loading the PVDF copolymer with ceramic nanoparticles, such as barium strontium titanate. However, the intrinsic ferroelectric properties of PVDF copolymers can cause hysteresis in the electrical behaviour of the OFET (Section 7.2.3).

In certain situations, for example using amorphous organic semiconductors such as polytriarylamines, the temperature activation of the field-effect mobility can increase when high permittivity insulators are used. This has been attributed to a broadening of the density of states at the semiconductor/insulator interface [46]. Figure 8.14 illustrates the phenomenon. The polarity of the gate insulator affects the electronic nature of the semiconductor/insulator interface. Disorder of the polar groups can result in a broadening of $S(E)$ at the semiconductor surface and increased trapping. In these cases, effects such as increased hysteresis can be avoided by using lower permittivity gate insulators. High carrier mobilities can also be achieved [46].

8.5 Organic Integrated Circuits

A significant step in the microelectronics industry was the ability to manufacture a complete circuit, an integrated circuit, on the same silicon substrate. Integrated circuits are now routinely fabricated using organic materials [47, 48]. Figure 8.15 shows the characteristics of an early inverter (a circuit that produces logic '1' for logic '0' input and vice-versa) based on two interconnected pentacene OFETs [49]; V_{dd} represents the supply voltage. One of the transistors acts as the load for the switching device. The resulting inverter has sufficiently large gain, but due to the positive turn-on voltage, the input and output levels do not match. Also evident is a hysteresis of a few hundred millivolts in the inverter characteristic, which is possibly due to mobile charges in the gate dielectric. This situation can be improved by the incorporation of a voltage-level-shifting network in the integrated

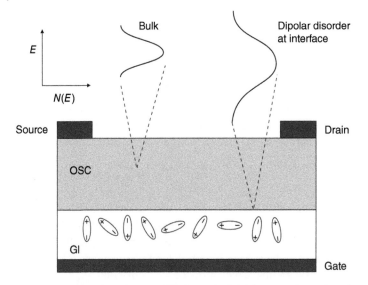

Figure 8.14 Density of states, $S(E)$, in the bulk of the organic semiconductor and at the semiconductor/insulator interface for an OFET. Local polarization in the dielectric may lead to a broader density of states function. OSC = organic semiconductor, GI = gate insulator. Source: Veres et al. [46].

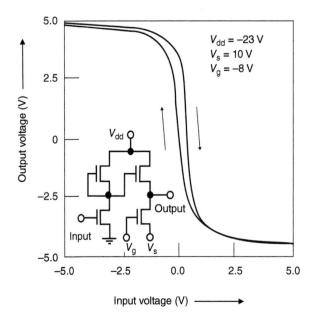

Figure 8.15 Output versus input voltage of an inverter device based on pentacene. The circuit of the structure is shown inset. Source: Klauk et al. [49].

circuit. These simple inverter structures have now led to the development of high-performance oscillator circuits, e.g. [50].

Thin film transistors or integrated circuits based on organic semiconductors are unlikely, in the immediate future, to compete with single crystal inorganic semiconductors, such as Si or GaAs, for fabricating very fast switching devices. However, these may find roles in niche areas, such as

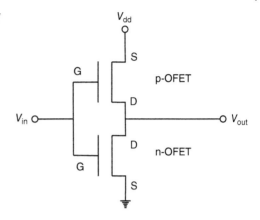

Figure 8.16 Inverter circuit based on complementary (p-type and n-type) OFETs. V_{dd} = supply voltage, V_{in} = input voltage, V_{out} = output voltage.

components of plastic circuitry for use as display drivers in televisions, portable computers, and pagers (Chapter 10), and as memory elements in transaction cards and identification tags.

Radiofrequency identification (RFID) is a method of remotely storing and retrieving data using devices called RFID tags [13]. These are small objects, such as adhesive stickers, which can be attached to, or incorporated into, a commercial product. The tags contain aerials to enable them to receive and respond to radio-frequency queries from an RFID transceiver. Identification tags can operate at different frequencies, ranging from about 100 kHz to the microwave region, around 2 GHz. Such systems have been based on pentacene integrated circuits and operate up to 6.5 MHz [51].

A particularly useful device for digital applications is a combination of n-channel and p-channel MOS transistors on adjacent regions of an integrated circuit. This complementary MOS, referred to as CMOS, circuitry dominates the market for applications such as microprocessors (Section 1.3.1). Complementary circuits are also of interest for less sophisticated applications, especially those that are battery-powered since they can provide very low static power dissipation and therefore extend battery life. Figure 8.16 shows the circuit for a p-channel and an n-channel FET connected as a complementary pair in an inverter configuration. This circuit consumes little DC power; when the input V_{in} is high or low, one of the transistors in series is off so that there is only a small steady-state current (subthreshold current).

By combining low-voltage p-channel and n-channel organic thin-film transistors in a complementary arrangement, simple integrated circuits such as digital-to-analogue converters may be achieved [52]. An integrated 13.56 MHz RFID tag in a printed complementary organic transistor technology has also been successfully developed [53]. This circuit is relatively complex, with more than 250 organic transistors on the same substrate, and operates at a supply voltage of 24 V.

Substantial and rapid progress continues to be made on OFET device architectures and materials. The yield from manufacturing and long-term reliability are now the key factors for further progress [50, 54].

8.6 Nanotube and Graphene FETs

Carbon nanotubes and graphene are unique materials (Sections 1.5.5 and 1.5.6).

Both experiment and theory have demonstrated that single-wall carbon nanotubes can be either metallic or semiconducting. The remarkable electrical properties of SWNTs originate from the unusual electronic structure of the two-dimensional material graphene. When wrapped to form a

SWNT, the momentum of the electrons moving around the circumference of the tube is quantized. The result is either a one-dimensional metal or a semiconductor, depending on how the allowed momentum states (wavevector or k states) compared with the preferred directions for conduction.

The physics and technology of electron transport in nanodevices are fully explained by the seminal work of Landauer, who derived Eq. (3.24) for the conductance, G, of a one-dimensional ballistic conductor. In the case of an SWNT, considering the sublattice degeneracy of graphene and assuming that the contacts are perfect (transmission probability $T = 1$), gives [55, 56]:

$$G = \frac{4e^2}{h} = 155 \ \mu S \tag{8.34}$$

This corresponds to a resistance of about 6.5 kΩ. Experiments reveal that the resistance of SWNTs varies considerably, from approximately 6 kΩ to many megohms, the differences mostly due to contact resistances between the electrodes and the nanotubes. When contact resistances are eliminated, the measurements reveal a resistance per length of 4 kΩ μm^{-1}, a mean free path of 2 μm and a room temperature resistivity of approximately 10^{-6} Ω cm. The conductivity of metallic nanotubes can therefore be equal to, or even exceed, the conductivity of metals like copper at room temperature. Their small dimensions, the lack of carrier scattering (ballistic transport) and the potential for doping to form both p- and n-type semiconductors make carbon nanotubes ideal candidates for high frequency transistor devices. Two main challenges have hindered technological progress: the exclusive growth of semiconducting SWNTs and the ability to position these precisely in integrated circuits.

The first carbon nanotube FETs were bottom gate devices and the SWNTs were deposited directly onto high-k dielectrics such as HfO$_2$ [57]. By adjusting the concentration of nanotubes in solution, it was possible to fabricate field effect transistors that operated with a single SWNT. The electrical behaviour of such devices revealed output and transfer characteristics typical for p-type MOSFETs. However, the device operation was controlled by Schottky barriers that existed at the interfaces between the three-dimensional source and drain electrodes and the one-dimensional nanotube. This resulted in the subthreshold slope not decreasing with temperature, as expected for a conventional MOSFET (Eq. (8.12)).

Improved results were achieved with top gate structures in which a gate metal is formed, by atomic layer deposition, on part of the SWNT, at a distance (≈ 0.5 μm) from both the Mo drain and source electrodes [58]. Figure 8.17 shows a schematic diagram for such a device, with a thin film (≈ 8 nm) of ZrO$_2$ ($\varepsilon_r \approx 25$) as the gate insulator, together with its output characteristics. The nanotube FET exhibits both the linear and saturation regions of a conventional MOSFET. A transconductance of 12 μA V^{-1} (or 12 μS) was obtained from the saturation region ($V_d = -1.2$ V). It should be noted that it is common to normalize carbon nanotube transistor parameters to the width of the nanotube. In this case, using a value of 2 nm, provides a normalized transconductance of 6000 S m^{-1}. However, a fair comparison between nanotube FETs and Si-based devices is non-trivial due to the quasi-one-dimensional nature of SWNTs. Subthreshold swing values in the range 70–100 mV per decade were achieved for the ZrO$_2$ FETs.

A further notable development has been the demonstration of transistors with self-aligned geometries, so that the edges of the source, drain, and gate electrodes are precisely positioned with no overlapping or significant gaps between them [59]. In this case, Pd is used to form low resistance source and drain electrodes. Carbon nanotube transistors with sub-10 nm channel lengths have also been fabricated [60]. Such devices outperform silicon transistors with more than four times the diameter-normalized current density (2.41 mA μm^{-1}) at a low operating voltage of 0.5 V.

High performance n-type carbon nanotube FETs have also been reported [61]. This was achieved by doping the source and drain regions with potassium, leading to transconductance figures of

Figure 8.17 (a) Schematic diagram of top gate single-walled carbon nanotube field effect transistor; ZrO_2 forms the gate dielectric, Mo forms source and drain electrodes, while Ti/Au is used as the gate electrode. (b) Drain current versus drain voltage behaviour for the SWNT FET. Source: Javey et al. [58].

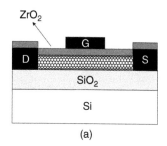

(a)

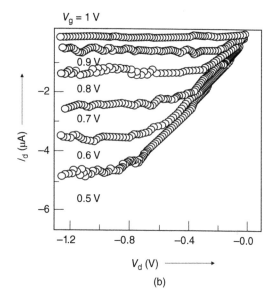

(b)

20 μS with substhresold swings of 70–80 mV per decade. Despite the promising performance of individual SWNT devices, the control of nanotube diameter, chirality, density, and placement remains inadequate for microelectronics production, particularly where large area coverage is required. Transistors incorporating tens of thousands of individual nanotubes are more practical. Devices with charge carrier mobilities of 10s of $cm^2\ V^{-1}\ s^{-1}$ can be produced, some on flexible supports [62]. These are attractive for driving organic light emitting devices (OLEDs – Chapter 10) and displays as they possess a higher mobility than amorphous silicon and can be fabricated by low temperature, non-vacuum methods.

Wafer-scalable FETs using aligned nanotube bundles can work at GHz frequencies [63]. The individual transitors contain 2000–3000 nanotubes operating in parallel, with source to drain separations of 160 nm. The average transconductance for such devices at 1.5 GHz is 330 mS mm^{-1}. The extrinsic (as-measured) cut-off frequencies exceed 100 GHz, which is on a par with high-frequency transistor designs based on GaAs pHEMTs (pseudomorphic high electron mobility transistors) and Si CMOS devices (at similar gate lengths) [63]. Figure 8.18 provides a comparison, in the form of the extrinsic cut-off frequency versus gate length for different transistor technologies. Although further research and development is needed to establish carbon nanotube transistors in the commercial arena, the fabrication and operation of a 16-bit microprocessor, comprising more than 14 000 complementary transistors demonstrates that many of the earlier hurdles for the technology have been largely overcome [64].

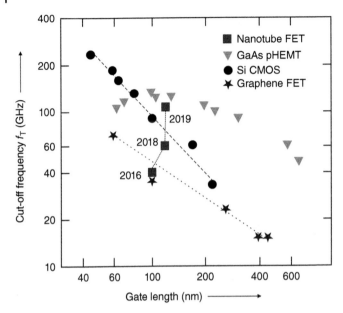

Figure 8.18 Extrinsic cut-off frequency, f_T, versus gate length for various electronic transistors: nanotube FET, GaAs pHEMT, Si CMOS and graphene FET. Source: Rutherglen et al. [63].

Graphene is also under intense scrutiny as a material for beyond-silicon electronic systems [65, 66]. However, its exploitation is difficult, as the material does not possess a band gap (Section 1.5.5). The graphene energy band structure leads to both low On/Off current ratios and the inability to switch the device to an Off state. Further challenges include the deposition of graphene over large areas with controlled grain size, thickness and crystallographic orientation, and the development of a suitable gate insulator with a high-quality interface with graphene. Several research strategies are being targeted to open the band gap such as using graphene nanoribbon and chemical modification. However, these approaches could lead to degradation in the carrier mobility. The problem of low On/Off ratio may be resolved using new transistor architectures, which exploit the modulation of the graphene work function, allowing control over vertical (rather than in-plane) carrier transport [67]. Despite all these difficulties, graphene transistors operating with low contact resistances, high values of transconductance (>700 mS mm^{-1}), and high cut-off frequencies (>400 GHz) have been reported, and transistor fabrication procedures have been developed that allow a graphene integrated circuit to perform practical wireless communication functions, receiving and restoring digital text transmitted on a 4.3 GHz carrier signal [68–70]. Some of these data are shown in Figure 8.18. Generally, the performance of graphene devices falls short of that for carbon nanotube FETs [63].

8.7 Single-Electron Transistors

The possibility of constructing nanodevices in which the manipulation of single (or small numbers) of electrons can be exploited was introduced in Section 3.8.3. The principle of Coulomb blockade enables electrons to be localized on an isolated island (or quantum dot). A simple device that illustrates this phenomenon is a double tunnel junction structure. If the voltage across junction 1 exceeds $e/2C$, an electron enters the island; it is then energetically favourable for another electron to tunnel through junction 2 out of the island. Therefore, an electron will almost immediately exit the island after the first electron has entered. This is a space-correlated tunnelling of electrons and establishes charge neutrality on the island. However, if the transmission coefficients of the tunnel junctions are different, carriers may enter the island through junction 1 but are restricted by the high resistance of junction 2 from immediately leaving it. Eventually, the electron will, due to the high bias, tunnel out of the island, which allows another electron to enter through junction 1. For most of the time the island is charged with one excess elementary charge. If the bias is increased, more electrons will populate the island. This leads to a step-like current versus voltage characteristic – the so-called Coulomb staircase.

Experimental data, at 77 K, for the current versus voltage behaviour of such a double tunnel junction are shown in Figure 8.19 [71]. The device was fabricated by depositing a Pd metallic nanocluster onto an insulating self-assembled decanethiol monolayer on a polycrystalline Au substrate. The tip of a scanning tunnelling microscope formed the second metallic electrode. Consequently, this double tunnel junction arrangement was highly asymmetric. The tunnelling of an electron to the metal particle is blocked until the applied voltage reaches a value equal or higher than the charging energy of the particle. Thereafter, the current versus voltage curve shows the expected Coulomb staircase, depicted in Figure 8.19a. Measurements taken with uncoated 'reference' monolayer exhibit no such features. From the derivative of the I–V data, dI/dV – Figure 8.19b, the fractional charge on the cluster can be determined. As seen in the curves, the experiment and the theory are in good agreement. Coulomb blockade has been suggested as a mechanism that might explain the low temperature I–V behaviour of monolayers of conductive polymers [72].

Metrologists have pursued electron-counting standards for capacitance and current using single-electron devices [73]. Although a single-electron box can control the number of electrons in the island, it does not have the properties of a switching device. Single-electron transistors (SETs) are three-terminal switching devices that can transfer single electrons from source to drain. A schematic diagram of such a device is shown in Figure 8.20 [1]. Figure 8.20a shows the three-electrode SET arrangement while Figure 8.20b depicts the electrical equivalent circuit. The minimum energy needed to transport a single electron to and from the island is $e^2/2C_t$, where C_t is the total capacitance:

$$C_t = C_g + C_s + C_d \tag{8.35}$$

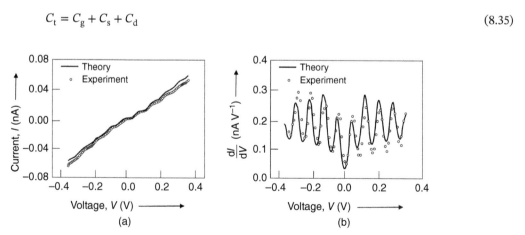

Figure 8.19 (a) Current versus voltage characteristic showing Coulomb staircase. (b) Derivative dI/dV for the data presented in (a). In each case, the experimental points are shown as open circles, while the full curves represent the theoretical fits. The experimental I versus V curve is taken for an Au/monolayer/Pd cluster/STM tip architecture at 80 K. Source: Oncel et al. [71].

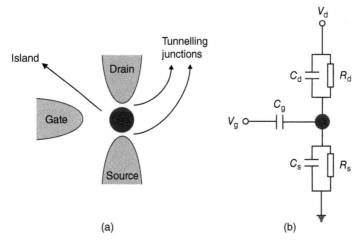

Figure 8.20 (a) Electrode configuration for a single-electron transistor (SET). (b) Equivalent circuit for SET: V_g and V_d are the gate and drain voltages, respectively. C_g, C_d, and C_s are the capacitances of the gate, drain, and source junctions, and R_d and R_s are the resistances of the drain and source junctions, respectively.

and C_g, C_s, and C_d are the gate, source, and drain capacitances, respectively, as shown in Figure 8.20b. These are all absolute values of capacitance, in contrast to capacitance per unit area values encountered elsewhere in this book. The minimum energy for electron transfer must also be significantly larger than $k_B T$, the thermal energy ($>100\,k_B T$). Therefore, at room temperature, C_t needs to be the order of aF (10^{-18} F), which requires a single-electron island size of less than 2 nm. Commonly, metal clusters or nanoparticles are used as the island, but fullerenes and carbon nanotubes have also been exploited (see below). If semiconductors are used, then additional quantization effects are expected to be observed (this aspect not included in the following discussion).

The capacitances between the island and the source and drain of the SET will have resistances associated with them – R_s and R_d in Figure 8.20b. As discussed in Section 3.8.3, the lower limiting values of these will be constrained by the Uncertainty Principle to 25.8 kΩ.

For current to flow between the source and drain, there are two junctions through which the electrons must tunnel. However, at any one time only one of these junctions will control the electron flow. Figure 8.21 shows potential energy band diagrams for the source/island/drain configuration [1]. The initial assumption is that the gate voltage is zero. If the source/island junction provides the bottleneck to current flow, an electron will tunnel if the voltage across this junction exceeds $e/2C_t$ (note that Figure 8.21 shows energy rather than potential). This is identified as process 1 in Figure 8.21a. Subsequently, the potential of the island decreases, and the electron can tunnel to the drain, indicated by process 2. However, if the tunnelling from the island to the drain is the current-controlling step, then Figure 8.21b more appropriately depicts the energy configurations. In both diagrams, process 1 occurs before process 2. The island potential will change by e/C_t after each tunnelling event.

The potential of the island will also be affected by the voltage applied to the gate electrode; this will alter the energy levels of the island shown in Figure 8.21. As the gate voltage is increased, the island voltage will also increase, but scaled down by a factor of C_g/C_t. In the case of the source/drain controlling the current (Figure 8.21a), the electron will start to tunnel if the voltage across the

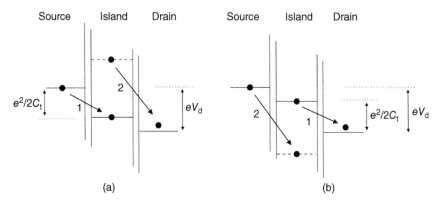

Figure 8.21 Potential energy diagrams illustrating the tunnelling events that can take place in a double-junction source/island/drain structure, with an applied voltage V_d. (a) The island is unoccupied and, first, an electron tunnels from the source to the island (process 1); the energy of the island changes and the electron subsequently tunnels to the drain (process 2). (b) The island is occupied and the electron tunnels first to the drain (process 1) allowing another electron to tunnel to the island from the source (process 2). Source: Sze and Ng [1].

junction exceeds $e/2C_t$. This corresponds to:

$$\frac{V_d C_d}{C_t} + \frac{V_g C_g}{C_t} \geq \frac{e}{2C_t} \tag{8.36}$$

which gives a minimum value of V_d for tunnelling or Coulomb blockade voltage:

$$V_d = \frac{e}{2C_d} - \frac{V_g C_g}{C_d} \tag{8.37}$$

As V_g is increased, a point will be reached where Coulomb blockade will be removed, allowing electrons to tunnel into and out of the island. The Coulomb blockade will be lifted when the gate charges the island potential to $e/2C_t$. The variation of V_d with gate voltage is:

$$\frac{dV_d}{dV_g} = -\frac{C_g}{C_d} \tag{8.38}$$

which has a negative slope.

On the other hand, for the situation where the island/drain controls the current (Figure 8.21b), electrons will flow from the island to the drain if:

$$V_d - \left(\frac{V_d C_d}{C_t} + \frac{V_g C_g}{C_t}\right) \geq \frac{e}{2C_t} \tag{8.39}$$

Using Eq. (8.35), this gives a further expression for the Coulomb blockade voltage:

$$V_d = \frac{(e/2) + V_g C_g}{C_g + C_s} \tag{8.40}$$

and will vary with gate bias according to:

$$\frac{dV_d}{dV_g} = \frac{C_g}{C_g + C_s} \tag{8.41}$$

which has a positive slope. The SET will therefore possess both positive and negative transconductance values, depending on the gate voltage range. A clear way of visualizing the regions of Coulomb blockade is to plot V_d as a function of V_g to reveal Coulomb blockade 'diamonds' [1], as shown in

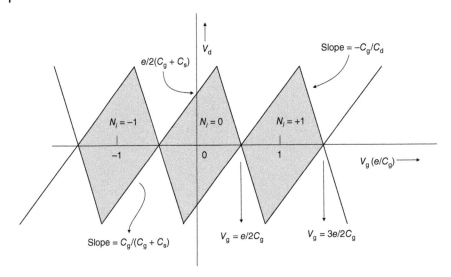

Figure 8.22 Coulomb diamonds resulting from a plot of V_d versus V_g. The shaded regions correspond to Coulomb blockade. N_i is the number of excess electrons in the island. C_g, C_s, and C_d are the gate, source and drain capacitances, respectively (absolute values).

Figure 8.22 (Problem 8.5). The shaded regions correspond to stable conditions – bias values where charge transport is blocked. The number of excess electrons on the island, N_i, is indicated.

From the above discussion, it is evident that the output and transfer curves for SETs differ from those of the conventional FETs described earlier in this chapter. The output characteristic (I_d versus V_d) will exhibit a knee below which the current will be suppressed by Coulomb blockade. However, for $V_g = e/2C_g$ a current will flow. A Coulomb staircase is anticipated as the drain voltage is increased. The transfer characteristics (I_d versus V_g) will show oscillations, corresponding to the transfer of an electron to the island. Both these sets of characteristics are illustrated in Figure 8.23 (output behaviour in (a); transfer in (b)).

Many SETs have been demonstrated in the laboratories, mostly using metallic clusters as the islands [see, for example, 74, 75]. Figures 8.24 and 8.25 show the output and transfer characteristics for a SET for a nano-granular Pt island positioned between prefabricated source and drain electrodes [75]. Both sets of data were recorded at 270 mK. The I_d versus V_d curves in Figure 8.24 reveal the expected Coulomb blockade for low values of V_d (as depicted in Figure 8.23a) followed by a drain current rise in the form of a Coulomb staircase. As the temperature is increased, thermally assisted tunnelling (Section 6.3) leads to a shrinkage of the Coulomb blockade features in the device current versus voltage characteristics [75]. Coulomb blockade oscillations in the drain current versus gate voltage characteristics are evident in Figure 8.25. As discussed above, these features occur whenever the Coulomb blockade is suppressed by the gate voltage.

For applications, the SET can perform logic. Since it can possess both positive and negative g_m values, a complementary type of logic can be devised using only one type of device, operated in different bias regimes. However, the role of SETs as practical switches in electron systems is unclear. Organic molecular materials should be ideally suited to provide components in these quantum structures – both as the tunnelling insulator (as in Figure 8.19) or as the electron island (or both). Reports of SETs using nanotubes [76, 77], C_{60} [78, 79] and graphene [80, 81] may be found in the literature. In 2002, the feasibility of exploiting a macromolecule that uses a single atom bonded within insulating organic regions has also been demonstrated [82]. Such devices are difficult to

Figure 8.23 Theoretical electrical behaviour of a SET. (a) Output characteristics (I_d versus V_d) for two values of V_g. (b) Transfer characteristics (I_d versus V_g). C_g = gate capacitance (absolute value).

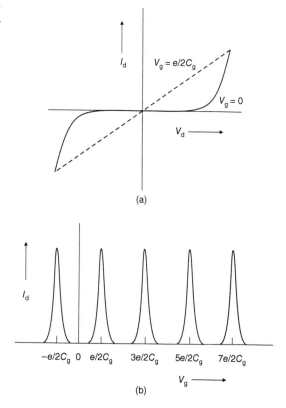

(a)

(b)

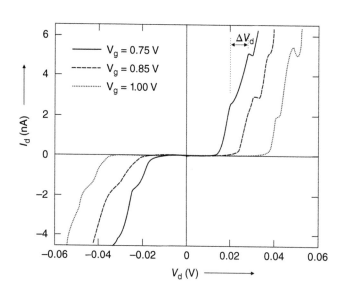

Figure 8.24 I_d versus V_d characteristics of a SET based on a nano-granular Pt island. Measurements at 270 mK. Data are shown for different values of gate voltage. ΔV_d indicates one step in the Coulomb staircase. Source: Di Prima et al. [75]. Licenced under CC BY-3.0.

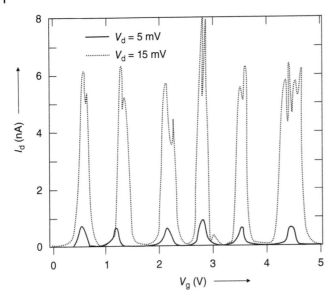

Figure 8.25 Coulomb oscillations in I_d versus V_g characteristics for a SET based on a nano-granular Pt island. Measurements at 270 mK. Source: Di Prima et al. [75]. Licenced under CC BY-3.0.

fabricate reproducibly and in the form of the large arrays required by computer processors and memories. The small currents and parasitic capacitances associated with the single-electron island also provide major challenges for the commercialization of SET technology.

8.8 Transistor-Based Chemical Sensors

A noteworthy feature of many organic (and biological) materials is their ability to interact with molecular species in the environment. The sensitivity of the electrical resistance of organic semi-conductors, such as phthalocyanines (Section 1.5.2), to the presence of oxidizing and reducing gases can be high [13]. This trait can cause problems if the organic semiconductor is to be used as a logic element in electronic circuitry and necessitates meticulous device encapsulation. However, the emerging field of organic electronics offers the possibility of exploiting this property and integrating the chemical sensing ability of organic thin films (gaseous or aqueous species) with the amplification provided by an OFET. This research field is vibrant, with many different approaches under investigation [2, 10, 13, 78, 83–86]. There are several components of an OFET that might be influenced by molecular species: the semiconductor layer, the gate insulator or even the source, drain and gate electrodes. Two rather different examples of chemically sensitive OFETs are provided below: one based on changes to the gate electrode and the other exploiting modification to the gate metallization. Both structures allow easy interaction of the outer part of the OFET with the environment under investigation.

8.8.1 Ion-Sensitive FETs

Electrochemical sensors are probably the most versatile and reliable chemical transducers. Some of these measure the voltage (potentiometric) while others monitor electric current (amperometric),

Section 12.8. Such detectors are used where a chemical reaction takes place or when the charge transport is modulated by the interaction of the chemical species under observation. Electrochemical gas sensors ionize gas molecules at a three-phase boundary layer (atmosphere/electrode of a catalytically active material/electrolyte).

In potentiometric cells, the potential difference generated between the electrodes by different partial pressures on either side of the cell provides the output. An example is the Gas-FET, a chemically sensitive field effect transistor [13]. These devices use a field effect transistor with a gate metallization exposed to the surrounding atmosphere. In the case of a Pd-gate device, hydrogen gas becomes adsorbed on the metal surface and dissociates into atoms, which subsequently diffuse to the oxide/metal interface where a dipole layer is created. This effective change in the work function of the gate electrode results in a shift in the threshold voltage, with a consequent change in the output characteristics of the transistor.

For solution work, ion-selective electrodes are used. These devices consist of membranes that respond to one ionic species in the presence of others. A well-known example is the pH electrode. The output is in the form of a voltage V that is related to the concentration of an ion by the Nernst equation:

$$V = V_0 + 2.3 \frac{RT}{nF} \log \left(\frac{a_{red}}{a_{ox}} \right) \tag{8.42}$$

where V_0 is a constant to account for the other potentials in the system, R is the gas constant (8.31×10^3 J kmole^{-1} K^{-1}), n is the number of electrons transferred, F is Faraday's constant (F = electronic charge x Avogadro's number = 9.65×10^7 C kmole^{-1}) and a_{red} and a_{ox} are the activities of the reduced and oxidized species, respectively. The activity is related to the ion concentration c by the relationship:

$$a = \gamma c \tag{8.43}$$

where γ is the activity coefficient. This parameter is essentially a correction factor, introduced to take into account non-ideal behaviour. In dilute solutions, γ approaches unity and the activity can then be identified with concentration.

For a solid electrolyte in equilibrium with ions in solution, a plot of the measured potential versus the logarithm of the ion concentration will give a straight line, with a slope of about 60 mV (=$2.3RT/F$) per decade at room temperature for a reaction involving the transfer of one electron, e.g. a pH sensor. This is the same value as that for the subthreshold swing of an ideal MOSFET (Eq. (8.12)) as the underlying physical principles are similar.

The complete measurement system consists of an ion-selective electrode, an internal reference electrode and an external reference electrode. Commercial ion-selective electrode systems often combine the two electrodes into one unit.

Ion-sensitive field effect transistors, or ISFETs, have been developed since the 1970s [87, 88]. In essence, the ISFET is a MOSFET in which the gate connection is separated from the silicon substrate in the form of a reference electrode. This is inserted in an aqueous solution that is in contact with the gate insulator, as shown in Figure 8.26, which compares the architecture of (a) an FET to that of (b) an ISFET. A change in the interfacial potential at the liquid/gate insulator interface will produce a change in the threshold voltage of the transistor. Gate insulators such as SiO_2, Si_3N_4, Al_2O_3 and Ta_2O_5 are responsive to hydrogen ions. The mechanism of operation is associated with the formation of a thin hydrated surface layer (4–5 nm). In the case of quartz, it is known that surface silanol groups dissociate:

$$\equiv SiOH \Leftrightarrow SiO^- + H^+ \tag{8.44}$$

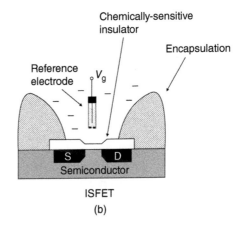

Figure 8.26 Comparison of the structure of (a) FET (field effect transistor) and (b) ISFET (ion-sensitive field effect transistor) devices. S, D represent the source and drain electrodes, respectively. V_g = gate voltage.

Sensitivity to ions other than H^+ can be achieved by the deposition of suitable gate materials. Here, organic or biological thin films can be useful. For example, polymeric membranes containing the ionophore valinomycin (Section 12.3.4) can provide a response to potassium ions. Layers of immobilized enzymes are often used to convert a substance for which no sensor is available into a substance for which a chemical sensor exists. A well-known example is the urea sensor that makes use of immobilized urease. This converts urea into ammonia ions, carboxyl ions, and hydroxyl ions and provides the option of monitoring the pH as an indirect method for detecting urea. Such devices are called enzymeFETs or ENFETs. ImmunoFETs or IMFETs make use of highly specific immunological reactions between antigens and antibodies (Section 12.8.2).

Although most pH ISFETs have been based on the semiconductor silicon combined with an inorganic insulator, e.g. silicon nitride or tantalum oxide, to provide hydrogen ion selectivity, there are increasing reports of device architectures exploiting both organic semiconductors and insulators [14, 48]. Furthermore, other organic layers (e.g. ionophores such as valinomycin) may extend the selectivity of OFET-based sensors [89, 90]. Such devices offer additional benefits over their inorganic counterparts, such as mechanical flexibility, and, possibly, incorporation with textiles and biological structures.

One advantage of ISFET technology is that it offers integration with the signal processing required for a complete instrumentation system. Organic circuitry can readily be combined with the chemically sensitive transistor devices to provide improved signal-to-noise ratio, a reduction in drift and some compensation for temperature changes [90].

8.8.2 Charge-Flow Transistor

A chemiresistor sensor exploits the resistance change of a thin layer of a gas-sensitive material. Although the most common type of chemiresistor uses semiconducting metal oxides, such as tin oxide, as the sensing element (so-called Taguchi sensors), devices can be based on organic materials [13]. The latter respond to a broad range of gases and vapours and may operate at room temperature (chemiresistors based on inorganic semiconductors usually have to be operated at elevated temperatures). Furthermore, a wide range of compounds can be synthesized, which offers the prospect of tailoring devices to respond to particular analytes. A chemiresistor can be fabricated by depositing

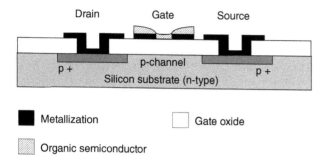

Figure 8.27 Schematic diagram of a charge-flow transistor [91]. Part of the gate metallization in the silicon MOSFET is replaced by a semiconductive organic material. Source: Modified from Senturia et al. [91].

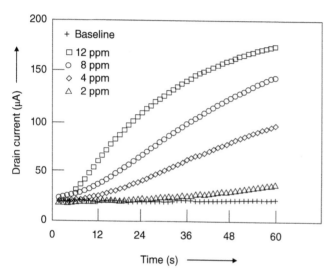

Figure 8.28 Drain current response for a charge flow transistor when a voltage step of −5 V is applied to the gate electrode. Data are shown for the device exposed to different concentrations of nitrogen dioxide. Width of hole in the gate metallization = 35 μm. A phthalocyanine derivative is spin coated to fill the gate hole. Source: Barker et al. [92].

a thin film of the active (gas-sensitive) material onto an interdigitated microelectrode structure. This allows a large surface area of the sample to be exposed to the analyte gas or vapour.

One problem associated with many chemiresistor devices is that the current outputs are low (typically picoamperes) requiring elaborate detection electronics and careful shielding and guarding of components. This difficulty may be overcome by incorporating the organic sensing layer into a field effect transistor. A schematic diagram of one type of structure is shown in Figure 8.27 [91, 92]. Note that configuration is quite different from that described for the Gas-FET described above, which is a potentiometric device based on the change in work function of a gate metal on exposure to a gas. The FET device shown in Figure 8.27 is a hybrid FET, comprising of an organic semiconducting layer deposited onto the gate area of a conventional silicon MOSFET. In principle, an OFET might be used to form the transistor architecture. The hybrid device has an opening in its gate metal that is filled with an organic semiconductor. The principle of operation of this 'charge-flow' transistor is as follows [91]. When a voltage is applied to the gate of the FET, the capacitor that is formed between the gate and the silicon substrate charges in two stages. First, the metallic part of the gate charges very rapidly to the applied voltage. Charge then gradually flows through the organic semiconductor (which will have a lower conductivity than the gate metal) until the organic part of the capacitor is uniformly charged to the applied voltage. The time required for this charging process to be complete will depend on the sheet resistance (the resistivity divided by the film thickness – Section 2.4) of the semiconductor. Therefore, exposure to gas will affect the turn-on time of the FET. Figure 8.28 shows data for such an FET incorporating a film of a spin-coated phthalocyanine derivative in the gate (gate 'hole' = 35 μm) [92]. Increasing the concentration of NO_2 increases the conductivity of the phthalocyanine and reduces the turn-on time of the transistor. Because of the amplification provided by the transistor configuration, the current flowing in this FET device is significantly higher than that for a chemiresistor. Of course, it is also possible to use a chemically sensitive organic film as the semiconductive layer in a diode or a transistor. However, in these cases the device must be fabricated so that the gas can interact readily with the organic material.

There are many other examples of OFET-based sensing devices in the literature. But here is a note of caution. It is often easy (perhaps, too easy) to demonstrate a chemical influence on the operation of an OFET in the laboratory. It is more challenging to develop a reproducible and reliable device that has the required sensitivity and specificity to conform to the stringent demands for process control and environmental monitoring. Nature's sensing apparatus, such as the olfactory system, is a combination of highly developed sensing elements and complex signal processing [13], suggesting that a sophisticated approach will be needed to emulate the successful chemical sensing machinery found in the natural world (as discussed in Section 12.9). Nonetheless, chemical and biochemical sensing is destined to become an important area of endeavour for organic and molecular electronics, from both scientific and commercial perspectives.

Problems

8.1 A p-type OFET (hole mobility = 0.15 cm² V⁻¹ s⁻¹) has the following dimensions: channel length = 10 μm, channel width = 80 μm. The gate insulator has a relative permittivity = 2.8, thickness = 100 nm.

    (a) Assuming that the device is operating in the saturation regime and that the threshold voltage $V_0 = 0$ V, calculate the drain current for an applied gate voltage $V_g = -2.0$ V.

    (b) The organic semiconductor layer thickness = 150 nm, and its doping density = 10^{12} cm⁻³. What is the On/Off drain current ratio at saturation?

    (c) If the organic semiconductor now includes a positive fixed charge density of 10^{-8} C cm⁻² at the semiconductor/insulator interface, calculate the new drain current.

8.2 Derive an expression for the cut-off frequency f_T for an OFET at saturation in terms of the material parameters and device dimensions. Compare this with Eq. (8.17) for the transit time for carriers in the channel. Calculate f_T for the OFET described in Q2, for a drain voltage of 5 V. How might this figure be increased?

8.3 The output characteristics of a p-type OFET with an Ohmic leakage resistance R_{dg} between the gate and drain (see equivalent circuit shown in Figure 8.11a) reveals a positive drain current at low values of negative drain voltage. As the magnitude of V_d is further increased, the magnitude of I_d reduces to zero. Thereafter, the drain current becomes negative, increases in magnitude and saturates, as expected (e.g. as shown in Figure 8.4). Explain this phenomenon. Derive an equation for the approximate value of V_d for $I_d = 0$. Assume $V_0 = 0$ V. If this condition is observed at $V_d = -1.1$ V for a particular device, calculate the value of R_{dg}. OFET parameters: hole mobility $= 0.30\,\mathrm{cm^2\,V^{-1}\,s^{-1}}$, channel length $= 5\,\mu\mathrm{m}$, channel width $= 50\,\mu\mathrm{m}$; gate insulator, relative permittivity $= 2.8$, thickness $= 80\,\mathrm{nm}$; $V_0 = 0\,\mathrm{V}$.

8.4   An OFET has its channel length L channel width W and gate insulator thickness d reduced by a common scaling factor K ($K > 1$). Assuming that the supply voltage and other voltages are also scaled by the same amount (so-called constant field scaling), calculate the effect on (i) the device capacitance, (ii) the device transconductance, and (iii) the power dissipation in the transistor. How are these results affected if the scaling is done so that the voltages remain constant?

8.5   Derive an expression for the drain voltage corresponding to the maximum value of the Coulomb blockade diamonds shown in Figure 8.22. In a SET experiment, this maximum value corresponds to $V_d = 0.05\,\mathrm{V}$. If the diamonds appear periodically in gate voltage at values of $0.75\,\mathrm{V}$, calculate the gate, source and drain capacitance values. Assume that $C_d = C_s$.

References

1 Sze, S.M. and Ng, K.K. (2007). *Physics of Semiconductor Devices*, 3e. Hoboken, NJ: Wiley.

2 Bao, Z. and Locklin, J. (eds.) (2007). *Organic Field Effect Transistors*. Boca Raton, FL: CRC Press.

3 Stallinga, P. (2009). *Electrical Characterization of Organic Electronic Materials and Devices*. Chichester: Wiley.

4 Klauk, H. (2010). Organic thin-film transistors. *Chem. Soc. Rev.* 39: 2643–2666.

5 Sirringhaus, H. (2014). 25th anniversary article: organic field effect transistors: the path beyond amorphous silicon. *Adv. Mater.* 26: 1319–1335.

6 Sandberg, H.G.O. (2008). Polymer field effect transistors. In: *Introduction to Organic Electronic and Optoelectronic Materials and Devices* (eds. S.-S. Sun and L.R. Dalton), 319–350. Boca Raton, FL: CRC Press.

7 Watkins, N.J., Yan, L., and Gao, Y. (2002). Electronic structure symmetry of interfaces between pentacene and metals. *Appl. Phys. Lett.* 80: 4384–4386.

8 Kymissis, I. (2009). *Organic Field Effect Transistors: Theory, Fabrication and Characterization*. New York: Springer.

9 Stutzmann, N., Friend, R.H., and Sirringhaus, H. (2003). Self-aligned, vertical-channel, polymer field-effect transistors. *Science* 299: 1881–1884.

10 Uno, M., Nakayama, K., Soeda, J. et al. (2011). High-speed flexible organic field-effect transistors with a 3D structure. *Adv. Mater.* 23: 3047–3051.

11 Xu, B., Dogan, T., Wilbers, J.G.E. et al. (2017). Fabrication, electrical characterization and device simulation of vertical P3HT field-effect transistors. *J. Sci. Adv. Mater. Devices* 2: 501–514.

12 Kleemann, H., Krechan, K., Fischer, A., and Leo, K. (2020). A review of vertical organic transistors. *Adv. Funct. Mater.* 1907113.

13 Petty, M.C. (2019). *Organic and Molecular Electronics: From Principles to Practice*, 2e. Chichester: Wiley.

14 Tang, W., Huang, Y., Han, L. et al. (2019). Recent progress in printable field effect transistors. *J. Mater. Chem. C* 7: 790–808.

15 Yun, Y., Pearson, C., Cadd, D.H. et al. (2009). A cross-linked poly(methyl methacrylate) gate dielectric by ion-beam irradiation for organic thin-film transistors. *Org. Electron.* 10: 1596–1600.

16 Necliudov, P.V., Shur, M.S., Gundlach, D.J., and Jackson, T.N. (2000). Modeling of organic thin film transistors of different designs. *J. Appl. Phys.* 88: 6594–6597.

17 Street, R.A. and Salleo, A. (2002). Contact effects in polymer transistors. *Appl. Phys. Lett.* 81: 2887–2889.

18 Benor, A. and Knipp, D. (2008). Contact effects in organic thin film transistors with printed electrodes. *Org. Electron.* 9: 209–219.

19 Ante, F., Kälblein, D., Zaki, T. et al. (2012). Contact resistance and megahertz operation of aggressively scaled organic transistors. *Small* 8: 73–79.

20 Horowitz, G., Hajlaoui, M.E., and Hajlaoui, R. (2000). Temperature and gate voltage dependence of hole mobility in polycrystalline oligothiophene thin film transistors. *J. Appl. Phys.* 87: 4456–4463.

21 Levinson, J., Shepherd, F.R., Scanlon, P.J. et al. (1982). Conductivity behaviour in polycrystalline semiconductor thin film transistors. *J. Appl. Phys.* 53: 1193–1202.

22 Street, R.A., Knipp, D., and Völkel, A.R. (2002). Hole transport in polycrystalline pentacene transistors. *Appl. Phys. Lett.* 80: 1658–1660.

23 Park, J., Jeong, Y.-S., Park, K.-S. et al. (2012). Subthreshold characteristics of pentacene field effect transistors influenced by grain boundaries. *J. Appl. Phys.* 111: 104512.

24 Marinov, O., Deen, M.J., Zschieschang, U., and Klauk, H. (2009). Organic thin film transistors: part I – compact DC modelling. *IEEE Trans. Electron Devices* 56: 2952–2961.

25 Kim, C.-H., Bonnassieux, Y., and Horowitz, G. (2014). Compact DC modelling of organic field-effect transistors: review and perspectives. *IEEE Trans. Electron Devices* 61: 278–287.

26 Groves, C. (2017). Simulating charge transport in organic semiconductors and devices: a review. *Rep. Prog. Phys.* 80: 026502.

27 Meijer, E.J., Matters, M., Herwig, P.T. et al. (2000). The Meyer-Neldel rule in organic thin-film transistors. *Appl. Phys. Lett.* 76: 3433–3435.

28 Jahinuzzaman, S.M., Sultana, A., Sakariya, K. et al. (2005). Threshold voltage instability of amorphous silicon thin-film transistors under constant current stress. *Appl. Phys. Lett.* 87: 023502.

29 Park, J., Do, L.M., Bae, J.-H. et al. (2013). Environmental effects on the electrical behaviour of pentacene thin-film transistors with a poly(methyl methacrylate) gate insulator. *Org. Electron.* 14: 2101–2107.

30 Kalb, W.L. and Batlogg, B. (2010). Calculating the trap density of states in organic field-effect transistors from experiment: a comparison of different methods. *Phys. Rev. B* 81: 035327.

31 Streetman, B.G. and Banerjee, S. (2000). *Solid State Electron Devices*, 5e. Hoboken, NJ: Prentice Hall.

32 Torsi, L., Dodabalapur, A., and Katz, H.E. (1995). An analytical model for short-channel organic thin-film transistors. *J. Appl. Phys.* 78: 1088–1093.

33 Austin, M.D. and Chou, S.Y. (2002). Fabrication of 70 nm channel length polymer organic thin-film transistors using nanoimprint lithography. *Appl. Phys. Lett.* 81: 4431–4433.

34 Gundlach, D.J., Zhou, L., Nichols, J.A. et al. (2006). An experimental study of contact effects in organic thin film transistors. *J. Appl. Phys.* 100: 024509.

35 Paterson, A.F., Singh, S., Fallon, K.J. et al. (2018). Recent progress in high-mobility organic transistors. *Adv. Mater.* 30: 1801079.

36 Zschieschang, U. and Klauk, H. (2019). Organic transistors on paper: a brief review. *J. Mater. Chem. C* 7: 5522–5533.

37 Anthony, J.E. (2008). The larger acenes: versatile organic semiconductors. *Angew. Chem. Int. Ed.* 47: 452–483.

38 Braga, D. and Horowitz, G. (2009). High-performance organic field-effect transistors. *Adv. Mater.* 21: 1473–1486.

39 Lüssem, B., Tietze, M.L., Kleeman, H. et al. (2013). Doped organic transistors operating in the inversion and depletion regime. *Nat. Commun.* 4: 2775.

40 Yi, H.T., Payne, M.M., Anthony, J.E., and Podzorov, V. (2012). Ultra-flexible solution-processed organic field-effect transistors. *Nat. Commun.* 3: 1259.

41 Zhang, W., Smith, J., Watkins, S.E. et al. (2010). Indacenodithiophene semiconducting polymers for high-performance, air-stable transistors. *J. Am. Chem. Soc.* 132: 11437–11439.

42 Zhao, Y., Guo, Y., and Liu, Y. (2013). 25th anniversary article: recent advances in n-type and ambipolar organic field effect transistors. *Adv. Mater.* 25: 5372–5391.

43 Niazi, M.R., Li, R., Li, E.Q. et al. (2015). Solution-printed organic semiconductor blends exhibiting transport properties on par with single crystals. *Nat. Commun.* 6: 8598.

44 Chen, K., Gao, W., Emaminejad, S. et al. (2016). Printed carbon nanotube electronics and sensor systems. *Adv. Mater.* 28: 4397–4414.

45 Newman, C.R., Frisbie, C.D., da Silva Filho, D.A. et al. (2004). Introduction to organic thin film transistors and design of n-channel organic semiconductors. *Chem. Mater.* 16: 4436–44561.

46 Veres, J., Ogier, S., Lloyd, G., and de Leeuw, D. (2004). Gate insulators in organic field effect transistors. *Chem. Mater.* 16: 4542–4555.

47 Taylor, D.M. (2016). Progress in organic integrated circuit manufacture. *Jpn. J. Appl. Phys.* 55: 02BA01.

48 Matsui, H., Takeda, Y., and Tokito, S. (2019). Flexible and printed organic transistors: from materials to integrated circuits. *Org. Electron.* 75: 105432.

49 Klauk, H., Halik, M., Zschieschang, U. et al. (2003). Pentacene organic transistors and ring oscillators on glass and on flexible polymeric substrates. *Appl. Phys. Lett.* 82: 4175–4177.

50 Ogier, S.D., Matsui, H., Feng, L. et al. (2018). Uniform, high performance, solution processed organic thin-film transistors integrated in 1 MHz frequency ring oscillators. *Org. Electron.* 54: 40–47.

51 Baude, P.F., Ender, D.A., Haase, M.A. et al. (2003). Pentacene-based radio-frequency identification circuitry. *Appl. Phys. Lett.* 82: 3964–3966.

52 Xiong, W., Guo, Y., Zschieschang, U. et al. (2015). A 3-V, 6-bit C-2C digital-to-analog converter using complementary organic thin-film transistors on glass. *IEEE J. Solid-State Circuits* 45: 1380–1388.

53 Fiore, V., Battiato, P., Abdinia, S. et al. (2015). An integrated 13.56-MHz RFID tag in a printed organic complementary TFT technology on flexible substrate. *IEEE Trans. Circuits Syst. I Regul. Pap.* 62: 1668–1677.

54 Sirringhaus, H. (2009). Reliability of organic field-effect transistors. *Adv. Mater.* 21: 3859–3873.

55 McEuen, P.L. and Park, Y.J. (2004). Electron transport in single-walled carbon nanotubes. *MRS Bull.* 29: 272–275.

56 Anantram, M.P. and Léonard, F. (2006). Physics of carbon nanotube electronic devices. *Rep. Prog. Phys.* 69: 507–561.

57 Appenzeller, J., Knoch, J., Martel, R. et al. (2002). Carbon nanotube electronics. *IEEE Trans. Nanotechnol.* 1: 184–189.

58 Javey, A., Kim, H., Brink, M. et al. (2002). High-K dielectrics for advanced carbon-nanotube transistors and logic gates. *Nat. Mater.* 1: 241–246.

59 Javey, A., Guo, J., Farmer, D.B. et al. (2004). Self-aligned ballistic molecular transistors and electrically parallel nanotube arrays. *Nano Lett.* 4: 1319–1322.

60 Franklin, A.D., Luisier, M., Han, S.-J. et al. (2012). Sub-10 nm nanotube transistor. *Nano Lett.* 12: 758–762.

61 Javey, A., Tu, R., Farmer, D.B. et al. (2005). High performance n-type carbon nanotube field-effect transistors with chemically doped contacts. *Nano Lett.* 5: 345–348.

62 De Volder, M.F.L., Tawfick, S.H., Baughman, R.H., and Hart, A.J. (2013). Carbon nanotubes: present and future commercial applications. *Science* 339: 535–539.

63 Rutherglen, C., Kane, A.A., Marsh, P.F. et al. (2019). Wafer-scalable, aligned carbon nanotube transistors operating at frequencies over 100 GHz. *Nat. Electron.* 2: 530–539.

64 Hills, G., Lau, C., Wright, A. et al. (2019). Modern microprocessor built from complementary carbon nanotube transistors. *Nature* 572: 595–602.

65 Novoselov, K.S., Geim, A.K., Morozov, S.V. et al. (2004). Electric field effect in atomically thin carbon films. *Science* 306: 666–669.

66 Warner, J.H., Schäffel, F., Bachmatiuk, A., and Rümmeli, M.H. (2013). *Graphene: Fundamentals and Emergent Applications*. Waltham, MA: Elsevier.

67 Novoselov, K.S., Fal'ko, V.I., Colombo, L. et al. (2012). A roadmap for graphene. *Nature* 490: 192–200.

68 Han, S.-J., Garcia, A.V., Oida, S. et al. (2013). Graphene radio frequency integrated circuit. *Nat. Commun.* 5: 3086.

69 Yu, C., He, Z.Z., Song, X.B. et al. (2017). Improvement of the frequency characteristics of graphene field-effect transistors on SiC substrate. *IEEE Electron Device Lett.* 38: 1339–1342.

70 Urban, F., Lupina, G., Grillo, A. et al. (2020). Contact resistance and mobility in back-gate graphene transistors. *Nano Express* 1: 010001.

71 Oncel, N., Hallbäck, A.-S., Zandvliet, H.J.W. et al. (2005). Coulomb blockade of small Pd clusters. *J. Chem. Phys.* 123: 044703.

72 Akai-Kasaya, M., Okuaki, Y., Nagano, S. et al. (2015). Coulomb blockade in a two-dimensional conductive polymer monolayer. *Phys. Rev. Lett.* 115: 196801.

73 Stewart, M.D. Jr., and Zimmerman, N.M. (2016). Stability of single electron devices: charge offset drift. *Appl. Sci.* 6: 187.

74 Bitton, O., Gutman, D.B., Berkovits, R., and Frydman, A. (2017). Multiple periodicity in a nanoparticle-based single electron transistor. *Nat. Commun.* 8: 402.

75 Di Prima, G., Sachser, R., Trompenaars, P. et al. (2019). Direct-write single electron transistors by focused electron beam induced deposition. *Nano Futures* 3: 025001.

76 Postma, H.W.C., Teepen, T., Yao, Z. et al. (2001). Carbon nanotube single-electron transistors at room temperature. *Science* 293: 76–79.

77 Seike, K., Kanai, Y., Ohno, Y. et al. (2015). Carbon nanotube single-electron transistors with single-electron storages. *Jpn. J. Appl. Phys.* 54: 06FF05.

78 Nasri, A., Boubaker, A., Hafsi, B. et al. (2018). High-sensitivity sensor using C_{60}-single molecule transistor. *IEEE Sens. J.* 18: 248–254.

79 Khademhosseini, V., Dideban, D., Ahmadi, M.T. et al. (2018). Single electron transistor scheme based on multiple quantum dot islands: carbon nanotube and fullerene. *ECS J. Solid State Sci. Technol.* 7: M145–M152.

80 Puczkarski, P., Gehring, P., Lau, C.S. et al. (2015). Three-terminal graphene single-electron transistor fabricated using feedback-controlled electroburning. *Appl. Phys. Lett.* 107: 133105.

81 Kim, G., Kim, S.-S., Jeon, J. et al. (2019). Planar and van der Walls heterostructures for vertical tunnelling single electron transistors. *Nat. Commun.* 10: 230.

82 Park, J., Pasupathy, A.N., Goldsmith, J.L. et al. (2002). Coulomb blockade and the Kondo effect in single-atom transistors. *Nature* 417: 722–725.

83 Elkington, D., Cooling, N., Belcher, W. et al. (2014). Organic thin-film transistor (OTFT)-based sensors. *Electronics* 3: 234–254.

84 Lee, M.Y., Lee, H.R., Park, C.H. et al. (2018). Organic transistor-based chemical sensors for wearable electronics. *Acc. Chem. Res.* 51: 2829–2838.

85 Li, H., Shi, W., Song, J. et al. (2019). Chemical and biomolecule sensing with organic field-effect transistors. *Chem. Rev.* 119: 3–35.

86 Surya, S.G., Raval, H.N., Ahmad, R. et al. (2019). Organic field effect transistors (OFETs) in environmental sensing and health monitoring: a review. *Trends Anal. Chem.* 111: 27–36.

87 Janata, J. (2009). *Principles of Chemical Sensors*, 2e. New York: Springer.

88 Jimenez-Jorquera, C., Orozco, J., and Baldi, A. (2010). ISFET based microsensors for environmental monitoring. *Sensors* 10: 61–83.

89 Ritjareonwattu, S., Yun, Y., Pearson, C., and Petty, M.C. (2012). An ion sensitive organic field effect transistor incorporating the ionophore valinomycin. *IEEE Sens. J.* 12: 1181–1186.

90 Shiwaku, R., Matsui, H., Nagamine, K. et al. (2018). A printed organic amplification system for wearable potentiometric electrochemical sensors. *Sci. Rep.* 8: 3922.

91 Senturia, S.D., Sechen, C.M., and Wishneusky, J.A. (1977). The charge-flow transistor: a new MOS device. *Appl. Phys. Lett.* 30: 106–108.

92 Barker, P.S., Petty, M.C., Monkman, A.P. et al. (1996). A hybrid phthalocyanine/silicon field effect transistor sensor for NO_2. *Thin Solid Films* 284–285: 94–97.

Further Reading

De Volder, M.F.L., Tawfick, S.H., Baughman, R.H., and Hart, A.J. (2013). Carbon nanotubes: present and future commercial applications. *Science* 339: 535–539.

Klauk, H. (ed.) (2006). *Organic Electronics: Materials, Manufacturing and Applications*. Weinheim: Wiley-VCH.

Klauk, H. (ed.) (2012). *Organic Electronics II: More Materials and Applications*. Weinheim: Wiley-VCH.

Nicollian, E.H. and Brews, J.R. (1982). *MOS (Metal Oxide Semiconductor) Physics and Technology*. New York: Wiley-Interscience.

Valizadeh, P. (2016). *Field Effect Transistors: A Comprehensives Overview*. NJ: Wiley.

9

Electronic Memory

9.1 Introduction

It is evident from the previous chapter that, over the past 20 years, research progress in the field of organic transistors has been impressive. Field effect mobility values in the range 0.1–1 cm^2 V^{-1} s^{-1} are routinely demonstrated and arrays of OFETs can conveniently be fabricated using cost-effective techniques such as inkjet printing. The availability of n-type devices opens up the possibility of an all-organic technology equivalent to silicon CMOS. These signal processing systems will not compete directly with single crystal Si and GaAs technology in terms of operation speed but are likely to form key components in display drivers and as components in transaction cards and RFID tags. Integration into textiles ('smart' clothing) is also envisaged.

For organic electronics to realize its full potential, it is necessary that a further basic circuit element is developed – a memory cell. Although memories represent by far the largest part of conventional (silicon-based) electronic systems, work on organic memories is lagging behind the development of OFETs. Many different concepts have been proposed, but there is no consensus on the way forward [1–7]. Although many memories are entirely electronic in nature (i.e. these are both written to and read from by electrical means), some devices, such as Compact Discs (CDs), Digital Video Discs (DVDs) and hard discs, require additional transducers (drivers) to convert optical, magnetic, or other physical quantities to electronic signals. Proposed memories also exploit other physical effects such as photochromism, chemical hole burning or molecular chirality [8]. This chapter will focus on electronic memories and their physical principles of operation.

9.2 Memory Types

The fundamental requirement for a memory is that the individual memory elements, or bit-cells, possess at least two stable states (or metastable states, but stable for a period appropriate for the data storage) and that these can be switched by an external stimulus (the writing process). The two binary states are usually indicated as a '0' and a '1'. It is also important that the states can be distinguished by applying a further external signal (the reading process). Reading the memory cell may leave the stored information unchanged (non-destructive readout) or may alter it in some way, so that the memory cell has to be re-written (destructive readout). Information storage devices are commonly grouped into random access devices and sequential access devices. In the former, the storage cells are organized in an array of memory elements (an early example is the magnetic core memory). This architecture provides short access times, which are independent of the location of a particular memory cell. In contrast, the access time for a cell in a sequential device will depend on

Electrical Processes in Organic Thin Film Devices: From Bulk Materials to Nanoscale Architectures, First Edition. Michael C. Petty.
© 2022 John Wiley & Sons Ltd. Published 2022 by John Wiley & Sons Ltd.
Companion Website: www.wiley.com/go/petty/organic_thin_film_devices

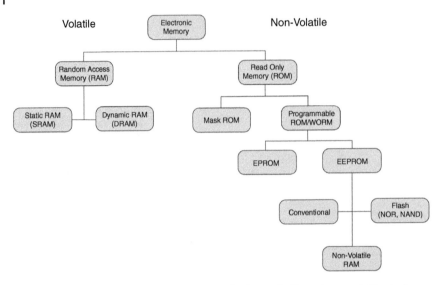

Figure 9.1 Classification of electronic memories into volatile and non-volatile devices.

its physical location relative to the position of the read and write units. Such mass storage devices (a common example is the hard disc drive found in computer systems) are usually used as secondary memory for information archiving and retrieval.

Electronic memory devices are inherently volatile or non-volatile. In the former case, the information is lost if the electrical power is removed. Figure 9.1 shows how these broad memory classifications are further subdivided. Examples of silicon volatile memory are the static random access memory (SRAM) and the dynamic random access memory (DRAM) [9, 10]. The overall organization of these two device types is rather similar. Each memory cell is located in a two-dimensional array, at the intersection of word lines and bit lines. These memories are all random access in the sense that the cells can be addressed for write or read operations in any order, depending on the word line and bit line addresses that are selected.

Currently, DRAM is used as the main memory in most computers. This technology exploits the charge stored on the plates of a capacitor; two different charge storage levels represent the binary information. A DRAM cell, as depicted in Figure 9.2a, comprises one capacitor and a single access transistor (to select a particular memory element during the read and write operations). This memory is relatively efficient in its use of area on a silicon chip. However, due to charge leakage from the capacitor, the data stored in DRAM cells must periodically be refreshed (Problem 9.1). The minimum required cell capacitance is about 50 fF (50×10^{-15} F). This is determined by so-called soft errors resulting from the earth's bombardment by cosmic rays, which can lead to charged particles affecting the memory cell. Maintaining a capacitance of 50 fF as the cell dimensions are reduced from one chip generation to the next presents an enormous challenge (but less demanding than the aF capacitances needed for single-electron transistors, Section 8.7). There is a limit to how much the gate insulator thickness can be reduced. Alternative approaches focus on increasing the device area (e.g. using trenches and multi-capacitor stacks) and exploiting high permittivity insulators (e.g. ferroelectrics).

SRAM is a memory technology that uses flip-flop logic gates. This memory is also volatile as the information stored (as a result of the logic gates being locked in a particular configuration) will be lost if the power is removed. A single SRAM cell, shown in Figure 9.2b, consists of two

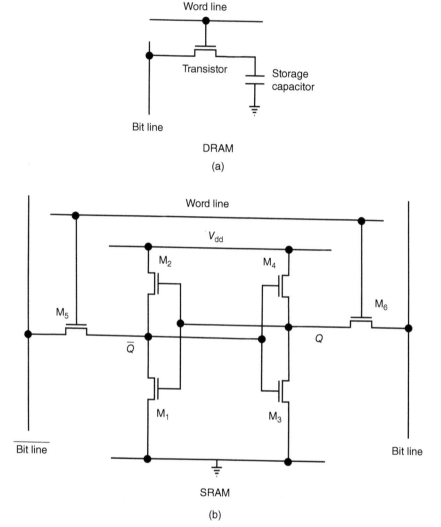

Figure 9.2 (a) Circuit diagram of a Dynamic Random Access Memory (DRAM) cell. The memory element is located, in a two-dimensional array, at the intersection of word lines and bit lines. (b) Circuit diagram of a Static Random Access Memory (SRAM) cell. Transistors are identified $M_1 - M_6$. The outputs of the two inverters are designated Q and $\overline{Q}$. The operating voltage is V_{dd}.

cross-coupled inverters and two access transistors, resulting in six transistors ($M_1 - M_6$) for each cell. Each memory element is bistable: if the output of one inverter is high (Q), then the output from the other inverter will be low ($\overline{Q}$). Such memory takes up additional space on the silicon wafer and is more expensive than DRAM. However, SRAM has faster access times than DRAM and is used as cache (temporary) memory to store frequently used instructions and data for quicker access by the system processor.

Non-volatile read only memories (ROMs) can be further divided into those that are programmed once and devices that are re-programmable, as depicted in Figure 9.1. For the first group, the programming can be achieved during chip manufacturing by appropriate layout of the final metal-lization mask (mask ROM). This may also be accomplished by the user in an initial programming step in which metal bridges can be fused or left intact to represent the binary information (programmable ROMs or PROMs). Using organic materials, write-once/read-many-times (WORM) memory can be realized by burning polymer 'fuses' [11]. A thin conductive polymer layer, or fuse, of poly(3,4-ethylenedioxythiophene) (PEDOT) is sandwiched between two electrodes. The as-deposited device exhibits high conductance due to the polymer, but when a burning-voltage pulse is applied (the write process), the PEDOT fuses 'blow' and causes the device to be in an open-circuit condition. The write process can be as short as microseconds, depending on the thickness of the PEDOT layer and the amplitude of the voltage pulse. Other organic WORM memories exist, based on the irreversible change in resistance of a material on application of a voltage [8, 12].

A nanomechanical sequential data storage concept based on an AFM – called the 'millipede' – that has potentially ultrahigh density has been developed by the company IBM [13, 14]. Indentations, 4–8 nm, are made in a thin (100 nm) polymer layer. Demonstrations of endurance and retention performed at a storage density of 1 Tbit in.$^{-2}$ show 10^8 write cycles using the same AFM tip, 10^3 erase cycles and 3×10^5 read cycles, and extrapolated to 10 years of retention at 85 °C. However, this type of memory is largely of academic interest; because of competing data storage technologies, no commercial product has been developed.

Electronically programmable ROMs or EPROMs can exploit charge storage on a floating gate of a field effect transistor. In the well-established configuration, shown in the top diagram of Figure 9.3, the gate electrode of a silicon MOSFET is replaced by a floating (electrically isolated) gate, usually made from polycrystalline silicon (polysilicon). The memory is programmed by exciting electrons over or through the potential energy barrier associated with the silicon dioxide layer. As a result, the gate of the MOSFET becomes negatively charged and is no longer floating.

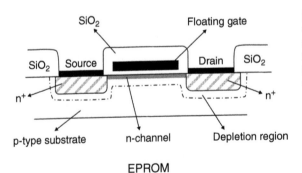

EPROM

Figure 9.3 Schematic diagrams of electronic programmable memories based on silicon. Top: electronic programmable memory (EPROM). Bottom: electronic programmable and erasable memory (EEPROM or E^{2}PROM).

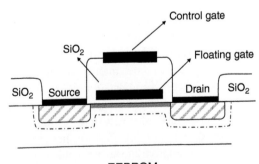

EEPROM

Once sufficient charge has been built up, the transistor is turned off and there is no current flow between the drain and the source electrodes. To remove the charge, the electrons located on the polysilicon gate need to acquire large energies to pass back over the polysilicon/SiO_2 barrier. This can be achieved by illuminating the memory device with UV light. Thus, the EPROM is electrically programmable and optically erasable.

If a MOSFET has both a conventional (contacted or control gate) and a floating gate, it has the potential for both electrical programming and erasing. The structure of such an EEPROM (or E^2PROM) is also shown in the bottom diagram of Figure 9.3. Figure 9.4 shows the effect of placing charge on the floating gate on the transfer characteristics of the transistor. The voltage V_t represents the threshold, or turn-on, voltage of the transistor (Section 8.2). The drain current, I_d, versus gate voltage, V_g, characteristic is displaced along the voltage axis as the charge accumulates on the floating gate. The voltage shift, ΔV_g, is related to the stored charge, ΔQ_{fg}:

$$\Delta V_g = \frac{d \Delta Q_{fg}}{\varepsilon_i \varepsilon_0} = \frac{\Delta Q_{fg}}{C_i} \tag{9.1}$$

where d is the oxide thickness beneath the floating gate of the EEPROM, ε_i is its relative permittivity, and C_i is the gate insulator capacitance per unit area (i.e. the capacitance associated with the dielectric between the control gate and the floating gate, which will determine the voltage that has to be applied to the control gate to compensate for ΔQ_{fg}). This equation may be compared to that for the effect of trapped charge in a field effect transistor, Eq. (8.30). The voltage shift can be measured from the change in drain conductance.

In a floating gate memory device, charge is injected to the floating gate either by hot-carrier injection or by Fowler–Nordheim (FN) tunnelling (Section 6.7). Both processes are depicted in the potential energy versus distance diagrams shown in Figure 9.5. The energy corresponding to the potential barrier that the electrons must surmount is shown as ΔE. In the hot-carrier process, the high longitudinal field in the pinch-off region of the transistor channel accelerates electrons towards the drain. If the electrons acquire sufficient energy from the field, they become 'hot' and can surmount the 3.1 eV energy barrier ($=\Delta E$) that exists between the conduction bands of the silicon and the SiO_2. At the same time, the high field also induces impact ionization. The secondary electrons that are generated by this process can also be injected to the floating gate. Hot-carrier injection currents result in an effective gate current in a conventional MOSFET; this peaks at $V_{fg} \approx V_d$, where V_{fg} is the potential of the floating gate.

Figure 9.4 Effect of charge storage on the floating gate of an EEPROM. The charge results in a shift in the drain current, I_d, versus gate voltage, V_g, characteristic by an amount ΔV_g. The voltage V_t is the threshold voltage.

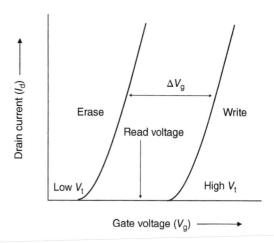

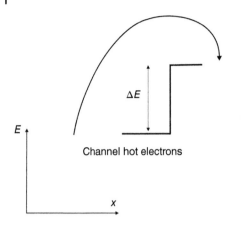

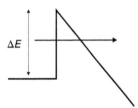

Fowler–Nordheim (FN) tunnelling

Figure 9.5 Charging/discharging mechanisms in floating gate memories. Top: excitation of hot carriers over a potential energy barrier ΔE. Bottom: Fowler–Nordheim (FN) tunnelling through the barrier.

The Fowler–Nordheim mechanism is a quantum mechanical tunnelling process. The application of a strong electric field (in the range $8–10\,\mathrm{MV\,cm^{-1}}$) across the oxide layer results in a significant tunnelling current (current transport *through* rather than *over* the potential energy barrier) without destroying the dielectric properties. For an n-channel MOSFET, application of a positive voltage across the two insulating layers will establish electric fields in the two insulators. If $\mathcal{E}_1$ and $\mathcal{E}_2$ represent the fields in the bottom and top insulators, respectively, then using Gauss' law:

$$\varepsilon_1 \mathcal{E}_1 = \varepsilon_2 \mathcal{E}_2 + Q_{\mathrm{fg}} \tag{9.2}$$

and

$$V_{\mathrm{g}} = V_1 + V_2 = d_1 \mathcal{E}_1 + d_2 \mathcal{E}_2 \tag{9.3}$$

where $\varepsilon_1, \varepsilon_2$ are the relative permittivity values and d_1, d_2 are the thicknesses of the bottom and top insulators, respectively; V_1 and V_2 represent the voltages across these layers (Problem 9.2).

For both the writing and erasing operations, the floating gate potential is affected by the voltage applied to the control gate. The efficiency of this is determined by the coupling ratio, K, which determines the portion of control gate voltage that is capacitively coupled to the floating gate:

$$K = \frac{C_2}{C_1 + C_2} \tag{9.4}$$

where C_1 and C_2 are the capacitances of the bottom and top insulator layers, respectively. In practice, the areas are not the same (e.g. the top control gate may wrap around the floating gate, unlike the arrangement depicted in Figure 9.3). The floating gate potential is therefore given by:

$$V_{\mathrm{fg}} = K V_{\mathrm{g}} \tag{9.5}$$

Coupling ratios 0.5–0.6 are typical in silicon floating gate memories.

Following programming, a long retention time, t_R, is required for non-volatile memories. Charge on the floating gate can decrease by thermal assisted tunnelling to the semiconductor (Section 6.3). This process can also be influenced by trapped charge in the tunnel insulator (trap-assisted tunnelling), which itself will change with time. Experimentally, the retention time is taken as the time for the memory charge to reduce by a particular fraction of its initial value (e.g. 15% or 50%). There are a number of theoretical models that can be used to describe t_T, which is a temperature-dependent parameter and often described by a classical Arrhenius law [15]:

$$t_R = t_0 \exp\left(e\phi_i/k_B T\right) \tag{9.6}$$

where t_0 is a constant and ϕ_i is the energy barrier of the floating gate to the insulator (Problem 9.3). An alternative model, which provides a good fit to experimental data is [16]:

$$t_R = t_0 \exp\left(-T/T_{DR}\right) \tag{9.7}$$

where T_{DR} is a characteristic temperature of data retention. The physical explanation for this model resides in the exponential temperature dependence of the thermally assisted Fowler–Nordheim current (Section 6.7).

Flash memory is a simple, cost-effective EEPROM architecture, in which the devices are re-programmed in blocks of information rather than by individual memory cells (the term 'flash' refers to this erase process). In addition, flash memory offers fast read access times (although not as fast as volatile DRAM memory) and better shock resistance than hard disk memories. The memory is categorized by two basic designs, NOR and NAND, dependent on how the individual cells are interconnected. In NOR gate flash, each cell has one end connected directly to ground, and the other end connected to a bit line. The memory operates in a similar fashion to a NOR logic gate. In contrast, the transistors in NAND flash are connected in a manner that resembles a NAND gate, with several devices connected in series. NOR flash possesses a fast initial access for high read performance, while NAND flash has slow initial access and high write performance. This configuration also permits a denser layout and greater storage capacity per chip and is the main form of universal serial bus (USB) flash devices that are used for high-capacity data storage. One advantage, however, of the NOR approach is that the technology provides full address and data buses, allowing random access to any memory location. Hot-carrier injection is usually used to write data to NOR cells, whereas Fowler–Nordheim tunnelling is exploited to write and erase NAND cells, and to erase NOR cells.

Multilevel flash technology allows more than one bit to be stored per memory cell. For example, rather than a 0 and a 1 being associated with the bit states, values may be interpreted as four distinct combinations, 00, 01, 10, and 11 (in this particular case, providing two bits of information). However, in these devices, it is necessary to be able to distinguish between multiple charge levels on the floating gate.

Key issues for any electronic memory technology include the operating voltage, the time for changing the memory state, the read time, the retention time, the amount of energy to change the state, and the number of memory cycles. At present (2021), Moore's Law continues to drive transistor-based memory scaling, but the complexity is increasing. The company Intel started production of three-dimensional transistors, known as FinFETs and based on a 22 nm lithography process, in late 2011. The distinguishing characteristic of the FinFET is that the gate wraps around a silicon 'fin', which forms the source and the drain. The thickness of the fin (measured in the direction from source to drain) determines the effective channel length of the device. Tri-gate FinFETs, introduced in 2014, were based on 14 nm processing. Currently 7 nm devices are in production (giving gate pitches of around 50 nm), while 5 nm structures are expected to be commonplace in the early 2020s. A prediction by IBM is that devices based on 2 nm processing will be in market by late 2024.

The potential for scaling the technology for multiple generations beyond this is therefore important. Current memory technologies, such as DRAM, SRAM, and NAND flash, are approaching very difficult issues related to their reduction in size. Flash scaling is easier due to its relatively simple structure. However, scaling will eventually become difficult due to the high electric fields required for the programming and erase operations, and the stringent requirements for long-term charge storage. These demands are imposing fundamental limitations on the memory cell operating voltages and on the physical thickness of the tunnelling dielectric.

The International Technology Roadmap for Semiconductors has identified a number of promising candidates for new memory technologies [4]: ferroelectric (FET and tunnel junction memory), metal oxide filamentary, Mott, carbon, macromolecular, and molecular. Some of these technologies listed are interrelated. Many of the devices can be classified as non-volatile random RAM rather than re-programmable ROM. The distinction is blurred, but is generally based on write times, which are in the nanosecond to microsecond range for the non-volatile RAMs and of the order of seconds for the re-programmable ROM. Non-volatile RAM represents the ideal memory. The following sections focus of some of these ideas, in as far as these have been developed using organic materials.

9.3 Resistive Memory

Resistive memory devices rely on an appropriate physical mechanism to switch the resistance of a material between high resistance and low resistance states. The absolute value of the On/Off ratio need not be that high – it has been suggested that a factor of 10 or more is sufficient [17] – so long as the distributions of the On/Off values over a large number of memory cells is narrow. Resistive RAM is a very broad classification of memory technologies based on different physical phenomena [4, 7, 18]. The intense research activity on such devices is motivated by their very simple structure and ability for scaling to small dimensions. There is also the possibility that arrays of the memory cells can be stacked in multiple levels in three dimensions.

Organic resistive memory structures are formed by interposing thin layers of organic molecules between two electrodes, as shown in Figure 9.6. The crossed-bar (or crossed-point) architecture permits the closest packing of bit-cells, with each occupying an area of $4F^2$, where F is the minimum feature size (the line width and spacing of the electrodes). A variation uses the tip of a scanning probe microscope instead of a (thermally evaporated) metallic top electrode, leading to ultrahigh data storage density.

A very wide range of organic materials that exhibit resistive switching has been reported. At least six different types of switching characteristics have been identified, according to the nature of the current, I, versus voltage, V, behaviour [17]. These are illustrated in Figure 9.7. Figure 9.7a shows relatively symmetric hysteresis; there is no distinct threshold voltage, or stable states, which might be used as the basis for a memory device. In contrast, the $I–V$ characteristics in Figure 9.7b reveal a sharp increase in current at a particular positive applied voltage; reversing the polarity of the bias can then turn off the device. Similar behaviour is observed in the characteristics of Figure 9.7c; however, the device turns off as the positive voltage is reduced and before zero bias is reached. The $I–V$ curves in Figure 9.7d show a similar sharp turn-on at a positive voltage, but the device does not turn off again. This type of characteristic is typical of a WORM memory. A negative differential

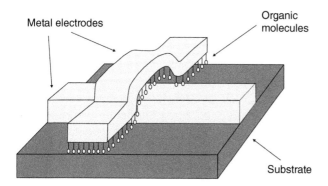

Figure 9.6 Crossed-bar architecture in which organic molecules are sandwiched between orthogonal metallic electrodes.

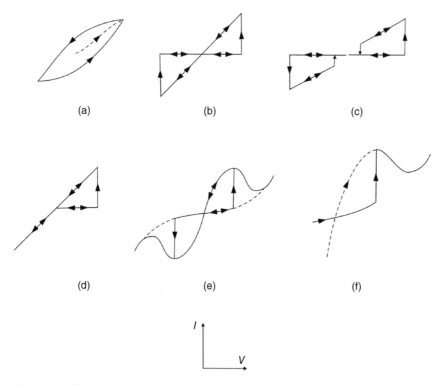

Figure 9.7 Current, *I*, versus voltage, *V*, behaviour reported for resistive switching/memory devices. (a) symmetric hysteresis; (b) sharp increase in current at a positive applied voltage; (c) similar to (b) but device turns off as the positive voltage is reduced; (d) sharp turn-on at a positive voltage, but the device does not turn off again; (e) and (f) negative differential resistance (NDR) regions are observed. See text for details. Source: Scott and Bozano [17].

resistance (NDR) region is noted in the characteristics shown in Figure 9.7e,f. Here, as the voltage is increased, the current decreases (but, the absolute resistance value, of course, remains positive).

Memory devices based on a change in resistance are frequently referred to as memristors. This is a name that was originally introduced in 1971 to suggest the existence of a fourth fundamental passive electrical circuit element (in addition to the resistor, capacitor and inductor) [19]. However, this assertion is controversial [20]. The use of the term memristor is still somewhat unclear in the literature but generally describes a resistor exhibiting both nonlinear current versus voltage characteristics and non-volatile memory function [7].

Over the past 60 years or so, there have been many reports of switching and memory effects in thin films. The film thicknesses are generally less than 1 μm and the phenomena are observed in different types of material (inorganic compounds, such as silicon dioxide and metal oxides; and organic compounds, such as polymers and charge-transfer complexes). Furthermore, the thin films have been formed using a variety of techniques (e.g. spin-coating, thermal evaporation). The only experimental parameter that is common to all the structures studied is the presence of metallic electrodes (e.g. Al) below and on top of the thin film. The switching characteristics of organic memory devices may be identified with one of the *I–V* curves shown in Figure 9.7. However, it is important to note that researchers undertake their experiments in different ways, e.g. different voltage ranges and/or different measurement protocols. It is often difficult to compare directly the data reported by individual groups.

Figure 9.8 shows some typical resistive switching data, like those encountered in the literature [21]. The organic compound used in this study was a polyfluorene derivative, 7-{4-[5-4-*tert*-butylphenyl-1,3,4-oxadiazol-2-yl]phenyl}-9,9-dihexyl-*N*,*N*-diphenyl-fluoren-2-amine, although similar effects are found in a range of organic materials. The device was formed by sandwiching a thin film of this compound (approximately 50 nm in thickness and formed by spin-coating) between aluminium electrodes. The current versus voltage behaviour, shown in the

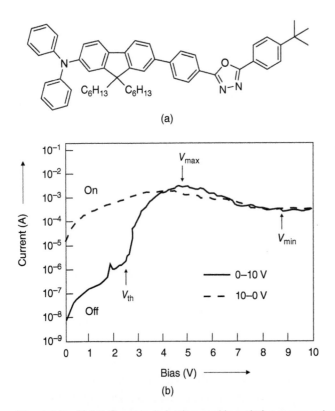

(a)

(b)

Figure 9.8 (a) Polyfluorene derivative used in resistive memory device. (b) Current versus voltage characteristics of resistive memory device based on the polyfluorene molecule. Source: Dimitrakis et al. [21].

figure (this was reported to be reasonably symmetric to the polarity of the applied voltage) exhibits both NDR and memory effects, i.e. the features shown by the *I–V* characteristics of Figure 9.7e. For the data shown in Figure 9.8, the On state is obtained by applying a voltage V_{max} close to the current maximum (just before the NDR region ∼ 4.5 V) and reducing it to zero (write). In contrast, switching from the high conductivity On state to the Off state is accomplished by selecting a voltage V_{min} corresponding to the current minimum in the NDR region (∼8 V) and reducing this rapidly to zero (erase). The state of the device (On or Off) is determined by measuring the current at a voltage of 1 V (read).

There is currently no consensus on how resistive thin film memories, such as those described above, operate. The explanations generally fall into two categories: (i) the injection and storage of charge in the thin film and (ii) metallic filament formation. A device 'formation' step is often required before switching effects are observed, but details of the process are sometimes sketchy. There is some confusion in the literature between the electrical behaviour of the unformed (virgin) device and that of the On and Off states. Three different states for resistive memory devices are often reported, each exhibiting different current versus voltage behaviour: unformed; formed On state; and formed Off state [21, 22].

Many of the recipes for switching are based on the incorporation of metallic nanoparticles within an organic thin film [23]. The presence of the nanoparticles is thought to influence the transfer of charge (e.g. by a trapping/hopping process). It is now clear that the intentional addition of such nanoparticles is not necessary to observe memory effects, although these may improve the switching performance [17, 24]. Bistable devices that do not include nanoparticles are preferable, as their presence will restrict the device scaling, i.e. when the cell dimensions become comparable to those of the nanoparticles.

The direct observation of metallic filaments by electron microscopy and energy dispersive X-ray spectroscopy suggests that these may be responsible for some of the electrical characteristics reported for organic thin films [25, 26]. Figure 9.9 shows the essential features of a simple model for a device fabricated by sandwiching an organic material between aluminum electrodes [24]. This is similar to that suggested to explain the switching in thin film structures with a silver electrode [25]. During film formation, aluminium ions, formed by oxidation, are injected from

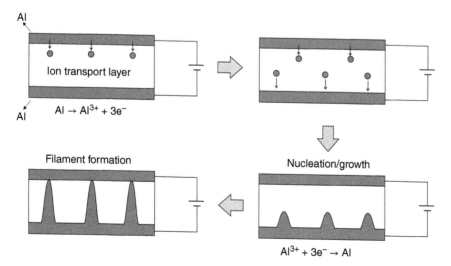

Figure 9.9 Model for filament nucleation and growth. Source: Lee et al. [24].

the top, positively biased, electrode, drift through the organic film and are subsequently reduced at the bottom electrode to form aluminium, which grows in the form of filaments into the film. When these reach the top electrode, the device switches into a higher conductivity regime (follow the arrows in Figure 9.9). The creation of a network of conductive filaments throughout the organic film constitutes the device formation. After formation, the application of an increased voltage results in excessive heat dissipation within these filamentary regions, leading to their destruction (Problem 9.4). However, filaments can subsequently reform by the drift of Al^{3+} ions from either electrode (depending on the polarity of the applied bias). There will be a dynamic equilibrium between filament formation and destruction, as first suggested in 1970 [27]. The voltage-controlled, 'N-shaped' current versus voltage characteristic (Figure 9.8b) results from non-regenerative filament formation, i.e. the resistance of the device is only slightly perturbed by the creation of a filament and the conditions are retained for the development of further filaments.

The above model also explains the almost Ohmic conductivity observed for the On state and is consistent with the observation that the switching appears to be related to the electric field in the organic film [24]. The role of the organic thin film in the resistive memory device is to provide a suitable environment for the transport of metal ions, which allow the redox processes involved in filament nucleation and growth to occur. Nanoparticles, if present, provide nodes within the organic film where filaments can terminate and/or grow. Such effects stabilize the network of filaments, providing more reproducible switching electrical behaviour.

Figure 9.10 shows how the two-terminal resistive elements can be incorporated into a two-dimensional array (random access memory [RAM]). There is one memory cell at the intersection of a word line and a bit line. The most common way to read the resistive elements is to apply a voltage to the selected word line and to couple a sense amplifier to each bit line, as shown in Figure 9.11a. The voltage increase at each sense amplifier depends on the resistance of the element and the (parasitic) capacitance at the input. Thus, the time taken for the voltage signal to rise above a threshold level is directly proportional to the resistance. The crossed-bar architecture will have many parallel paths connecting each word line to each bit line; only one of these passes directly through the addressed node. These additional currents can easily overwhelm the desired signal. However, if the nodes are rectifying, for example, by the addition of a diode to each resistive memory element, the excess current can be minimized. Figure 9.11b illustrates this effect: one of the additional diodes (top left in the figure) will be reverse biased, thereby reducing

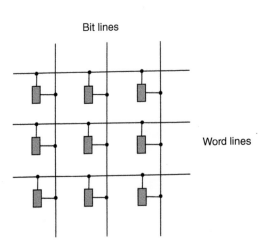

Bit lines

Word lines

Figure 9.10 Incorporation of two-terminal resistive memory elements (shown as shaded rectangles) in a two-dimensional array.

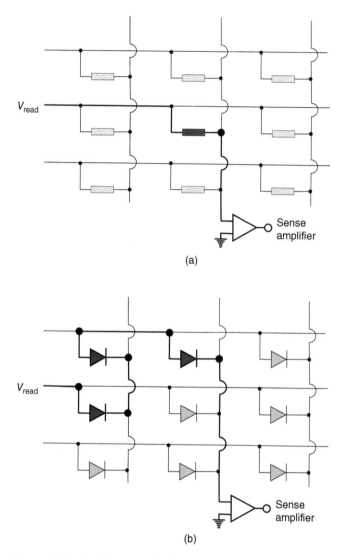

(a)

(b)

Figure 9.11 Reading state of resistive memory element in an array. (a) Basic array. By selecting a particular memory element, some parallel current paths are also available via other circuit elements. (b) Resistive memory elements with rectification properties. In the parallel path, one of the diodes will be reverse biased. Source: Scott and Bozano [17].

significantly the current flowing though the unwanted path. It is envisaged that high rectification ratios (approaching 10^8 for a 10 MB memory) will be needed for practical devices [17].

Resistive memory devices are not restricted to solid supports. Foldable and disposable (e.g. on paper) memory arrays based on resistive switching can be fabricated [28]. Furthermore, a three-dimensional 8×8 crossed-bar array of polymer resistive devices based on a composite of polyimide and 6-phenyl-C_{61} butyric acid methyl ester has been reported [29]. The particular architecture that was fabricated possessed three active layers, providing a total of $64 \times 3 = 192$ memory cells. Five-layer crossed-bar arrays of self-rectifying resistors based on Si/SiO$_2$/Si structures have been developed [30]; these possess On/Off ratios of up to 10^4. Such results suggest that three-dimensional arrays may provide a viable technology for the high-density integration of memory cells. A number of companies and research organizations are pursuing resistive random access memory (ReRAM or RRAM) or memristor technologies. Crossbar ReRAM technology uses a silicon-based switching material as the host for a metallic filament formation; the ReRAM cell is capable of withstanding temperature swings from −40 to 125 °C, over 10^6 write cycles, and a retention of 10 years at 85 °C [31]. Further research and development work will be required to transfer this technology to organic materials.

Resistive memory devices can also be based on single-walled carbon nanotubes. One scheme uses a crossed-bar arrangement of rows and columns of nanotubes separated by supporting blocks [32, 33]. The application of an appropriate voltage between the desired column and row bends the top tube into contact with the bottom nanotube. Van der Waals forces maintain the contact, even after the voltage has been removed. Separation of the SWNTs is then achieved by the application of a voltage pulse of the same potential. Assuming a minimum cell size to be a square of 5 nm length, a packing density of 10^{12} elements cm^{-2} can be achieved. The inherent switching time is estimated to be in the 100 GHz range. The memory is as fast as, and denser than, DRAM and has essentially zero power consumption in standby mode.

9.4 Organic Flash Memory

As noted in the Section 9.2, charge storage is the basis of the operation of flash memory. A flash device is similar in structure to a MOSFET, except that it has two gate electrodes, one above the other. The top electrode forms the control gate, below which a floating gate is capacitively coupled to the control gate and the underlying semiconductor. The memory cell operation involves putting charge on the floating gate or removing it, corresponding to two logic levels.

Nanoflash devices can exploit single or multiple nanoparticles (or nanocrystals) as the charge storage elements. These are usually embedded in the gate oxide of a field effect transistor and located in close proximity to the transistor channel. In a simple version of this device, a very thin (<3 nm) insulating capping layer surrounding gold nanoparticles (Au-NPs) allows charge (electrons or holes) to move between an organic semiconductor and the nanoparticles by the process of quantum mechanical tunnelling [34]. The charge retention time is key to the successful development of any flash memory. Equation (9.6) provides a measure of t_R based on the energy barrier associated with the floating gate. The retention time will also be dependent on the leakage current that can flow through the gate insulator. These currents must be less than 10^{-22} A if a device is to lose less than 50% of its charge over a 10-year period.

Figure 9.12 depicts the architecture of a flash memory OFET based on pentacene/Au-NP/PMMA [35]. The dependence of I_d on V_d (OFET output characteristics) is shown for the memory device and also for control device (with no Au-NPs) in Figure 9.13. In each measurement, the forward and reverse voltage scans are shown. Whereas both the Au-NP and control devices exhibit typical transistor behaviour, the addition of the nanoparticles produces a distinct hysteresis in the output characteristics. This is attributed to the charging and discharging of the Au-NPs with the applied voltage. Figure 9.14 shows the effect of programming pulses on the value of the drain current when a voltage is applied to the gate electrode. The write state of the memory is achieved by the application of a pulse of −30 V for one seconds while, for the erase state, a voltage pulse of 30 V is used for the same period of time. With a gate voltage of 10 V, it is possible to distinguish whether the device is in the write or erase state from the value of the drain-to-source current, evident from Figure 9.14.

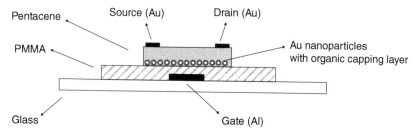

Figure 9.12 Schematic diagram of pentacene organic flash memory transistor. Source: Modified from Mabrook et al. [35].

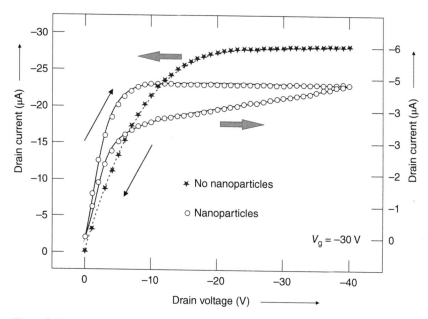

Figure 9.13 Transistor output characteristics for pentacene memory transistor. Data are shown for devices with and without gold nanoparticles. No hysteresis is evident for the reference (no nanoparticles) device. Source: Mabrook et al. [35].

Pentacene memory transistor arrays (26 × 26 elements) on plastic substrates and operating at relatively low voltages have been achieved using thermally evaporated aluminium as the floating gate and a 2 nm thick alkyl-phosphonic acid-based self-assembled monolayer as the dielectric [36]. The small thickness of the dielectric allows very low programming and erase voltages (≤6 V) to produce a large, non-volatile, reversible threshold-voltage shift. Furthermore, the transistors endured more than 1000 program and erase cycles. Other notable achievements in this category of organic flash memory include the development of solution processing to build up memory devices on flexible supports [37, 38]. The challenge now is to exploit manufacturing routes such as inkjet printing, screen printing, and spray coating to achieve low-cost memories compatible with flexible substrates, which can be integrated into devices such as display drivers, radio RFID tags, and electronic paper (e-paper).

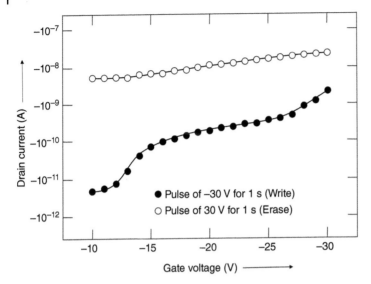

Figure 9.14 Effect on drain current versus gate voltage characteristics of applying write and erase pulses to pentacene organic memory transistor. Source: Mabrook et al. [35].

9.5 Ferroelectric RAMs

Longer retention times in DRAM-type memory cells can be achieved if the dielectric layer has ferroelectric properties [2, 39, 40]. A ferroelectric has an in-built electric polarization; furthermore, this can be reversed by an applied electric field. Such materials have been introduced in Section 7.2.3. One important polymer that exhibits such behaviour is poly(vinylidene difluoride) (PVDF). The chemical structure has been shown in Figure 7.2a, consists of the mer unit CH_2CF_2, and has at least four different crystalline phases. The α-phase is the most common and the other forms can be obtained from this parent phase by applications of mechanical stress, heat, and electric field. The polymer chain in the α-phase results in the dipole moments associated with the carbon-fluorine bonds arranged in opposite directions, so that there is no net polarization within the crystal.

When the α-phase of PVDF is mechanically deformed by stretching or rolling at temperatures below 100 °C, the β-phase of PVDF is formed. The unit cell of this structure (Figure 7.2b) has a net dipole moment (normal to the chain direction). However, because of the random orientation of the crystallites, there is no net polarization in the material. The application of a strong electric field (poling) is needed to confer the PVDF with an overall dipole moment. Furthermore, the material can be taken though a hysteresis loop, shown in Figure 7.5, in the form of a plot of the polarization as a function of the applied electric field.

Co-polymers of PVDF with trifluoroethylene (TrFE) and tetrafluoroethylene (TFE) have also been shown to exhibit strong ferroelectric effects and do not require mechanical stretching. These can be preferable for use in memory (and other) devices.

Writing a '1' or a '0' to the ferroelectric memory element requires the application of a high voltage across the ferroelectric capacitor. Depending on the polarity of the voltage, this polarizes the capacitor in one particular state. To read the cell, a further voltage is applied. If the polarity of this voltage matches the polarization direction in the ferroelectric material, nothing happens. However, in the case that the applied voltage polarity opposes the dipole direction, the ferroelectric will be repolarized, and charge will flow to the sensing circuit. The reading process – destructive read-out – results in the loss of data, which must be rewritten. Complex addressing circuitry is therefore required.

An issue that relates to the reliability of ferroelectric memories is that of fatigue – the tendency of the polarization to decrease as a result of repeated switching. This led to the failure to commercialize this memory technology in the 1950s [41]. The fatigue phenomenon can have a number of physical origins including the 'pinning' of ferroelectric domain walls as a result of the migration and trapping of free electron carriers, the inhibition of opposite domain wall nucleation by charge injection and the formation of a passive surface layer that reduces the electric field in the ferroelectric film. The relatively large number of programming cycles required writing data to and reading data from ferroelectric memories (see above) exacerbates the fatigue problem. A further issue of exploiting ferroelectric memories relates to scaling; a reduction in the capacitor size will reduce the average current during a switching event.

Some of the problems outlined above may be circumvented by combining a ferroelectric polymer with a transistor – to form a ferroelectric FET or FeFET [2]. The different polarization states of the ferroelectric gate insulator lead to different values of the drain current of the transistor. Reading the memory state is therefore a non-destructive process, so the device lifetime is determined only by the number of times that the memory is written.

There have been recent attempts to re-introduce polymer ferroelectric printed memory [1]. The memory elements are addressed with n-channel and p-channel organic transistors. Although this approach is unlikely to challenge mainstream semiconductor memories (e.g. based on NAND flash), the technology could be useful within the flexible electronics arena; for example applications in food and medicine spoilage, by recoding time–temperature doses [42–44].

A related technology is the ferroelectric memory diode [45–49]. The basic idea relies on the variation of the quantum mechanical tunnelling current in an ultrathin ferroelectric film and was proposed in 1971 [41, 48]. Figure 9.15 shows the potential energy diagrams (Figure 9.17a) and operation (Figure 9.17b) of such a device. Different Thomas–Fermi screening lengths (Section 4.4.1) in the two dissimilar metal electrodes (M_1 and M_2) give rise to a tunnelling barrier that is dependent on the polarization direction of the ferroelectric (F). The barrier height is, on average, higher for one direction of P than for the other. As a consequence, the device resistance depends on the polarization vector. A semiconductor may also be used instead of a metal as a second electrode [48]. The piezoelectric effect can also play a role as it will modulate the tunnelling distance. The relatively high readout current densities for these devices suggest possible applications as high-density ferroelectric memories with non-destructive readout (the reading process uses only a low voltage and the original polarization in the material is unaffected). The early ferroelectric diodes were based on polycrystalline inorganic materials (e.g. barium titanate). However, it is quite difficult to achieve stable ferroelectric behaviour in layers of only 2–3 nm in thickness. Double tunnel junctions, in which the ferroelectric is combined with a very thin tunnelling layer such as SiO_2 or Al_2O_3 provide a practical alternative [45].

Organic ferroelectric devices are commonly based on films that are significantly thicker than tunnel diodes. In these cases, simple tunnelling between the electrodes is not the dominant conductivity mechanism. The interplay between ferroelectric/metal interface properties and the device current can be complex [47]. Ferroelectric polymers, such as poly(VDF-TrFE) are often used. Mixtures of the ferroelectric polymer with an organic semiconductor can improve the device current. Simple crossed-bar arrays and a 1 kbit reconfigurable memory fabricated from such blends on plastic foil have been demonstrated [50, 51].

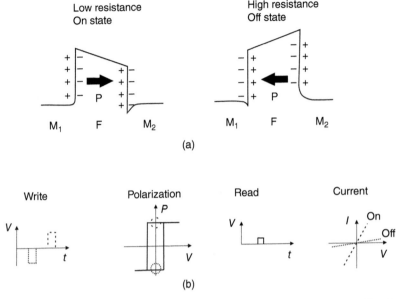

Figure 9.15 Operation of a ferroelectric tunnel diode. (a) Energy barrier profile of device in On and Off states. The device architecture is that of a ferroelectric ultrathin film (F) sandwiched between two metal electrodes (M_1 and M_2). **P** = polarization vector. (b) Write and read operations in which a write pulse is first applied to switch **P**. The device is read by a small pulse, which does not affect the polarization. Source: Wen and Wu [48].

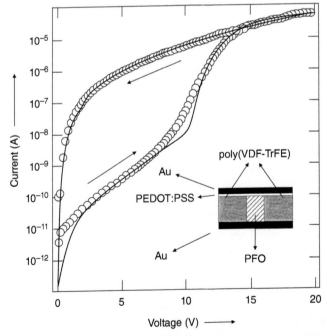

Figure 9.16 Current versus voltage data for an organic ferroelectric diode. The device structure is inset and consists of a phase separated blend of the ferroelectric polymer poly(VDF-TrFE) and the semiconducting polymer PFO. This organic layer (thickness = 265 ± 10 nm) is sandwiched between Au electrodes, one of which is coated with a layer of PEDOT:PSS. Points represent experimental data measured at ambient temperature. Lines are simulated fits. Source: Ghittorelli et al. [47]. Licenced under CC BY-4.0.

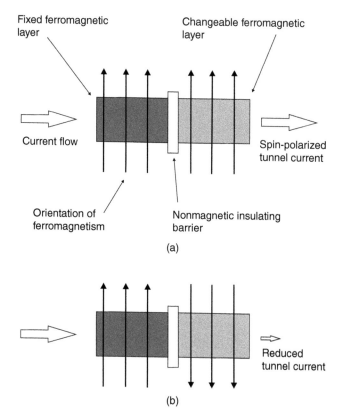

Fixed ferromagnetic layer

Changeable ferromagnetic layer

Current flow

Spin-polarized tunnel current

Orientation of ferromagnetism

Nonmagnetic insulating barrier

(a)

Reduced tunnel current

(b)

Figure 9.17 Schematic diagram of a magnetic tunnel junction. (a) The fixed ferromagnetic layer polarizes the spins of current-carrying electrons, which cross the barrier to the second layer by quantum mechanical tunnelling when the magnetism of both layers is aligned. (b) When the direction of magnetism in the second layer is reversed, the tunnelling current is reduced.

Figure 9.16 shows the current versus voltage behaviour (the points represent the experimental data measured at ambient temperature) for a ferroelectric memory in which the organic material (thickness = $265 \pm 10\,\text{nm}$) is a phase-separated blend of poly(VDF-TrFE) and the semiconducting polymer poly(9,9-dioctylfluorene) (PFO) [47]. The organic layer is sandwiched between Au electrodes, one of which had been coated with a layer of a mixture of the conductive polymer PEDOT and the water-soluble polyelectrolyte, poly(styrene sulfonic acid) (PSS). The solid line is the result of a numerical simulation taking into account the polarization of the ferroelectric component, charge injection at the electrodes and transport through the organic semiconductor. Charge transport in the organic semiconductor is described by means of variable range hopping (Section 6.4), while charge flow at the metal/semiconductor contact is modelled using drift-diffusion, energy disorder, thermionic emission, tunnelling, and image force barrier lowering. An intriguing result is the formation of a strongly accumulated hole density along the entire interface between the semiconducting and ferroelectric polymers. The diode operation resembles the channel current in a field effect transistor operated in saturation at pinch-off.

9.6 Spintronics

Spintronics (an acronym for SPIN TRansport electrONICS) exploits the spin of an electron to process and store digital information [52–55]; the subject is also called magnetoelectronics. The simplest method of generating a spin-polarized current is to inject the current through a ferromagnetic material. Much of the interest in the area started with IBM exploiting magnetically induced resistance, or magnetoresistance, in the early 1990s. Magnetic random access memory (MRAM) is the first mainstream spintronic non-volatile RAM. The most successful device to date is the spin valve. This uses a layered structure of thin films of magnetic materials, which changes electrical resistance depending on applied magnetic field direction. When the magnetization vectors of the ferromagnetic layers are aligned, a relatively large current will flow, whereas if the vectors are antiparallel then the resistance of the system is much higher. The magnitude of the change ((antiparallel resistance – parallel resistance)/parallel resistance) is called the giant magnetoresistance (GMR) ratio. Devices have been demonstrated with GMR ratios as high as 200%, with typical values greater than 10%. This is a vast improvement over the anisotropic magnetoresistance effect in single layer materials, which is usually less than 3%.

Figure 9.17 shows a schematic diagram for a non-volatile memory based on a magnetic tunnel junction. The device consists of two ferromagnetic thin films separated by a thin insulating barrier. The first layer polarizes the spins of current-carrying electrons, which then cross the barrier to the second layer by quantum mechanical tunnelling (Section 6.3). In the case that both layers are magnetically aligned, the resulting current is relatively high. However, when the magnetism in the second layer is reversed, the tunnelling current is reduced. The magnitude of the current can be exploited in a memory cell to indicate a 0 or a 1.

The first ever MRAM commercial products were launched in 2006. Such devices possessed read and write speeds of tens of nanoseconds. Early devices were based on 180 nm MOSFET technology. The spin-momentum-transfer effect [52] exploits the net angular momentum that is carried by a spin-polarized current and the transfer of this momentum to the magnetization of the second layer. It offers the potential of orders of magnitude lower switching currents and therefore a much lower energy for writing. The spin-momentum-transfer effect becomes important when the minimum dimension of the memory cell is less than 100 nm and is more efficient as the cell size is reduced (the opposite to what occurs with the use of conventional magnetic field switching). The related memory technology is called Spin Transfer Torque Random Access Memory, or STT-RAM.

MRAM has similar performance to SRAM, a comparable density to DRAM but lower power consumption, is much faster and suffers no degradation over time in comparison to flash memory. It is this combination of features that some suggest make it the 'universal memory', able to replace SRAM, DRAM, EEPROM, and flash. However, to date, MRAM has not been widely adopted in the market. This may simply be related to the reluctance of companies to invest enormous amount of money in new technologies.

There are proposals for a transistor that works by switching electrons between spin states. In one option, a gate controls whether most of the electrons have the same spin state. It is also possible to exploit magnetic domain wall motion to form computational logic elements. Information is stored through the presence or absence of a domain wall in a linear array of domain walls in a magnetic thin film loop confined to a channel in a silicon chip. Such a memory might eventually provide a solid-state replacement for the magnetic hard drive used in computers.

Although most of the materials developed for spintronic devices are inorganic, a case has been made for exploiting organic compounds [52–59]. This is based on the weak spin-orbit and hyperfine interactions in organic molecules, which leads to the possibility of preserving spin-coherence over times and distances much longer than in conventional metals or semiconductors [60].

Although reports on all-organic spintronic devices have yet to appear, spin-valves using a tris-(8-hydroxyquinoline) aluminium (Alq_3) layer sandwiched between ferromagnetic electrodes have been shown to possess a GMR value of about 5% at low temperatures [61].

9.7 Molecular Memories

There are many proposals to develop memories that exploit changes in individual molecules or groups of molecules. Bistable rotaxane molecules have been used in several of these studies; an example is shown in Figure 9.18 [63, 64]. This molecule is amphiphilic, and the ring component can move between the polar (right-hand side of the molecule) and nonpolar (central sulfur-rich TTF group) regions of the main part of the molecule. The molecules can be assembled on an electrode using the Langmuir–Blodgett approach (which allows single molecular layers to be built-up on solid supports) and a top electrode then deposited to form a crossed-bar structure [62]. The switching mechanism involves oxidation of the TTF site followed by translation of the ring component from the TTF site (as depicted in Figure 9.18) to the dioxynapthalene site (double benzene ring moiety to the right in Figure 9.18). The TTF is reduced back to form the metastable state co-conformer, which is the high-conductance, or '1' state. The metastable state relaxes back to the ground state with a half-life of about an hour.

In 2007, a 160 000 bit molecular electronic memory circuit, at a density of 10^{11} bits cm^{-2} was fabricated from such bistable rotaxanes [65]. Although the circuit possessed a large number of defects, these could be readily identified through electronic testing, and isolated using software coding. The working bits were then configured to form a fully functional RAM (a fault-tolerant technique adopted for Si systems, Section 12.5).

The idea of using single molecules as key components in computers has been around since the 1950s [8]. This would offer the ultimate in miniaturization (Problem 9.5). Certainly, at the start of the twenty-first century, with progress such as that noted in the previous paragraph, this vision was seemingly becoming a reality. There have been intriguing examples of single molecule diodes demonstrated in the laboratory, for example [66]. Section 4.5 has provided an overview of devices, such as the molecular rectifier. Some of these might form the basis for a single switching element in an electronic memory. DNA (Section 12.3.2) is arguably the ultimate molecular storage molecule. It has been estimated that 1 kg of DNA could encode all the data in the world [67]. Moreover, the data density would be around 10^{19} bits cm^{-3} with a retention time of greater than 100 years. Although DNA can be used for computation (Section 12.4.5), the practical realization of DNA digital storage architectures is likely to be some way into the future.

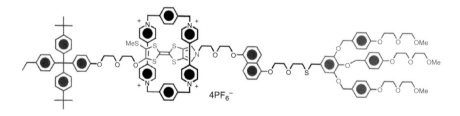

Figure 9.18 Bistable rotaxane structure. The 'ring' component of the molecule can move along the molecule between the hydrophobic 'stopper' on the left end of the molecule and the hydrophilic stopper region on the far right of the molecule. The switching mechanism involves oxidation of TTF followed by translation of the ring to the adjacent dioxynapthalene site. Source: Petty [62].

The rationale for developing organic memory devices is that some memory capability is essential if plastic/organic electronics is to become a fully integrated and significant technology. Clearly, options for organic memory are to develop the organic equivalents of silicon DRAM, SRAM, and flash, i.e. based around organic transistors. Some success in this direction has already been achieved, with the demonstration of arrays of organic semiconductor memory transistors. As organic electronics is not likely to compete directly, in terms of operational speed and device density, with inorganic-based technologies, the specifications for the memory elements can, to some extent, be relaxed. However, the devices must be robust and operate for a useful number of read and write cycles ($>10^{12}$ cycles, with a lifetime of >10 years).

The International Technology Roadmap for Semiconductors reviews progress on emerging memory devices – those based on both inorganic and organic materials. Table 9.1 shows some data that are relevant to organic devices [4]. Carbon memory includes devices based on nanotubes, graphene, and amorphous carbon. Macromolecular memory includes organic and polymeric thin film resistive switches.

Some of the technologies outlined in Table 9.1 (e.g. devices exploiting resistive switching) can be realized with either inorganic or organic materials. Inorganic compounds are generally favoured because of their better stability (e.g. temperature and time) and compatibility with standard semiconductor processing methods. However, one category in which organic memories will excel is those devices exploiting individual molecules or small groups of molecules. This field presents formidable scientific and technological challenges. The problems of fabricating reproducible arrays of devices on a commercial scale should not be underestimated. However, the availability of memory elements on the molecular scale would revolutionize not only organic electronics but also solve a potential 'bottleneck' Moore's Law scaling issue for inorganic-based technologies.

A different approach to developing organic memories is not to assemble piecemeal signal processing and memory elements but to look beyond the current von Neumann architectures and, thereby, to produce systems that emulate the processing and memory devices found in nature. Such ideas are explored in Chapter 12.

Table 9.1 Memory parameters for emerging research memory devices [4].

Parameter	Ferroelectric FET	Ferroelectric tunnel junction	Carbon memory	Macromolecular memory	Molecular memory
Best projected feature size	Same as CMOS transistor	<10 nm	<5 nm	5 nm	5 nm
Best demonstrated feature size	28 nm	50 nm	22 nm	100 nm	30 nm
Write/erase time	20 ns/10 ns	10 ns	10 ns	15 ns	10 s/ 0.2 s
Retention time	2.5×10^3 s (3 d)	3 d	168 h (250 °C)	10^5 s	2 mo
Write cycles	10^{12}	4×10^6	5×10^7	10^5	2×10^3
Write operating voltage	±5 V	2–3 V	5–6 V	1.4 V	4 V
Read operating voltage	0.5 V	0.1 V	1.5 V	0.2 V	0.5 V
Write energy per bit	1 fJ	10 fJ	—	10 fJ	—

Carbon memory includes devices based on nanotubes, graphene, and amorphous carbon. Macromolecular memory includes organic and polymeric thin film resistive switches.
Source: Modified from ITRS (International Technology Roadmap for Semiconductors) [4].

Problems

9.1 A DRAM has a refresh time of 5 ms. Each memory cell has a capacitance of 60 fF and is fully charged at 5 V. If the memory cannot tolerate a reduction in its charge by more than 50% before it is refreshed, estimate the maximum leakage current from each cell.

9.2 An EEPROM has a lower insulator of relative permittivity 3.5 and thickness 5 nm and a top insulator (i.e. above the floating gate) of relative permittivity 7.0 and thickness 40 nm. The memory cell is programmed by applying 10 V to the control gate. This results in a stored floating gate charge density of 10^{-6} C cm^{-2}. Calculate the electric fields and voltages associated with both the lower and upper insulating layers. What change in the device threshold voltage will the charged floating gate produce?

9.3 If a non-volatile EEPROM memory has a charge retention time of 100 years at 125 °C and 1 year at 170 °C, calculate the floating gate to tunnel insulator barrier height. Assume that the charge retention time follows a thermally activated Arrhenius model.

9.4 A resistance memory operates by the destruction and formation of aluminium filaments. If the thin film memory operates at 5 V has a thickness of 40 nm and the filaments each have a diameter of 10 nm, estimate the time taken for a filament to melt following formation. You will need to look up appropriate material constants. State any assumptions that you make.

9.5 It is proposed to construct a molecular memory based on closed packed molecules of an electro-active organic compound with a relative permittivity of 3.0. Each molecule is cylindrically shaped, with a length of 4 nm and a cross-sectional area of 0.5 nm^2. Assuming, it is possible to make electrical contacts to the ends of each molecule, calculate the associated capacitance of each memory bit. How many typical books could be stored on 1 cm^2 area of the molecular memory? Comment on your answers.

References

1 Petty, M.C. (2019). Organic electronic memory devices. In: *Handbook of Organic Materials for Optical and Optoelectronic Devices*, 2e (ed. O. Ostroverkhova), 843–874. Oxford: Woodhead.
2 Heremans, P., Gelinck, G.H., Müller, R. et al. (2011). Polymer and organic non-volatile memory devices. *Chem. Mater.* 23: 341–358.
3 Pershin, Y.V. and Di Ventra, M. (2011). Memory effects in complex materials and nanoscale systems. *Adv. Phys.* 60: 145–227.
4 ITRS (International Technology Roadmap for Semiconductors) (2015). *ITRS-ERD Meeting*, 2015. Beyond CMOS. http://www.itrs2.net/ (accessed 17 March 2021).
5 Chen, A., Hutchby, J., Zhirnov, V., and Bourianoff, G. (eds.) (2015). *Emerging Nanoelectronic Devices*. Chichester: Wiley.
6 Vourkas, I. and Sirakoulis, G.C. (2016). *Memristor-Based Nanoelectronic Computing Circuits and Architectures*. Cham: Springer.
7 Goswami, S., Goswami, S., and Venkatesan, T. (2020). An organic approach to low energy memory and brain inspired electronics. *Appl. Phys. Rev.* 7: 021303.

8 Petty, M.C. (2019). *Organic and Molecular Electronics: From Principles to Practice*, 2e. Chichester: Wiley.

9 Streetman, B.G. and Banerjee, S. (2014). *Solid State Electronic Devices*, 7e. London: Pearson.

10 Prince, B. (2000). *High Performance Memories: New Architecture DRAMs and SRAMs – Evolution and Function*. New York: Wiley.

11 Möller, S., Perlov, C., Jackson, W. et al. (2003). A polymer/semiconductor write-once read-many-times memory. *Nature* 426: 166–169.

12 Hümmeigen, I.A., Coville, N.J., Cruz-Cruz, I., and Rodrigues, R. (2014). Carbon nanostructures in organic WORM memory devices. *J. Mater. Chem. C* 2: 7708–7714.

13 Vettiger, P., Despont, M., Dreschsler, U. et al. (2000). The 'Millipede' – more than one thousand tips for future AFM data storage. *IBM J. Res. Dev.* 44: 323–340.

14 Gotsmann, B., Knoll, A.W., Pratt, R. et al. (2010). Designing polymers to enable nanoscale thermomechanical data storage. *Adv. Funct. Mater.* 20: 1276–1284.

15 Sze, S.M. and Ng, K.K. (2007). *Physics of Semiconductor Devices*, 3e. Hoboken, NJ: Wiley.

16 De Salvo, B., Ghibaudo, G., Pananakakis, G. et al. (1999). Experimental and theoretical investigation of non-volatile memory data-retention. *IEEE Trans. Electron Devices* 46: 1518–1524.

17 Scott, J.C. and Bozano, L.D. (2007). Nonvolatile memory elements based on organic materials. *Adv. Mater.* 19: 1452–1463.

18 Ieimini, D. (2016). Resistive switching memories based on metal oxides: mechanisms, reliability and scaling. *Semicond. Sci. Technol.* 31: 063002.

19 Chua, L. (1971). Memristor – the missing circuit element. *IEEE Trans. Circuit Theory* 18: 507–519.

20 Abraham, I. (2018). The case for rejecting the memristor as a fundamental circuit element. *Sci. Rep.* 8: 10972.

21 Dimitrakis, P., Normand, P., Tsoukalas, D. et al. (2008). Electrical behaviour of memory devices based on fluorene-containing organic thin films. *J. Appl. Phys.* 104: 044510.

22 Cölle, M., Büchel, M., and De Leeuw, D.M. (2006). Switching and filamentary conduction in non-volatile organic memories. *Org. Electron.* 7: 305–312.

23 Yang, Y., Ouyang, J., Ma, L. et al. (2006). Electrical switching and bistability in organic/polymeric thin films and memory devices. *Adv. Funct. Mater.* 16: 1001–1014.

24 Lee, M.-W., Pearson, C., Moon, T.J. et al. (2014). Switching and memory characteristics of thin films of an ambipolar organic compound: effects of device processing and electrode materials. *J. Phys. D: Appl. Phys.* 47: 485103.

25 Cho, B., Yun, J.-M., Song, S. et al. (2011). Direct observation of Ag filamentary paths in organic resistive memory devices. *Adv. Funct. Mater.* 21: 3976–3981.

26 Pearson, C., Bowen, L., Lee, M.-W. et al. (2013). Focused ion beam and field-emission microscopy of metallic filaments in memory devices based on thin films of an ambipolar organic compound consisting of oxadiazole, carbazole and fluorine units. *Appl. Phys. Lett.* 102: 213301.

27 Dearnaley, G., Morgan, D.V., and Stoneham, A.M. (1970). A model for filament growth and switching in amorphous oxide films. *J. Non-Cryst. Solids* 4: 593–612.

28 Lee, B.-H., Lee, D.-I., Bae, H. et al. (2016). Foldable and disposable memory on paper. *Sci. Rep.* 6: 38389.

29 Song, S., Cho, B., Kim, T.-W. et al. (2010). Three-dimensional integration of organic resistive memory devices. *Adv. Mater.* 22: 5048–5052.

30 Li, C., Han, L., Jiang, H. et al. (2017). Three-dimensional crossbar arrays of self-rectifying $Si/SiO_2/Si$ memristors. *Nat. Commun.* 8: 15666.

31 Crossbar Inc. (2021). ReThink IoT with ReRAM. http://www.crossbar-inc.com/ (accessed 17 March 2021).

32 Rueckes, T., Kim, K., Joselevich, E. et al. (2000). Carbon nanotube-based nonvolatile random access memory for molecular computing. *Science* 289: 94–97.

33 Nantero (2021). The future of memory is now. http://nantero.com/ (accessed 17 March 2021).

34 Mabrook, M.F., Yun, Y., Pearson, C. et al. (2009). Charge storage in pentacene/polymethylmethacrylate memory devices. *IEEE Electron Device Lett.* 30: 632–634.

35 Mabrook, M.F., Yun, Y., Pearson, C. et al. (2009). A pentacene-based organic thin film memory transistor. *Appl. Phys. Lett.* 94: 173302.

36 Sekitani, T., Yokota, T., Zschieschang, U. et al. (2009). Organic nonvolatile memory transistors for flexible sensor arrays. *Science* 326: 1516–1519.

37 Leong, W.L., Mathews, N., Tan, B. et al. (2011). Solution processed non-volatile top-gate polymer field-effect transistors. *J. Mater. Chem.* 21: 8971–8974.

38 Tho, L.V., Baeg, K.-J., and Noh, Y.-Y. (2016). Organic nano-floating gate transistor memory with metal nanoparticles. *Nano Convergence* 3: 10.

39 Ishiwara, H. (2012). Ferroelectric random access memories. *J. Nanosci. Nanotechnol.* 12: 7619–7627.

40 Li, H., Wang, R., Han, S.-T., and Zhou, Y. (2020). Ferroelectric polymers for non-volatile memory devices: a review. *Polym. Int.* 69: 533–544.

41 Mikolajick, T., Schroeder, U., and Slesazeck, S. (2020). The past, the present, and the future of ferroelectric memories. *IEEE Trans. Electron Devices* 67: 1434–1443.

42 Ng, T.N., Schwartz, D.E., Mei, P. et al. (2015). Printed dose-recording tag based on organic complementary circuits and ferroelectric non-volatile memories. *Sci. Rep.* 5: 13457.

43 Ng, T.N., Schwartz, D.E., Mei, P. et al. (2017). Printed organic circuits for reading ferroelectric rewritable memory capacitors. *IEEE Trans. Electron Devices* 64: 1981–1984.

44 Street, R.A., Ng, T.N., Schwartz, D.E. et al. (2015). From printed transistors to printed smart systems. *Proc. IEEE* 103: 607–618.

45 Wen, Z., Li, C., Wu, D. et al. (2013). Ferroelectric-field-effect-enhanced electroresistance in metal/ferroelectric/semiconductor tunnel junctions. *Nat. Mater.* 12: 617–621.

46 Garcia, V. and Bibes, M. (2014). Ferroelectric tunnel junctions for information storage and processing. *Nat. Commun.* 5: 4289.

47 Ghittorelli, M., Lenz, T., Dehsari, H.S. et al. (2017). Quantum tunnelling and charge accumulation in organic ferroelectric memory diodes. *Nat. Commun.* 8: 15741.

48 Wen, Z. and Wu, D. (2019). Ferroelectric tunnel junctions: modulations on the potential barrier. *Adv. Mater.* 31: 1904123.

49 Xiao, C., Sun, H., Cheng, L. et al. (2020). Temperature dependence of transport mechanisms in organic multiferroic tunnel junctions. *J. Phys. D* 53: 325301.

50 Asadi, K., Li, M., Stingelin, N. et al. (2010). Crossbar memory array of organic bistable rectifying diodes for non-volatile data storage. *Appl. Phys. Lett.* 97: 193308.

51 Van Breemen, A.J.J.M., Van der Steen, J.-L., Van Heck, G. et al. (2014). Crossbar arrays of non-volatile, rewritable polymer ferroelectric diode memories on plastic substrates. *Appl. Phys. Express* 7: 031602.

52 Wolf, S.A., Lu, J.W., Stan, M.R. et al. (2010). The promise of nanomagnetics and spintronics for future logic and universal memory. *Proc. IEEE* 98: 2155–2168.

53 Geng, R., Daugherty, T.T., Do, K. et al. (2016). A review of organic spintronic materials and devices: I. Magnetic field effect on organic light emitting diodes. *J. Sci.: Adv. Mater. Devices* 1: 128–140.

54 Joshi, V.K. (2016). Spintronics: a contemporary review of emerging electronics devices. *Eng. Sci. Technol.* 19: 1503–1513.

55 Vardeny, Z.V. (ed.) (2017). *Organic Spintronics*. Boca Raton, FL: CRC Press.

56 Rocha, A.R., García-Suárez, V.M., Bailey, S.W. et al. (2005). Towards molecular spintronics. *Nat. Mater.* 4: 335–339.

57 Sun, D., Ehrenfreund, E., and Vardeny, Z.V. (2014). The first decade of organic spintronics research. *Chem. Commun.* 50: 1781–1793.

58 Majumdar, S., Majumdr, H.S., and Österbacka, R. (2011). Organic spintronics. In: *Comprehensive Nanoscience and Technology*, vol. 1 (eds. D.L. Andrews, G.D. Scholes and G.P. Wiederrecht), 109–142.

59 Liu, H., Zhang, C., Malissa, H. et al. (2018). Organic-based magnon spintronics. *Nat. Mater.* 17: 308–312.

60 Dediu, V.A., Hueso, L.E., Bergenti, I., and Taliani, C. (2009). Spin routes in organic semiconductors. *Nat. Mater.* 8: 707–716.

61 Wang, F.J., Xiong, Z.H., Wu, D. et al. (2005). The case for $Fe/Alq_3/Co$ spin-valve devices. *Synth. Met.* 155: 172–175.

62 Petty, M.C. (1996). *Langmuir-Blodgett Films*. Cambridge: Cambridge University Press.

63 Chen, Y., Ohlberg, D.A.A., Li, X. et al. (2003). Nanoscale molecular-switch devices fabricated by imprint lithography. *Appl. Phys. Lett.* 82: 1610–1612.

64 Yang, W., Li, Y., Liu, H. et al. (2012). Design and assembly of rotaxane-based molecular switches and machines. *Small* 4: 504–505.

65 Green, J.E., Choi, J.W., Boukai, A. et al. (2007). A 160-kilobit molecular electronic memory patterned at 10^{11} bits per square centimetre. *Nature* 445: 414–417.

66 Capozzi, B., Xia, J., Adak, O. et al. (2015). Single-molecule diodes with high rectification ratios through environmental control. *Nat. Nanotechnol.* 10: 522–527.

67 Extance, A. (2016). Digital DNA. *Nature* 537: 22–24.

Further Reading

Tour, J.M. (2003). *Molecular Electronics: Commercial Insights, Chemistry, Devices, Architecture and Programming*. Singapore: World Scientific.

Woulters, D.J., Tokumitsu, E., Auciello, O., and Panagiotis, D. (2011). *New Functional Materials and Emerging Device Architectures for Nonvolatile Memories*, 2011 MRS Symp. Proc., vol. 1337 (ed. Y. Fujisaki). New York: Cambridge University Press.

10

Light-Emitting Devices

10.1 Introduction

Organic light-emitting devices (or diodes) (OLEDs) represent a significant research and development activity [1–5]. These require no backlighting and currently compete with liquid crystal displays (LCDs) in the consumer (e.g. television and mobile phone) market. OLED displays can be manufactured in large areas, in flexible form, and by low-cost methods such as inkjet printing. A further area of interest concerns the industrial and domestic lighting sector. There is also potential for other applications, for example as chemical sensors and in the medical field [6–8].

The inorganic counterpart to the OLED is the light-emitting diode (LED). These devices, which were discovered in the early twentieth century and developed in the 1960s, are based on junctions (e.g. between n- and p-type material) of compound semiconductors such as InAsGaP and InGaN. Reports of light emission from organic materials on the application of an electric field also go back many years. In 1987, scientists at the Kodak company reported efficient low voltage electroluminescence in an organic thin film incorporating dye molecules [9]. Interest intensified in the 1990s following the publication of a paper concerning light-emitting devices that incorporated the conjugated polymer poly(p-phenylenevinylene) (PPV) [10]. The technology can now be considered as mature, in as far as commercial products incorporating OLEDs are popular and widely available.

10.2 Light Emission Processes

Luminescence is a term that is used to describe the emission of light by a substance caused by any process other than a rise in temperature. Molecules may emit a photon of light when they decay from an electronically excited state to the ground state. The excitation leading to this emission may be caused by a photon (photoluminescence), an electron (electroluminescence or cathodoluminescence), or a chemical reaction (chemiluminescence).

Most organic molecules possess an even number of electrons with all their electrons paired. Within each pair, the opposing spins cancel, and the molecule has no net electron spin. Such an electronic structure is called a singlet (S) state. The processes leading to light emission can be considered in terms of the energy levels in the molecule. Figure 10.1, known as a Jablonski diagram, shows a very simplified energy scheme (omitting internal conversion) illustrating some processes that may occur. The states are arranged vertically by energy and grouped horizontally by spin multiplicity. Each molecular state in the diagram corresponds to a bonding or anti-bonding molecular orbital. The orbitals associated with a carbon–carbon bond can be either σ or π type, with corresponding antibonding orbitals σ^* and π^*, respectively (Section 1.5). Additionally, there

Electrical Processes in Organic Thin Film Devices: From Bulk Materials to Nanoscale Architectures, First Edition. Michael C. Petty.
© 2022 John Wiley & Sons Ltd. Published 2022 by John Wiley & Sons Ltd.
Companion Website: www.wiley.com/go/petty/organic_thin_film_devices

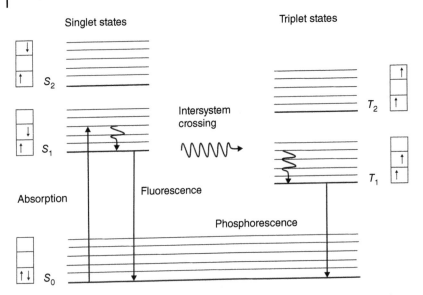

Figure 10.1 Jablonski diagram showing the radiative and nonradiative decay of an excited molecule. The orientations of the electron spins are shown in the boxes next to each state. S_n = singlet state, T_n = triplet state.

are valence-shell electrons that do not participate in the formation of molecular bonds. These non-bonding orbitals are designated as n.

When one electron is excited from its ground state to a higher energy level, either a singlet or a triplet (T) can form. Each of these electronic energy levels (shown by the darker horizontal lines in Figure 10.1) is subdivided into vibrational sublevels (lighter horizontal lines). In the excited S state, the electron is promoted in the same spin orientation as it was in the ground state (paired). In a T excited stated, the electron that is promoted has the same spin orientation (parallel) to the other unpaired electron.

Following promotion to an electronically excited singlet excitonic state (S_1 or S_2 in the figure), the molecule usually relaxes to the lowest vibrational level of S_1, the energy being lost as heat via inter-molecular collisions. This process takes about 10^{-11} seconds, or approximately 10^2 vibrations of the molecule. The lifetime of S_1 in its lowest vibrational state is longer, 10^{-8} to 10^{-7} seconds. This state may then decay to the ground state with the emission of a photon.

The emission is termed fluorescence. Alternatively, the energy may be transferred by a process called intersystem crossing (ISC) to a triplet excitonic state. The spins of the two electrons are now parallel and a transition to the ground state, with the emission of a photon, involves a change of spin. This is, in theory, a spin-forbidden transition. Quantum mechanical selection rules forbid transitions in which the electron spin changes, i.e. only singlet to singlet or triplet to triplet transitions are allowed. Triplet states are long lived, with lifetimes of greater than 10^{-5} seconds. The triplet state is typically at a lower energy than the corresponding singlet state.

In practice, the triplet state can decay to S_0, emitting a photon. This process is known as phosphorescence and may persist for seconds or even longer after the incident excitation has ceased. Nonradiative transitions from the excited singlet and triplet states to the ground state are also possible. In these instances, the original light quantum is converted into heat.

Triplet excitons in aromatic molecules are more effective energy conductors than singlet excitons (Section 1.5.2). Their lifetimes are longer and times to move between sites in the material (hopping times) are not much shorter than those in the singlet state. Triplet excitons can also combine their energy to form a singlet exciton, resulting in delayed fluorescence or triplet annihilation. The former gives rise to the emission of fluorescence light emitted with a delay (determined by the lifetime of the triplet excitons – Section 10.7.4), whereas triplet annihilation provides reduced phosphorescent emission as two triplet states have been combined into one.

The different emission processes outlined above may all be exploited in OLEDs [5]. In an electroluminescence process, the population of excited states occurs via a recombination of negatively charged electrons and positively charged holes. These attract each other by Coulomb interaction and form excitons in the emission layer of an OLED. Since both holes and electrons possess spins, four different spin combinations are possible. According to quantum mechanics, there is one combination of antiparallel spins, giving a singlet, and three combinations of parallel spins, resulting in a triplet state [1, 11]. From a statistical viewpoint, 25% of the excitons represent singlets and 75% triplets.

The upper limit of the internal quantum efficiency (ratio of the number of generated photons to the number of injected electrons) of OLEDs exploiting conventional fluorescent emitter is therefore around 25%. However, the maximum external efficiency is significantly less than this because of other factors, as discussed below in Sections 10.3 and 10.8. In the case of triplet-triplet annihilation (TTA), the maximum internal efficiency improves to 62.5% [5]. However, designing TTA molecules is challenging because of strict requirements for the energy levels. Phosphorescent light-emitting materials generally possess complex metal structures (e.g. based on Ir, Au, Os, Re, Ru, or Cu) that enable singlet to triplet and triplet to singlet energy transfer to occur. As noted above, these processes are theoretically prohibited; however, the rules of quantum mechanics can be broken because both singlet and triplet excitations are formed as a result of highly efficient spin-orbital coupling caused by the heavy metal centre of the complex. As triplet to singlet energy transfer is also allowed by the heavy metal effect, triplet exciton decay occurs faster than normal.

10.3 Operating Principles

The simplest OLED is an electroluminescent (EL) compound, such as a polymer or dye, sandwiched between metals of high and low work function, as depicted in Figure 10.2. One electrode must be semi-transparent (e.g. ITO) to allow light to escape from the structure. Like all other electronic devices based on organic compounds, OLEDs are fabricated in an inert atmosphere, e.g. nitrogen, and are encapsulated to prevent the inclusion of water and oxygen, both of which can lead to short operating lifetimes [12].

On application of a voltage, electrons are injected from the low work function cathode into the LUMO level (or conduction π^* band in the case of an organic compound possessing a delocalized electron system) of the organic compound, and holes are injected from the high work function anode into the HOMO level (or valence π band). The recombination of these oppositely charged carriers then causes the emission of light. As the electrons and holes are usually bound by a few meV in the form of excitons, their radiative recombination leads to emission energies slightly lower than the HOMO–LUMO separation. Efficient injection of both electrons and holes is a prerequisite to obtain high performance devices. Current displays exploit both low-molecular-weight organic molecules and polymers as EL materials [2, 3]. The chemical formulae of some of the types of materials that have been used as the emissive compounds in OLEDs are shown in Figure 10.3. Their HOMO and LUMO levels and colours are indicated [13, 14]. However, the reader should

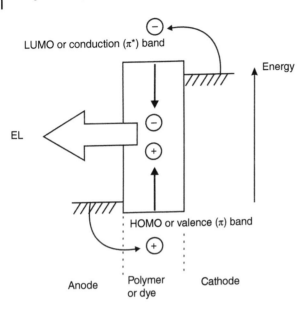

Figure 10.2 Schematic energy band structure of an OLED. The LUMO (or conduction π^* band) and HOMO (or valence π band) levels of the organic semiconductor are indicated. The recombination of electrons and holes results in the emission of electroluminescence (EL). The emission energy is slightly less than the HOMO–LUMO separation (see text for details).

treat the energy level values as approximate because these are obtained from different sources, some experimental and some theoretical.

The ability to fabricate thin films from such compounds is an important practical issue. For example, the addition of the side chain to the PPV monomer, to form poly[2-methoxy-5-(2-ethylhexyloxy)-1,4-phenylenevinylene] (MEH-PPV), allows the latter compound to be easily spin-coated onto a variety of substrates. The non-polymeric compounds, such as tris-(8-hydroxyquinoline) aluminium (Alq_3), are usually formed into thin films by the process of thermal evaporation.

As discussed above, radiative emission in OLEDs may be characterized as either fluorescence or phosphorescence, depending on the relaxation path. Many of the fluorescent dyes that have been incorporated in OLEDs were originally developed as laser dyes, since these materials were designed to possess high photoluminescence quantum efficiencies in dilute solution, along with good stability. In general, their solid-state fluorescence is extremely weak due to concentration quenching; this reduces the EL efficiency. A way to circumvent this is to add bulky side groups to the molecule, which prevents aggregation by increasing the steric hindrance. Unfortunately, this often leads to poor charge transport through the material. A further remedy is to dope the emissive dye into an organic matrix, which dilutes the concentration of the emissive dye (the dopant), thereby preventing aggregation. So long as the dopant is red shifted compared to the host (it has its emission at a lower energy), excitons formed in the host material will tend to migrate to the dopant prior to relaxation. This results in emission that is predominantly from the dye.

Phosphorescent materials, such as rare earth complexes, can improve the OLED efficiency [13]. Both singlet and triplet states can be used to harvest the light emission. One approach uses a triplet-emitting, heavy atom containing, (electro)phosphorescent metal–organic complexes, as typified by bis[2-(2-pyridinyl-*N*)phenyl-C](2,4-pentanedionato-O^2,O^4)iridium(III) (Ir(ppy)$_2$(acac)), the structure of which is shown in Figure 10.3. These 'dopants' (in this context, the term should not be confused with that used when a material is added to alter the electrical conductivity of an organic semiconductor – Section 1.5.7) must have high luminescence quantum yields and relatively short lifetimes, of the order 1 μs, to avoid exciton-ion quenching.

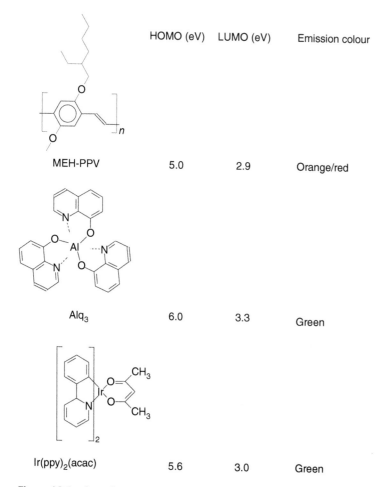

	HOMO (eV)	LUMO (eV)	Emission colour
MEH-PPV	5.0	2.9	Orange/red
Alq$_3$	6.0	3.3	Green
Ir(ppy)$_2$(acac)	5.6	3.0	Green

Figure 10.3 Organic compounds used as the emissive layer in OLEDs. The HOMO and LUMO energies and emission colours are indicated. The materials are: poly[2-methoxy-5-(2-ethylhexyloxy)-1, 4-phenylenevinylene] (MEH-PPV); tris-(8-hydroxyquinoline) aluminium (Alq$_3$); and bis[2-(2-pyridinyl-*N*) phenyl-C](2,4-pentanedionato-O^2,O^4)iridium(III) (Ir(ppy)$_2$(acac)).

Figure 10.4 shows the forward current versus voltage and the EL power versus voltage characteristics of an OLED based on a fluorescent polymer. The device structure is ITO/PPV/Al, and the ITO electrode is positively biased to obtain these data [15]. Negligible current and light emission result with the opposite polarity of applied voltage. The OLED therefore acts as a rectifying device. The steep rise in the forward current between 1 and 2 V can be modelled with the equation for a Schottky diode (Eq. (4.39)) with an ideality factor typically in the range 1.6–2.4 for different devices. The deviation from this exponential behaviour above 2 V may be explained by the limiting series resistance associated with the device. Electroluminescence is observed as the forward voltage is increased beyond about 1.6 V. The luminance (defined in Section 10.5) from an OLED is generally proportional to current density over a wide range [16].

OLEDs based on other materials exhibit rectifying electrical characteristics, but the results sometimes indicate other conductivity mechanisms at work. Space-charge limited conduction and Fowler–Nordheim conduction (Sections 6.6 and 6.7) are often reported [16, 17]. Different electrical behaviour has also been noted for, supposedly, the same OLED architectures. For example, the

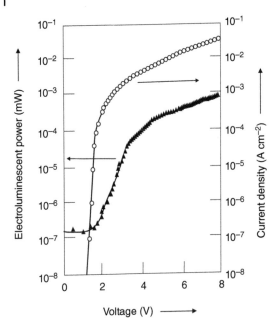

Figure 10.4 Current density versus voltage and electroluminescence power versus voltage characteristics for an ITO/MEH-PPV/Al OLED. Source: Karg et al. [15].

current versus voltage behaviour for another ITO/PPV/Al device has been given in Figure 6.13 [18]. In this case, the forward current characteristics were of the form of the power law of Eq. (6.43) suggesting a model in which the injection of space charge was influenced by an exponential distribution of trapping states:

$$I \propto L \left(\frac{V^m}{L^{2m+1}} \right) \tag{10.1}$$

where L is the electrode separation and m ranges from about 7 at 290 K to about 18 at 11 K. The different conductivity mechanisms reported by different research groups are probably related to differences in the organic materials (purity), device fabrication (electrodes, organic film thickness), or to the measurement protocols (voltage and/or temperature ranges). In these respects, the technique of impedance spectroscopy, as described in Section 7.4, can prove useful in determining the electrical equivalent circuit of OLEDs (e.g. to identify resistances associated with electrodes).

10.4 Colour Measurement

By judicious choice of compound, EL at different wavelengths across the visible spectrum is possible using an OLED. The human eye with normal vision has three kinds of cells, called cones, that sense light with peaks of spectral sensitivity at blue (420–440 nm), green (530–540 nm), and red (560 –580 nm) wavelengths. These cells determine human colour perception in conditions of medium and high brightness. Mathematical functions – colour matching functions – may be written to correspond to the eye sensitivity curves for the red, green, and blue cones. Under very low light conditions, colour vision diminishes, and the monochromatic 'night vision' receptors, known as rod cells, become effective. The sensation of colour varies slightly among individuals. Since the human eye has three types of colour sensors, a full plot of all visible colours is a three-dimensional figure. However, chromaticity, or the concept of colour, can be divided into two parts: hue and saturation. Hue describes the most dominant wavelength in a colour (a particular blue may be described

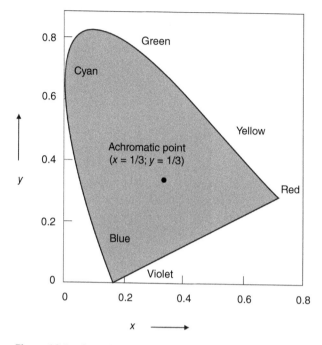

Figure 10.5 Commission International de l'Eclairage (CIE) chromaticity diagram. Any colour can be expressed in terms of two colour coordinates *x* and *y*.

using the adjectives such as 'sky', 'navy', or 'pastel'), whereas saturation is a measure of wavelength purity (a highly saturated colour will contain a narrow set of wavelengths).

A convenient way in representing the colour output of an OLED (or any other light source) is to use the CIE (Commission International de l'Eclairage) chromaticity diagram, as shown in Figure 10.5 [1, 19]. Any colour can be expressed in terms of two colour coordinates, *x* and *y*. The tongue-shaped figure forms the limit of all perceivable hues. The curved edge of the diagram is called the spectral locus and corresponds to monochromatic light, each point representing a pure hue of a single wavelength. The colour purity of many inorganic LEDs (e.g. red and blue devices) is high, with coordinates very close to the spectral locus. The straight edge on the lower part of the figure is called the line of purples. These colours have no counterpart in monochromatic light; no monochromatic light source can generate them. Less saturated colours appear in the interior of the figure with white at the centre. The calculation of the CIE chromaticity coordinates for an OLED requires the multiplication of its spectral power at each wavelength by a weighting factor from each of three colour matching functions. Light with a flat power spectrum in terms of wavelength (equal power in every 1 nm interval) corresponds to the point $(x, y) = (1/3, 1/3)$; this achromatic point defines 'white' light on the CIE diagram.

A further characteristic of a light source, particularly relevant to white lighting, is its colour-rendering index (CRI). This is a measure of the ability of the light source to show object colours 'realistically' compared to a familiar reference source, either incandescent light or daylight. The CRI is calculated from the differences in the chromaticities of eight CIE standard colour samples when illuminated by the light source and by a reference; the smaller the average difference in chromaticities, the higher the CRI. A CRI of 100 represents the maximum value. Lower CRI values indicate that some colours may appear unnatural when illuminated by the lamp. Fluorescent lamps (discrete line spectral output) have lower CRI values than incandescent lights (continuous spectral output), which can have a CRI above 95. A related feature of a white light source is its colour temperature. This is the temperature of an ideal black-body radiator (in K) that radiates light of a colour comparable to that of the light source. Hence, a 'warm' white lamp may have a specified colour temperature of 2700 K, whereas a 'cool' white lamp may possess a value of 4000 K (its emission spectrum is shifted to shorter wavelengths).

10.5 Photometric Units

A brief introduction to the SI units relevant to display technology is provided in this section.

First, it is necessary to make a distinction between photometric and radiometric units [1, 20]. Radiometry is the measurement of optical radiation, which is EM radiation within the frequency range 3×10^{11} to 3×10^{16} Hz. This corresponds to wavelengths between 0.01 and 1000 μm, and includes the ultraviolet, visible, and infrared regions. Photometry on the other hand is the measurement of light, defined as EM radiation that is detectable by the human eye. It is therefore restricted to the wavelength range from about 380 to 780 nm. Photometry is essentially radiometry except that everything is normalized to the spectral response of the eye, known as the photopic response. Radiometry uses the familiar unit of the watt as a measure of power (rate of energy emission) or radiant flux. In the following discussion, the subscripts 'rad' and 'phot' are used to distinguish between radiometric and photometric units, respectively.

The photometric equivalent of the watt is the luminous flux, Φ_{phot}, measured in lumens (lm). The luminous intensity, I_{phot}, measured in candelas (cd) refers to the flux emitted into a unit solid angle in space or steradian. Figure 10.6a illustrates the relationship between luminous flux and luminous intensity. If the luminous flux originating from a point source, and contained within a solid angle ω, traverses the surfaces A and B (the fluxes crossing the two surfaces will be identical), the relationship between the intensity and flux is:

$$I_{\text{phot}} = \frac{\Phi_{\text{phot}}}{\omega} \tag{10.2}$$

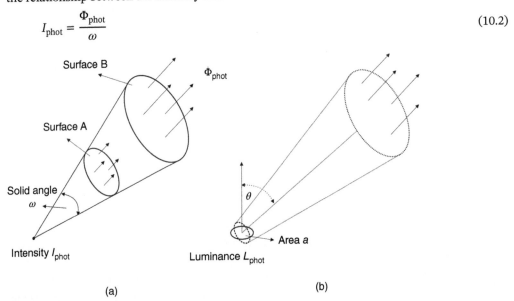

(a) (b)

Figure 10.6 Diagrams showing relationship between luminous intensity, I_{phot}, luminous flux, Φ_{phot}, and luminance, L_{phot} of a light source. (a) Point source radiating isotropically within a solid angle ω. (b) Area source with Lambertian emission within the same solid angle ω, but at an angle θ to area normal.

Since there are 4π steradians in a sphere, for a light source emitting equally in all directions (isotropic), the relationship between lumens and candelas is $1\,\text{cd} = 4\pi\,\text{lm}$. Hence, an isotropically emitting light source with a luminous intensity of 1 cd has a luminous flux of 12.57 lm.

The formal definition of the candela (which derived from an older unit, the candlepower) has changed over the years. Currently, the candela is defined as the luminous intensity, in a given direction, of a source that emits monochromatic radiation of frequency $540 \times 10^{12}\,\text{Hz}$ (555 nm) and that has a radiant intensity in that direction of $1/683\,\text{W}\,\text{sr}^{-1}$. The candela is the only basic SI unit that is associated with human perception.

An OLED will have a finite surface area and will take the form of a small area source, rather than a point source (to appear as a point source, a light emitting element will subtend an angle of some two minutes or arc or less at the point of observation). The amount of light leaving such an extended area source is the luminance, L_{phot}, measured in cd m^{-2}; these units are often called nits (from the Latin 'nitere' = to shine). A typical computer screen has a luminance of about 100 cd m^{-2}, while the luminance of an average clear sky will be around 8000 cd m^{-2}. Figure 10.6b depicts a light source of area a emitting light within a solid angle ω, but at an oblique angle to the area normal. If the angle between the normal to the face of the area source and the direction of emission is θ, then the luminance can be expressed:

$$L_{phot} = \frac{I_{phot}}{a\cos\theta} = \frac{\Phi_{phot}}{a\omega\cos\theta} \tag{10.3}$$

This equation shows that the luminous flux emitted from an area emitter will be a maximum for the surface normal and will then decrease with the angle θ, in contrast to the isotropically emitting point source discussed above. For example, when $\theta = 60°$, the luminous flux from a small-area emitter will have fallen to one-half of its maximum value and will vanish as θ approaches 90°. This dependence of L_{phot} on Φ_{phot} is characteristic of a Lambertian emitter. A further property is that a Lambertian emitter will have the same brightness (luminance), as perceived by the human eye, when viewed from any angle. This appears to contradict Eq. (10.3). However, the cosine law is relevant if a photodetector such as a photodiode is used to measure the radiation. In the case where the detector is some kind of camera with a lens and focal plane image sensor (e.g. CMOS image sensor or the human eye), then the amount of light captured by the camera is reduced according to Eq. (10.3). At the same time, the image of the emitting surface on the sensor becomes smaller (fewer pixels of the sensor are illuminated). Although the emitted power from a given area element is reduced by the cosine of the emission angle, the solid angle, subtended by surface visible to the viewer, is reduced by the same very amount. Because the ratio between power and solid angle is constant, the radiance or brightness (power per unit solid angle per unit projected source area) stays the same.

For a Lambertian source, it can be shown that the following relationship holds between the luminance and the total flux emitted into hemisphere above the emitter [1]:

$$\Phi_{phot} = \frac{L_{phot}}{\pi} \tag{10.4}$$

Therefore, a surface with a luminance of 100 cd m^{-2} (the typical PC monitor) and an area of 0.1 m^2 ($\approx$19-in. screen) will produce a total luminous flux (total light emitted) of 31.8 lm.

The stacked layer architecture of an OLED (Section 10.7) can introduce significant interference effects that modulate the device efficiency; this can change the emission pattern from that of an ideal Lambertian emitter.

A further photometric unit is the illuminance, E_{phot}. This is the luminous flux incident per unit area and is a measure of the amount of light incident on a surface. The unit of illuminance is the lux

Table 10.1 Common radiometric and photometric units.

Property	Description	Units
Radiant power (Φ_{rad})	Rate of energy emitted from a source	Watt (W)
Luminous flux (Φ_{phot})	Rate of energy emitted from source as perceived by the human eye	Lumen (lm)
Luminous intensity (I_{phot})	Flux emitted from a point source per unit solid angle	Candela (cd) $\equiv$ Lumen per steradian
Luminance (L_{phot})	Flux emitted per unit surface area of an extended source per unit solid angle	cd m^{-2}
Illuminance (E_{phot})	Incident luminous flux per unit area	Lux = lm m^{-2}

($1\,\text{lx} = 1\,\text{lm m}^{-2}$). A full moon with have an illuminance of about 1 lx and direct sunlight will possess a value of about 10^5 lx. Thus, illuminance is measure of the incident light, while luminance is what's leaving the surface. Luminance is the parameter used to describe the brightness of a display. Table 10.1 provides a list of the photometric quantities encountered in OLED work (Problem 10.1).

A key issue is to link the radiometric and photometric worlds. The CIE has adopted as a standard an average eye with a predictable response to light at various frequencies. As a result, the photosensitivity of the 'standard' human eye has a peak value of 683 lm W^{-1} at a wavelength of 555 nm (green). The luminous flux Φ_{phot} is obtained from the radiometric light power using the equation:

$$\Phi_{phot} = 683 \int_{\lambda} V(\lambda)\, \Phi_{rad}(\lambda) d\lambda \; \text{lm W}^{-1} \tag{10.5}$$

where $\Phi_{rad}(\lambda)$ is the radiometric power spectral density (light power emitted per unit wavelength) and $V(\lambda)$ is the eye sensitivity function or photopic response [21].

According to the above definition, there are 683 lm W^{-1} for 555 nm light that is propagating in a vacuum. Hence, for a monochromatic light source ($\Delta\lambda \to 0$), it is straightforward to convert from watts to lumens; the power in watts is multiplied by the appropriate $V(\lambda)$ value, and the conversion factor applied from the definition for a candela. For example, the photometric power of a 5 mW red (=650 nm) laser pointer, which corresponds to $V(\lambda) \approx 0.1$, is $0.1 \times 0.005\,\text{W} \times 683\,\text{lm W}^{-1} = 0.34\,\text{lm}$, whereas the value for a 5 mW green laser pointer (=532 nm, $V(\lambda) \approx 0.89$) is 3.0 lm. Although both pointers have identical radiometric power, the green laser pointer will appear approximately nine times brighter than the red one (assuming both have the same beam diameter). For lights with a broad spectral output, and especially for white light sources, the integration over the appropriate wavelength range (Eq. (10.5)) must be undertaken to calculate Φ_{phot}.

10.6 OLED Efficiency

Several different efficiency figures are used in OLED work. For display applications, a useful parameter its external quantum efficiency, η_{ext}. This is the ratio of the number of photons emitted by the OLED into the viewing direction to the number of electrons injected. For devices emitting monochromatic light, the energy emitted can be taken approximately as equal to the band gap, E_g, or HOMO–LUMO separation, and the external efficiency is:

$$\eta_{ext} = \frac{\Phi_{rad}/E_g}{I/e} = \frac{\lambda e \Phi_{rad}}{Ihc} \tag{10.6}$$

where E_g (in J) $= hf = hc/\lambda$ and f, λ are the frequency and wavelength of the light, respectively. Most OLEDs emit radiation over a range of frequencies; hence, Eq. (10.6) must be integrated over the wavelength limits [20]. In this case, the wavelength dependence of the photodetector (e.g. photodiode) must be considered to avoid large errors. While η_{ext} could also be defined as the ratio of the total number of photons emitted from the device (in all directions) to the number of electrons injected, this definition is not useful for display devices. A large fraction of the light can be waveguided by the organic layer(s)/substrate combination, ultimately emerging out at the edge of the device (Section 10.8). Thus, the total amount of light emitted from the device will be significantly higher than the light emitted in the viewing direction. The efficiency based on total light emitted can be up to four or five times larger than η_{ext}.

As noted in Section 10.2, the internal quantum efficiency, η_{int} is the ratio of the total number of photons generated within the structure to the number of electrons injected. The internal and external efficiencies therefore differ by the fraction of light coupled out of the structure in the viewing direction; this is called the extraction efficiency (or coupling coefficient), η_c:

$$\eta_{ext} = \eta_{int}\eta_c \tag{10.7}$$

However, it has been pointed out that there is an ambiguity with this definition since a photon emitted within an OLED can be re-absorbed, and in some cases re-emitted at a longer wavelength [20]. Moreover, η_c may vary with the device operating conditions (e.g. operating voltage and viewing angle).

A convenient measure of the light output of an OLED is its luminous (or current) efficiency, η_l, measured in cd A^{-1}:

$$\eta_l = \frac{aL_{phot}}{I} = \frac{L_{phot}}{J} \tag{10.8}$$

where L_{phot} is the luminance of the OLED (in cd m^{-2}), a is the device active area (not necessarily equal to the area of light emission), I is the current, and J the current density.

In many respects, η_l is equivalent to η_{ext}, with the exception that η_l weights all incident photons according to the photopic response of the eye. Another frequently used efficiency unit is the luminous power efficiency (or luminosity or luminous efficiency), η_p, measured in lm W^{-1}. This is the ratio of luminous power emitted in the viewing direction (in lm) to the total electrical power required to drive the OLED at a particular voltage:

$$\eta_p = \frac{\Phi_{phot}}{IV} \tag{10.9}$$

The luminous power efficiency is often used as the figure of merit for an OLED white light source. The most common OLED efficiency parameters are contrasted in Table 10.2 (Problems 10.2 and 10.3).

The external quantum efficiency of an OLED can be measured using a calibrated photodetector (e.g. a photodiode), as shown in Figure 10.7a [20]. The main objective is to ensure that the light output from the OLED is completely coupled into the photodetector. A convenient way of doing this is to use a detector with an area that is significantly larger than that of the OLED. It is also necessary to block photons emerging from the substrate edges (waveguided light) from the detector, for example by coating the these with black paint.

To measure η_{int}, the OLED and substrate may be placed inside an integrating sphere, as depicted in Figure 10.7b. This is an optical instrument consisting of a hollow spherical cavity with its interior covered with a diffuse white reflective coating. Its distinguishing feature is a uniform scattering or

Table 10.2   Efficiencies used in OLED work.

Parameter	Description	Symbol	Units
Internal quantum efficiency	Ratio of photons generated within the OLED to the number of injected electrons	η_{int}	—
External quantum efficiency	Ratio of the number of photons emitted into viewing direction to the number of electrons injected	η_{ext}	—
Extraction efficiency	Fraction of light coupled out of OLED	η_c	—
Luminous efficiency (or current efficiency)	Luminance divided by current density	η_l	$cd\,A^{-1}$
Luminous power efficiency	Ratio of the luminous flux emitted into viewing direction to the electrical power input	η_p	$lm\,W^{-1}$

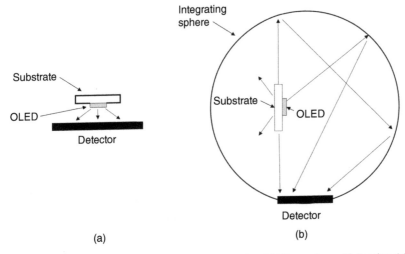

(a)   (b)

Figure 10.7   Experimental geometries for measuring OLED quantum efficiencies. (a) Measurement of η_{ext} with a large area calibrated photodetector; the back and edges of the substrate are coated with black paint. (b) Measurement of η_{int} using an integrating sphere; all the emitting photons are collected by the detector. Source: Forrest et al. [20].

diffusing effect. Light incident on any point on the inner surface is, by multiple reflections, distributed equally to all other points. The effects of the original direction of light are minimized. An integrating sphere may be thought of as a diffuser that preserves power but destroys spatial information. This allows all the photons, including those from the substrate, to be collected.

Table 10.3 provides luminous efficiency and T_{95} lifetime (the time taken for the light output to fall to 95% of its initial value) data for some commercial polymer-based OLEDs (J. Burroughes, personal communication March 2021).

Other up-to-date specifications for commercial devices can be found on a dedicated OLED website [22] and on the webpages of major OLED display manufacturers, such as JOLED [23]. The degradation of OLEDs usually implies some undesirable internal processes such as chemical reactions, morphological phenomena (phase transformations, crystallization, and delamination processes), and other physical (e.g. charge accumulation) changes [24]. Such effects are discussed

Table 10.3 Performance of polymer-based organic light emitting devices].

	Red	Green	Blue
Efficiency $(cd\,A^{-1})$	23	76	5.7
Colour (CIE: x, y)	0.66, 0.34	0.32, 0.63	0.14, 0.09
T_{95} lifetime (h) @ $1000\,cd\,m^{-2}$	7000	25 000	1100

Device structure: ITO (45 nm)/soluble HIL (35–65 nm)/interlayer (20 nm)/light emitting polymer (60–90 nm)/cathode.
Source: J. Burroughes, personal communication.

further in Section 10.8.4. Over recent years, the convention of how lifetime data are reported and compared has changed, with the requirements that device parameters meet industrial standards. The device lifetime was originally quoted at luminance of $100\,cd\,m^{-2}$, a previously assumed display brightness. Lifetime values are now commonly published at a luminance of $1000\,cd\,m^{-2}$, a standard appropriate for display and simple lighting applications. A particular challenge, evident from the data in Table 10.3, has been to produce efficient blue OLEDs with extended lifetimes [25]. Nonetheless, the availability of the current generation of OLEDs has allowed manufactures to produce large displays (55- or 77-in. televisions) with expected lifetimes of over 100 000 hours.

It is evident that the efficiency of an OLED is an important gauge of its performance. However, as discussed, there are different efficiency parameters and different ways to measure these. Efficiency will vary with wavelength and OLED operating conditions (e.g. drive current) and will not, generally, remain constant over long periods of time.

10.7 Device Architectures

A breakthrough in terms of improved OLED performance was achieved when the functions of charge transport and light emission were separated. The overall structure of a state-of-the-art OLED can be quite complex, as shown in Figure 10.8 (Problem 10.4). The two electrodes can each be modified to improve the carrier injection (electron- and hole-injection layers – EIL and HIL); furthermore, carrier-transporting layers (electron transport layer [ETL] and hole transport layer [HTL]) can provide efficient transfer of electrons and holes to the layer that emits the light (EML). The role of the injection and transport layers can be combined. For example, the carrier transport layers may be doped (Section 1.5.7) to provide low resistance contact regions. Common p-type dopants are strong electron acceptors such as F_4-TCNQ or transition metal oxides, while alkali metals such as Li or Cs can be effective as n-type dopants. The resulting p-i-n devices can exhibit high efficiencies and operating voltages that are close to the thermodynamic limit, i.e. the photon energy divided by the electronic charge [5, 26]. The efficiency gain from the multilayer structure depicted in Figure 10.8 must, of course, be offset against the increased fabrication costs.

10.7.1 Top- and Bottom-Emitting OLEDs

The light output from an OLED can be directed through either its bottom or top surfaces. Figure 10.9 contrasts bottom-emitting and top-emitting configurations. In the case of bottom emission, the light passes through the transparent or semi-transparent bottom electrode and

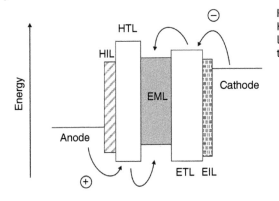

Figure 10.8 Multilayer OLED architecture. HIL = hole injection layer. HTL = hole transport layer. EML = emissive layer. ETL = electron transport layer. EIL = electron injection layer.

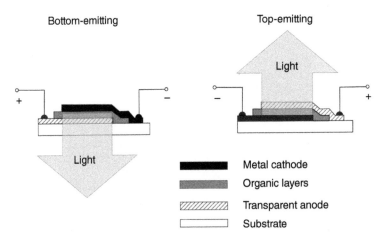

Figure 10.9 Bottom- and top-emitting OLED configurations.

substrate on which the device was manufactured. For top emission, the OLED device is built-up on a non-transparent cathode and the light exits through a top semi-transparent electrode (or via a lid that is added following device fabrication). Top-emitting OLEDs are better suited for display applications in which individual pixels need to be connected to transistors (for switching the OLED on and off). The thin film transistor array (backplane) is attached to the bottom substrate on which OLEDs are manufactured and is typically non-transparent (although transparent devices are under development), resulting in a considerable reduction of transmitted light if the device follows a bottom emitting scheme.

Other variations on the basic OLED architecture use transparent or semi-transparent contacts on both sides of the device to create displays that can be made to be both top and bottom emitting [5]. These devices can greatly improve contrast, making it much easier to view displays in bright sunlight. The technology can be used in head-up displays, smart windows, or augmented reality applications. Mechanically flexible OLEDs are a further emerging technology, offering applications in wearable electronics, bendable displays, and electronic newspapers [27]. The requirement here is for the devices to be fabricated on flexible (e.g. plastic) supports rather than the more conventional solid glass or semiconductor substrates.

10.7.2 Electrodes

The work function of the anode in an OLED must be matched to the HOMO level of the organic semiconductor for effective hole injection. This electrode is typically ITO, since this is transparent and highly conductive. This material is commonly treated with a thin layer of a hole conductor (see the following section) such as PEDOT:PSS, phthalocyanine, or polyaniline to improve both the contact to the emissive layer and the device stability. Another advantage of these layers is that they smooth out the relatively rough surface of the ITO, preventing any local short-circuiting that would otherwise cause the device to fail. In many OLEDs, the hole current can far exceed the electron current. This results in significant energy wastage since the excess holes cannot combine with electrons to generate light. Efficiency improvements can be achieved by coating the anode with a thin film that reduces the hole current (hole blocking layer). Graphene and graphene-silver nanowire composites offer alternatives to ITO as semi-transparent anodes. The mechanical compliances of such materials make these very attractive for flexible OLED displays [28, 29].

A low work function metal or alloy is used for the cathode in OLEDs (to match with the LUMO level of the organic semiconductor). Unfortunately, metals with very low work functions, such as Li and Ca, are reactive and require careful encapsulation. A thin inorganic insulating layer such as LiF, or an organic monolayer, may be inserted between the cathode and the emissive material. The introduction of such an interfacial layer can modify the electronic band structure in several ways. For example, the layer may introduce fixed charge, affecting the height of the barrier to electron injection (i.e. modifying the work function of the metal), Section 5.3. An important difference between MIS and MS structures is that, in the former case, the metal Fermi level is no longer 'tied' to the energy band structure of the semiconductor. This means that, as the forward applied voltage is increased, the cathode Fermi level can move with respect to the LUMO level in the organic emissive layer, thereby aligning filled electron states in the metal with vacant states in the organic semiconductor. If the interfacial layer is 'transparent' to electrons, the EL will increase. This, of course, imposes some restrictions on the nature of the interfacial layer: it must support a direct current (DC) voltage, but at the same time be sufficiently thin that the electron current can pass through, e.g. by tunnelling.

10.7.3 Hole- and Electron-Transport Layers

The HTL in an OLED serves two purposes. First, it provides a path for holes to be injected into the emitting layer. It also acts as an electron-blocker to confine electrons within the emissive material. Figure 10.10 shows the chemical formulae of three organic compounds that have been used as hole transport materials, and to modify the anode electrode (see above) [2, 3]. Hole transport materials generally possess a shallow LUMO level to prevent electrons entering from the emissive layer, and a HOMO level matching that of the emissive layer and the work function of the anode. Typical hole-injecting electrodes, such as ITO or PEDOT:PSS, have work functions between 4 and 5 eV and compounds that have HOMO levels matching these values are air-stable weak electron donors. One of the most common classes of compound that are used are the arylamines, of which N,N'-bis(3-methylphenyl)-N,N'-diphenylbenzidine (TPD, Figure 10.10) is one example. However, this compound has a glass transition (softening) temperature, T_g, of only 63 °C, which can result in recrystallization and delamination from the ITO anode. A significantly higher T_g (200 °C) is found in the hole transport material poly(vinylcarbazole) (PVK), an organic semiconductor used in photocopiers (the chemical structure is provided in Figure 10.10)

The roles of the ETL are to transport injected electrons to the emitting layer and to serve as a hole-blocking layer to confine holes in the emitting layer. Some of the electron transport

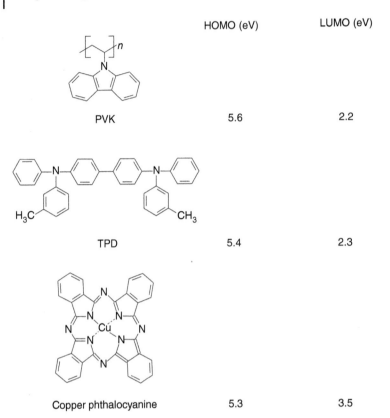

	HOMO (eV)	LUMO (eV)
PVK	5.6	2.2
TPD	5.4	2.3
Copper phthalocyanine	5.3	3.5

Figure 10.10 Hole-transporting molecules used in OLEDs. The HOMO and LUMO levels are indicated. The materials are: poly(vinyl carbazole) (PVK); *N,N*′-bis(3-methylphenyl)-*N,N*′-diphenylbenzidine (TPD); and copper phthalocyanine.

materials that have been used in OLEDs are depicted in Figure 10.11 [13, 30]. Shown are 2-(4-*tert*-butylphenyl)-5-[3-[5-(4-*tert*-butylphenyl)-1,3,4-oxadiazol-2-yl]phenyl]-1,3,4-oxadiazole (OXD-7), 3-(4-*tert*-butylphenyl)-4-phenyl-5-(4-phenylphenyl)-1,2,4-triazole (TAZ), and 1,3,5-tris(*N*-phenylbenzimidizol-2-yl)benzene (TPBI). Many of these contain oxadiazole groups (five-membered rings containing one oxygen and two nitrogens); these are very electron deficient, which allows them to block holes and transport electrons effectively.

10.7.4 Triplet Management

As noted earlier, the external quantum efficiency of devices fabricated using fluorescent materials is generally limited to 25% because of the 1 : 3 singlet to triplet ratio. To transfer energy to a phosphorescent dopant, both the singlet- and triplet-excited states of the host material must be higher in energy than the triplet-excited state of the dopant. This wide band gap requirement for the host can make it difficult to optimize simultaneously the charge injection and light emitting properties of the device. The host material must also be able to transport electrons and holes effectively and be energy-level matched to the adjacent transport layers. Figure 10.12 shows an example of an OLED architecture based on the phosphorescent emitter Ir(ppy)$_2$(acac) [13, 31]. In this case, the emitter is doped into a host material TAZ (Figure 10.11). The HTL is *N,N,N*′,*N*′-tetrakis(3-methylphenyl)-3,3′-dimethylbenzidine (HMTPD) (a similar molecule to

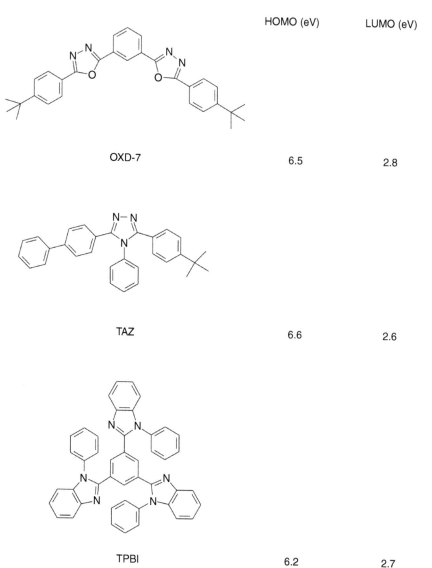

	HOMO (eV)	LUMO (eV)
OXD-7	6.5	2.8
TAZ	6.6	2.6
TPBI	6.2	2.7

Figure 10.11 Electron-transporting molecules used in OLEDs. The HOMO and LUMO levels are indicated. The materials are 2-(4-*tert*-butylphenyl)-5-[3-[5-(4-*tert*-butylphenyl)-1,3,4-oxadiazol-2-yl]phenyl]-1,3, 4-oxadiazole (OXD-7), 3-(4-*tert*-butylphenyl)-4-phenyl-5-(4-phenylphenyl)-1,2,4-triazole (TAZ), and 1,3,5-tris(*N*-phenylbenzimadazol-2-yl)benzene (TPBI).

TPD, which shown in Figure 10.10) while the ETL is Alq_3 (Figure 10.3). Device efficiencies can be relatively large for phosphorescent OLEDs, e.g. η_{ext} approaching 18% for blue emitting devices [32]. However, very high values are only achieved at low current densities. With increasing currents, the efficiency gradually decreases due to a growing influence of different quenching effects, of which triple-triplet annihilation is regarded as being of particular importance (Section 10.2).

A further strategy to improve OLED efficiency, which does not rely on heavy metal complexes, is to exploit thermally activated delayed fluorescence (TADF). In a TADF emitter, the S_1 and T_1

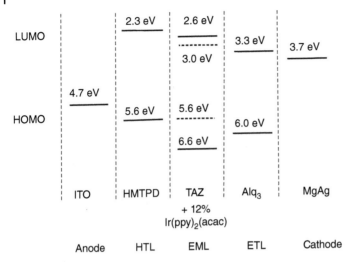

Figure 10.12 Energy level diagram for multilayer OLED based on the phosphorescent molecule Ir(ppy)$_2$(acac). The HOMO and LUMO levels of the emitter are indicated by dashed lines. Source: Yersin and Finkenzeller [13].

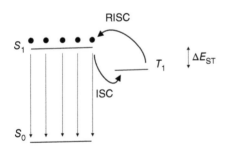

Figure 10.13 Schematic representation of thermally activated delayed fluorescence (TADF). S, T are singlet and triplet energy levels, respectively. ΔE_{ST} = singlet to triplet energy difference. ISC = intersystem crossing. RISC = reverse intersystem crossing.

levels are strongly coupled, which allows ISC between the two levels (Figure 10.1). In addition, the molecule is designed so that the energy difference between the S_1 and T_1 (ΔE_{ST}) is much smaller than in typical organic molecules; for efficient TADF, ΔE_{ST} is less than 200 meV [5, 33, 34]. The process of reverse intersystem crossing (RISC) can then occur, whereby excitons in the T_1 are converted to S_1 in a thermally activated process. The RISC constant, K_{RISC}, can be expressed using the Boltzmann distribution function (Eq. (1.9)):

$$K_{RISC} \propto \exp\left(-\frac{\Delta E_{ST}}{k_B T}\right) \tag{10.10}$$

Once in the S_1 state, the excitons can decay back to the S_0 ground state via fluorescence. The TADF process is illustrated in Figure 10.13. Since RISC is a slow process, the fluorescence from the originally produced triplet excitons occurs later than the fluorescence from the excitons created directly in the S_1 state – hence the name delayed fluorescence. Very high external quantum efficiencies are achievable with suitably designed emitters: ≈28% for red (maximum $\eta_l = 40\,\mathrm{cd\,A^{-1}}$) [35]; ≈37% for green (maximum $\eta_l = 140\,\mathrm{cd\,A^{-1}}$) [36]; and ≈38% for blue (maximum $\eta_l = 64\,\mathrm{cd\,A^{-1}}$) [37]. However, work is required to improve the long-term stability of this type of OLED [5].

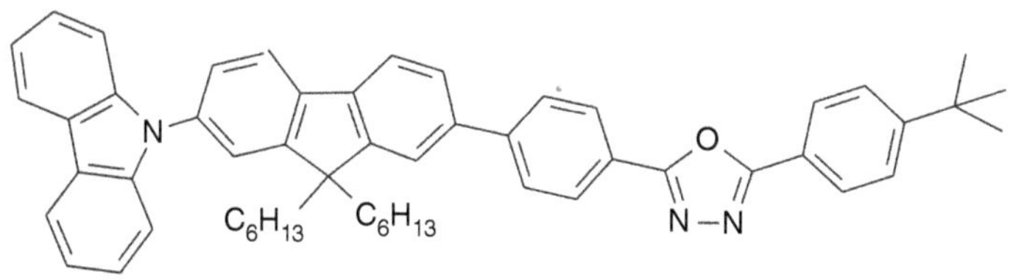

Figure 10.14 Donor-acceptor dyad for blue emission. The carbazole group (far left) is the donor while the oxadiazole group (far right) is the acceptor. The fluorene group (centre) is used as a spacer. Source: Linton et al. [40].

10.7.5 Blended-Layer and Molecularly Engineered Devices

Devices incorporating blended single layers of emissive and charge transport materials may be fabricated as an alternative to multilayer OLEDs [14, 38]. These invariably possess lower efficiencies than multilayer architectures, but they have the advantage of ease of manufacture. Electron-transport and hole-transport moieties can also be covalently incorporated into the emissive compound [39]. Figure 10.14 shows an example of such a molecule – a donor–acceptor dyad [40]. Carbazole (the chemical group to the left of the molecule as depicted) is a popular hole-transporting chemical group, while the oxadiazole unit (the right hand of the molecule) acts as an electron transporter. A fluorene chemical group (a five-carbon ring with a benzene ring on each side) is used as a spacer between the donor and acceptor units.

A simple device architecture based on this molecule (ITO/PEDOT:PSS/carbazole-oxadiazole compound/Ca/Al) was found to exhibit a deep-blue EL close to the CIE spectral locus ($x = 0.16$, $y = 0.079$), with an external quantum efficiency of 4.7%. The colour was thought to arise from the presence of the carbazole group.

10.8 Increasing the Light Output

This section examines different strategies that have been used to optimize the efficiency of OLEDs. These techniques can, broadly, be divided into methods to redirect light that is lost within the device to the emitting hemisphere and those to eliminate degradation effects and preserve a luminance output over a long time. First, the origins of the major losses of out-coupling efficiency are described.

10.8.1 Efficiency Losses

The standard OLED architecture consists of two or more layers of organic material sandwiched between two electrodes (Figure 10.8). Although the internal quantum efficiency of the EL material can be high, only a fraction of the light generated finds its way out of the device structure. For example, consider an optical point source (e.g. a fluorescent molecule) emitting radiation isotropically within a material. When the light reaches an interface, such as the semitransparent electrode, some of this will arrive at angle greater than the critical angle and will be reflected. Only the rays inside a circular cone, defined by the critical angle, will be transmitted. Ray optics models, based on Snell's Law, can be used to predict the amount of light that will eventually emerge into the air [41, 42]. A simple estimate gives η_c, the extraction efficiency as:

$$\eta_c = \frac{1}{n_{eff}^2} \tag{10.11}$$

where n_{eff} is the effective refractive index of the organic layer stack.

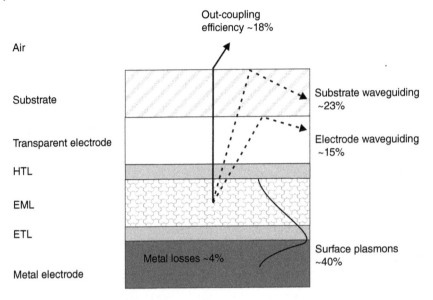

Figure 10.15 Main efficiency losses in an OLED. Source: Hong and Lee [43].

Figure 10.15 provides more details on the origin of the main out-coupling losses in an OLED [43]. Due to total internal reflections caused by the mismatch in the refractive index between each layer, as well as plasmon coupling at the metal cathode, less than 20% of the emitted light may leave the device. The metal cathode is a focus of significant loss, as approximately 40% of the total emitted light can couple to surface plasmons. The latter are collective oscillations of the free charges at a metal boundary that propagate along the interface [44]. The intensity of these waves is a maximum at the surface and reduces exponentially perpendicular to the surface. Total internal reflection can occur at the substrate/transparent anode and the substrate/air interfaces and light can be confined within the OLED layer architecture by waveguiding. Furthermore, a small proportion of the EL (4%) is absorbed at the electrodes.

Extracting light from the surface plasmon modes, thereby increasing the OLED external efficiency, is challenging. Three main strategies have been investigated: index coupling, prism coupling, and grating coupling [45]. In the first method, emission from a green dye can excite an orange dye across a thin silver layer via the plasmon modes. The prism technique exploits an optical coupling technique in which dye-emission plasmon modes propagate across a thin silver layer. Rather than exciting another dye molecule, the plasmon mode emits into an SiO_2 substrate. This emission is subsequently released into air by a prism matched to the substrate. The third and perhaps most useful technique is to scatter the surface plasmon modes by incorporating a periodic structure onto the organic metal interface such that the subsequent scattering recovers light from some of the surface plasmon modes. Other methods exploit dielectric materials with high refractive index values (e.g. WO_3), in an attempt to suppress the plasmon modes.

10.8.2 Microlenses and Shaped Substrates

Substrate waveguided light can be relatively easily extracted from the OLED by any means that disturbs the planar nature of the substrate waveguide. This disruption can be achieved by either

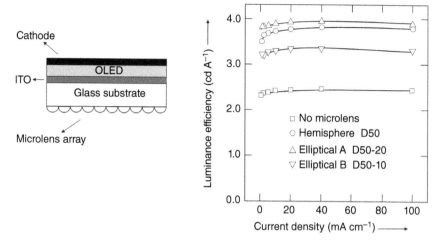

Figure 10.16 Luminance efficiency of OLEDs based on Alq$_3$ with and without microlens arrays. The device architecture is depicted on the left. Light outputs measured at a drive current density of 20 mA cm^{-2}. The data refer to different shape microlenses: D50 – hemispherical; D50-20 and D50-10 – elliptical (see text for details). Source: Yang et al. [47].

modification to the substrate/air interface or by incorporating scattering centres into the substrate itself. Early work involved attaching a small lens directly above the OLED; microlens arrays can also be utilized. Increases in the light output by a factor of approximately two can be achieved using such approaches [46]. Figure 10.16 shows an example of this strategy [47]. The OLED is based on the molecule Alq$_3$ (Figure 10.3). Plastic microlens arrays were fixed to the underside of the OLED substrate, as shown in the diagram. Three types of microlens were used in the study: a hemispherical microlens (50 μm in diameter, referred to as D50 in the figure); a short elliptical microlens (50 μm in transverse diameter, 20 μm in conjugate diameter, D50-20); and a long elliptical microlens (50 μm in transverse diameter, 10 μm in conjugate diameter, D50-10). All the lenses increased the output of the OLED, with the maximum luminance efficiency (an enhancement by a factor of 1.6) obtained for the D50-20 lens. The experimental results are in good agreement with theoretical predictions.

Increases in efficiency can be achieved by shaping the substrate to ensure that the maximum possible fraction of light emitted from the OLED leaves the device in the forward-scattered direction. Patterning of the substrate can enhance the effective emission angle by scattering additional light energy into the viewing direction. One particularly useful structure is a preformed glass mesa, an example of which is shown schematically in Figure 10.17 [46].

Compared with the substrate modes, it is significantly more difficult to recover the EL lost by waveguided modes propagating through the anode/organic layer (e.g. ITO/HTL). Strategies that have been used include: (i) use of a substrate with a refractive index higher than the emitting layer (or reducing the index of the organic layer below that of the substrate), thereby combining the anode/organic waveguided modes with the substrate modes; (ii) disrupting the planar nature of the anode/organic waveguide by introducing scattering centres or internal reflecting surfaces; and (iii) employing photonic band gap structures within the OLED [46]. Method (ii) offers the best combination of potential gain in light enhancement, process simplicity and cost-effectiveness.

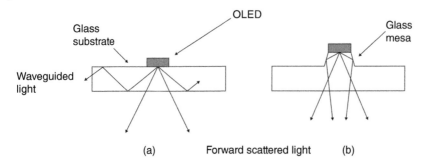

(a)   Forward scattered light   (b)

Figure 10.17 Schematic cross section (not to scale) of an OLED fabricated on a shaped substrate designed to increase the proportion of light emitted by the OLED in the forward direction. (a) A proportion of the emitted light is waveguided in the substrate. (b) Internal reflection re-directs the waveguided light into the viewing direction. Source: Burrows et al. [46].

10.8.3 Microcavities

The molecules responsible for light emission exist in a medium bounded on both sides by reflective surfaces and interfaces. This optical layer is usually composed of organic layers with a total thickness of several hundred nanometres. Therefore, the OLED stack constitutes a microcavity. Interference effects from the reflection of light from the upper and lower surfaces of a thin film will give rise to a series of maxima and minima in the transmitted (or reflected) electromagnetic wave. By using a stack of thin films, formed from alternate materials of different refractive indexes, these interference fringe maxima become very sharp. The arrangement in which the dielectric stack is formed between two mirrors is called a Fabry–Perot resonator (the structure is an example of a one-dimensional photonic crystal). The term optical microcavity is also used widely to refer to such a structure with dimensions of the order of light. These have been used successfully with OLEDs to both increase their EL output and to tune their emission wavelength. The multilayer architecture of one arrangement is shown in Figure 10.18. The SiO_2 and TiO_2 layers possess different refractive indices (1.4 and 2.3, respectively) and are alternated to provide a total thickness of one quarter of a wavelength. The OLED is then assembled on top of this stack. The EL output of a microcavity OLED based on Alq_3 is contrasted with that from a reference device in Figure 10.19 [48]. The electroluminescence spectrum for the microcavity comprises narrow and intense peaks. The wavelengths of these can be predicted using calculations based on classical optics. By varying the stack, the emission colour of the OLED can be altered, providing a useful way to produce a range of colours from the same device.

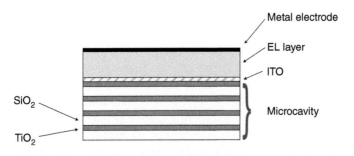

Figure 10.18 OLED fabricated on an optical microcavity.

Figure 10.19 Electroluminescent output for an Alq$_3$ OLED fabricated on a microcavity compared to a reference device. Source: Nakayama [48].

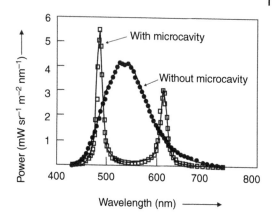

10.8.4 Device Degradation

Over the last 30 years of OLED research, much effort has been devoted to the understanding and prevention of failure and degradation processes, which limit device lifetimes [24, 49, 50]. There are a few definitions of lifetime found in the literature. Often, this parameter is expressed as the time taken for the luminance to drop to one-half of its original value at a constant current density. This is denoted as T_{50} or $T_{1/2}$. More challenging metrics, such as the T_{95} lifetime used in Table 10.3, are quoted by display panel manufacturers. Device lifetimes are usually measured by monitoring the luminance over time at a fixed current density and with the OLED in a controlled environment (e.g. constant temperature and humidity). The decrease in the luminance over time is often accompanied by a voltage increase. This is attributed to an increased injection barrier at one of the interfaces within the device architecture and/or an increase in resistance of one or more of the transport layers (e.g. due to accumulation of trapped charge) [24].

Lifetimes will depend on the precise OLED configuration. For example, top emission structures with transparent top contacts often exhibit lower lifetimes than their bottom emission counterparts (Figure 10.9). This can be caused by both the differences in charge carrier injection and damage that may occur during device fabrication, especially by particular deposition techniques (e.g. sputtering) used for transparent top contacts. The influence of different operating condition schemes (i.e. alternating/pulsed current [AC] or direct current [DC]) may also influence the lifetime [24].

An empirical scaling law is often found to relate the initial luminance L_0 to the OLED lifetime [51]:

$$L_0^n T_{1/2} = C \tag{10.12}$$

where C is a constant and n is an 'acceleration' factor dependent on the device architecture and materials.

The degradation of OLEDs may be influenced by many independent mechanisms. The lifetime can exhibit an initial rapid decay followed by a long-term decline in the luminance. The behaviour can be fitted with a series of exponential functions, with different time constants. Alternatively, a stretched exponential decay function can be fitted to the luminance versus time behaviour [51]:

$$L(t) = L_0 \exp\left(-\left(\frac{t}{\tau}\right)^\beta\right) \tag{10.13}$$

where τ is the decay constant and β is the 'stretching' factor. Neither parameter has a clear physical meaning [22]. However, the above expression may be associated with the idea that degradation loss mechanisms are linked to the accumulation of defects created within the OLED [24, 51].

Degradation phenomena that were observed in early OLED work have now been largely identified and eliminated. One example was the instantaneous breakdown of the EL behaviour (catastrophic failure), a feature of sub-optimal device fabrication processes (leading to high electric fields and electrical short circuits). A further commonly observed effect was the growth of non-emissive areas, mainly known as 'dark-spots'. These have been shown to be related to the delamination of the cathode electrode [1, 24].

The degradation of OLEDs can be divided between and extrinsic and intrinsic processes. The former category is associated with device fabrication and includes the incorporation of undesirable impurities and deficient encapsulation. A relatively well-known problem is the ingress of water and oxygen. Both these molecular species can initiate deleterious chemical or electrochemical processes within the OLED. Temperature-induced degradation is a further issue [49]. Relatively high current densities are used to achieve the required luminance for commercial displays and lighting modules. The resulting temperature increase within a device may be excessive. Adequate heat sinking and the avoidance of organic materials with low melting points or glass transition temperatures are therefore key device design strategies.

Intrinsic degradation phenomena are much more difficult to identify, and therefore to eliminate. A wide range of physical effects and chemical reactions are found to affect OLED lifetime [24]. The device architecture is complex (Figure 10.8), and a variety of process can occur within any single layer or across an interface. Diffusion and drift have been discussed in Section 2.7.2. Entities being transported will include electrons/holes, ions and excitons. Charge accumulating in deep traps may act as nonradiative recombination centres (Section 3.4) and quench the OLED luminescence. Over time, internal electric fields can re-orient electric dipoles, modifying the polarization (Section 7.2). If these processes involve the emitter molecules, the outcoupling efficiency of the OLED will be changed.

Electronic charges (either holes or electrons) can initiate chemical or electrochemical reactions, particularly in the presence of water and/or oxygen. These include chemical processes at the electrodes and oxidation of organic materials. Finally, morphological changes such as crystallization of the various layers of organic compounds may be induced by contaminants and can lead to decreases in luminance.

In some instances, chemical reactions have been identified that can be linked quantitatively to efficiency losses over time [24]. However, it impossible to provide definitive rules for preventing degradation in particular devices. OLED development is a very competitive commercial activity and many fabrication 'recipes' include proprietary information. Nonetheless, there are some empirical guidelines, which can be followed to provide a significant improvement in the device lifetime. Factors that will enhance OLED stability include the use of highly purified materials and substrates, meticulous cleanliness throughout all the processing steps, adequate encapsulation and device operation at low heat dissipation.

10.9 Full-Colour Displays

An important application area for OLEDs is in flat panel displays [1]. A display consists of a matrix of contacts made to the bottom and top surfaces of each organic light-emitting element, or pixel. The individual pixels may be addressed either passively or actively. In the former case, the display is addressed one line at a time, so that if a display has 2140 lines, then an individual pixel can only be emitting for 1/2140th of the time. High drive currents are needed, leading to heating problems and to expensive electrical driver circuits. These issues are to some extent offset by the simplicity of the technique.

Active addressing schemes involve using a device, such as a thin film transistor, attached to each pixel. Consequently, the pixel can remain emitting for the entire frame rather than for a small fraction of it. Organic EL technology is suited to active matrix addressing since it is a low voltage technology and OFETs are likely, eventually, to provide the drive circuitry.

To generate a full-colour image, it is necessary to vary the relative intensities of three closely spaced, independently addressed pixels, each emitting one of the three primary colours of red, green or blue. Some of the different approaches that have been used are shown in Figure 10.20 [46]. The simplest scheme, Figure 10.20a, is the side-by-side-positioned red, green, and blue (R, G, B) subpixels. This arrangement requires each of the three closely spaced OLEDs to be sequentially grown and patterned. Alternatively, optical filtering of white OLEDs can produce red, green, and

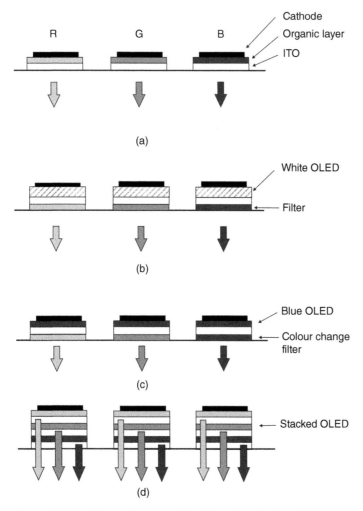

Figure 10.20 Schemes for generating full-colour displays. (a) Separate red, green, and blue emitters (R, G, and B) providing pixels side-by-side. (b) The light from white-emitting OLEDs is filtered to provide R, G, and B emission. (c) The light from blue-emitting OLEDs is used to generate R, G. and B emission through colour changing filters. (d) Stacked OLEDs emit R, G, and B. Source: Modified from Burrows et al. [46].

blue emission, Figure 10.20b. This method is not particularly efficient as a significant amount of light is absorbed by the filters. A further option, shown in Figure 10.20c, exploits a single blue or ultraviolet OLED to excite organic fluorescent wavelength down converters. Each of these 'filters' consists of a material that efficiently absorbs the blue light and re-emits the energy as either green or red light.

An elegant way of achieving fill-colour displays is to stack the red, green and blue pixels on top of each other, as shown in Figure 10.20d. Since there are no subpixels, this has the advantage of increasing the resolution of the display by a factor of three. However, the method requires semi-transparent electrodes that are compatible with high current densities. One problem encountered with stacking numerous transparent organic layers is the formation of unwanted optical cavities whose resonances alter the emission spectra of OLEDs (Section 10.8.3). Such effects can be eliminated by careful control of layer thickness and composition. With appropriate device design, three independently addressable electrodes may be used to drive these devices [52].

10.10 Organic Semiconductor Lasers

The light from an OLED will generally be incoherent, which means that there is no ordered phase relationship between the EM waves that are emitted. The electron in the excited state E_2 spontaneously falls back into the ground state E_1, with the emission of a photon of energy $hf = E_2 - E_1$. If there are many such excited electrons, these will emit photons at random times and so generate incoherent light. The normal processes of light absorption and spontaneous emission are illustrated in Figure 10.21a,b, respectively. However, it is possible for an incoming photon to trigger the emission process. This is called stimulated emission, as shown in Figure 10.21c. The emitted photon is in phase with the incoming photon; it is travelling in the same direction and has the same frequency (because it has an identical energy, $E_2 - E_1$). The result is coherent radiation. Stimulated emission is the basis for the operation of the laser (an acronym for *l*ight *a*mplification by *s*timulated *e*mission of *r*adiation). This device acts as a photon amplifier since one incoming photon results in two outgoing photons. An important requirement is that there are more electrons in energy state E_2 than in E_1, a situation referred to as population inversion. It is not possible to achieve this condition with only two energy levels as, in the steady state, the incoming photon flux will cause as many upward transitions as downward stimulated transitions (Problem 10.5). To create a population inversion, an additional level(s) is required, and energy is channelled into the lasing medium by some process such as the passage of a current, the creation of an electrical discharge or illumination with EM radiation.

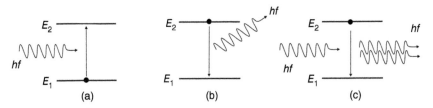

Figure 10.21 Absorption and emission processes associated with two electron energy levels E_1 and E_2. (a) Absorption of a photon with energy hf promotes an electron from level E_1 to E_2. (b) Spontaneous emission, the photons are emitted randomly. (c) Stimulated emission, the emitted photons are coherent with the stimulating photons.

In a three-level laser, the material is first excited to a short-lived high-energy state that spontaneously drops to a lower-energy metastable state with a long lifetime. The metastable state is important because it traps and holds the excitation energy, building up a population inversion that can be further stimulated, and finally allowing the species to return to the ground state with the emission of radiation. An example of a three-level laser material is ruby ($Cr^{3+}:Al_2O_3$). The three-level laser works only if the ground state is depopulated. As atoms or molecules emit light, they accumulate in the ground state, where they can absorb the stimulated emission and constrain laser action. Consequently, most three-level lasers can only generate pulses. This difficulty is overcome in the four-level laser, where an extra transition state is located between metastable and ground states.

A schematic diagram for a four-level system is shown in Figure 10.22a. Light excites a molecule from the ground state at energy E_1 to an excited state E_2; the molecule rapidly relaxes to another energy level E_3. The lasing transition then occurs between E_3 and a level E_4, which is above the ground state. Finally, there is a rapid return to the ground state. From the figure, it is evident that the emission will occur at a longer wavelength than the absorption. The advantage of a four-level system is that there can be a population inversion between levels E_3 and E_4, even when most molecules are in the ground state, so lasing is achieved for a very low rate of excitation, i.e. the threshold for lasing is low. A popular example of a four-level solid state laser medium is neodymium-doped yttrium aluminium garnet (Nd:YAG). Lasers based on Nd:YAG can operate in both pulsed and continuous mode.

The distribution of energy levels in a typical organic semiconductor lends itself, in principle, to four-level lasing [53]. Figure 10.22b shows the ground state and first excited singlet state (depicted in more detail in Figure 10.1). Each of these electronic energy levels is subdivided into vibrational sublevels. Typical vibrational energy quanta are in the range of 100–300 meV, implying that at room temperature only the lowest vibrational level is occupied. Light can excite the molecule from its ground, S_0, state to an excited vibrational level of the singlet S_1 level. This will be followed by rapid relaxation to the bottom of the S_1 state. Lasing can then take place by transition to a vibrationally excited level of the S_0, followed by vibrational relaxation to the ground state.

A second essential element for a practical laser is an optical feedback system that repeatedly passes resonant light through the gain medium to establish a very intense, coherent optical field

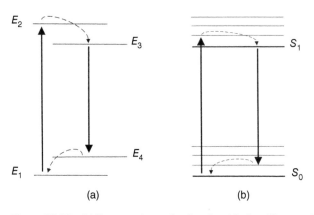

(a)   (b)

Figure 10.22 (a) Energy scheme for four-level lasing. The transition from level E_1 to E_2 represents an optical absorption while that between E_3 and E_4 is the emission. The transitions from E_2 to E_3 and from E_4 to E_1, shown as dashed lines, are thermal relaxations. (b) Energy levels for lowest two singlet states (S_0 and S_1) in an organic semiconductor showing corresponding optical and thermal transitions to those in (a).

inside the laser. In the very simplest case, this optical cavity may comprise only two mirrors, configured as a Fabry–Perot interferometer, between which the amplifying gain medium is situated (e.g. Figure 10.18). Other useful types of resonators for organic semiconductor lasers are diffractive structures. These do not use either mirrors or total internal reflection for feedback, but instead exploit periodic, wavelength-scale microstructures that diffract, or Bragg-scatter, the light.

There have been several different diffractive structures explored for organic semiconductor lasers, including simple diffraction gratings that form so-called distributed feedback (DFB) lasers. These can be readily incorporated into planar organic semiconductor waveguides and avoid the need for good-quality end facets to provide the feedback [53]. An example of this approach is provided in Figure 10.23, which shows a semiconducting polymer film deposited onto a corrugated substrate (e.g. silica) of period Λ. Light propagating in a waveguide mode of the organic film is scattered by the corrugations and the scattered light combines coherently to create a Bragg-scattered wave propagating in some new direction. The angle through which the light is diffracted is highly dependent on wavelength. For a given period of the corrugation, there is a particular set of wavelengths that will be diffracted from a propagating mode of the waveguide into the counter-propagating waveguide mode. This situation will arise when the Bragg condition is satisfied:

$$m\lambda = 2n_{\text{eff}}\Lambda \tag{10.14}$$

where λ is the wavelength of the light and m is an integer that represents the order of the diffraction [54]. The parameter n_{eff} is the effective refractive index of the waveguide. The DFB resonators provide optical feedback in the plane of the polymer film via second-order Bragg scattering at a wavelength determined by $n_{\text{eff}}\Lambda$.

Organic dyes have been exploited as dopants in optically pumped (excited) lasers. However, organic semiconductors offer several advantages over dyes [53]. The former can possess high photoluminescence quantum yields, which offer stronger absorption and higher gain than dyes. Importantly, organic semiconductor films are capable of charge transport, opening the possibility of electrical excitation. This would lead to the organic equivalent of the solid-state LED laser, in which population inversion is achieved, not by using another optical source (optical pumping), but by forward biassing a p-n junction diode.

An optically pumped organic semiconductor microcavity laser was first shown in 1996 [55]. This consisted of a 100 nm thick layer of the polymer PPV between a pair of mirrors and was excited by a Nd:YAG laser. The demonstration of laser action provided direct evidence for a model in which the main photoexcitation in PPV is an emissive intrachain species (rather than a non-emitting

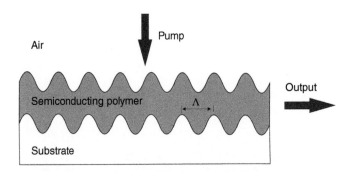

Figure 10.23 Schematic diagram of a polymer distributed feedback (DFB) laser with corrugations of period Λ. Second-order diffraction induces the optical feedback necessary for lasing operation.

interchain species). Lasing in other compounds has since been observed [53]. The use of inorganic LEDs to pump organic semiconductor lasers is also possible [56].

Much research effort has focused on the realization of an electrically driven organic semiconductor laser. However, this goal remains elusive. A fundamental problem concerns the nature of electrical pumping. Optical excitation mainly leads to the formation of singlet excitons – the states required to provide the necessary gain. In contrast, electrical excitation involves charge injection followed by capture of charges to form an exciton, which can be singlet or triplet. Both the injected charges and triplet excitons have associated absorptions, leading to losses within the material. High current densities would be required (over a few hundred amperes per square centimetre) to suppress these losses. But there have been encouraging results. One of the most promising molecules for the realization of an electrically driven organic semiconductor laser is 4,4′-bis[(N-carbazole)styryl]biphenyl (BSBCz) [57]. This compound appears to possess an appropriate combination of optical and electrical properties such as a low amplified spontaneous emission and the ability to withstand the injection of current densities as high as $2.8\,kA\,cm^{-2}$ under pulse operation.

10.11 OLED Lighting

There are keen commercial and environmental influences that are driving the development of white light organic displays for lighting applications. A large-area white light-emitting device will provide a solid-state light source that should compete with conventional technologies [58–61]. OLED specifications for lighting applications are somewhat different to those for displays. Although the brightness, lifetime, efficiency, and CRI are all critical factors, the ability to make large area panels and their colour stability are also crucial features. The CIE co-ordinates required are approximately 0.31, 0.32, but the interpretation of 'white' does vary throughout the world.

A white light-emitting OLED is commonly referred to by the acronym WOLED. For lighting, WOLEDs need to deliver significantly higher luminance values than displays. In the case of the latter, values of $200\,cd\,m^{-2}$ are adequate, whereas a luminance of at least 10 times this figure is required if the OLED is to be used in lighting. This presents a challenge in terms of device operating lifetime. The luminous power efficiencies for incandescent light bulbs are typically less than $20\,lm\,W^{-1}$, while the figures for fluorescent tubes are in the 50–$100\,lm\,W^{-1}$ range. Clearly, OLED lights need to target the higher efficiencies of fluorescent tubes. Individual white OLEDs with power efficiencies of around $100\,lm\,W^{-1}$ have already been demonstrated [61] and efficiencies greater than $150\,lm\,W^{-1}$ are thought to be achievable [62]. For a white light source to be human eye-friendly, it should possess a CRI > 80.

Different methods of making WOLEDs by blending emissive species, either in single or multiple layers (tandem devices) have been studied [61]. Alternatively, a blue OLED can be used with one or more down-conversion layers. Many white OLEDs exploit phosphorescent emitters, as the fraction of excitons that can be radiative can approach 100%. All-phosphorescent white devices can be fabricated using multiple dopants in the same emitting layer. A blue fluorescent emitter can be combined with phosphorescent emitters of other colours, creating a hybrid white OLED. In this case, the management of singlet and triplet excitons is important in order to channel most of the triplet energy to the phosphorescent molecules, retaining the singlet energy on the blue fluorescent dopant [60].

The roadmap for WOLED development is generally following that for OLEDs, for example by synthesizing more efficient materials, optimizing the device structures, and improving the light outcoupling using methods outlined in Section 10.8.2. A desirable feature is the ability for users to vary the WOLED luminance (dimming). This can be achieved either by adjusting the amount of current used to drive the device or by changing the duty cycle using a pulse-width modulated driver. Unfortunately, as WOLEDs are fabricated using multiple emitters, and the proportion of photons from the individual emitters can change as the current level changes, the output colour will generally vary with dimming. If the colour temperature of a lighting source shifts with the driving voltage, the CRI will inevitably tend to change, leading to distortion in the actual colour of objects under the lighting source.

10.12   Light-Emitting Electrochemical Cells

In a light-emitting electrochemical cell (LEC), the light emitting layer consists of a blend of an EL organic compound (polymer, small molecule or an ionic transition metal complex) with an electrolyte [63–65]. Under applied bias, the mobile anions and cations of the electrolyte accumulate at the electrodes, forming electric double layers. The effect is to reduce the charge injection barriers at the electrodes. One of the advantages of LECs over OLEDs is that they have relatively low operating voltages, which can approach E_g/e, where E_g is the band gap of the organic EL material (as for the p-i-n doped structures discussed in Section 10.7). Unlike OLEDs, LECs do not require specific work-function-selected electrodes, but rather are able to make use of more stable and inert metallic materials. Air-stable cathodes avoid complicated packaging processes and further improve the device lifetime.

The earliest LECs were based on the emissive polymer PPV. This was blended with a polymer electrolyte, consisting of an ion-solvating/transporting polymer and a molecular salt. The polymer electrolyte used in the original LECs was poly(ethylene oxide) (PEO, with a repeat unit CH_2CH_2O) complexed with lithium trifluoromethanesulfonate ($LiCF_3SO_3$) [63]. Figure 10.24 depicts the arrangement. Before the application of a voltage, the solvated free cations (Li^+) and anions ($CF_3SO_3^-$) in the LEC are randomly distributed between the metal electrodes as shown in Figure 10.24a. When a sufficiently large voltage bias (typically 2–4 V) is applied, electrons are injected from the negatively biased cathode, and holes are injected from the positively biased anode. As a result, the luminescent polymer is oxidized near the anode and reduced near the cathode. The injected electronic charges are compensated by the oppositely charged ionic species, which are abundant throughout the polymer film. This leads to electrically neutral, but highly conductive, doped polymer layers near the electrode/polymer interfaces. The luminescent polymer is p-doped near the anode, and n-doped near the cathode (Figure 10.24b). Under a constant bias, the in-situ electrochemical doping process continues with the injection of electronic charges and the redistribution of ionic charges, or counter-ions, causing the doping regions to expand. Eventually, the p- and n-doping fronts meet within the bulk of the LEC film to form a semiconductor p-n junction. EL is then observed as a result of the radiative recombination of the injected electrons and holes in the vicinity of the junction (Figure 10.24c).

The most used electrolyte groups in LECs are alkali metal salts dissolved in ether-based ion transporters and ionic liquids. Ionic liquids (or 'molten salts') are electrolytes with a low melting point (in some cases below room temperature), which can be hydrophobic, and highly conductive. These do not require an ion transporter for ion solvation and transport. The first ionic liquids employed were based on tetraalkylammonium cations, but other ionic liquid LECs incorporating imidazolium or phosphonium cations have been investigated [64]. Polymerizable electrolytes and

PPV:PEO:LiCF$_3$SO$_3$

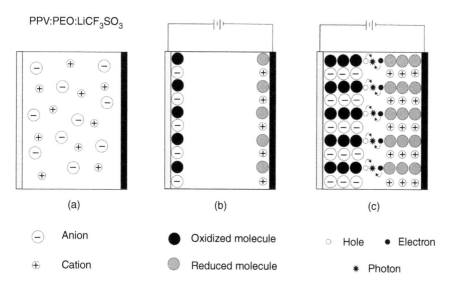

(a)	(b)	(c)

⊖ Anion ⬤ Oxidized molecule ○ Hole ● Electron

⊕ Cation ◓ Reduced molecule ✳ Photon

Figure 10.24 Operation of a light-emitting electrochemical cell. (a) Without applied bias. (b) Initiation of doping with bias applied. (c) Electroluminescence from a fully formed p-n junction. The material between the electrodes is a blend of the polymer PPV with an electrolyte comprising PEO and LiCF$_3$SO$_3$. The symbols representing the various species in the cell are indicated. Source: Gao [63].

mixed ion and electron conductors have also been used to address specific problems such as a slow turn-on time and phase separation.

A wide range of EL polymers and small molecules have now been exploited in LEC configurations [65]. Cationic iridium complexes offer relatively high efficiency devices colour tuning over a wide wavelength range. White-light LECs can be fabricated using mixtures of emitting species. Compared to conventional OLEDs and WOLEDs, LECs devices only achieve a moderate performance, in terms of efficiencies and lifetimes. The electrochemical devices also possess slow response times, as their operation involves the redistribution of ions and the doping of the luminescent organic compound. However, the low fabrication costs may offer advantages for certain device architectures, for example flexible and nonconventional substrates (paper, fibres and complex-shaped surfaces).

10.13 Light-Emitting Transistors

Organic light-emitting transistors (OLETs) combine the functionalities of OFETs and OLEDs [66–69]. In the case of the OFET, the injection of either electrons or holes from the source and drain contacts is sufficient for successful device operation. In contrast, the simultaneous injection of electrons and holes, followed by their subsequent recombination, is required in an OLET. Figure 10.25a depicts a typical OLET architecture. Electrons and holes are injected from the source and drain contacts and move towards each other under suitable bias conditions. In practice, a good match is needed between the work functions of the electrodes and the LUMO, HOMO levels of the organic semiconductor.

The materials requirements for OLETs also differ from those in OLEDs. For the latter devices, the carrier mobilities in the light-emitting layer are not generally important parameters since the 'channel' length of an OLED is typically a few 10s of nanometres. However, the electrons and holes must travel significantly further distances in the transistor structure and high carrier mobilities are

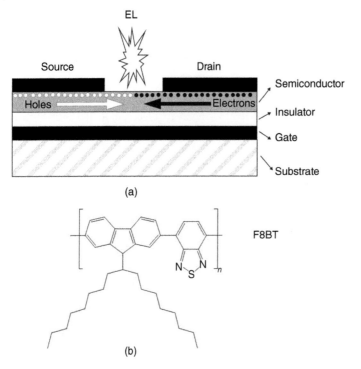

Figure 10.25 (a) Architecture of an organic light-emitting transistor (OLET). (b) Chemical structure of the green-emitting polymer, poly(9,9-di-*n*-octylfluorene-*alt*-benzothiadiazole) (F8BT) used in OLETs.

needed. Organic semiconductors that have been used successfully in OLET architectures include: fluorene derivatives, PPV-based polymers, acenes, oligothiophenes, furan-incorporated oligomers, spirobifluorenes, and various phosphorescent compounds [66].

Both unipolar and ambipolar device operation have been demonstrated. The majority of OLETs operate in the unipolar (p-channel) regime in which the hole density extends all the way across the transistor channel such that the hole accumulation layer functions as the anode for the OLED with electron injection by tunnelling from the drain electrode. Early work exploited acenes such as tetracene to achieve light emission. For unipolar OLETs, the light generation is restricted to a region close to the minority carrier injection electrode; this can reduce the light emission due to metal contact-induced exciton quenching. Improvements in the carrier injection may be achieved by using asymmetric work function source and drain contacts, for example Al for electrons and Au for holes.

The first ambipolar OLETs were fabricated using a bulk heterojunction as the active component, by mixing two materials with complementary properties. In such devices, the position of the exciton recombination zone can be controlled by using appropriate (gate) biassing conditions. Efficient photoluminescent organic compounds, which also possess balanced electron and hole mobilities, have now been developed. An example is the green-emitting polymer, poly(9,9-di-*n*-octylfluorene-*alt*-benzothiadiazole) (F8BT), the chemical structure of which is depicted in Figure 10.25b. Bilayer OLETs can be fabricated; these have separate layers for the charge transport and EL [69]. Other modifications to improve the operation the basic OLET architecture shown in Figure 10.25a include the use of a split-gate architecture, non-planar electrodes and a vertical device configuration [66]. High-*k* dielectrics (Section 8.4.5), such as based on PVDF, can result in larger channel currents and lower operating voltages for OLETs [70]. Solid electrolytes may also be used as the gate electrode [71]. Again, the relatively high capacitance of the electrolytic layer leads to low operating voltages. Hybrid organic–inorganic OLETs can combine the transport features of inorganic semiconductors with the optical properties of organic dyes. Devices based on In_2O_3-ZnO semiconducting layer and a yellow-emitting polymer and operating at drain and gate voltages of 10 V have been shown to produce a luminance of 700 cd m^{-2} [72].

As high current densities can be obtained and the light emission can be chosen to occur far away from absorbing metal electrodes, OLETs offer a versatile and attractive platform not only for low-loss light signal transmission in optoelectronic integrated circuits but also for the realization of electrically pumped organic lasers. OLETs may constitute a key element for the development of next generation organic active-matrix display technology (Section 10.9) since these devices combine electrical switching with light emission.

Problems

10.1 A commercial LED (based on an inorganic semiconductor) 'warm white' light bulb is rated at 10.5 W and has a luminous flux of 1055 lm. Assume that light is emitted isotropically from the bulb:

- (i) What is the luminous intensity of the light bulb?
- (ii) Calculate the illuminance on a desk located 1.5 m below the bulb.
- (iii) What is the luminous power efficiency of the light bulb? How does this value compare to that from a 60 W incandescent light bulb with a luminous flux of 1000 lm?
- (iv) Research the origin of the (1/683) factor involved in the definition of the candela (Eq. (10.2)).

10.2 Consider red (625 nm) and amber (590 nm) OLEDs. If both devices possess external quantum efficiencies $\eta_{ext} = 0.2$, estimate their luminous power efficiencies. Assume that the device operating voltages are given by $V = E_g/e = hf/e$, where E_g is the band gap between the conduction and valence bands. A few values of the eye sensitivity function, $V(\lambda)$, are provided in the table below:

Wavelength (nm)	$V(\lambda)$	Wavelength (nm)	$V(\lambda)$
580	0.87	610	0.50
585	0.82	615	0.44
590	0.76	620	0.38
595	0.70	625	0.32
600	0.63	630	0.27
605	0.57	635	0.22

10.3 An OLED with a diameter of 0.2 mm area is viewed at a distance of 1 m. It emits light at a wavelength of 600 nm and has an external quantum efficiency of 5%. If the device is operated at 20 mA, calculate (i) the radiant flux, (ii) the luminous flux, and (iii) the luminous intensity at normal incidence. Use the table given in Q2 above for values of the eye sensitivity function.

10.4 A multilayer OLED is based on the architecture $Al/Alq_3/NPB/MDATA/ITO$ (you will need to research some of these compounds). What is the function of each layer? The thicknesses of the Alq_3 and 4,4′,4″-tris[phenyl(m-tolyl)amino]triphenylamine (MTDATA) are both fixed at 70 nm while that of the N,N'-di(1-naphthyl)-N,N'-diphenyl-(1,1′-biphenyl)-4,4′-diamine (NPB) layer is varied. To achieve a fixed current density of 80 mA cm^{-2}, the voltage applied to the OLED (drive voltage) increases with the thickness of the NPB layer as shown in the table below.

NPB layer thickness (nm)	Driving voltage (V)
10	12.0
20	12.8
40	14.4
60	16.0

Assuming that the voltage drop in the MTDATA layer is negligible, calculate the electric field in the NPB layer. Estimate the electric field in the Alq_3 layer.

10.5 Estimate the relative electron populations in thermal equilibrium of two energy levels such that a transition from the higher to the lower will give visible radiation. Comment on the suitability of this system to provide the basis for lasing.

References

1 Tsujimura, T. (2012). *OLED Displays: Fundamentals and Applications*. Hoboken, NJ: Wiley.

2 Gaspar, D.J. and Polikarpov, E. (eds.) (2015). *OLED Fundamentals: Materials, Devices, and Processing of Organic Light-Emitting Diodes*. Boca Raton, FL: CRC Press.

3 Li, Z.R. (ed.) (2015). *Organic Light-Emitting Materials and Devices*, 2e. Boca Raton, FL: CRC Press.

4 Kalyan, N.T., Swart, H., and Dhoble, S.J. (2017). *Principles and Applications of Organic Light Emitting Diodes (OLEDs)*. Duxford: Woodhead Publishing.

5 Zou, S.-J., Shen, Y., Xie, F.-M. et al. (2020). Recent advances in organic light-emitting diodes: towards smart lighting and displays. *Mater. Chem. Front.* 4: 788–820.

6 Shinar, J. and Shinar, R. (2008). Organic light-emitting devices (OLEDs) and OLED-based chemical and biological sensors: an overview. *J. Phys. D: Appl. Phys.* 41: 133001.

7 Elze, T., Taylor, C., and Bex, P.J. (2013). An evaluation of organic light emitting diode monitors for medical applications: great timing, but luminance artifacts. *Med. Phys.* 40: 92701.

8 Noctura Sleep Mask. https://noctura.com (accessed 19 March 2021).

9 Tang, C.W. and VanSlyke, S.A. (1987). Organic electroluminescent diodes. *Appl. Phys. Lett.* 51: 913–915.

10 Burroughes, J.H., Bradley, D.D.C., Brown, A.R. et al. (1990). Light-emitting diodes based on conjugated polymers. *Nature* 347: 539–541.

11 Monkman, A.P. (2013). Singlet generation from triplet excitons in fluorescent organic light-emitting diodes. *ISRN Mater. Sci.* 2013: 670130.

12 Moro, L., Boesch, D., and Zeng, X. (2015). OLED encapsulation. In: *OLED Fundamentals: Materials, Devices, and Processing of Organic Light-Emitting Diodes* (eds. D.J. Gaspar and E. Polikarpov), 25–65. Boca Raton, FL: CRC Press.

13 Yersin, H. and Finkenzeller, W.J. (2008). Triplet emitters for organic light-emitting diodes: basic properties. In: *Highly Efficient OLEDs with Phosphorescent Materials* (ed. H. Yersin), 1–98. Weinheim: Wiley-VCH.

14 Oyston, S., Wang, C., Hughes, G. et al. (2005). New 2,5-diaryl-1,3,4-oxadiazole-fluorene hybrids as electron transporting materials for blended-layer organic light emitting diodes. *J. Mater. Chem.* 15: 194–203.

15 Karg, S., Meier, M., and Riess, W. (1997). Light-emitting diodes based on poly-*p*-phenylene-vinylene: I charge-carrier injection and transport. *J. Appl. Phys.* 82: 1951–1966.

16 Burrows, P.E., Shen, Z., Bulovic, V. et al. (1996). Relationship between electroluminescence and current transport in organic heterojunction light-emitting devices. *J. Appl. Phys.* 79: 7991–8006.

17 Gong, X. and Wang, S. (2008). Polymer light-emitting diodes: devices and materials. In: *Introduction to Organic Electronic and Optoelectronic Materials and Devices* (eds. S.-S. Sun and L.A. Dalton), 373–400. Boca Raton, FL: CRC Press.

18 Campbell, A.J., Bradley, D.D.C., and Lidzey, D.G. (1997). Space-charge limited conduction with traps in poly(phenylene vinylene) light emitting diodes. *J. Appl. Phys.* 82: 6326–6342.

19 Coaton, J.R. and Marsden, A.M. (eds.) (1977). *Lamps and Lighting*, 4e. London: Arnold.

20 Forrest, S.R., Bradley, D.D.C., and Thompson, M.E. (2003). Measuring the efficiency of organic light-emitting devices. *Adv. Mater.* 15: 1043–1048.

21 Schubert, E.F. (2003). *Light-Emitting Diodes*. Cambridge: Cambridge University Press.

22 OLED Info (2021) Web pages dedicated to up-to-date OLED information. http://www.oled-info.com (accessed 19 March 2021).

23 JOLED (2021) Web pages of JOLED company. https://www.j-oled.com/eng/company (accessed 19 March 2021).

24 Scholz, S., Kondakov, D., Lüssem, B., and Leo, K. (2015). Degradation mechanisms and reactions in organic lighting devices. *Chem. Rev.* 115: 8449–8503.

25 Lee, J.-H., Chen, C.-H., Lee, P.-H. et al. (2019). Blue organic light-emitting diodes: current status, challenges, and future outlook. *J. Mater. Chem. C* 7: 5874–5888.

26 Loeser, F., Tietze, M., Lüssem, B., and Blochwitz-Nimoth, J. (2015). Conductivity doping. In: *OLED Fundamentals: Materials, Devices, and Processing of Organic Light-Emitting Diodes* (eds. D.J. Gaspar and E. Polikarpov), 189–233. Boca Raton, FL: CRC Press.

27 Liu, Y.-F., Feng, J., Bi, Y.-G. et al. (2019). Recent developments in flexible organic light-emitting devices. *Adv. Mater. Technol.* 4: 1800371.

28 Kwon, O.E., Shin, J.-W., Oh, H. et al. (2020). A prototype active-matrix OLED using graphene anode for flexible display application. *J. Inf. Disp.* 21: 49–56.

29 Li, H., Liu, Y., Su, A. et al. (2019). Promising hybrid graphene-silver nanowire composite electrode for flexible organic light-emitting diodes. *Sci. Rep.* 9: 17998.

30 So, F. and Shi, S. (2008). Organic molecular light-emitting materials and devices. In: *Introduction to Organic Electronic and Optoelectronic Materials and Devices* (eds. S.-S. Sun and L.A. Dalton), 351–372. Boca Raton, FL: CRC Press.

31 Adachi, C., Baldo, M.A., Thompson, M.E., and Forrest, S.R. (2001). Nearly 100% internal phosphorescence efficiency in an organic light emitting device. *J. Appl. Phys.* 90: 5048–5051.

32 Klimes, K., Zhu, Z.-Q., and Li, J. (2019). Efficient blue phosphorescent OLEDs with improved stability and color purity through judicious triplet exciton management. *Adv. Funct. Mater.* 29: 1903068.

33 Uoyama, H., Goushi, K., Shizu, K. et al. (2012). Highly efficient organic light-emitting diodes from delayed fluorescence. *Nature* 492: 234–240.

34 Penfold, T.J., Dias, F.B., and Monkman, A.P. (2018). The theory of thermally activated delayed fluorescence for organic light emitting diodes. *Chem. Commun.* 54: 3926–3935.

35 Zhang, Y.-L., Ran, Q., Wang, Q. et al. (2019). High-efficiency red organic light-emitting diodes with external quantum efficiency close to 30% based on a novel thermally activated delayed fluorescence emitter. *Adv. Mater.* 31: 1902368.

36 Wu, T.-L., Huang, M.-J., Lin, C.-C. et al. (2018). Diboron compound-based organic light-emitting diodes with high efficiency and reduced efficiency roll-off. *Nat. Photonics* 12: 235–240.

37 Ahn, D.H., Kim, S.W., Lee, H. et al. (2019). Highly efficient blue thermally activated delayed fluorescence emitters based on symmetrical and rigid oxygen-bridged boron acceptors. *Nat. Photonics* 13: 540–546.

38 Oyston, S., Wang, C., Perepichka, I.F. et al. (2005). Enhanced electron injection and efficiency in blended-layer organic light emitting diodes with aluminium cathodes: new 2,5-diaryl-1,3,4-oxadioazole-fluorene hybrids incorporating pyridine units. *J. Mater. Chem.* 15: 5164–5173.

39 Kamtekar, K.T., Wang, C., Bettington, S. et al. (2006). New electroluminescent bipolar compounds for balanced charge-transport and tuneable colour in organic light emitting diodes: triphenylamine-oxadiazole-fluorene triad molecules. *J. Mater. Chem.* 16: 3823–3835.

40 Linton, K.E., Fisher, A.L., Pearson, C. et al. (2012). Colour tuning of blue electroluminescence using bipolar carbazole-oxadiazole molecules in single-active-layer organic light emitting devices (OLEDs). *J. Mater. Chem.* 22: 11816–11825.

41 Greenham, N.C., Friend, R.H., and Bradley, D.D.C. (1994). Angular dependence of the emission from a conjugated polymer light-emitting diode: implications for efficiency calculations. *Adv. Mater.* 6: 491–494.

42 Gu, G., Garbuzov, D.Z., Burrows, P.E. et al. (1997). High-external-quantum-efficiency organic light-emitting devices. *Opt. Lett.* 22: 396–398.

43 Hong, K. and Lee, J.-L. (2011). Recent developments in light extraction technologies of organic light emitting diodes. *Electron. Mater. Lett.* 7: 77–91.

44 Petty, M.C. (2019). *Organic and Molecular Electronics: From Principles to Practice*, 2e. Chichester: Wiley.

45 Lu, M.-H.M. (2015). Microcavity effects and light extraction enhancement. In: *OLED Fundamentals: Materials, Devices, and Processing of Organic Light-Emitting Diodes* (eds. D.J. Gaspar and E. Polikarpov), 299–337. Boca Raton, FL: CRC Press.

46 Burrows, P.E., Gu, G., Bulović, V. et al. (1997). Achieving full-colour organic light-emitting devices for lightweight, flat-panel displays. *IEEE Trans. Electron. Devices* 44: 1188–1202.

47 Yang, J.P., Bao, Q.Y., Xu, Z.Q. et al. (2010). Light out-coupling enhancement of organic light-emitting devices with microlens array. *Appl. Phys. Lett.* 97: 223303.

48 Nakayama, T. (1997). Organic luminescent devices with a microcavity structure. In: *Organic Electroluminescent Materials and Devices* (eds. S. Miyata and H.S. Nalwa), 359–389. Amsterdam: Gordon and Breach.

49 Tyagi, P., Srivastava, R., Giri, L.I. et al. (2016). Degradation of organic light emitting diode: heat related issues and solutions. *Synth. Met.* 216: 40–50.

50 Zhao, C. and Duan, L. (2020). Review on photo- and electrical aging mechanisms for neutral excitons and ions in organic light-emitting diodes. *J. Mater. Chem. C* 8: 803–820.

51 Meerheim, R., Walzer, K., Pfeiffer, M., and Leo, K. (2006). Ultrastable and efficient red organic light emitting diodes with doped transport layers. *Appl. Phys. Lett.* 89: 061111.

52 Fröbel, M., Fries, F., Schwab, T. et al. (2018). Three-terminal RGB full-color OLED pixels for ultrahigh density displays. *Sci. Rep.* 8: 9684.

53 Samuel, I.D.W. and Turnbull, G.A. (2007). Organic semiconductor lasers. *Chem. Rev.* 107: 1272–1295.

54 Ghafouri-Shiraz, H. (2003). *Distributed Feedback Laser Diodes and Optical Tunable Filters*. Chichester: Wiley.

55 Tessler, N., Denton, G.J., and Friend, R.H. (1996). Lasing from conjugated polymer microcavities. *Nature* 382: 695–697.

56 Tsiminis, G., Wang, Y., Kanibolotsky, A.L. et al. (2013). Nanoimprinted organic semiconductor laser pumped by a light-emitting diode. *Adv. Mater.* 25: 2826–2830.

57 Sandanayaka, A.S.D., Matsushima, T., Bencheikh, F. et al. (2019). Indication of current-injection from an organic semiconductor. *Appl. Phys. Express* 11: 061010.

58 Sasabe, H. and Kido, J. (2013). Development of high performance OLEDs for general lighting. *J. Mater. Chem. C* 1: 1699–1707.

59 Tyan, Y.-S. (2015). Design considerations for OLED lighting. In: *OLED Fundamentals: Materials, Devices, and Processing of Organic Light-Emitting Diodes* (eds. D.J. Gaspar and E. Polikarpov), 399–436. Boca Raton, FL: CRC Press.

60 Sun, Y., Giebink, N.C., Kanno, H. et al. (2006). Management of singlet and triplet excitons for efficient white organic light-emitting devices. *Nature* 440: 908–912.

61 Yin, Y., Ali, M.U., Xie, W. et al. (2019). Evolution of white organic light-emitting devices: from academic research to lighting and display applications. *Mater. Chem. Front.* 3: 970–1031.

62 Universal Display Corporation (2021) White OLEDs. https://oled.com/oleds/white-oleds-woleds (accessed 19 March 2021).

63 Gao, J. (2008). Polymer light-emitting electrochemical cells. In: *Luminescent Materials and Applications* (ed. A. Kitai), 161–205. Chichester: Wiley.

64 Mindemark, J. and Edman, L. (2016). Illuminating the electrolyte in light-emitting electrochemical cells. *J. Mater. Chem. C* 4: 420–432.

65 Su, H.-C., Chen, Y.-R., and Wong, K.-T. (2019). Recent progress in white light-emitting electrochemical cells. *Adv. Funct. Mater.* 29: 1906898.

66 Zhang, C., Chen, P., and Hu, W. (2016). Organic light-emitting transistors: materials, device configurations, and operations. *Small* 12: 1252–1294.

67 Muccini, M. and Toffanin, S. (2016). *Organic Light-Emitting Transistors: Towards the Next Generation Display Technology*. Hoboken, NJ: Wiley.

68 Liu, C.-F., Liu, X., Lai, W.-Y., and Huang, W. (2018). Organic light-emitting field-effect transistors: device geometries and fabrication techniques. *Adv. Mater.* 30: 1802466.

69 Yuan, D., Sharapov, V., Liu, X., and Yu, L. (2020). Design of high-performance organic light-emitting transistors. *ACS Omega* 5: 68–74.

70 Nam, S., Chaudhry, M.U., Tetzner, K. et al. (2019). Efficient and stable solution-processes organic light-emitting transistors using a high-*k* dielectric. *ACS Photonics* 6: 3159–3165.

71 Liu, J., Zhao, F., Li, H., and Pei, Q. (2018). Electrolyte-gated light-emitting transistors: working principle and applications. *Mater. Chem. Front.* 2: 253–263.

72 Chaudry, M.U., Tetzner, K., Lin, Y.-H. et al. (2018). Low-voltage solution-processes hybrid light-emitting transistors. *ACS Appl. Mater. Interfaces* 10: 18445–18449.

Further Reading

Chang, Y.L. (2015). *Efficient Organic Light Emitting-Diodes (OLEDs)*. Boca Raton, FL: CRC Press.

Eisberg, R. and Resnick, R. (1985). *Quantum Physics*, 2e. New York: Wiley.

Klauk, H. (ed.) (2006). *Organic Electronics: Materials, Manufacturing and Applications*. Weinheim: Wiley-VCH.

Klauk, H. (ed.) (2012). *Organic Electronics II: More Materials and Applications*. Weinheim: Wiley-VCH.

Köhler, A. and Bässler, H. (2015). *Electronic Processes in Organic Semiconductors*. Weinheim: Wiley-VCH.

Ostroverkhova, O. (ed.) (2019). *Handbook of Organic Materials for Optical and Optoelectronic Devices*, 2e. Oxford: Woodhead.

Schwoerer, M. and Wolf, H.C. (2006). *Organic Molecular Solids*. Berlin: Wiley-VCH.

Waser, R. (ed.) (2005). *Nanoelectronics and Information Technology*, 2e. Weinhein: Wiley-VCH.

11

Photoconductive and Photovoltaic Devices

11.1 Introduction

This chapter concerns the photoconductive and photovoltaic properties of organic materials and their associated electronic devices. The photoconductivity of anthracene was discovered at the beginning of the twentieth century [1]. However, widescale application did not follow until the 1960s when organic photoreceptors for electrophotography were introduced [2]. The field of organic photovoltaics (OPVs) dates to 1959 when it was found that anthracene could also be used to make a solar cell [3]; the early research is described by Chamberlain [4]. The first devices were single-layer OPVs. The donor/acceptor bilayer configuration, or planar heterojunction, was subsequently introduced by Tang in 1986, using copper phthalocyanine (Figure 10.10) in combination with a perylene derivative (two materials that are still under study). This solar cell exhibited an efficiency of about 1% [5]. As noted in Chapter 10, the same author reported on the OLED in the following year [6]. This chapter will introduce the physical backgrounds to photoconductivity and the photovoltaic effect and then review the exploitation of organic materials in these fields.

11.2 Photoconductivity

For inorganic semiconductors, the phenomenon of photoconductivity involves the optical absorption in a material and the subsequent generation of free carriers, which increases the electrical conductivity. These free carriers can subsequently recombine or be captured at a trapping centre, as described in Sections 2.6.2 and 3.4. An equilibrium between generation and recombination of the carriers is eventually reached. In the case of low permittivity organic semiconductors (i.e. lower than values found in inorganic materials), the absorption of light generally produces excitons, which must be dissociated in order to produce free electrons and holes.

11.2.1 Optical Absorption

Optical absorption is described quantitatively through the absorption coefficient (or constant) α (units m^{-1}). In the simplest case, neglecting reflection or interference effects, if light of power density P_i ($W\,m^{-2}$) is incident on a material of thickness d with absorption coefficient α, the power density of the transmitted light P is given approximately by Beer's law [7]:

$$P = P_i \exp(-\alpha d) \tag{11.1}$$

Electrical Processes in Organic Thin Film Devices: From Bulk Materials to Nanoscale Architectures, First Edition. Michael C. Petty.
© 2022 John Wiley & Sons Ltd. Published 2022 by John Wiley & Sons Ltd.
Companion Website: www.wiley.com/go/petty/organic_thin_film_devices

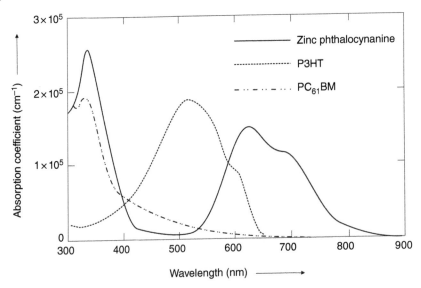

Figure 11.1 Absorption coefficient versus wavelength for organic semiconductors: zinc phthalocyanine; poly(3-hexylthiophene) (P3HT); and [6,6]-phenyl-C_{61}-butyric acid methyl ester (PC$_{61}$BM). Source: Marinova et al. [8].

This is often written:

$$A = -\log_{10}\left(\frac{P_i}{P}\right) \tag{11.2}$$

where A is the absorbance of the sample and is given by $A = (\log_{10}e)\alpha d$. The absorption coefficient of a sample will vary with the wavelength of the incident light, increasing when this corresponds to an optical transition (e.g. equal to the energy gap of an inorganic semiconductor or the LUMO–HOMO separation in an organic semiconductor). Figure 11.1 shows the absorption coefficient (in common units of cm^{-1}) versus photon wavelength for three organic semiconductors: zinc phthalocyanine, the polymer poly(3-hexylthiophene) (P3HT), and [6,6]-phenyl-C_{61}-butyric acid methyl ester (PC$_{61}$BM), a fullerene derivative used in OPVs [8]. Absorption coefficients of the order of 10^5 cm^{-1} are observed in the wavelength regions corresponding to optical transitions, indicating that most of the incident radiation can be absorbed in a very thin film (Problem 11.1).

11.2.2 Carrier Lifetime

For a semiconductor with free electrons and/or holes, the electrical conductivity in the dark, σ_D, is given by Eq. (2.40):

$$\sigma_D = e\left(n\mu_n + p\mu_p\right) \tag{11.3}$$

Under illumination, the conductivity, σ_L, is increased by the photoconductivity $\Delta\sigma$:

$$\sigma_L = \sigma_D + \Delta\sigma \tag{11.4}$$

In the case that the conductivity is dominated by one carrier, for example holes:

$$\sigma_L = e\left(p_D + \Delta p\right)\left(\mu_D + \Delta\mu\right) \tag{11.5}$$

where p_D and μ_D refer to the hole concentration and mobility in the dark, respectively. This equation allows for the possibility that absorption of photons may change both the carrier density and the carrier mobility.

It is generally true that Δp is proportional to the carrier generation rate G:

$$\Delta p = \tau_p G \tag{11.6}$$

where τ_p is the hole lifetime (Eq. (2.74)).

Therefore,

$$\Delta \sigma = e\mu_D \tau_p G + pe\Delta \mu \tag{11.7}$$

It should be noted that the lifetime may itself be a function of the excitation rate. There are a number of important consequences that may arise from Eq. (11.7) [9]. If the carrier concentration in the dark is low (the material is insulating) and/or the photoexcitation rate is high:

$$\Delta \sigma = e\mu_D \tau_p G \tag{11.8}$$

This equation must be modified if the carrier lifetime depends on the photoexcitation intensity. There is also the situation where the increase in the carrier mobility (second term in Eq. (11.7)) dominates:

$$\Delta \sigma = pe\Delta \mu \tag{11.9}$$

Different physical processes might account for this, such as changes to the density of scattering centres under photoexcitation (Section 2.5) or the reduction of surface or intergrain potential barriers. Alternatively, the photoexcitation may result in a carrier being excited from a band characterized by one mobility to a band with a different mobility (similar to the transferred electron effect described in Section 2.5).

A useful figure of merit for a single-carrier photoconductor may be defined:

$$\frac{\Delta \sigma}{Ge} = \mu \tau \tag{11.10}$$

11.2.3 Photosensitivity

There are a several measures of photosensitivity, depending on whether a material or a device is under consideration. The large-signal sensitivity for insulators ($\Delta \sigma \gg \sigma_D$) can be expressed in terms of the specific sensitivity S^* (units $m^2\ \Omega^{-1}\ W^{-1}$). This the material's sensitivity in terms of its $\mu \tau$ product [9]:

$$S^* = \left(\frac{e}{h\nu}\right)\mu\tau \tag{11.11}$$

In contrast, the specific detectivity, D^* (units $m\ Hz^{1/2}\ W^{-1}$), is commonly used for devices such as infrared detectors (discussed in Section 11.9), where $\Delta \sigma \leq \sigma_D$, and it is necessary to detect small photoexcited signals in the presence of background noise:

$$D^* = \frac{(A\Delta f)^{1/2}}{\text{NEP}} \tag{11.12}$$

where A is the area of the photodetector, Δf is its bandwidth, and NEP is the noise equivalent power (units W), which represents the radiation power required to produce a signal equal to the noise. In this case, the signal to noise, or S/N, ratio is unity. Electrical noise has been introduced in Section 7.6. However, the NEP of a photodetector will also depend on noise resulting from statistical fluctuations of the device temperature [10]. A related parameter, called the detectivity, D

(in contrast to the specific detectivity) is sometimes encountered. This has the units of W^{-1} and is equal to $(NEP)^{-1}$. The specific detectivity is independent of the detector area, A, and bandwidth, as is the case for many infrared detectors, and is a useful parameter for specifying device performance [11].

A further measure of photosensitivity is the photoconductive gain, G^* [9, 12]. This combines material and device parameters and can be defined as the number of charges collected in the external circuit for each photon absorbed. It can be expressed as the ratio of the carrier lifetime to the carrier transit time, t_T:

$$G^* = \frac{\tau}{t_T} \tag{11.13}$$

If the lifetime of the carrier is greater than its transit time, it will effectively make several transits through the material between the electrodes, provided that the contacts are Ohmic and are able to replenish carriers drawn off at the opposite contact (Section 4.3). The photoconductive gain can vary continuously across the value of unity without any change in the physical processes at work. A change in either the voltage applied to the photoconductor or in the electrode spacing can affect the gain. If both the carrier mobility and lifetime are voltage independent, the measured photocurrent should increase linearly with applied voltage (Problem 11.2). However, at very high values of voltage, space charge injection (Section 6.6) will become important. The carrier transit time then becomes equal to the dielectric relaxation time (Section 7.3.2) and both will decrease with increasing voltage [12]. The effect of traps can complicate the photoconductor response [9, 12]. Under certain conditions, negative photoconductivity can even be observed (i.e. a decrease in current to a value less than that measured in thermal equilibrium).

11.3 Xerography

Organic photoconductive materials are widely exploited in office photocopiers using the process of xerography (from the Greek 'xeros' for dry and 'graphos' for writing). This was invented in 1938 and patented in 1940 by Chester F. Carlson (see, for example [2]). There are a number of variations of the technology, but the most common is based on the development of a latent electrostatic image formed on the surface of a pre-charged insulating photoconductor. The five 'fundamental' stages of the xerography process are shown in Figure 11.2.

Initially, an electrostatic charge is uniformly distributed over the surface of the photoconductor by a corona discharge (Step I. Charging). For a typical organic photoreceptor with a thickness of $25\,\mu m$, relative permittivity of 3 and an applied voltage of $500\,V$, the capacitance per unit area is $10^{-10}\,F\,cm^{-2}$ (using the formula for a parallel plate capacitor, Eq. (7.25)), corresponding to about 10^{11} charges cm^{-2}. Assuming the surface is composed of molecules, each with an area of $1\,nm^2$ $(10^{-14}\,cm^2\,molecule^{-1})$, then less than 1% of the surface molecules are associated with the surface charge.

The photoconductor is invariably deposited onto the surface of a drum; this rotates as the image of the document or photograph is scanned. The charge is built up on the material's surface by the discharge from a fine wire held at a high potential close to the drum. An image of the document to be copied is projected onto the photoconductor. Electron–hole pairs are generated in the photoconductor at points where the light is incident. For a positively charged surface, electrons will drift to the surface, neutralizing the positive charge. Hence, the initial surface potential is reduced locally and the optical image is converted into a latent electrostatic image (Step II. Exposure). The ideal

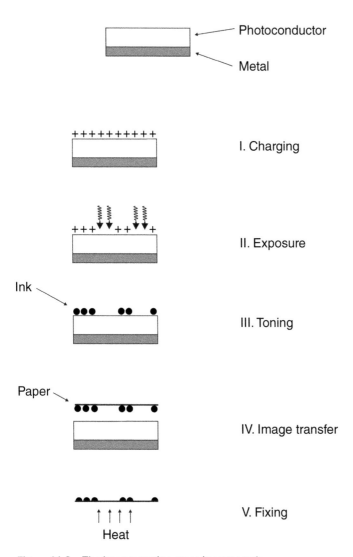

Figure 11.2 The key processing steps in xerography.

photoconductor will have a high dark resistivity and a panchromatic photoconductive response (i.e. sensitive to all wavelengths of visible light) with high quantum efficiency.

The latent image is converted into a visible one by bringing charged toner particles (the 'ink') close to the surface of the photoconductor (Step III. Toning). The toner particles are charged by agitated contact with carrier particles, which also serve to transport the toner to the surface of the photoconductor. The toner neutralizes the charge on the photoconductor so that the image can be transferred to a charged piece of paper (Step IV. Image transfer). Finally, the paper is heated to fix the image onto the paper (Step V. Fixing). In two further processes, cleaning and erase, the drum is first cleaned of any remaining toner, which did not transfer to the paper, by a rotating brush or a wiper blade under suction. Any residual electrostatic image is then removed by flooding the entire photoreceptor with uniform illumination.

The properties needed for the photoreceptor are a high dark surface resistivity, a reasonable level of photoconductivity when illuminated with either incandescent lamps or semiconductor laser

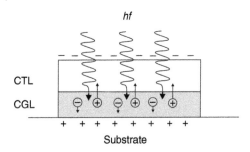

Figure 11.3 A two-layer xerographic material. CTL = charge-transport layer. CGL = charge-generation layer. Electrons and holes are generated in the CGL by the incoming radiation of energy hf and transported (holes in the example shown) in the CTL.

diodes and good abrasion resistance. The latter ensures adequate photoreceptor lifetime by minimizing damage during the image transfer, cleaning, and discharging. A photoreceptor resistivity of greater 10^{13} Ω cm is required [2]. For a material with a relative permittivity of 3, this will provide a dielectric relaxation time ($\varepsilon_r\varepsilon_0\rho$) of about 3 seconds. This is adequate as development typically occurs at a fraction of a second to one second after charging.

The first practical copiers used photoreceptors fabricated by the vacuum deposition of selenium and selenium alloys. Amorphous silicon has also been employed by plasma decomposition of volatile silicon and dopant compounds. Most commercial photocopiers now use semiconducting polymers. These offer simple fabrication techniques such as dip or spray coating of an aluminium drum or solution coating of metallized poly(ethylene terephthalate) (PET) fabric to make photoreceptor belts for high speed copiers.

Typically, as illustrated in Figure 11.3, polymeric photoreceptors have multilayer structures comprising an outer charge-transport later (CTL) adjacent to a charge-generation layer (CGL). An additional layer between the CGL and the metal substrate can be used to prevent charge injection from the metal. The CTL must have a high carrier mobility and be transparent at wavelengths that generate electrons and holes in the underlying CGL. There are two important parameters that characterize charge motion through the CTL. These are the carrier drift mobility, a measure of how fast the carrier moves per unit applied field, and the normalized carrier range $\mu\tau$ (Eq. (11.10)), which is how far the injected carrier moves per unit field before becoming immobilized in a trap. For this layer, much work has focused on polymers with pendent conjugated groups, such as PVK or molecular materials, such as TPD mixed in a polymer binder, e.g. polycarbonate. The chemical structures of PVK and TPD have been given in Figure 10.10 [2, 13]. Many of the organic compounds developed as the CTL in xerography are now used as transport layers in organic light emitting devices.

The CGL usually comprises of fine crystals of pigment dispersed in a polymer binder. Materials with suitable properties include azo dyes, perylene, and phthalocyanines (Figure 10.10).

11.4 Photovoltaic Principles

The sun is the earth's most abundant source of energy. The annual energy potential of sunlight incident at the earth's surface, estimated at 23 000 TW-years, is over 13 times greater than the combined energy available from all known reserves of coal, natural gas, petroleum, and uranium, along with wind and other types of renewable energy sources [14]. Figure 11.4 reveals how the energy output of the sun changes across the electromagnetic spectrum from the ultraviolet to the infrared. The diagram is in the form of graph of spectral irradiance (units W m^{-2} nm^{-1}) as a function of wavelength. The data show the variation between the sunlight spectrum at the top of the earth's atmosphere and that measured at sea level. The differences result mainly from the absorption of the sunlight

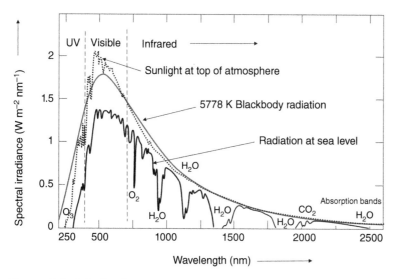

Figure 11.4 Solar irradiance spectra at top of the atmosphere (dotted curve) and at sea level (dark full curve). The spectrum for a blackbody radiator at 5778 K, predicted by Planck's law, is also shown (light full curve). Absorption bands for atmospheric molecular species are indicated.

by atmospheric molecular species, such as H_2O, O_2, O_3, and CO_2. Also shown in Figure 11.4 is the theoretical spectrum from a black body radiator at 5778 K, according to Planck's law. The precise intensity of radiation reaching the earth's surface will depend on how much of the earth's atmosphere the sunlight has to penetrate. The spectrum after travelling through the atmosphere to sea level with the sun directly overhead is referred to, by definition, as AM1 (air mass 1). This is a useful illumination condition for estimating the performance of solar cells in equatorial and tropical regions of the earth. Most terrestrial solar cells are tested under the AM1.5G (global) condition. (The figure of 1.5 refers to a relative atmosphere thickness of 1.5 and corresponds to a solar zenith angle of approximately 48°.) On a cloudless day, 1000 W m^{-2} (100 mW cm^{-2}) of solar radiation reach the earth's surface according to the AM1.5G standard.

The first generation of solar cells was base on crystalline silicon. This technology has developed since the 1940s and offers conversion efficiencies of around 24% for off-the-shelf commercial devices. Higher efficiencies, around 27%, are reported (usually for relatively small area devices) in the laboratory [15]. The success of silicon photovoltaics is not only due to their high efficiencies but also because their manufacturing has benefited from the experience of crystalline silicon processing by the microelectronics industry. Although crystalline silicon solar cells are highly efficient, these require expensive and energy intensive fabrication. The energy payback period is therefore significant for silicon photovoltaics. As a consequence, new solar cell technologies have been developed using thin-film materials. These include amorphous silicon, cadmium telluride, and copper indium gallium diselenide (CIGS). Organic solar cells and, more recently, perovskite cells can now be added to this list.

Photovoltaics based on organic compounds, such as polymers or dyes, offer the possibility of large-scale manufacture at low temperature coupled with low cost. Moreover, the use of flexible and lightweight substrates allows the devices to be applied to non-planar surfaces while also reducing installation costs. As noted above, efficiencies of OPVs were modest until the end of the twentieth century, with typical values of 1%. The availability of new conductive organic materials and different OPV designs has significantly improved on this figure. Till 2020, conversion efficiencies of over 18% have been reported [16].

11.4.1 Electrical Characteristics

The purpose of any photovoltaic device is to convert photons into useful electrical energy. In an inorganic solar cell, the photons are absorbed in or close to (within a diffusion length of) the depletion layer of a device such as a p–n junction or a Schottky barrier. The high associated electric field serves to separate photogenerated electron–hole pairs, which can be extracted at the device contacts [17]. Figure 11.5 shows the current versus voltage characteristics for a simple solar cell based on these principles, in the dark and under illumination. These features are preserved for OPV devices, although the process of charge separation (discussed later) is more complex.

In the dark, Figure 11.5a, the device is rectifying with an $I–V$ relationship of a similar form to that of the Schottky diode described in Eq. (4.39). Under illumination, the $I–V$ characteristic is moved down the current axis as a photocurrent is generated. The dependence of current on applied voltage now becomes:

$$I = I_0 \left(\exp\left(\frac{eV}{nk_B T} \right) - 1 \right) - I_l(V) \tag{11.14}$$

where $I_l(V)$ is the voltage-dependent photocurrent and n is the diode ideality factor (Section 4.4). The photocurrent flows in the same direction as the reverse bias dark current. Two important device parameters, shown in Figure 11.5a, are V_{oc}, the open-circuit voltage, and I_{sc}, the short-circuit current. The former represents the voltage appearing across the terminals of the solar cell when no current is drawn from the device; this is the maximum attainable voltage that the cell can provide. When the device terminals are short-circuited, I_{sc} represents the maximum current that can be extracted. The $I–V$ characteristics under illumination are shown in more detail in Figure 11.5b. The actual operating point on a given solar cell $I–V$ characteristic is determined by the load resistance that is connected between its terminals. To maximize the output power, it is desirable to choose a value for the load resistance such that the operating condition of the cell is the maximum power point, MPP, in Figure 11.5b. Here, the product of the current and voltage outputs of the cell is a maximum and their corresponding values are denoted I_{max} and V_{max}.

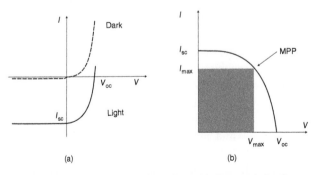

(a)   (b)

Figure 11.5 (a) Current I versus voltage V characteristics for a semiconductor photovoltaic diode in the dark and in the light. V_{oc} = open-circuit voltage. I_{sc} = short-circuit current. (b) Expanded region for the fourth quadrant of the $I–V$ curve showing the maximum voltage V_{max} and the maximum current I_{max} available from the solar cell for optimized power output. The shaded area represents the maximum power available from the cell. MPP = maximum power point.

11.4.2 Efficiency

The power conversion efficiency η_{PV} of a photovoltaic cell is the ratio of the electrical power extracted from the cell to the radiant input power:

$$\eta_{PV} = \frac{I_{max} V_{max}}{P_i A} = \frac{I_{sc} V_{oc} F}{P_i A} \tag{11.15}$$

where P_i is the incident solar power density (W m^{-2}) and A is the area of the solar cell. The ratio of the product $(I_{max} V_{max})$ to $(I_{sc} V_{oc})$ is defined as the fill-factor, F, of the solar cell. It is clearly desirable to produce devices with F values as close to unity as possible.

Other efficiency measures for a photovoltaic cell are its external quantum efficiency, η_{ext}, and its internal quantum efficiency, η_{int}:

$$\eta_{ext} = \frac{\text{electrons extracted}}{\text{incident photons}} \tag{11.16}$$

$$\eta_{int} = \frac{\text{electrons extracted}}{\text{photons absorbed by active layers}} \tag{11.17}$$

The definition of the external quantum efficiency is based on the number of photons incident on the device and therefore incorporates optical losses, such as reflection from the substrate and absorption in nonactive layers. The external and internal efficiencies differ by the absorption efficiency of the active layers in the solar cell:

$$\eta_{int} = \frac{\eta_{ext}}{\eta_a} \tag{11.18}$$

where η_a is the photoabsorption efficiency. This parameter will depend on the wavelength, whereas η_{ext} and η_{int} will be functions of both wavelength and the applied voltage.

Figure 11.6 shows the equivalent circuit of the photovoltaic cell connected to a resistive load, R_{load}. Imperfections in the device (formed during fabrication) leading to current leakage are represented by a shunt resistance, R_{shunt}, and parasitic series resistance effects are signified by R_{series}. The value of the latter includes the resistances of the various layers (organic films and electrodes) and interfaces within the OPV architecture. The current generated by the incoming EM radiation is taken into account by the incorporation of a constant current generator of value I_1. Note that a forward dark current I_d will flow through the rectifying junction (represented by the diode symbol in the figure). The voltage and current are given by V_{load} and I_{load}, respectively, in Figure 11.6.

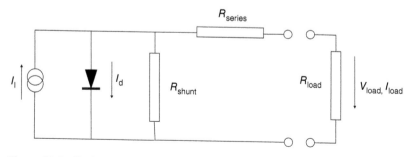

Figure 11.6 Equivalent circuit of a photovoltaic solar cell. I_l is the current generated by the solar radiation. I_d = forward biased current through rectifying barrier. R_{series}, R_{shunt}, and R_{load} represent the series, shunt, and load resistors. V_{load} and I_{load} are the voltage and current, respectively, associated with the load resistor.

Using simple circuit theory, it can be shown that (Problem 11.3):

$$I_{load} \left(1 + \frac{R_{series}}{R_{shunt}} \right) = I_1 - I_d - \frac{V_{load}}{R_{shunt}} \tag{11.19}$$

This equation can be solved numerically to ascertain the effects of R_{series} and R_{shunt} on the cell output characteristics. It is important for R_{series} to be small (ideally zero) and for R_{shunt} to be large (ideally infinite). For example, the fill-factor has the following dependence on R_{series} and R_{shunt} [14]:

$$F \left(R_{series}, R_{shunt} \right) = F(0, \infty) \left(1 - \frac{I_{sc} R_{series}}{V_{oc}} - \frac{V_{oc}}{I_{sc} R_{shunt}} \right) \tag{11.20}$$

It is evident that F decreases as R_{shunt} decreases, or as R_{series} increases. The shunt and series resistances also affect the values of V_{oc} and I_{sc} through their relationship with F. The I–V relationship of Eq. (11.14) becomes:

$$I = \frac{R_{shunt}}{R_{series} + R_{shunt}} \left(I_0 \left(\exp \left(\frac{eV - IR_{series}}{nk_B T} \right) - 1 \right) + \frac{V}{R_{shunt}} - I_1(V) \right) \tag{11.21}$$

In the case where the shunt resistance is infinite and the series resistance is zero, Eq. (11.14) may be solved to obtain an expression for V_{oc} (i.e. by putting $I = 0$):

$$V_{oc} = \frac{nk_B T}{e} \ln \left(\frac{I_1 (V_{oc})}{I_0} \right) + 1 \tag{11.22}$$

The open-circuit voltage therefore depends on the reverse bias saturation current.

The optical absorption of the photoactive layer must also be tuned to match the incoming photon flux from the solar irradiation. The peak in the solar spectrum corresponds to a wavelength of approximately 600 nm (depending on location on the earth's surface) and, as shown in Figure 11.4, it has a useful range of about 0.3–2 μm.

There is a fundamental thermodynamic limit to efficiency that can be achieved in a photovoltaic cell, such as described above [18]. The first study on this topic was by Shockley and Queisser in 1961; the maximum efficiency of a single semiconductor solar cell under solar illumination is referred to as the Shockley–Queisser limit [19]. The theoretical model requires that the solar cell is in equilibrium with its surroundings and must therefore emit blackbody radiation by virtue of its non-zero temperature. This radiation is an unavoidable source of energy loss. However, there is also a band-gap-dependent maximum efficiency that can be obtained with a solar cell fabricated by a single semiconductor as absorber. The underlying assumption of this limit is that every photon incident on the solar cell with an energy exceeding the band gap is absorbed and converted into an electron and a hole, while photons with energies below the bandgap do not contribute. However, the potential energy of photogenerated carriers is independent of the photon energy. Photon energies above the band gap only provide 'one-band-gap's-worth' of energy. There is thermalization down to the band edge for all absorption events. The maximum efficiency is then found by considering the solar spectrum and the underlying thermodynamics, which is why the limit is often also referred to as the detailed balance limit. Using the Shockley–Queisser model, with AM1.5G radiation, the maximum power conversion efficiency of a single junction inorganic solar cell is 33% and coincides with a band gap of 1.3 eV [20]. For higher band gaps, the energy per generated charge is higher, but a significant amount of incident light cannot be absorbed. On the other hand, for smaller bandgaps most of the sunlight is converted, but thermalization losses reduce the overall power conversion efficiency.

A further important issue is that, under solar illumination at the earth's surface, there is only a very narrow angle distribution of incident light (sunlight is virtually parallel), while the black body radiation of the device is isotropic. The theoretical device efficiency can be enhanced when using light concentration, which broadens the angle distribution of the excitation light. Sunlight from a large area can be focused onto a smaller solar cell using a variety of lens arrangements (Section 11.8).

An efficient solar cell is generally an efficient light-emitting device [21]. When all non-essential charge recombination paths are eliminated, such as trap-assisted recombination at defect or impurity sites, the charges present in the active material can only recombine through radiation of light. Thus, bright electroluminescence from a photovoltaic cell operating in forward bias, typically measured when the dark recombination current equals the short-circuit photocurrent under sunlight, is an indication that a small fraction of recombination occurs via non-radiative channels.

The efficiencies of photovoltaic cells that contain more than one type of absorber (e.g. two semiconductors with different band gaps) can differ from those described above for single band gap inorganic semiconductors. The efficiency of OPVs is also reduced from that for devices based on inorganic materials as the excitons produced by the absorption of sunlight must first be dissociated. These effects are considered in Sections 11.5.1, 11.5.2, and 11.5.4 of this chapter.

It is essential for researchers and technologists to be able to compare the photovoltaic parameters for the different types of photovoltaic device that are now available, both in the form of research laboratory devices and as commercial solar panels. Research devices have smaller active areas (fewer defects) and possess higher efficiencies than the module and submodule commercial solar panels. Up-to-date parameters for all these devices are available on the web pages of the US National Renewable Energy Laboratory (NREL) [22]. Moreover, since January 1993, the journal 'Progress in Photovoltaics' has published six-monthly listings of the highest confirmed efficiencies for a range of photovoltaic cell and module technologies [15]. One important criterion for inclusion of results into the tables is that they must have been independently measured by a recognized solar cell test centre. Table 11.1 shows some recent data for single semiconductor material solar cells. These are not necessarily the highest efficiencies reported in the literature. A separate listing in [15] provides data for 'notable exceptions', i.e. parameters for photovoltaic devices that do not conform strictly to the main criteria. For example, Table 11.1 reveals an efficiency for a single layer OPV with an active area of 1.02 cm^2 of 15.2%, while an efficiency of 18.2% (device area 0.032 cm^2) is given in reference [15] under notable exceptions. The data in Table 11.1 show an efficiency of over 29% for a GaAs cell. This inorganic semiconductor has a band gap of 1.4 eV, close to the figure of 1.3 eV predicted

Table 11.1 Confirmed solar cell efficiencies measured under global AM1.5G spectrum.

Classification	Efficiency (%)	Area (cm^2)	V_{oc} (V)	J_{sc} (mA cm^{-2})	Fill factor (%)
Single crystal Si	26.7	79.0	0.738	42.7	84.9
GaAs thin film	29.1	0.998	1.13	29.8	86.7
CdTe thin film	21.0	1.06	0.876	30.3	79.4
CIGS thin film	23.4	1.04	0.734	39.6	80.4
Amorphous Si	10.2	1.00	0.896	16.4	69.8
Microcrystalline Si	11.9	1.04	0.550	29.7	75.0
Perovskite	21.6	1.02	1.19	21.6	83.6
Organic	15.2	1.02	0.845	24.2	74.3
Dye sensitized	11.9	1.00	0.744	22.5	71.2

Source: Modified from Green et al. [15].

above for optimum efficiency. Gallium arsenide is also a direct band gap semiconductor (Figure 1.7) with a high absorption coefficient for incident photons.

11.5 Organic Solar Cells

As discussed in Section 1.5.2, excited states in organic materials are strongly bound with energies of at least a few 100 meV. This energy is high compared to the thermal energy at room temperature ($k_B T$), which is approximately 25 meV. Thermal splitting of these excitons in organic materials is therefore unlikely. The earliest OPVs were based on a simple metal/organic/metal thin film architecture that relied on the built-in electric field between metal electrodes, or the field associated with a Schottky barrier at one of the metal/organic contacts, to dissociate the photogenerated excitons into free carriers. Figure 11.7 shows a schematic diagram of such a structure. As is the case for OLEDs, one of the electrodes must be semitransparent – for OPVs this enables the light to get *into* the device. A common choice is ITO, which has a relatively high work function and can therefore be used to collect the holes. A low work function electrode, such as aluminium, then collects the electrons.

11.5.1 Carrier Collection

The single-layer OPV has a very low device efficiency. A significant improvement can be achieved by splitting the excitons at an interface between donor and acceptor molecules [5]. A photon incident on a bilayer donor/acceptor solar cell can be absorbed by one of the two active materials. Figure 11.8 shows a series of potential energy diagrams representing the interface between a donor and an acceptor. In this example, the magnitudes, the LUMO and HOMO energy levels of the donor are both lower (closer to the vacuum level) than the equivalent values for the acceptor. Assuming that an incident photon of sufficient energy (equal or greater than the HOMO–LUMO separation or band gap) is absorbed in the donor layer, then an exciton will be formed, shown in Figure 11.8a. The binding energy of this entity will be slightly less than the band gap. The exciton will diffuse from its point of creation and if it reaches an interface with an acceptor within its lifetime, a charge-transfer state can be formed. This is an intermediate state represented by a geminate pair (i.e. originating

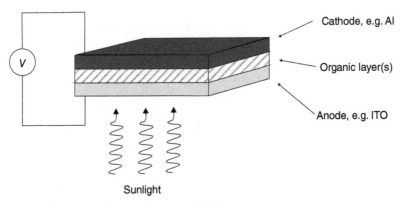

Figure 11.7 Schematic diagram showing the structure of an organic photovoltaic (OPV) cell based on a single-layer organic film.

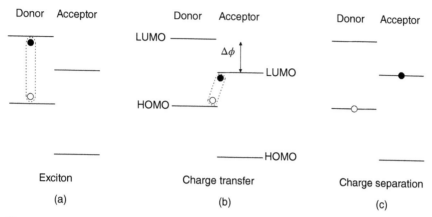

Figure 11.8 Formation of an electron and hole at the interface between organic donor and acceptor molecules. (a) Optical absorption in donor produces an exciton in donor layer. (b) The exciton diffuses to interface with acceptor to produce a charge-transfer state. (c) The charge-transfer state can be dissociated if the occupation of the final charge-separated state (free electron and hole in the acceptor and donor, respectively) is energetically favourable. $\Delta\phi$ = energy separation between donor and acceptor LUMO levels.

from the same photoabsorption event) of opposite charges, with the electron located in the acceptor and the hole in the donor, as illustrated in Figure 11.8b. To dissociate the exciton, the acceptor material has to offset the binding energy of the exciton in the donor. The charge-transfer state at the interface can only be formed if the following condition is satisfied:

$$X_A - X_D \geq U_D \tag{11.23}$$

where X_A, X_D are the electron affinities (Figure 1.21) of the acceptor and donor, respectively, and U_D is the binding energy of the exciton in the donor. The probability of this is increased if the energy offset between the LUMO of the donor and the LUMO of the acceptor, $\Delta\phi$ in Figure 11.8b, is increased.

The next step is to dissociate the charge-transfer state into free carriers, as depicted in Figure 11.8c. The formation of the free electrons and holes from a pair of charges held together by Coulomb forces has been explained by a classic theory by Onsager, subsequently extended by Braun [23–26]. The charge-transfer state can be considered as a strongly bound electron–hole pair, in which the hole is located on the donor side of the interface and the electron on the acceptor side. Initial separation of the electron and hole increases the Coulomb radius between the two species, resulting in a lower binding energy for the charge-transfer state compared to the initial exciton formed in the donor. This binding energy E_b can be given [27]:

$$E_b = \frac{e^2}{4\pi\varepsilon_r\varepsilon_0 r_{DA}} \tag{11.24}$$

where ε_r is the relative permittivity of the donor/acceptor interface and r_{DA} is the initial separation between the charges in the donor and acceptor immediately following transfer at the interface. Since the charge-transfer state is obtained by mixing the HOMO of the donor and the LUMO of the acceptor, its energy is also significantly reduced compared to the equilibrium exciton energy in the donor. Therefore, the charge-transfer state can be thermally split into an electron residing in an acceptor molecule close to the interface and a hole in the donor material (it is also theoretically possible for the charge-transfer state to be regenerated). Depending on the operation conditions (external bias) of the device, the free carriers can then migrate away from the interface towards the external contacts, resulting in the generation of a photocurrent.

For a photovoltaic cell under illumination, excess carriers are generated, and the open-circuit voltage is determined by the difference between the hole and electron quasi-Fermi levels (Section 2.7). In single-layer photovoltaic cells, the maximum quasi-Fermi level splitting is restricted to the band gap of the absorber, but in practice the maximum value of V_{oc} is limited to the difference in work functions of the two electrodes. For bilayer solar cells, V_{oc} is related to the energy separation between the HOMO level of the donor and the LUMO level of the acceptor and can be estimated from the following [28] (Problem 11.4):

$$V_{oc} = 1/e \left(E_{LUMO(acceptor)} - E_{HOMO(donor)} \right) - 0.3 \tag{11.25}$$

The empirical factor 0.3 V arising in the above equation is related partly to the ideality factor and the reverse current of the diode (≈ 0.2 V loss) and partly (≈ 0.1 V loss) from the limited electric field across the active layer. Additionally, the value of V_{oc} can be dependent on recombination processes at the donor/acceptor interface [14].

For the case where the acceptor HOMO level is lower than that of the acceptor, then it is possible for the entire exciton to transfer to the acceptor without dissociation [29].

The foregoing discussion has assumed that the photoexcitation occurs in the donor layer (Figure 11.8a). On the other hand, if the acceptor is excited, a hole is injected into the HOMO of the donor, while the electron remains in the LUMO of the acceptor. Hybrid organic/inorganic photovoltaic structures are also possible. If either the organic donor or acceptor is replaced by an inorganic semiconductor, the hole is injected into the valence band or the electron into the conduction band of the material, respectively (Section 11.7.1).

The difference between the LUMO energy levels between the materials in donor-acceptor OPVs ($\Delta\phi$ in Figure 11.8b) creates the driving force whereby dissociation by rapid electron transfer from the donor to the acceptor can occur. Following exciton dissociation, the free carriers created at the interface are ideally transported to their respective electrodes under the influence of the internal electrical field. Two important loss mechanisms limit this. The first, termed geminate recombination, is that the electron–hole pair created at the interface recombines by the transfer of the electron back from the LUMO level of the acceptor to the HOMO level of the donor. The other possibility, called bimolecular recombination occurs when the dissociated free carriers recombine before reaching the electrodes; this process may be trap assisted as in Shockley–Read–Hall recombination (Section 3.4.2). Since the recombining charge carriers originate from different photoexcitations, this mechanism is also referred to as non-geminate recombination. The rate of non-geminate recombination depends on both the carrier concentration and the effective lifetime. Usually, the higher the charge carrier concentration, the faster the recombination. For most state-of-the-art organic solar cells, the non-geminate recombination of separated electrons and holes is the dominant loss mechanism [30].

The internal efficiency of an OPV as defined in Eq. (11.18) may be broken down into further components:

$$\eta_{int}(\lambda, V) = \eta_{ed}(\lambda, V)\eta_{ct}(\lambda, V)\eta_{cc}(\lambda, V) \tag{11.26}$$

where $\eta_{ed}(\lambda, V)$, $\eta_{ct}(\lambda, V)$, and $\eta_{cc}(\lambda, V)$ represent the efficiencies of the exciton diffusion, charge transfer, and charge collection processes, respectively.

11.5.2 Bulk Heterojunction Solar Cells

For soluble organic materials such as polymers, the exciton diffusion lengths are relatively small, on the order of 10 nm. Although significantly longer diffusion lengths (e.g. 1 μm) are reported in

Figure 11.9 Bulk heterojunction (BHJ) material morphologies. (a) Bilayer or planar heterojunction structure. (b) Theoretical ordered blend of donor and acceptor regions. (c) Random domain network.

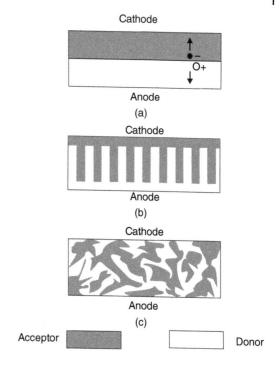

some single crystalline organic compounds, processing in the form of thin films (as required for OPVs) invariably results in many defect states that reduce the exciton diffusion length to a few nanometres.

If the exciton does not encounter a donor/acceptor interface during its lifetime, it can undergo non-radiative decay in a loss mechanism for the cell. For OPVs, film thicknesses of 100–200 nm are required to absorb all the incident light (thicker layers result in an unacceptable series resistance). For a simple bilayer architecture, illustrated in Figure 11.9a (a flat heterojunction), this would result in most of the excitons being created away from the donor/acceptor interface and they would not be harvested. The answer to this problem is to mix the donor and acceptor species in a so-called bulk heterojunction (BHJ) arrangement. Figure 11.9b illustrates such a theoretical blend of donor and acceptor materials. A practical way to achieve the desired material morphology is to exploit solution processing, in which rapid drying leads to a random network of donor and acceptor domains; this is illustrated in Figure 11.9c. Continuous percolation pathways are required through each phase to transport the electrons and holes to their respective electrodes. This charge transport will be strongly influenced by the local order and crystallinity within each phase, with crystalline morphologies generally leading to the highest charge carrier mobilities.

One method of controlling the nanoscale morphology of a BHJ device is to add a small fraction of a higher boiling point co-solvent to the blend solution before spin-coating the thin film. This can allow the BHJ morphology to continue to evolve in a solvent-rich environment. Post deposition methods can also be used to enhance the phase separation between the donor and acceptor species. These include thermal annealing and solvent annealing. In the latter technique, the organic material is exposed to solvent vapours that can dissolve selectively in one of the components of the thin film, mobilizing it sufficiently to change the morphology. With appropriate control of the nanoscale morphology, all excitons can be created within a diffusion length of a donor/acceptor interface, and hence be harvested [26, 31, 32].

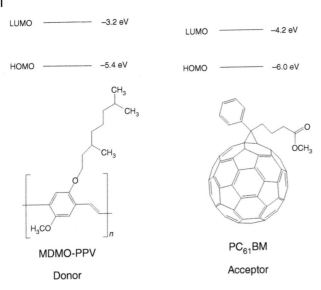

LUMO ———— −3.2 eV

LUMO ———— −4.2 eV

HOMO ———— −5.4 eV

HOMO ———— −6.0 eV

MDMO-PPV

Donor

PC$_{61}$BM

Acceptor

Figure 11.10 Chemical structures of organic molecules used OPV devices. Left: poly(2-methoxy-5-(3′,7′-dimethyloctyloxy)-1,4-phenylenevinylene) (MDMO-PPV) – donor. Right: [6,6]-phenyl-C$_{61}$ butyric acid methyl ester (PC$_{61}$MB) – acceptor. HOMO, LUMO levels. Source: Thompson and Fréchet [33].

The material requirements for the BHJ structures include the ability for the donor and acceptor species to form the required nano-morphology and the resulting domains to be sufficiently conductive to allow the charges to be transported efficiently to the electrodes. The optical absorption of the photoactive layer must also be tuned to match the incoming photon flux from the solar radiation. Figure 11.10 shows two materials that are commonly used in BHJ OPV cells: poly(2-methoxy-5-(3′,7′-dimethyloctyloxy)-1,4-phenylenevinylene) (MDMO-PPV) copolymer as the donor blended with [6,6]-phenyl-C$_{61}$ butyric acid methyl ester (PC$_{61}$BM – also known as PCBM) as the acceptor. The HOMO and LUMO levels of these compounds are also given [33]. A related acceptor, PC$_{71}$BM, [6,6]-phenyl-C$_{71}$ butyric acid methyl ester, is popular as it is more absorbing than PC$_{61}$BM. However, numerous other compounds have been studied [14, 34–37]. Polymer donors include MEH-PPV, polythiophenes, fluorene-containing polymers, carbazole-based polymers, and low-band gap compounds such as polymers containing benzothiadiazole moieties. Acceptor molecules are generally based on fullerenes, such as C$_{60}$ and C$_{70}$. Some work has also been devoted to the use of carbon nanotubes [29]. Their relatively high surface area provides an opportunity for the formation of many heterojunctions with polymers. The success of fullerenes can be attributed to their deep HOMO and LUMO levels and relatively high electron mobilities. However, most of these materials possess narrow absorption spectra, which do not overlap well with visible light; the data for PC$_{61}$MB shown in Figure 11.1 are a good example. Many non-fullerene acceptor materials are now available [38] and OPVs incorporating nonfullerene acceptors are some of the highest-performing OPVs [39].

The two-phase model for the internal morphology for a donor–acceptor BHJ solar cell implied by Figure 11.9c is an oversimplification. While this may be the case for small molecule systems, polymer-fullerene BHJs are invariably much more complex. This results from crystalline behaviour of the different materials and the potential presence of mixed phases. For example, there are at least five different phases present in most polymer–fullerene BHJs, and where all pairs of two phases can be in direct contact [26]. These are: crystalline regions of pure donor; less-ordered and amorphous regions of pure donor; crystalline regions of pure acceptor; less-ordered and amorphous regions of pure acceptor; and mixed phases of donor and acceptor. The latter are commonly characterized by a fine intermingling of the donor and acceptor, resulting in a rather amorphous molecular arrangement. Excitons preferentially separate in mixed phases due to the large donor/acceptor interfacial area. Subsequent transport of electrons and holes towards their respective collection electrodes is more efficient in crystalline phases, which will possess higher charge carrier mobilities.

The donor–acceptor BHJ OPVs described above have a limited optical absorption range. Ternary blend devices are designed to broaden the absorption spectrum and light-harvesting abilities of the solar cell. In these devices, a third component with complementary absorption, usually in the infrared part of the EM spectrum, is added to the dominant donor–acceptor blend. Typical compositions include either two donors and one acceptor or one donor and two acceptors. These devices can be produced in the same manner as conventional BHJ blends [14, 40]. In the case of the P3HT:PC$_{61}$BM BHJ, an appropriate near infrared sensitizer is the low band gap copolymer poly[2,6-(4,4-bis-(2-ethylhexyl)-4H-cyclopenta[2,1-b;3,4-b']-dithiophene)-alt-4,7-(2,1,3-benzothiadiazole)] (PCPDTBT) [41]. The molecular structure of PCPDTBT is shown in Figure 11.11, together with its absorption spectrum [42, 43]. The donor part of the molecule is the cyclopentadithiophene unit on the left-hand side of the polymer repeat unit. A comparison with the absorption spectrum for donor polymer P3HT in Figure 11.1 reveals complementary absorption for P3HT and PCPDTBT over the visible part of the EM spectrum.

Charge transport and energy transfer processes can both occur between the components in a ternary blend photovoltaic cell. The former have been discussed above. Förster energy transfer is a long-range process (e.g. distances exceeding the van der Waals' radii of the molecules) that can take

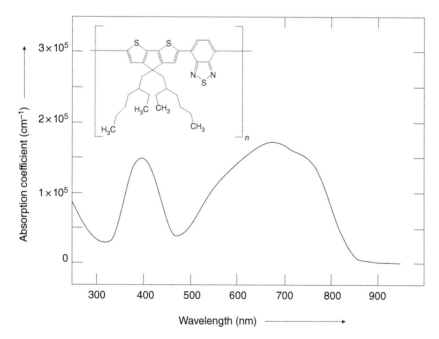

Figure 11.11 Absorption spectrum for a thin film of the donor copolymer poly[2,6-(4,4-bis-(2-ethylhexyl)-4H-cyclopenta[2,1-b;3,4-b']-dithiophene)-alt-4,7-(2,1,3-benzothiadiazole)] (PCPDTBT). The chemical structure is inset. Source: Vezie et al. [42].

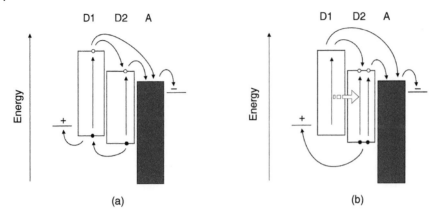

Figure 11.12 Ternary OPV architectures based on two donor materials (D1 and D2) and one acceptor (A). (a) Charge-transfer operation. (b) Energy-transfer operation.

place between donor and acceptor molecules for which the acceptor has an absorption spectrum that overlaps with the emission spectrum of the donor [25].

Figure 11.12 shows the specific example of the ternary arrangement containing two donors, D1 and D2, and one acceptor, A, with the LUMO and HOMO levels of the additional D2 material located between those of the host components D1 and A. Charge transfer between the molecules is illustrated in Figure 11.12a, whereas Figure 11.12b includes an energy transfer process. Excitons can be formed in both donor materials. In the case of Figure 11.12a, charge-transfer interfaces are formed at both the D1/D2 and D2/A interfaces. The photoexcited host donor D1 may transfer an electron to either the acceptor or to the additional donor D2. At the same time, the photoexcited D2 molecules can transfer a hole to D1. The open-circuit voltage of this arrangement is limited to the smallest energy difference between the LUMO of the acceptor and the HOMO of the donors. It is also possible for the charge transfer to operate in parallel rather than as a cascade process [40]. Here, excitons generated in each individual donor would migrate to their respective donor/acceptor interface where they would dissociate into free electrons and holes.

For the situation shown in Figure 11.11b, excitons formed on the larger band-gap donor D1 migrate to D2 via an energy transfer mechanism provided there is an overlap between the emission spectrum of D1 and the absorption spectrum of D2. Subsequently, the excitons in D2, either formed directly of via transfer from D1, dissociate at the D2/A interface. In this example, it is important that the HOMO levels of the two donors are aligned to avoid excessive reduction in V_{oc}. Photoluminescence experiments can be used to elucidate energy transfer processes.

Most ternary BHJs are based on polymeric donor components. However, the use of small molecules to extend the spectral response of the solar cell can offer certain advantages [40]. Such materials can possess relatively high charge-carrier mobilities and are easy to purify. Nanoparticles represent a further class of compounds that has attracted some interest. As well as offering increased light absorption from surface plasmon resonance [30], these materials can reflect and scatter light, thereby increasing the optical path length in BHJ architectures. Quaternary BHJ cells can also be fabricated (e.g. two donors and two acceptors). In one example, the quaternary structure outperforms the corresponding binary and ternary devices with an efficiency of 13% compared to 12% for the ternary and 11% for the binary solar cell [44]. As might be imagined, the morphology and energy/charge transfer processes in such architectures are complex [45].

BHJs are probably the most versatile category of OPV structure. Efficiencies of over 10% are now routinely reported [14, 15, 20] for BHJ solar cells, with the highest values exceeding 18% [16]. Ternary and quaternary cells can achieve figures of over 14% and over 13%, respectively [44, 46–48]. This all represents remarkable progress since the first report of a donor–acceptor cell with a 1% efficiency [5]. However, the absolute efficiency of a photovoltaic device is not the only important parameter. As noted above in Section 11.4.1, the device areas and measurement protocols are also significant factors that must be taken into consideration. Furthermore, additional features such as cost of manufacture and cell lifetime will also determine if a particular photovoltaic device has the potential for commercial development. An important metric here is the cost per unit power generated by the solar cell, which can be expressed in units such as dollars per watt [49].

11.5.3 Electrodes and Device Architectures

The electrodes are critical parts of any electronic device structure [31]. In the previous chapter, the use of additional thin films to control the injection and transport of charge into OLEDs (Section 10.7) was described. For OPVs, such layers are used to extract efficiently charge from the cell to the external circuit. It is imperative that the resistances of these layers are kept to a minimum to reduce electrical losses.

The most common anode material in an OPV device (and OLED) is ITO. This can be produced in thin film form by sputtering onto a transparent substrate such as glass or onto a flexible material, such as PET. Alternative oxide anode materials, including fluorine-doped tin oxide (FTO), indium-doped CdO (CIO), and aluminium-doped zinc oxide (AZO), have also been explored. Double-layer electrodes, e.g. CIO/ITO, can also be used to good effect as these can exhibit superior electrical and optical properties over the individual layers [50]. However, environmental considerations (e.g. supply of indium and toxicity of cadmium) can be just as important as the electrical properties. There has also been interest in the use of carbon nanotubes and graphene as transparent conductive electrode materials [7, 29]. These may offer advantages for solar cells based on flexible supports.

ITO is often coated with a thin layer of the conductive polymer PEDOT:PSS to smooth the relatively rough oxide surface and provide better electrical contact between the anode and the organic layers in the solar cell. A similar approach is used for OLEDs (Section 10.7.2). However, PEDOT:PSS is hygroscopic and can absorb water from the atmosphere leading to device degradation (see Section 11.5.6 below). The use of such materials increases the requirement of water barrier layers for the encapsulation of OPVs. Strategies to improve the wettability of FTO and AZO surfaces include the chemisorption of self-assembled monolayers (SAMs). However, the ultimate choice of electrode materials will also depend on the OPV configuration.

Additional buffer films are often incorporated between the electrodes and the active layers of a planar organic solar cell [14]. These can perform many functions, including their use as sacrificial layers to prevent damage to the organic components during the thermal deposition of metallic electrodes. Thin films of high band gap materials also reduce exciton quenching effects in the vicinity of the electrodes. A common buffer material is 2,9-dimethyl-4,7-diphenyl-1,10-phenanthroline, also known as bathocuproine (BCP), which can result in enhanced power conversion efficiency when inserted between the acceptor and metallic top electrode. It has been suggested that the BCP can form a passivating layer for the C_{60} acceptor that prevents the creation of donor states by diffusion from the Al electrode [51].

One significant development in organic solar cells has been the shift from the 'normal' geometry of the OPV stack, where electrons exit from the top electrode (e.g. Al) and holes at the bottom

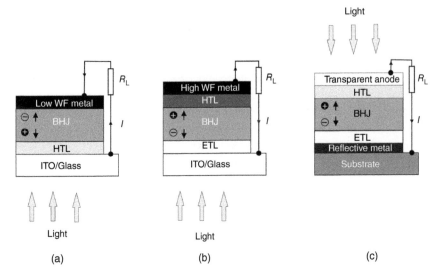

Figure 11.13 Organic thin film solar cell architectures. (a) Conventional device structure. (b) Inverted configuration. (c) Top illumination structure. WF = work function. ETL = electron-transport layer. HTL = hole-transport layer. Source: Wang et al. [52].

(e.g. ITO) to an inverted configuration where electrons exit at the bottom and holes at the top [52]. Figure 11.13 shows some of the device architectures that have been used. The conventional OPV is shown in Figure 11.13a. Illumination through the ITO is preserved for the inverted structure depicted in Figure 11.13b. However, the polarity of charge collection is reversed; this allows the use of more ambient stable (e.g. resistant to O_2 and H_2O) and high work function metals such as Au or Ag as the top anode. These materials can be deposited by relatively low-cost printing technologies. The inverted OPV eliminates the somewhat problematic PEDOT:PSS buffer layer. However, suitable electron-transport layers (ETLs) (e.g. ZnO, TiO_2) and hole-transport layers (HTLs) (e.g. semiconducting transition metal oxides such as MoO_3, WO_3) must be deployed [53, 54]. Reported efficiencies for inverted OPVs are usually less than those of their conventional counterparts, with values below 10% [52, 53]. However, low processing costs and potential for long lifetimes warrant their further study.

A further variation on the OPV structure, shown in Figure 11.13c, is to exploit top illumination. The top electrode must be semi-transparent and may be an ultra-thin metal film/grid or a highly conductive organic material such as carbon nanotubes or graphene. The device configuration can take the form of an optical resonant cavity consisting of a photoactive layer sandwiched between two metallic electrodes, allowing more photons to be captured.

The active layer in an OPV needs to be ultra thin to dissociate the photogenerated excitons, but not add excessive series resistance. One way of increasing the absorption of light without increasing the thickness of the absorbing layer is to exploit plasmonic phenomena [7] to provide a light-trapping architecture [29, 55]. Such techniques include: embedding metallic nanoparticles within the surface of the solar cells; incorporating these within the active layer; and using a back contact in the form of a diffraction grating. For the first strategy, light is scattered into the higher permittivity organic absorbing layer. The scattered light will then acquire an angular spread in the dielectric that effectively increases the optical path length.

Fully embedded metallic nanoparticles will possess plasmon modes surrounding the nanoparticles (due to the different permittivities of the metal and organic host matrix). These will be associated with strong electric fields, which will couple to the organic semiconductor, increasing its effective absorption cross-section. In the third approach, the back contact in the form of a corrugated structure can couple the incoming light into a surface plasmons, which the light will scatter through 90° and propagate along the metal/organic interface, again resulting in an increased light absorption.

11.5.4 Tandem Cells

Another way to utilize more of the incoming EM spectrum is to connect OPVs in tandem. For BHJ-type OPV devices, the upper limit to V_{oc} is determined by the energy difference between the LUMO level of the acceptor and the HOMO level of the donor (Eq. (11.25)). Exciton binding and other factors further reduce it. Large band gap materials will possess higher V_{oc} values but will harvest fewer incoming photons than low band gap materials. Reported open-circuit voltages for simple BHJ devices are generally between 0.5 and 1 V. However, significantly higher figures (>1.5 V) are achieved using tandem cells [14, 29, 34]. These architectures use a wide band gap material as the first charge-separation layer and a narrow band gap material as the second charge-separation layer with a thinner front cell than back cell so that the photocurrents are balanced. An example is shown in Figure 11.14 [29]. It is possible to increase the absorption spectrum by adding more subcells in this type of architecture, but this adds to fabrication complexity and cost. It has been suggested that the limiting efficiency of a two-subcell tandem device is around 30% higher than the figure for a single junction cell [14].

The subcells of a tandem OPV can be connected in series, as depicted in Figure 11.14, or in parallel. In the former case, the interlayer between the two subcells acts as a recombination zone for currents originating from the individual devices. The output voltage will be the sum of the voltages from the two subcells, while the output current will be common. An important requirement, therefore, is that the currents generated by the two cells are closely matched. The parallel connection of cells in a tandem device is less common but removes the current matching criterion. In this case,

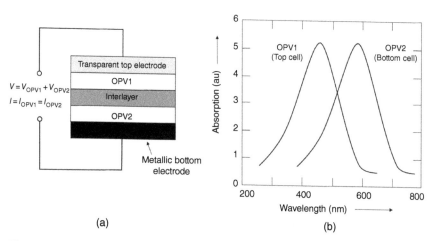

(a) (b)

Figure 11.14 Tandem OPV with two different subcells, OPV1 and OPV2. (a) Device architecture. For a series connection of devices, the output voltage will be the sum of those from the two subcells; the output current will be the same for the two subcells. (b) Optical absorption spectra for the two cells.

the interlayer must be both optically transparent and laterally conductive to allow the currents to be extracted from the two subcells.

The fabrication of tandem OPVs is challenging. It is essential to ensure that the material processing methodologies for the upper OPV do not affect the underlying device. However, efficiencies over 10% are reported for tandem architectures [14]. By combining a solution-processed non-fullerene-acceptor-based infrared absorbing subcell with a visible-absorbing fullerene-based subcell grown by vacuum thermal evaporation, an efficiency of 15% for $2\,mm^2$ cells has been achieved [56]. Moreover, this methodology resulted in a greater than 95% fabrication yield for devices with areas up to $1\,cm^2$.

11.5.5 Upconversion

As discussed earlier, a fundamental limit (Shockley–Queisser limit) on the efficiency of photovoltaic solar cells is that photons with energy below the band gap of the absorbers are transmitted while excitations that are created following absorbing photons with energy above the band gap undergo rapid thermalization. Spectral conversion aims to modify the incident solar spectrum to achieve a better match with the wavelength-dependent conversion efficiency of the solar cell. Upconversion (UC) combines two low-energy photons (with energies less than the band gap) to create one high-energy photon [57, 58]. The intermediate band solar cell (IBSC) introduces a narrow band of states within the band gap of the absorber that serves as a stepping-stone for the absorption of low energy photons, which would have otherwise been lost. There are many examples of inorganic semiconductor IBSC devices in which the intermediate band states are typically achieved by extreme lattice mismatch of dense sheets of quantum dots. However, charge recombination via the intermediate band is problematic [58].

For OPVs, a convenient means to achieve upconversion is to exploit triplet states [58–61]. Excitons are electron–hole pairs and are therefore pairs of half-integer spin particles, which can possess singlet or triplet character. Under certain conditions, two triplet states can recombine to form a singlet state in the processes of triplet–triplet annihilation (TTA) (Section 10.2). Hence, low-energy photons, which otherwise would be unabsorbed, can be harvested.

Figure 11.15 depicts an example of a phosphorescent sensitizer molecular adjacent to a donor (e.g. as part of the BHJ solar cell). Triplet states in the sensitizer are populated by intersystem crossing (ISC) (Figure 10.1). This leads to population of the triplet states in the adjacent donor absorber layer. TTA in the latter then produces high energy singlet excitons, which can subsequently dissociate via a charge-transfer mechanism at an interface with an acceptor material. Direct excitation of singlet excitons in the absorber also occurs.

For the upconversion to be successful, two conditions relating the material energy levels must be fulfilled. First, the singlet and triplet levels of the sensitizer should lie between those of the absorber to enable the triplet sensitization cycle to occur. The CT state of the donor–acceptor must also be higher in energy than the triplet level of the absorber. Otherwise, triplets that are generated in the absorber layer by the sensitization process would readily be dissociated at the donor/acceptor interface, bypassing TTA–UC [59].

In the reverse process to TTA, it is possible for one photoexcited high-energy singlet to be converted into a pair of low-energy triplets, a process called singlet fission (SF) [62]. If both triplets are then dissociated at the donor/acceptor interface, then, at certain wavelengths, the device can potentially generate two electron–hole pairs for every absorbed photon. This carrier multiplication process can be used to harvest the excess energy of above-band-gap photoexcitations that would otherwise be dissipated as heat.

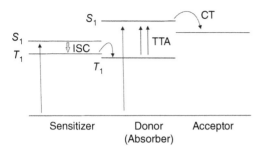

Figure 11.15 Upconversion organic solar cell. Optical absorption in a sensitizer molecule to the singlet S_1 states results in transfer to the triplet T_1 state in the sensitizer via intersystem crossing (ISC). Following transfer to the triplet state in the donor (absorber) molecule, triplet-triplet annihilation (TTA) provides excitation to the singlet state in the donor. Direct transfer to the singlet state in the donor molecule is also possible. As a result, photon absorption over an extended wavelength range is achieved. Excitons in the donor layer will dissociate at the donor/acceptor interface to electrons and holes via a charge-transfer (CT) state.

Both SF and TTA are elegant examples of molecular engineering. These have the potential to increase the limiting conversion efficiency of a solar cell by around 30% [63]. However, modest power conversion efficiencies are usually reported [14].

11.5.6 Device Degradation

Silicon-based solar cells may have a lifetime about 25 years. Organic materials are by nature more susceptible to chemical and physical degradation than their inorganic counterparts. While marked progress has been achieved with OPVs over the last decade, the stability of these devices remains a significant problem [64–70]. A physical understanding and the process control of the morphology of organic solar cells are central to providing further improvements in device performance.

Figure 11.16 indicates the origin of some of the main degradation pathways in polymer-fullerene BHJ cells. These can be grouped broadly into three categories: extrinsic degradation caused by chemical reactions with water and oxygen, intrinsic degradation in the dark, and intrinsic photo-induced degradation [65, 68]. However, degradation processes are complex, and there is overlap between these processes.

Many of the degradation effects in OPVs are common to those found in OLEDs, discussed in Section 10.8.4. Small amounts of oxygen and water can be introduced during device fabrication and these molecular species will also diffuse from the ambient into the manufactured device. This can cause physical and chemical degradation at the electrodes, in the carrier transport layers and within the organic active layers. Oxidation of metal electrodes such as aluminium can create an electrically insulating metal oxide layer. If grains of Al_2O_3 are located at the metal/active layer interface, as depicted in Figure 11.16, these will block carrier extraction. Following oxidation, defects, such as pinholes, can form at the interface between the electrodes and carrier transport layer. These can exacerbate the diffusion of oxygen and water into the OPV structure. Under certain conditions, it is even possible for atoms, or clusters of atoms, of the electrode material to diffuse into the inner part of the cell (Figure 9.9).

Some carrier transport layers are sensitive to ambient oxygen and water. As noted above (Section 11.5.3), the widely used HTL PEDOT:PSS is hygroscopic. The PSS chains also contain sulfonic acid groups, which facilitate etching of the ITO layer [71]. Indium can then diffuse into the device interior, as noted in Section 6.8.4. Strategies to avoid this problem include the addition

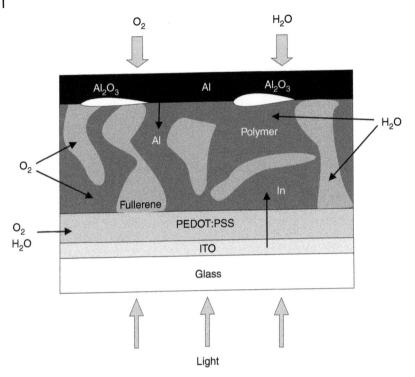

Figure 11.16 Schematic diagram of structure of bulk heterojunction polymer/fullerene solar cell indicating main pathways for degradation.

of a barrier between the PEDOT:PSS and ITO or coating the ITO with a self-assembled monolayer. The best long-term solution is to identify alternative electrode materials, as indicated above (Section 11.5.3).

External degradation processes are often activated by light. Some of these mechanisms are reasonably well understood [65]. For example, singlet oxygen is formed by energy transfer from a photoexcited polymer, e.g. PPV, to adsorbed ground state oxygen molecules. This is then thought to react with the vinylene groups in PPVs resulting in chain scission. Mixing a polymer with a fullerene, such as $PC_{61}BM$, as required in a BHJ device may improve the photooxidative stability [68].

Even when they are encapsulated and stable in the dark, many OPVs are observed to degrade rapidly for the first several hundred hours of illumination. This degradation rate decreases with time and the solar cells can then become relatively stable for many thousands of hours [68]. The initial period of fast degradation is called 'burn-in' and a burn-in loss of up to 60% is observed in some cells. Explanations for this phenomenon include the photo-induced dimerization of fullerene molecules and the creation of deep-trapping defect states. The former process decreases the exciton-harvesting efficiency leading to a loss in the short-circuit current. Changes to the electronic energy structures of the donors and acceptors following interactions with light can alter the energy offset at the donor/acceptor interfaces, affecting V_{oc}. Modifications to the density of states can also occur and defect levels within the HOMO–LUMO band gap may be introduced. These can decrease the mobility or increase the free carrier recombination (or both), thereby reducing J_{sc}.

Dense, crystalline BHJ film morphologies can improve the photochemical stability of the semiconductor materials themselves. The elimination of polymer side chains can help to achieve this. The use of higher adduct fullerenes (e.g. $PC_{71}BM$ rather than $PC_{61}BM$) can reduce photodimerization. An alternative is to utilize non-fullerene acceptors [38]. However, this strategy is not always successful in improving the solar cell lifetime [72]. Encapsulation can significantly slow degradation mechanisms caused by water and oxygen, although even perfectly encapsulated devices degrade over time. One interesting approach is to add an additional component to the BHJ cell that will absorb water molecules (i.e. acting as a desiccant). For example, the inclusion of poly(methyl-methacrylate) (PMMA) in $P3HT–PC_{61}BM$ devices has been shown to increase a cell's lifetime without unduly diminishing its electrical performance [73].

Dark degradation usually results from molecular rearrangement in the absorber layer or organic buffer layers. The interpenetrating network of the donor and acceptor materials is not necessarily a stable one. On a short time scale, molecules segregate and rearrange at material interfaces, forming layers that can hinder charge extraction. Over longer time scales, the two (or more) materials of the BHJ layer can phase separate, reducing the cell's ability to create free carriers from absorbed photons. These processes will be influenced by the operating temperature of the device and by the permeation of water and oxygen molecules. There is an ongoing search for new materials that can form stable morphologies. The thermal stability of OPVs can usually be improved by using materials with high glass transition temperatures. However, the optimum structure for device performance will, in all probability, not be the thermodynamically most stable.

There are also mechanical instability issues, induced by mechanical stresses. These can be introduced during device processing (e.g. roll-to-roll fabrication), solar cell module transportation, and installation. Operating weather conditions are also a significant factor.

Macroscopic manifestations of mechanical strains in the solar cell are delamination of the various layers, which will facilitate the ingress of oxygen and water. Strains within the active layers can produce changes in the material morphology.

As for OLEDs, there are some simple guidelines for extending OPV lifetimes. First and foremost, the devices must be fabricated under meticulously clean conditions and using the highest purity materials. Impurities can act as nucleation sites for degradation processes. It is imperative that the solar cells are encapsulated. Even though the stability of the OPV materials can be improved with ordered and dense film morphologies, the long-term rate of oxidation of electrode materials and absorber layers is limited by the diffusion rate of oxygen and water into the devices. Finally, the active materials themselves must possess high T_g values relative to the operating conditions to achieve stability in the dark. Low-molecular-weight-species, which can diffuse readily, should be avoided.

11.6 Dye-Sensitized Solar Cells

An artificial photosystem containing a light-harvesting antenna was manufactured by Grätzel and coworker in 1991 using a dye-sensitized titanium dioxide thin film with a redox electrolyte [74–76]. Grätzel's dye-sensitized photovoltaic device operates in the following manner, depicted schematically in Figure 11.17. Upon photon absorption, a dye electron is elevated to an excited state. The dye injects the excited electron directly into the conduction band of a semiconductor, which subsequently creates a current in an external circuit. An electrolyte then reduces the dye, regenerating it. Grätzel used a thin film of nanoparticles of titanium dioxide (a wide band gap semiconductor) and a

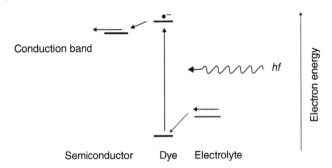

Figure 11.17 Carrier transport in dye-sensitized solar cell. Upon absorption of an incident photon of energy *hf*, a dye electron is elevated to an excited state. The dye then injects the excited electron directly into the conduction band of a semiconductor.

lithium ion liquid electrolyte. This combination of a nanoporous semiconductor with an electrolyte offers a high interfacial surface area, which causes rapid and efficient charge transfer.

Two factors account for the relatively high, 7%, conversion efficiency for Grätzel's cell. First, the ruthenium-based dye and the titanium oxide film absorbed 46% of the incoming solar flux. By using a light harvesting antenna, Grätzel was also able to separate light absorption and charge transport. This provided the cell with a superior fill-factor than a conventional silicon solar cell, since exciton recombination losses were minimized as there was no exciton travelling through the semiconductor. In 2015, conversion efficiencies of up to 13% were achieved for organic solvent-based liquid electrolyte cells [76]. However, a major problem is the contamination of the cell by moisture/water. Even the most robust encapsulation arrangements have been unable to eliminate this. Several alternatives to organic solvents have been proposed, for example solid-state conductors, plastic crystals, and gels. One development is the replacement of the traditionally used organic solvents with those based on water. Photoanode modifications, the introduction of additives and surfactants, the selection of specifically conceived redox couples, the preparation of suitable cathodes, and the stabilization of electrolytes have progressively led to the fabrication of 100% aqueous cells, with efficiencies of up to about 5% [76].

To make solid-state dye-sensitized solar cells, the liquid redox electrolyte can be replaced by a solid HTL. Most promising results have been achieved with molecular organic hole transporters such as 2,2′,7,7′-tetrakis(*N,N*-di-methoxyphenylamino)-9,9′-spirobifluorene (spiro-MeOTAD), conductive polymers like PEDOT, and using metal complexes [77, 78]. The main limitations of such devices arise from the very fast recombination between electrons in TiO_2 with holes in the HTL. Moreover, due to their solid nature, there is incomplete penetration of solid HTLs within nanoporous titanium dioxide layer, leading to poor electrical contact with the dye.

Dye-sensitized solar cells represent a low-cost solar cell technology. Confirmed efficiencies of 11.9% for a 1 cm^2 device and 8.8% for a submodule solar panel (399 cm^2) are reported [15], while a figure of 14.2% has been demonstrated in the laboratory [79]. Challenges are to produce large-area panels with a high degree of control over cell-to-cell reproducibility and to develop stable metal interconnects.

11.7 Hybrid Solar Cells

Many components make up an organic solar cell, such as a BHJ device. Each material has its own role to play (e.g. charge generation, charge extraction). Hybrid solar cells exploit the advantages of individual organic and inorganic compounds, with the aim of optimizing the overall device efficiency. The dye-sensitized solar cell described above may be considered as one example of such a hybrid. Arguably, the most successful example, in terms of power conversion efficiency, is the perovskite-based solar cell, which can routinely be fabricated with efficiency figures over 20%. This and other examples of hybrid solar cells containing organic materials are considered in this section [26, 31].

11.7.1 Polymer–Metal Oxide Devices

Hybrid organic–inorganic solar cells can take the form of either planar or BHJ architectures [31]. The operation of polymer–metal oxide devices is similar to the entirely organic BHJ cell. However, metallic oxide nanoparticles are used rather than fullerenes or other conjugated acceptor molecules. An example is ZnO, in which individual crystallites are 10–20 nm in diameter. This material is an n-type semiconductor (and acts as an electron acceptor) because of the presence of oxygen vacancies in the metal oxide lattice. Such metal oxides possess band gaps on the order of 3 eV. Consequently, the polymer is the only visible light absorber in the hybrid device. An exciton generated in the polymer will diffuse to an interface with a metal oxide particle, where a hybrid charge-transfer exciton is formed. This can be split via electron injection from the LUMO of the polymer into the conduction band of the polymer. Metal oxide–polymer BHJs are subject to the same loss mechanisms as organic BHJs. However, they can benefit from the relatively high electron mobility in the inorganic semiconductor.

11.7.2 Inorganic Semiconductor-Polymer Hole-Transporter Cells

Another type of hybrid BHJ solar cell combines organic hole transporters with inorganic absorbers. The rationale here is that inorganic nanoparticles or quantum dots can be used as the absorbing component. The nanoparticles provide stable structures on length scales of 2–100 nm that can be incorporated into an organic host matrix to produce the required exciton dissociation and charge-transport processes. Furthermore, the absorber coatings can be deposited using inexpensive solution coating methods [80, 81]. In such devices, n-type materials like CdTe, PbS, or CdSe are processed to form nanoparticles, which can then be blended with hole-transporting polymers such as P3HT [26].

11.7.3 Perovskite Solar Cells

Perovskite-based solar cells have seen dramatic progress, in terms of efficiency improvements, over recent years [8, 26, 82–84]. Confirmed efficiencies have been reported as 21.6% for a 1 cm^2 cell and 17.25% for a module of 804 cm^2 [15]. Perovskite compounds derive their name from L.A. Perovski, a mineralogist who first characterized the structure of the mineral calcium titanium oxide ($CaTiO_3$). A perovskite material has the chemical formula ABX_3, where A is a cation with a large atomic radius, B is a metal cation, and X is an anion. An example is shown in Figure 11.18. The atomic arrangement can be thought of as an AX_{12} cuboctahedron that shares its edges with a BX_6 octahedron [83]. A classic inorganic oxide perovskite that has found use in the electronics arena is barium titanate ($BaTiO_3$), which is known for its highly insulating and ferroelectric behaviour. In contrast, halide perovskites can be semiconducting. These may exhibit high carrier mobilities (of the order of a few cm^2 V^{-1} s^{-1}), balanced electron, and hole transport and long carrier diffusion lengths [85].

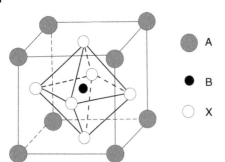

Figure 11.18 Crystal structure of perovskite with the ABX structure. A = cation with large ionic radius (e.g. methylammonium); B = metal cation (e.g. Pb); X = anion (e.g. I).

Moreover, some of the compounds are direct band gap semiconductors, processing high absorption coefficients (10^4–10^5 cm^{-1}).

Halide perovskites have the same perovskite structure as oxide perovskites, but monovalent halide ions occupy the sites populated by divalent oxygen ions. Therefore, these compounds can only host inorganic metal cations, such as Pb^{2+}, Sn^{2+}, and Ge^{2+}, with a valence state of 2^+ to satisfy charge neutrality. This limits the material diversity in terms of composition. However, this is compensated by the possibility of lower temperature synthesis. Significant hybridizations (Section 1.2) between cation s states and halogen p states (for example, overlap between Pb-6s and I-6p) contribute to the exceptional electronic properties of these materials.

Currently, the dominant material for perovskite solar cells is the organo-metal halide $CH_3NH_3MX_3$ (with M = Pb and X = I), which is based on the methylammonium cation, $CH_3NH_3^+$. The first attempt to exploit the materials in solar cells was to use these in a dye-sensitized device. Perovskite solar cells are of an excitonic nature, and the generation of excitons and their separation plays a major role in the device operation. However, the excitons are more of a Wannier–Mott type than a Frenkel type, i.e. the excitons are more delocalized (Figure 1.22). This leads to a smaller exciton-binding energy (less than 50 meV) than found in all-organic solar cells [26, 86].

Over time, several perovskite solar cell architectures have been studied. Many of these use nanoporous TiO_2 as an ETL, reflecting the fact that the devices evolved from the dye-sensitized solar cell. Figure 11.19 show two popular solid-state architectures built up on a semi-transparent FTO-coated glass substrate [26, 86]: Figure 11.19a is a structure using nanoporous TiO_2 permeated by perovskite as the active layer; and Figure 11.19b is a planar structure with the configuration FTO/ETL/intrinsic perovskite layer/HTL/Au. The nanoporous TiO_2 layer can be replaced by a nanoporous Al_2O_3 layer with no decrease in the solar cell performance, indicating that the

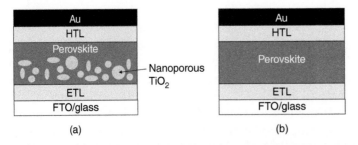

Figure 11.19 Two configurations for solid-state perovskite solar cells built up on a fluorine-doped tin oxide (FTO)/glass substrate. (a) structure using nanoporous TiO_2 permeated by perovskite as the active layer. (b) planar structure with the configuration FTO/ETL/intrinsic perovskite layer/HTL/Au. HTL = hole-transport layer. ETL = electron-transport layer. Source: Roy et al. [86].

perovskite is ambipolar and can conduct both electrons and holes. Spiro-MeOTAD is often used as the HTL, while a compact TiO_2 film can be used as an electron-selective contact.

The band gap of $CH_3NH_3PbI_3$ is around 1.55 eV, which is not ideal for optimum efficiency (absorption onset around 800 nm). However, the perovskite structure allows molecular engineering to tune the band gap without sacrificing the high absorption coefficient. One common replacement is to use a formamidinium $(HC(NH_2)_2{}^+)$ cation rather than methylammonium, which reduces the band gap by about 0.07 eV, extending the absorption wavelength by 40 nm. While $CH_3NH_3PbI_3$ shows n-type behaviour, $HC(NH_2)PbI_3$ is p-type [26].

Key advantages of perovskite solar cells are their high efficiencies and simple processing. For example, it is possible to produce a thin film in a single step by spin-coating a mixed solution of CH_3NH_3I and PbI_3 in appropriate solvent. Tandem cells (e.g. in conjunction with Si devices) are an attractive proposition [83]. If photovoltaics is to become a dominant technology, providing the terawatt electricity demands of the world, then the supply of raw materials becomes crucial. Here, lead perovskite technology will have an advantage. Lead is an abundant element, and it would only take a few days of current (2020) lead production capacity to scale to the TW level. This contrasts with about 500 years of gallium production capacity (for GaAs cells), 1000 years of tellurium (CdTe), or 400 years of indium (CIGS) [82]. However, challenges remain for both toxicity and degradation in lead-based perovskite solar cells [87].

11.8 Luminescent Solar Concentrator

The luminescent solar concentrator (LSC) is an interesting idea that has been under study since the 1970s [88–91]. The concept is somewhat different to classical imaging concentration systems that are based on the reflection and/or refraction of light, for example using a large area lens or parabolic mirror to focus sunlight onto a smaller photovoltaic cell.

Both strategies can eliminate the visual impact and minimize the amount of expensive semiconductor material required in photovoltaic solar cells. However, the dramatic cost reduction of photovoltaics over the past decades is making the latter motivation less important. The LSC approach is based on incorporating, in a transparent matrix, materials that are both absorbing and luminescent. The refractive index of the transparent matrix is chosen to be larger than the surrounding medium (air). As the luminescent materials emit light in all directions, part of the emitted light will be internally reflected at the matrix/air interface, as shown schematically in Figure 11.20. By designing geometries for which the length and width of the transparent matrix are larger than its thickness (as for a plate configuration), geometric concentration is accomplished. The emitted light is totally internally reflected and will be concentrated within the matrix where it can be harvested at the edges. There, it can be transformed into electrical energy by a photovoltaic cell.

The most evident feature of an LSC is its geometric factor or geometric gain, G, defined by the ratio of the area of the input surface, A_{in}, to the area of the edge where the concentrated emitted light is collected, A_{out}:

$$G = \frac{A_{in}}{A_{out}} \tag{11.27}$$

This gain represents the theoretical limit to the light concentration (which can be defined by the ratio of the flux of collected photons to the number of incident photons). Although increases in photon flux by a factor of one hundred, compared with exposing the same photocell to direct sunlight, have been predicted, only modest overall system efficiencies ($\approx 5\%$) are generally achieved [92].

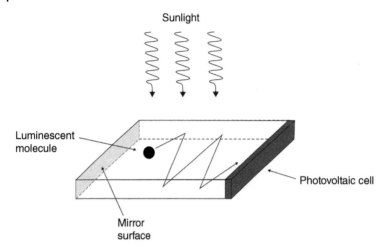

Figure 11.20 Schematic diagram of a luminescent plane concentrator. Direct and diffuse sunlight are absorbed by luminescent molecules. The emission is confined to the thin plate by total internal reflection. The radiation within the plate is collected by a photovoltaic cell at one end.

(It is important to note that the efficiency figures are somewhat deceptive as they are entirely dependent on the nature of the attached photovoltaic cell.) This is partly because for the organic luminescent materials used, the luminescent wavelengths are only slightly higher than the absorption wavelengths. Under these conditions, a significant part of the emitted light is lost by re-absorption.

The availability of newer emitter materials, based on quantum dots, photonic crystals, and even perovskites, is providing renewed interest in LSCs [92]. This renaissance may be associated with the emergence of building integrated photovoltaics, with niche areas of application, driven by environmental concerns and energy savings. One promising field of research lies in the considerable collection surface for solar light offered by the glazing of modern buildings, i.e. smart windows.

11.9 Organic Photodiodes and Phototransistors

The OPV configuration has also been used successfully to produce organic photodiodes for imaging technologies. Such structures offer cheaper processing methods and devices that are light, flexible, and compatible with large (or small) areas, and readily integrated on silicon substrates with CMOS components [93–95]. The devices may be grouped into those that absorb across a broad wavelength range (broadband or panchromatic) and those specifically designed for colour imaging in which a narrow wavelength band is targeted (narrowband or monochromatic). However, the approaches to photodiode design are similar. Organic photodiodes are suited to light detection in the visible to near-infrared regions of the EM spectrum.

The detector performance parameters are somewhat different to those for solar cells [11, 17, 93]. A key figure of merit is the responsivity, R. This is defined as the ratio of the photogenerated current, I_1, to the incoming radiant power:

$$R = \frac{I_1}{P_i A} \tag{11.28}$$

where P_i is the input radiant power density and A is the device area. The units of R are therefore $A\,W^{-1}$. Noting the definition of external quantum efficiency given by Eq. (11.16) (η_{ext} = electrons extracted/incident photons):

$$\eta_{ext} = \frac{I_1}{e}\left(\frac{hf}{P_i A}\right)$$

(11.29)

and

$$R = \frac{\eta_{ext} e}{hf} = \frac{\eta_{ext}\lambda(\mu m)}{1.24} \quad AW^{-1}$$

(11.30)

The responsivity will vary with the wavelength of the incident light. The spectral response of a photodetector is usually given as a plot of R or η_{ext} versus wavelength.

Responsivity should not be confused with sensitivity or detectivity, which refers to the lowest detectable light level of a device. This is related to the background noise and is therefore influenced by the detection bandwidth, Δf. The detectivity D and specific detectivity D^* for a photodetector have been discussed in Section 11.2.3 (Eq. 11.12). These parameters are both inversely related to the NEP, representing the radiation power needed to produce a signal equal to the noise (Problem 11.5).

Any current generated in the absence of light (signal) defines the noise level. Although in practical applications the dark current is subtracted from the measured signal (which includes the dark current and photocurrent), its random fluctuations cannot be mathematically eliminated. Noise sources have been discussed in Section 7.6. Of relevance to photodiodes are the thermal noise, shot noise (the result of random arrival of current quanta), $1/f$ noise and random telegraph noise (e.g. related to trapping and detrapping events). The total noise current, i_{noise}, is the root-mean-square (RMS) of the random fluctuations in the dark current at a detection bandwidth, Δf, and can be written in a general form:

$$i_{noise} = \left(i_{thermal}^2 + i_{shot}^2 + i_{1/f}^2 + i_{RTN}^2\right)^{1/2}$$

(11.31)

where $i_{thermal}$, i_{shot}, $i_{1/f}$, and i_{RTN} are the thermal, shot, $1/f$, and random telegraph noise components, respectively. While the first two terms are 'white' (frequency independent), $1/f$ noise and random telegraph noise will depend on frequency. However, the assumption is often made that these two latter noise components can be neglected in organic photodiodes [93, 96].

Using the expressions for thermal and shot noise power spectra from Section 7.6 (Eqs. (7.35) and (7.37)):

$$i_{noise} = \left(\frac{4k_B T \Delta f}{R_{shunt}} + 2eI_d\Delta f\right)^{1/2}$$

(11.32)

where I_d is the dark current of the photodiode. Here, the Fano factor associated with the shot noise is assumed to be unity and the thermal noise is assumed to be generated in the diode shunt resistor, R_{shunt}. The specific detectivity, Eq. (11.12), for the photodiode then becomes:

$$D^* = \frac{(A\Delta f)^{1/2} R_{shunt}}{i_{noise}}$$

(11.33)

In the case where shot noise dominates, this expression may be simplified further [96]:

$$D^* = \frac{R_{shunt}}{(2eJ_d)^{1/2}}$$

(11.34)

where J_d is the dark current density.

In principle, the D^* value of an organic photodiode is limited by two factors: the light harvesting capability of the photoactive layer, which controls the light response, and the dark current, which

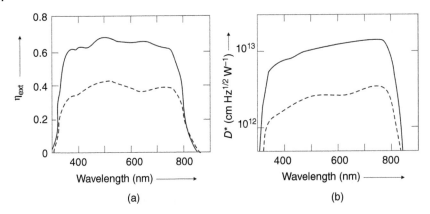

Figure 11.21 (a) External quantum efficiency η_{ext} versus wavelength, and (b) specific detectivity D^* versus wavelength data for devices based on high molecular weight PNTT–PC$_{71}$BM (full line) and low molecular weight PNTT–PC$_{71}$BM (dashed lines) BHJ photodiodes. Applied bias = −0.1. PNTT = a copolymer based on naphtho[1,2-c:5,6-c′]bis[1,2,5]-thiadiazole (NT) unit. Source: Zeng et al. [97].

can lead to excessive noise. An effective strategy to increase simultaneously the light harvesting capability and reduce the dark current is to increase the thickness of the photoactive layer. However, the film thickness is typically limited (to around 200 nm) because of the relatively low charge carrier mobility of organic semiconductors. High mobility organic semiconductors are therefore required.

The specific detectivity values for the best organic photodiodes are around 10^{13} cm Hz$^{1/2}$ W^{-1} (the unit of 'Jones' – in honour of Robert Clark Jones, who defined it – is also common in the literature) [93, 96, 97].

Figure 11.21 shows data for an organic photodiode detector using a high-mobility copolymer PNTT, which comprises a centrosymmetric naphtho[1,2-*c*:5,6-*c*′]bis[1,2,5]-thiadiazole (NT) unit [97]. This material allowed a relatively high film thickness to be used without unduly affecting the power conversion efficiency, which was over 9% for a photoactive layer thickness of 1 μm. Results are shown for the external quantum efficiency, Figure 11.21a, and the specific detectivity, Figure 11.21b, for PNTT:PC$_{71}$BM BHJ photodiodes with two different copolymer molecular weights (the full lines are data for the device fabricated using the higher molecular weight copolymer, while the dashed lines show results for the lower molecular weight photodiode).

Devices based on both the high- and low-molecular-weight samples exhibited broad responses from about 300 to 850 nm, although those based on the higher molecular weight material possessed greater values of η_{ext}. The hole mobility, as estimated from devices with high-molecular-weight PNTT was 1.08×10^{-3} cm^2 V^{-1} s^{-1}, which was about 2 orders of magnitude higher than the value for low molecular weight material. This was attributed to the more favourable film morphology for the former copolymer. The combination of a relatively high η_{ext} and a dark current provides a D^* figure of 1.39×10^{13} cm Hz$^{1/2}$ W^{-1} for the high-molecular PNTT:PC$_{71}$BM device (measured at 760 nm and a reverse bias of −0.1 V). These observations highlight the significance of film morphology on the operation of organic photodiodes and, more generally, the importance of thin film structure for organic devices based on blends of organic compounds.

In a similar approach to that used for the light-emitting transistor (Section 10.13) organic photoconductors can be integrated with transistors to realize organic phototransistors (OPTs) [98–103]. These devices benefit from the inherent gain of the transistor, leading to high values of η_{ext}. The active layer in an OPT can be a single-layer organic semiconductor (p-type, n-type, or ambipolar), in which case the exciton dissociation occurs because of the electric field in the channel. To improve the charge separation efficiency, BHJs may be used, either in isolation or in combination with other layers of organic semiconductors [100, 103]. OPTs can have similar architectures to OFETs. The favoured configuration is the bottom-gate top-contact structure depicted in Figure 8.6a, which allows the incident light to directly interact with the organic semiconductor layer. However, vertical transistors offer some advantages, in particular the relatively short channel length facilitates the separation and collection of photogenerated excitons [103].

Two different physical processes occur in the active-channel layers of OPTs: a photoconductive and a photovoltaic effect. The latter plays a dominant role for OPTs under illumination when the transistor is operating in the on state, i.e. $V_g < V_t$ for p-channel devices. Photogenerated holes will contribute to the channel current and drift to the drain electrode, while electrons will accumulate at the source electrode, effectively reducing the charge injection barrier and contact resistance at the source electrode. Consequently, for a p-channel device, a positive shift in the threshold voltage is produced. In addition, photogenerated electrons can be captured by electron-trapping sites within the organic layer or at the semiconductor/dielectric interface, resulting in a further positive shift in the transistor threshold voltage. The overall change in threshold voltage, ΔV_t, will produce an increase in the drain current, ΔI_d:

$$\Delta I_d = g_m \Delta V_t \tag{11.35}$$

where g_m is the transconductance of the transistor (Eq. (8.6)).

The photoconductive effect dominates when the transistor operates in the depletion mode ($V_g > V_t$ for p-channel devices). The transistor is turned off, leading to an increase in the photogenerated drain current with the optical power. The output current of the OPT will also depend on the transistor parameters such as capacitance, field-effect mobility, and the W/L ratio (Eq. (8.5)). Figure 11.22 shows the transfer characteristics for a p-channel OPT under illumination from a 650 nm LED [99]. The active layer was a semiconductor based on a diketopyrrolopyrrolethiazolothiazole copolymer (PDPPTzBT), which was deposited by inkjet printing. Both a shift in the threshold voltage and an increase in the off-current on exposure to light are evident in the figure.

The responsivity of the OPT can be calculated:

$$R = \frac{I_l}{P_i A} = \frac{I_{dl} - I_{dd}}{P_i A} \tag{11.36}$$

Figure 11.22 Drain current versus gate voltage (transfer characteristics) for an organic phototransistor based on a diketopyrrolopyrrolethiazolothiazole copolymer (PDPPTzBT). Data are shown for the device in the dark and under different illumination intensities from a 650 nm LED. ΔV_t indicates the threshold voltage change resulting from the light. Source: Wang et al. [99].

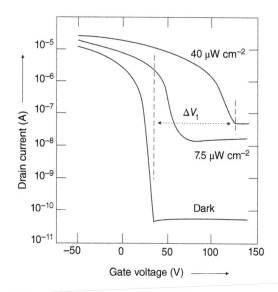

where I_{dl}, I_{dd} are the values of the drain current in the light and dark, respectively. OPTs can possess peak responsivities above 10^5 A W^{-1} [99, 100, 103] with D^* values up to 10^{17} cm Hz$^{1/2}$ W^{-1} [101]. Moreover, with low band gap organic semiconductors, the sensing range can extend from the visible part of the EM spectrum to 1500 nm.

An alternative approach to the detection of infrared radiation is to up-convert this to visible light, which can then be detected by the naked eye or a commercially available digital camera. This can be realized by the combination of a BHJ architecture and a phosphorescent OLED [102, 104–107]. Alternatively, hybrid organic–inorganic devices can be fabricated, exploiting the best attributes of the individual materials [105, 107].

Multiple organic photodiodes (OPDs) or OPTs can be organized in arrays to produce digital imaging cameras [98, 102, 103]. Passive or active addressing methodologies are possible, as for OLED displays (Section 10.9). In a passive matrix, the devices are assembled in a crossed-bar array. The architecture is simple, but there are readout problems, and it is susceptible to pixel-to-pixel crosstalk (Section 9.3, Figure 9.11). Active matrix solutions are more suitable for large-area applications. Here, the organic detector element is integrated with one or more transistors, which, depending on the chosen architecture, can act as simple device selectors or can also perform a signal processing function (e.g. amplification).

Problems

11.1 Using the experimental data from Figure 11.1, estimate the thickness of (a) a thin film of P3HT and (b) a thin film of PC$_{61}$BM to absorb 90% of incoming radiation at 500 nm.

11.2 Show that the current generated in a photoconductor under illumination is proportional to the applied voltage if both the carrier lifetime and mobility are voltage independent. (Hint: think about the relationship between the number of photogenerated carriers in the material and the resulting current that flows in an external circuit).

11.3 Derive Eq. (11.19) relating the load current and voltage, I_{load}, V_{load}, based on the solar cell equivalent circuit shown in Figure 11.6. (Hint: think about the current flowing through the shunt resistance, R_{shunt}).

11.4 A bulk heterojunction OPV cell uses P3HT (poly(3-hexylthiophene)) as the donor and PC$_{61}$BM ([6,6]-phenyl-C$_{61}$ butyric acid methyl ester) as the acceptor components. Research the HOMO and LUMO levels for these compounds. Draw a potential energy band diagram for the interface between the donor and acceptor. What wavelengths of incoming solar radiation will these materials absorb? Estimate the maximum open-circuit voltage for the heterojunction device. How might the HOMO and LUMO levels of the donor and acceptor compounds be modified to improve the device efficiency?

11.5 An organic infrared photodiode has a specific detectivity of 10^{12} cm Hz$^{1/2}$ W^{-1} at a wavelength of 1 μm. If the detector has a circular active area, with a diameter of 4 mm and a detection bandwidth of 5 kHz, calculate the minimum radiation intensity that the device will respond to at this wavelength. If the external quantum efficiency of the device is 0.5, what is its responsivity?

References

1 Pocchetino, A. (1906). Sul comportamento foto-elettrico dell'antracene. *Atti. Acad. Lincei. Rend.* 15: 355–368.

2 Weiss, D.S. and Abkowitz, M. (2017). Organic photoconductors. In: *Springer Handbook of Electronic and Photonic Materials*, 2e (eds. S. Kasap and P. Capper), 967–996. Cham: Springer.

3 Kallmann, H. and Pope, M. (1959). Photovoltaic effect in organic crystals. *J. Chem. Phys.* 30: 585–586.

4 Chamberlain, G.A. (1983). Organic solar cells: a review. *Sol. Cells* 8: 47–83.

5 Tang, C.W. (1986). Two-layer organic photovoltaic cell. *Appl. Phys. Lett.* 48: 183–185.

6 Tang, C.W. and VanSlyke, S.A. (1987). Organic electroluminescent diodes. *Appl. Phys. Lett.* 51: 913–915.

7 Petty, M.C. (2019). *Organic and Molecular Electronics: From Principles to Practice*, 2e. Chichester: Wiley.

8 Marinova, N., Valero, S., and Delgado, J.L. (2017). Organic and perovskite solar cells: working principles, materials and interfaces. *J. Colloid Interface Sci.* 488: 373–389.

9 Bube, R.H. (1992). *Photoelectronic Properties of Semiconductors*. Cambridge: Cambridge University Press.

10 Garn, L.E. (1984). Fundamental noise limits of thermal detectors. *J. Appl. Phys.* 55: 1243–1253.

11 Wilson, J. and Hawkes, J. (1998). *Optoelectronics: An Introduction*, 3e. Hemel Hempstead: Prentice-Hall.

12 Rose, A. (1978). *Concepts in Photoconductivity and Allied Problems*. New York: Robert E. Krieger Publishing.

13 Blythe, T. and Bloor, D. (2005). *Electrical Properties of Polymers*. Cambridge: Cambridge University Press.

14 Fusella, M.A., Lin, Y.L., and Rand, B.R. (2019). Organic photovoltaics (OPVs): device physics. In: *Handbook of Organic Materials for Optical and Optoelectronic Devices*, 2e (ed. O. Ostroverkhova), 665–693. Oxford: Woodhead.

15 Green, M., Dunlop, E., Hohl-Ebinger, J. et al. (2021). Solar cell efficiency tables (version 57). *Prog. Photovoltaics Res. Appl.* 29: 3–15.

16 Liu, Q., Jiang, Y., Jin, K. et al. (2020). 18% efficiency organic solar cell. *Sci. Bull.* 65: 272–275.

17 Sze, S.M. and Ng, K.K. (2007). *Physics of Semiconductor Devices*, 3e. Hoboken, NJ: Wiley-Interscience.

18 Nelson, J.A. (2003). *The Physics of Solar Cells*. London: Imperial College Press.

19 Shockley, W. and Queisser, H.J. (1961). Detailed balance limit of efficiency of *p-n* junction solar cells. *J. Appl. Phys.* 32: 510–519.

20 Rühle, S. (2016). Tabulated values of the Shockley–Queisser limit for single junction solar cells. *Sol. Energy* 130: 139–147.

21 Miller, O.D., Yablonovitch, E., and Kurtz, S.R. (2012). Strong internal and external luminescence as solar cells approach the Shockley–Queisser limit. *IEEE J. Photovoltaics* 2: 303–311.

22 Research photovoltaic cell and module efficiency records available from the National Renewable Energy Laboratory (NREL) at URL. https://www.nrel.gov/pv (accessed 24 March 2021).

23 Onsager, L. (1934). Deviations from Ohm's law in weak electrolytes. *J. Chem. Phys.* 2: 599–615.

24 Braun, C.L. (1984). Electric-field assisted dissociation of charge-transfer states as a mechanism of photocarrier production. *J. Chem. Phys.* 80: 4157–4161.

25 Köhler, A. and Bässler, H. (2015). *Electronic Processes in Organic Semiconductors*. Weinheim: Wiley-VCH.

26 Schmidt-Mende, L. and Weickert, J. (2016). *Organic and Hybrid Solar Cells*. Berlin: De Gruyter.

27 Rand, B.P., Burk, D.P., and Forrest, S.R. (2007). Offset energies at organic semiconductor heterojunctions and their influence on the open-circuit voltage of thin-film solar cells. *Phys. Rev. B* 75: 115327.

28 Scharber, M.C., Mühlbacher, D., Koppe, M. et al. (2006). Design rules for donors in bulk-heterojunction solar cells – towards 10% energy-conversion efficiency. *Adv. Mater.* 18: 789–794.

29 Abdulrazzaq, O.A., Saini, V., Bourdo, S. et al. (2013). Organic solar cells: a review of materials, limitations, and possibilities for improvement. *Part. Sci. Technol.* 31: 2427–2442.

30 Laquai, F., Andrienko, D., Deibel, C., and Neher, D. (2017). Charge carrier generation, recombination and extraction in polymer-fullerene bulk heterojunction organic solar cells. In: *Elementary Processes in Organic Photovoltaics*, Advances in Polymer Science, vol. 272 (ed. K. Leo), 267–291. Cham: Springer.

31 Brabec, C., Dyakonov, V., and Scherf, U. (eds.) (2008). *Organic Photovoltaics*. Weinheim: Wiley-VCH.

32 Li, Z., Ying, L., Zhu, P. et al. (2019). A generic green solvent concept boosting the power conversion efficiency of all-polymer solar cells to 11%. *Energy Environ. Sci.* 12: 157–163.

33 Thompson, B.C. and Fréchet, J.M.J. (2008). Polymer-fullerene composite solar cells. *Angew. Chem. Int. Ed.* 47: 58–77.

34 Jørgensen, M., Carlé, J.E., Søndergaard, R.R. et al. (2013). The state of organic solar cells – a meta analysis. *Sol. Energy Mater. Sol. Cells* 119: 84–93.

35 Kaur, N., Singh, M., Pathak, D. et al. (2014). Organic materials for photovotaic applications: review and mechanism. *Synth. Met.* 190: 20–26.

36 Moulé, A.J., Neher, D., and Turner, S.T. (2014). P3HT-based solar cells: structural properties and photovoltaic performance. In: *P3HT Revisited – From Molecular Scale to Solar Cell Devices*, Advances in Polymer Science, vol. 265 (ed. S. Ludwigs), 181–232. Berlin: Springer-Verlag.

37 Rwenyagila, E.R. (2017). A review of organic photovoltaic energy source and its technological designs. *Int. J. Photoenergy* 2107: 165612.

38 Cheng, P., Li, G., Zhan, X., and Yang, Y. (2018). Next-generation organic photovoltaics based on non-fullerene acceptors. *Nat. Photonics* 12: 131–142.

39 Perdigón-Toro, L., Zhang, H., Markina, A. et al. (2020). Barrierless free charge generation in the high-performance PM6:Y6 bulk heterojunction non-fullerene solar cell. *Adv. Mater.* 32: 1906763.

40 Ameri, T., Khoram, P., Min, J., and Brabec, C.J. (2013). Organic ternary solar cells: a review. *Adv. Mater.* 25: 4245–4266.

41 Koppe, M., Egelhaaf, H.-J., Dennier, G. et al. (2010). Near IR sensitization of organic bulk heterojunction solar cells: towards optimization of the spectral response of organic solar cells. *Adv. Funct. Mater.* 20: 338–346.

42 Vezie, M.S., Few, S., Meager, I. et al. (2016). Exploring the origin of high optical absorption in conjugated polymers. *Nat. Mater.* 15: 746–753.

43 Mühlbacher, D., Scharber, M., Morana, M. et al. (2006). High photovoltaic performance of a low-bandgap polymer. *Adv. Mater.* 18: 2884–2889.

44 Yan, D., Xin, J., Li, W. et al. (2019). 13%-efficiency quaternary polymer solar cell with non-fullerene and fullerene as mixed electron acceptor materials. *ACS Appl. Mater. Interfaces* 11: 666–773.

45 Makha, M., Schwaller, P., Strassel, K. et al. (2018). Insights into photovoltaic properties of ternary organic solar cells from phase diagrams. *Sci. Technol. Adv. Mater.* 19: 669–682.

46 Xiao, Z., Jia, X., and Ding, L. (2017). Ternary organic solar cells offer 14% power conversion efficiency. *Sci. Bull.* 62: 1562–1564.

47 Li, H., Xiao, Z., Ding, L., and Wang, J. (2018). Thermostable single-junction organic solar cells with a power conversion efficiency of 14.62%. *Sci. Bull.* 63: 340–342.

48 Cui, Y., Yao, H., Zhang, J. et al. (2019). Over 16% efficiency photovoltaic cells enabled by a chlorinated acceptor with increased open-circuit voltages. *Nat. Commun.* 10: 2515.

49 Gambhir, A., Sandwell, P., and Nelson, J. (2016). The future costs of OPV – a bottom-up model of material and manufacturing costs with uncertainty analysis. *Sol. Energy Mater. Sol. Cells* 156: 49–58.

50 Yang, Y., Wang, L., Yan, H. et al. (2006). Highly transparent and conductive double-layer oxide thin films as anodes for organic light-emitting diodes. *Appl. Phys. Lett.* 89: 051116.

51 Gommans, H., Verreet, B., Rand, B.P. et al. (2008). On the role of bathocuproine in organic photovoltaic cells. *Adv. Funct. Mater.* 18: 3686–3691.

52 Wang, K., Liu, C., Meng, T. et al. (2016). Inverted organic photovoltaic cells. *Chem. Soc. Rev.* 45: 2937–2975.

53 Tran, V.-H., Eom, S.H., Yoon, S.C. et al. (2019). Enhancing device performance of inverted organic solar cells with SnO_2/CS_2CO_3 as dual electron transport layers. *Org. Electron.* 68: 85–95.

54 Lee, K.S., Park, Y.J., Shim, J. et al. (2019). Effective charge separation of inverted polymer solar cells using versatile MoS_2 nanosheets as an electron transport layer. *J. Mater. Chem. A* 7: 15356.

55 Atwater, H.A. and Polman, A. (2010). Plasmonics for improved photovoltaic devices. *Nat. Mater.* 9: 205–213.

56 Che, X., Li, Y., Qu, Y., and Forrest, S.R. (2018). High fabrication yield organic tandem photovoltaics combining vacuum- and solution-processed subcells with 15% efficiency. *Nat. Energy* 3: 422–427.

57 Van Sark, W.G.J.H.M., De Wild, J., Rath, J.K. et al. (2013). Upconversion in solar cells. *Nanoscale Res. Lett.* 8: 81.

58 Chen, E.Y., Milleville, C., Zide, J.M.O. et al. (2018). Upconversion of low-energy photons in semiconductor nanostructures for solar energy harvesting. *MRS Energy Sustain.* 5: E16.

59 Simpson, C., Clarke, T.M., MacQueen, R.W. et al. (2015). An intermediate band dye-sensitised solar cell using triplet-triplet annihilation. *Phys. Chem. Chem. Phys.* 17: 24826–24830.

60 Lin, Y.L., Koch, M., Brigeman, A.N. et al. (2017). Enhanced sub-bandgap efficiency of a solid-state organic intermediate band solar cell using triplet-triplet annihilation. *Energy Environ. Sci.* 10: 1465–1475.

61 Bharmoria, P., Bildirir, H., and Moth-Poulsen, K. (2020). Triplet-triplet annihilation based near infrared to visible molecular photon upconversion. *Chem. Soc. Rev.* 49: 6529–6554.

62 Xia, J., Sanders, S.N., Cheng, W. et al. (2017). Singlet fission: progress and prospects in solar cells. *Adv. Mater.* 29: 1601652.

63 Schulze, T.F. and Schmidt, T.W. (2015). Photochemical upconversion: present status and prospects for its applications to solar energy conversion. *Energy Environ. Sci.* 8: 103–125.

64 Kawano, K., Pacios, R., Poplavskyy, D. et al. (2006). Degradation of organic solar cells due to air exposure. *Sol. Energy Mater. Sol. Cells* 90: 3520–3530.

65 Jørgensen, M., Norrman, K., and Krebs, F.C. (2008). Stability/degradation of polymer solar cells. *Sol. Energy Mater. Sol. Cells* 92: 686–714.

66 Krebs, F.C. (ed.) (2012). *Stability and Degradation of Organic and Polymer Solar Cells*. Chichester: Wiley.

67 Sing, V., Sharma, V., Arora, S. et al. (2016). Degradation analysis of organic solar cells under variable conditions. *Adv. Mater. Proceed.* 1: 71–74.

68 Mateker, W.R. and McGehee, M.D. (2017). Progress in understanding degradation mechanisms and improving stability in organic photovoltaics. *Adv. Mater.* 29: 1603940.

69 Duan, L. and Uddin, A. (2020). Progress in stability of organic solar cells. *Adv. Sci.* 7: 1903259.

70 Karakawa, M., Suzuki, K., Kuwabara, T. et al. (2020). Factors contributing to degradation of organic photovoltaic cells. *Org. Electron.* 76: 105448.

71 Cameron, J. and Skabara, P.J. (2020). The damaging effects of the acidity in PEDOT:PSS on semiconductor device performance and solutions based on non-acidic alternatives. *Mater. Horiz.* 7: 1759–1772.

72 Doumon, N.Y., Dryzhov, M.V., Houard, F.V. et al. (2019). Photostability of fullerene and non-fullerene polymer solar cells: the role of the acceptor. *ACS Appl. Mater. Interfaces* 11: 8310–8318.

73 AL-Busaidi, Z., Pearson, C., Groves, C., and Petty, M.C. (2017). Enhanced lifetime of organic photovoltaic diodes utilizing a ternary blend including an insulating polymer. *Sol. Energy Mater. Sol. Cells* 160: 101–106.

74 O'Regan, B. and Grätzel, M. (1991). A low-cost, high-efficiency solar cell based on dye sensitized colloidal TiO_2 films. *Nature* 353: 737–740.

75 Ye, M., Wen, X., Wang, M. et al. (2015). Recent advances in dye-sensitized solar cells: from photoanodes, sensitzers and electrolytes to counter electrodes. *Mater. Today* 18: 155–162.

76 Bella, F., Gerbaldi, C., Barolo, C., and Grätzel, M. (2015). Aqueous dye-sensitized solar cells. *Chem. Soc. Rev.* 44: 3349–3862.

77 Sharma, K., Sharma, V., and Sharma, S.S. (2018). Dye-sensitized solar cells: fundamentals and current status. *Nanoscale Res. Lett.* 13: 381.

78 Boschloo, G. (2019). Improving the performance of dye-sensitized solar cells. *Front. Chem.* 7: 77.

79 Ji, J.-M., Zhou, H., Eom, Y.K. et al. (2020). 14.2% efficiency dye-sensitized solar cells by co-sensitizing novel thieno[3,2-*b*]indole-based organic dyes with a promising porphyrin sensitizer. *Adv. Energy Mater.* 10: 2000124.

80 Chebrolu, V.T. and Kim, H.-J. (2019). Recent progress in quantum dot sensitizer solar cells: an inclusive review of photoanode, sensitizer, electrolyte, and the counter electrode. *J. Mater. Chem. C* 7: 4911–4933.

81 Lee, H., Song, H.-J., Shim, M., and Lee, C. (2020). Towards the commercialization of colloidal quantum dot solar cells: perspectives on device structures and manufacturing. *Energy Eviron. Sci.* 13: 404–431.

82 Liu, M., Johnston, M.B., and Snaith, H.J. (2013). Efficient planar heterojunction perovskite solar cells by vapour deposition. *Nature* 501: 395–398.

83 Snaith, H.J. (2018). Present status and future prospects of perovskite photovoltaics. *Nat. Mater.* 17: 372–376.

84 Seok, S.I. and Guo, T.-F. (eds.) (2020). Halide perovskite opto- and nanolectronic materials and devices. *MRS Bull.* 45: 427–484.

85 Motta, C., El-Mellouhi, F., and Sanvito, S. (2015). Charge carrier mobility in hybrid halide perovskites. *Sci. Rep.* 5: 12746.

86 Roy, P., Sinha, N.K., Tiwari, S., and Khare, A. (2020). A review on perovskite solar cells: evolution of architecture, fabrication techniques, commercialization issues and status. *Sol. Energy* 198: 665–688.

87 Yi, Z., Ladi, N.H., Shai, X. et al. (2019). Will organic-inorganic hybrid halide lead perovskites be eliminated from optoelectronic applications? *Nanoscale Adv.* 1: 1276–1289.

88 Weber, W.H. and Lambe, J. (1976). Luminescent greenhouse collector for solar radiation. *Appl. Opt.* 15: 2299–2300.

89 Debije, M.G. and Verbunt, P.P.C. (2012). Thirty years of luminescent solar concentrator research: solar energy for the built environment. *Adv. Energy Mater.* 2: 12–35.

90 Khamooshi, M., Salati, H., Egelioglu, F. et al. (2014). A review of solar photovoltaic concentrators. *Int. J. Photoenergy* 2014: 958521.

91 Zhao, Y., Meek, G.A., Levine, B.G., and Lunt, R.R. (2014). Near-infrared harvesting transparent luminescent solar concentrators. *Adv. Opt. Mater.* 2: 606–611.

92 Roncali, J. (2020). Luminescent solar collectors: quo vadis? *Adv. Energy Mater.* 10: 2001907.

93 Jansen-van Vuuren, R.D., Armin, A., Pandey, A.K. et al. (2016). Organic photodiodes: the future of full color detection and image sensing. *Adv. Mater.* 28: 4766–4802.

94 Jahnel, M., Thomschke, M., Ullbrich, S. et al. (2016). On/off-ratio dependence of bulk hetero junction photodiodes and its impact on electro-optical properties. *Microelectron. Eng.* 152: 20–25.

95 Jahnel, M., Thomschke, M., Fehse, K. et al. (2015). Integration of near infrared and visible organic photodiodes on a complementary metal-oxide-semiconductor compatible backplane. *Thin Solid Films* 592: 94–98.

96 Kim, I.K., Jo, J.H., Lee, J., and Choi, Y.J. (2018). Detectivity analysis for organic photodetectors. *Org. Electron.* 57: 89–92.

97 Zeng, Z., Zhong, Z., Zhong, W. et al. (2019). High-detectivity organic photodetectors based on a thick-film photoactive layer using a conjugated polymer containing a naphtho[1,2-*c*:5,6-*c*]bis[1,2,5]-thiadiazole unit. *J. Mater. Chem. C* 7: 6070–6076.

98 Baeg, K.-J., Binda, M., Natali, D. et al. (2013). Organic light detectors: photodiodes and photo-transistors. *Adv. Mater.* 25: 4267–4295.

99 Wang, H., Cheng, C., Zhang, L. et al. (2014). Inkjet printing short-channel polymer transistors with high-performance and ultrahigh photoresponsivity. *Adv. Mater.* 26: 4683–4689.

100 Ren, X., Yang, F., Gao, X. et al. (2018). Organic field-effect transistor for energy-related applications: low-power-consumption devices, near-infrared phototransistors, and organic thermoelectric devices. *Adv. Energy Mater.* 8: 1801003.

101 Ji, D., Li, T., Liu, J. et al. (2019). Band-like transport in small-molecule thin films toward high mobility and ultrahigh detectivity phototransistor arrays. *Nat. Commun.* 10: 12.

102 Wang, C., Zhang, X., and Hu, W. (2020). Organic photodiodes and phototransistors towards infrared detection: materials, devices, and applications. *Chem. Soc. Rev.* 49: 653–670.

103 Huang, X., Ji, D., Fuchs, H. et al. (2020). Recent progress in organic phototransistors: semiconductor materials, device structures and optoelectronic applications. *ChemPhotoChem* 4: 9–38.

104 Lv, W., Zhong, J., Peng, Y. et al. (2016). Organic near-infrared upconversion devices: design principles and operation mechanisms. *Org. Electron.* 31: 258–265.

105 Hany, R., Cremona, M., and Strassel, K. (2019). Recent advances with optical upconverters made from all-organic and hybrid materials. *Sci. Technol. Adv. Mater.* 20: 497–510.

106 Aderne, R., Strassel, K., Jenatsch, S. et al. (2019). Near-infrared absorbing cyanine dyes for all-organic optical upconversion. *Org. Electron.* 74: 96–102.

107 Motmaen, A., Rostami, A., and Matloub, S. (2020). Ultra high-efficiency integrated mid infrared to visible up-conversion system. *Sci. Rep.* 10: 9325.

Further Reading

Geoghegan, M. and Hadziioannou, G. (2013). *Polymer Electronics*. Oxford: Oxford.

Schwoerer, M. and Wolf, H.C. (2007). *Organic Molecular Solids*. Weinheim: Wiley-VCH.

Streetman, B.G. and Banerjee, S. (2000). *Solid State Electronic Devices*, 5e. Hoboken, NJ: Prentice Hall.

12

Emerging Devices and Systems

12.1   Introduction

The former chapters of this book have concerned the electrical behaviour of organic compounds that are finding use in electronics. OLEDs are currently in the marketplace, while other devices, including OPVs and FETs, are emerging as consumer products. Most of organic electronics is focused on emulating the operation of existing inorganic semiconductor components by using organic materials. There are key benefits to organic technologies, especially their relatively simple material processing methodologies, low cost, and the ability to manufacture flexible and bendable components. However, there are also major drawbacks, such as the low carriers mobilities found in organic semiconductors and the relatively poor physical and chemical stability of these materials. I therefore suggest that organic semiconductors will probably not displace silicon and gallium arsenide as major platforms for general computing in the immediate future.

In this chapter, I have taken a broader view of the field of organic electronics with some suggestions for developing areas. Scientists are notoriously unsuccessful in predicting what technologies are likely to dominate in the forthcoming decades. What follows is therefore tentative and very much a personal view (and elicits the occasional change in pronoun). It is evident that organic compounds can offer much more to the fields of materials science and electronics than as substitutes for their inorganic counterparts. Carbon and silicon are both found in Group IV of the Periodic Table and possess similar outer electronic configurations, but the absence of inner atomic orbitals allows carbon readily to form hybridized orbitals. A consequence is the richness and diversity of organic and biological compounds. Nature abounds with examples of electrical 'devices' – from those that can process information (the brain) to systems that can convert energy from one form to another (photosynthesis). It is therefore entirely fitting that the biological world influences our speculations. This approach should not be proscriptive. Learning from the natural world is more important that trying to copy it – after all, our aircraft are manufactured from metal rather than feathers.

This chapter begins by examining examples of electronic circuits based on individual molecules of groups of molecules. A brief introduction to the biological world then follows. Natures exploits both ions and electrons as charge carriers, and some of these processes are reviewed. The various computing architectures that are currently in use (von Neumann machines) and those that are contenders for future problem-solving and data manipulation (quantum computing and neuromorphic computing) are contrasted. Information processing in the natural world can cope with material defects (i.e. it is fault-tolerant), and systems are able to self-repair. Possible lessons for organic electronics are discussed. Finally, some relevant artificial biological systems are described.

Electrical Processes in Organic Thin Film Devices: From Bulk Materials to Nanoscale Architectures, First Edition. Michael C. Petty.
© 2022 John Wiley & Sons Ltd. Published 2022 by John Wiley & Sons Ltd.
Companion Website: www.wiley.com/go/petty/organic_thin_film_devices

12.2 Molecular Logic Circuits

In most of the experiments on molecular-scale electronics, discrete devices are studied, and the inorganic materials associated with silicon microelectronic devices are replaced by organic counterparts, albeit at the molecular scale. The designs for molecular rectifiers have been described in Section 4.5, while the prospects for developing memory devices using individual molecules or groups of molecules have been explored in Section 9.7. Such devices need to be connected to form the logic circuitry that is the foundation of digital computers. In the latter, information is encoded in electrical signals. Threshold values and logic conventions are established for each signal. In a positive logic convention, a '0' is used to represent a signal that is below the threshold value and a '1' indicates that a signal is above the threshold. The logic circuits of silicon microprocessor systems process such binary data through a sequence of logic gates. Although it is not necessarily true that the components of a molecular computer will have to operate in manner analogous to a silicon-based computer, much effort is being directed to the design, synthesis, and characterization of molecular systems that mimic conventional logical operations [1–3].

The three basic types of logic gates are the NOT, AND, and OR [4]. Gates that are a combination of these basic three functions also exist: examples are the NAND (AND gate plus a NOT gate) and the NOR (OR plus NOT). Each has been designed to perform according to a set of rules delineated in a truth table, which is a list of outputs that the gate should give in response to the complete range of input combinations. The NOT gate simply inverts the signal at its input, i.e. a logic 0 input results in a logic 1 output, and vice versa. In the chemical world, this function is quite common. For example a luminescence output can be quenched by a chemical input [5].

A two-input molecular AND gate is shown in Figure 12.1 [6]. This logic function has two inputs, A and B, and one output. Logic 1 is only obtained on the output when the input signals are both at 1, i.e. the output is 1 if, and only if, input A *and* input B are at 1. The gate exploits the chemistry of an anthracene derivative. In methanol, and in the presence of H^+ and Na^+, the fluorescence

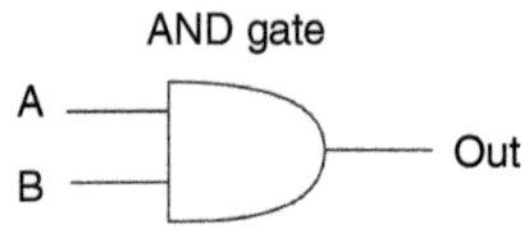

AND gate

(a)

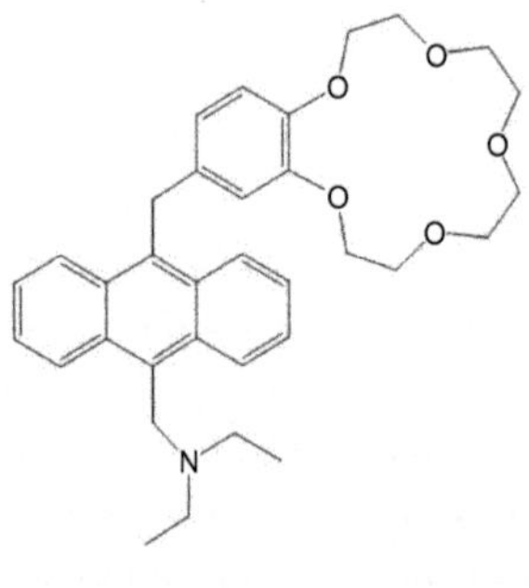

A (H^+)	B (Na^+)	Out (Fluorescence)
0	0	0
0	1	0
1	0	0
1	1	1

(b) (c)

Figure 12.1 A chemically-based AND logic gate. (a) Circuit symbol. (b) Anthracene derivative. (c) Truth table for AND function. Source: Balzani et al. [1].

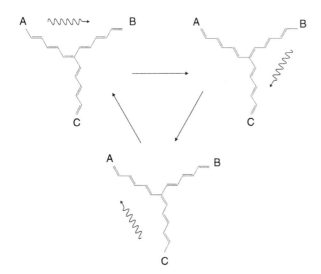

Figure 12.2 Soliton valving. The propagation of a soliton from chain A to chain B corresponds to a clockwise rotation by 120° of the upper left-hand configuration. Source: Carter [7].

quantum yield of this compound is high (output state 1 in the truth table). The other three output states 0 correspond to low fluorescence. The photo-induced electron-transfer process involves the amine moiety in the first two states of the truth table and the crown ether (the chemical group to the top right of the molecule depicted in Figure 12.1b) in the third. The crown ether alone cannot quench the anthracene fluorescence, but when the amine is protonated the process becomes thermodynamically allowed and does occur.

Other examples of chemically based logic operations can be found in the literature [1, 3]. In some of the early work, it was proposed to exploit conformational changes in molecules. The soliton switch, introduced by Carter in 1981 [7, 8], is perhaps the best-known example. A soliton in a conjugated polymer, such as polyacetylene, is a defect that separates chain segments with different bond alternation (Section 3.6.1). These segments may be considered as logical states and a passing soliton can 'switch' the chain from one state to another. Figure 12.2 shows how solitons in polymer networks might be switched and steered – soliton valving. The figure illustrates the change of state of a three-state network following the passage of a soliton. Progression of the soliton from segment A to B (or from B to A) moves the double bond at the branch carbon from the A chain to the B chain. In the upper right of the Figure 12.2, a soliton moving from B to C moves the double bond to the C chain.

There have been a few suggestions that utilize the properties of the hydrogen bond [9]. This is nearly unique in its range of energies (approximately 0.02–1.3 eV per atom) and is ubiquitous in biological systems. Figure 12.3 shows a proposed switch based on hydrogen bonding in an hemiquinone molecule [10]. The two isomers (molecules with identical formulae but distinct arrangements of atoms in space – Section 1.5.1) depicted differ only in the placement of the protons, and are called tautomers. The hydrogen atoms that are involved in the tautomerism are suspended in a perfectly symmetrical double well potential and are not completely localized in one well, or in one part of the molecule, but move to the left and right, giving rise to an oscillation of the structure shown in Figure 12.3. In the proposed operation of the device, an applied electric field perturbs the double well potential, which becomes asymmetric and therefore allows the two states of the molecule to be distinguished. The protons move between the potential wells by quantum mechanical tunnelling (Section 6.3).

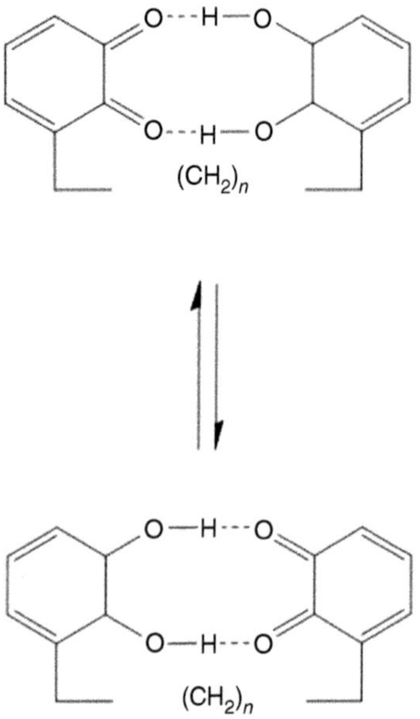

Figure 12.3 Tautomeric forms of a hemiquinone derivative proposed as a molecular switch. Source: Aviram et al. [10].

Many organic molecules, such the fullerene C_{60}, exhibit multiple reduction/oxidation (redox) states, offering the possibility of 'multibit' information storage [11]. One bit of data can be associated with each redox state. Other approaches to molecular logic exploit the electrical properties of organic molecules and propose to construct circuitry by chemically linking the various components (e.g. molecular rectifier molecules, Section 4.5). These and other suggestions have attracted their fair share of criticism for being impractical (or even impossible) to realize. The problems of interconnecting individual devices, be these molecular diodes, switches, or simple logic circuits, are formidable. There is also the issue of how to read-in and read-out data at the molecular level. In the case of soliton switching, to 'read' the state of a chain segment, two adjacent carbon atoms must be 'marked', and the bond between them needs to be 'inspected'. However, I suggest that the value of such ideas is to generate lateral thinking and develop new concepts for molecular information processing, which might one day turn out to be achievable.

There have also been many speculations about self-assembling, self-repairing, and fault-tolerant molecular systems; none have yet been well specified or fabricated. However, the steps needed have been outlined [12]. First, a high-level logic function associated with a molecule must be identified. A means is required to place such molecules with precision, and to make an interface in order to verify their position and provide a method for communication. It is then essential to be able to interconnect and/or isolate the molecules without destroying their functionality. Finally, an efficient assembly scheme for very large numbers of such molecules must be developed. These ideas are explored further in Section 12.5.

12.3 Inspiration from the Natural World

The biological cell is the fundamental structural and functional unit of all living organisms [13]. Many different molecules are found here. The detailed chemical and conformation of each compound determines in which chemical reactions it can participate, and therefore its role in the life of the cell. Important classes of biomolecules include nucleic acids, proteins, carbohydrates, and lipids. Other compounds perform functions such as transporting energy from one part of the cell to another or utilizing the sun's energy to drive chemical reactions. All these molecules, and the cell itself, are in a state of constant change. A cell cannot remain healthy unless it is continually forming and breaking down proteins, carbohydrates, and lipids, repairing damaged nucleic acids, and using and storing energy. Such energy-linked reactions are collectively known as metabolism.

Below is an introduction to some of the fundamental building blocks of the biological world. I have given prominence to those components that might have a role in the development of new electronic devices and systems.

12.3.1   Amino Acids, Peptides, and Proteins

Amino acids are a class of organic compounds that contain both the amino (NH_2) and carboxyl (COOH) chemical groups. The primary building blocks of all proteins, regardless of their species of origin, are the group of 20 different amino acids. The amino and carboxyl groups are both attached to a single carbon atom called the α-carbon. The α-amino group is free or unsubstituted in all the amino acids except one, proline. A further variable group, R, is attached to the α-carbon; it is in their R groups that the molecules of the 20 amino acids differ. The simplest of the acids, glycine, contains an R group that comprises a single hydrogen atom. Since the amino and carboxylic acid groups are basic and acidic, respectively, amino acids are zwitterions, with both negative and positive charges. The structures of the neutral (undissociated) and charged forms of glycine are shown in Figure 12.4a. Zwitterions are highly polar substances (Section 7.2) for which inter-molecular electrostatic attraction leads to strong crystal lattices. Apart from glycine, all the amino acids are chiral molecules, that is they exist as both right-handed and left-handed isomers. Amino acids containing aromatic rings, such as tyrosine, tryptophan, phenylalanine and histidine, are highly polarizable (i.e. electric dipoles can be induced by applied fields) because of their delocalized electron systems [14]. Tryptophan is the most polarizable amino acid, possessing a particular double ring (an 'indole

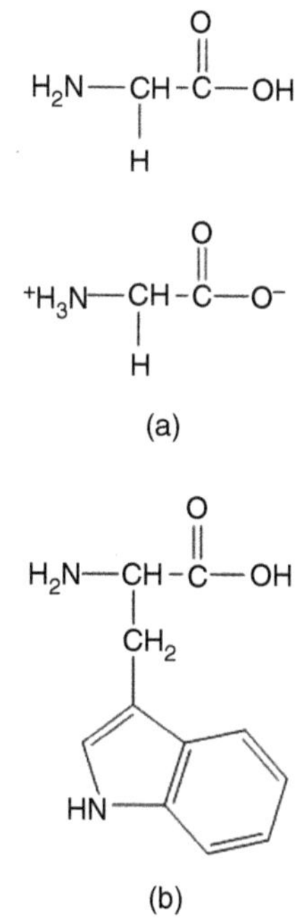

Figure 12.4   (a) Undissociated and zwitterionic forms of the amino acid glycine. The latter has a strong permanent dipole. (b) Chemical structure of tryptophan, a highly polarizable amino acid.

ring') – a six-carbon ring joined to a five-carbon ring with one nitrogen and four carbons; its molecular structure is depicted in Figure 12.4b.

Two amino acid molecules may be joined to yield a dipeptide through a peptide bond, formed by the removal of a water molecule from the carboxyl group of one amino acid and the α-amino group of the other by the action of strong condensing agents. Higher peptides are also possible; a tripeptide contains three amino acids, a tetrapeptide four, and so on. Peptides are named from the sequence of their constituent amino acids, beginning from the amino-terminated end. When many amino acids are joined in a long chain, this is called a polypeptide. Such compounds contain only one free α-amino acid group and one free α-carboxyl group at their ends. In addition to the amino acids that form proteins, many other amino acids have been found in nature, including some that have the carboxyl and amino groups attached to separate carbon atoms. These unusually structured amino acids are most often found in fungi and higher plants.

A protein is a complex, high molecular weight, organic compound consisting of amino acids joined by peptide bonds. These molecules are an abundant species in most cells and generally constitute 50% of their dry weight. A protein can be a single polypeptide chain, or it may consist of several such chains held together by weak molecular bonds. The R groups of the amino acid subunits determine the final shape of the protein and its chemical properties. Using only 20 different amino acids, a cell constructs thousands of different proteins, each of which has a highly specialized function.

Proteins are versatile cell components and serve two important functions. First, they can act as a structural material. The structural proteins tend to be fibrous in nature, i.e. the polypeptide chains are lined up more or less parallel to each other and are joined to one another by hydrogen bonds. These are physically tough and are normally insoluble in water. Depending on the actual three-dimensional arrangement of the individual protein molecule and its interaction with other similar molecules, a variety of structural forms may result. Typical examples of fibrous proteins are α-keratin, the major component of hair, feathers, nails and skin, and collagen, the key component of tendons.

The other important property of proteins is their role as biological regulators. Here, the proteins are responsible for controlling the speed of biochemical reactions and the transport of various materials throughout the organisms. The catalytic proteins (the enzymes) and transport proteins tend to be globular in nature. The polypeptide chain is folded around itself in such a way to give the entire molecule a rounded shape. Globular proteins are soluble in aqueous systems and diffuse readily.

The sequence of amino acids in the covalent backbone of a protein is called its primary structure. The secondary structure refers to the specific geometric arrangement of the polypeptide chain along one axis. Two common arrangements are the α-helix and the β-conformation, in which the polypeptide chains are in an extended zigzag configuration called a pleated sheet. These structures are depicted in Figure 12.5. In Figure 12.5a, the α-helix is right-handed with a pitch of 0.54 nm or 3.6 amino acid units. Three parallel protein chains in a β-pleated sheet are shown in Figure 12.5b. All the R groups project above or below the plane of the figure. Both the helix and pleated sheet are robust structures, held together by hydrogen bonding. The specific configurations of the polypeptide chains are stable because of particular amino acid sequences. For example, an α-helix tends to form spontaneously only in the case of polypeptide chains in which consecutive R groups are relatively small and uncharged, as in α-keratin.

A tertiary structure for proteins also exists. This term is used to refer to the three-dimensional structure of globular proteins, in which the polypeptide chain is tightly folded and packed into a compact spherical form. The molecule tends to orient itself so that the nonpolar side chains lie inside the bulk of the structure, where they attract each other by van der Waals forces. The polar

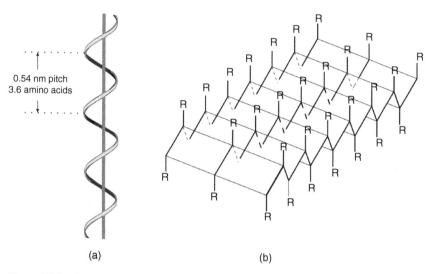

0.54 nm pitch
3.6 amino acids

(a)   (b)

Figure 12.5 Secondary structures of proteins. (a) Right-handed α-helix. (b) Three parallel protein chains in a β-pleated sheet. All the R groups project above or below the plane of the figure.

side chains are usually found on the surface of the molecule; consequently, these can hydrogen bond to the solvent molecules and confer the necessary water solubility. The process by which a linear protein chain forms its secondary and tertiary structure is called protein folding. A protein's shape is closely linked with its function, and the ability to predict this structure unlocks a greater understanding of what it does and how it works. Many current scientific challenges, such as developing treatments for diseases or discovering enzymes that break down industrial waste, are fundamentally linked to proteins and the role they play. The accurate prediction of protein structures from their amino-acid sequence is a formidable problem although approaches using artificial intelligence are beginning to prove successful [15].

12.3.2 Nucleotides, DNA, and RNA

Nucleotides are the recurring structural units of the nucleic acids. These molecules contain three characteristic components: a nitrogenous base, a five-carbon sugar, and phosphoric acid. The nucleotides are the compounds responsible for storing and transferring genetic information and are enormous molecules made up of long strands of subunits, called bases, arranged in a precise sequence. The bases are 'read' by other components of the cell and used as a guide in making proteins. Two important types of nucleic acid are ribonucleic acid (RNA) and deoxyribonucleic acid (DNA).

The bases found in nucleotides are of two types. These are derivatives of two parent heterocyclic compounds pyrimidine and purine, which are themselves not found in nature. Pyrimidine bases are heterocyclic compounds, which consist of a single carbon ring containing two nitrogen atoms. Purines are derived from pyrimidines by addition of an imidazole group and therefore contain a greater number of nitrogen atoms. Three pyrimidine bases are common in nucleic acids: uracil, thymine, and cytosine, universally abbreviated as U, T, and C. Uracil is generally found in RNA and thymine in DNA; cytosine is found in both RNA and DNA (Problem 12.1). There are two common purine bases, found in both RNA and DNA: adenine (A) and guanine (G). The pyrimidine

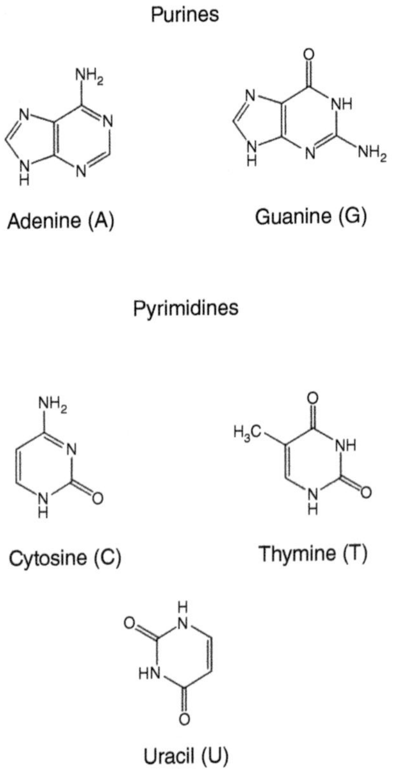

Purines

NH$_2$

N

N

N

N

H

Adenine (A)

O

N

NH

N

N

NH$_2$

H

Guanine (G)

Pyrimidines

NH$_2$

N

N

O

H

Cytosine (C)

O

H$_3$C

NH

N

O

H

Thymine (T)

H

O

N

HN

O

Uracil (U)

and purine bases are nearly flat molecules and are relatively insoluble in water; their chemical structures are provided in Figure 12.6.

In 1953, James Watson and Francis Crick established the structure of DNA, which in 1962 led to the award of the Nobel Prize for Medicine (together with Maurice Wilkins). The DNA molecule is composed of two long strands in the form of a double helix, as shown in Figure 12.7. The strands comprise alternating phosphate and sugar molecules. The nitrogen bases provide links between these strands, holding them together. Each base is attached to a sugar molecule and is linked by a hydrogen bond to a complementary base on the opposite strand. Adenine always binds to thymine (A to T), and guanine always binds to cytosine (G to C). To make a new, identical copy of the DNA molecule, the two strands need only unwind and separate at the bases (which are weakly bound); with more nucleotides available in the cell, new complementary bases can link with each separated strand, and two double helixes result. If the sequence of bases were AGATC on one existing strand, the new strand would contain the complementary, or 'mirror image', sequence TCTAG. In nature, the DNA backbone is tightly coiled up. Genes are sections of DNA that are used for coding information for the synthesis of proteins and other molecules. The information storage capacity of DNA is enormous. As noted in Section 9.7, it has been estimated that 1 kg of DNA can store all the world's data.

Translation of the genetic code from base sequences to amino acid sequences does not occur in a single step. This is the role of the other nucleic acid, RNA, which is a molecule that links the two worlds of DNA and proteins. DNA's letter Ts are made from RNA's letter Us. RNA makes up about 5–10% of the total weight of the cell. There are three major types of ribonucleic acids, messenger RNA (mRNA), ribosomal RNA (rRNA), and transfer RNA (tRNA). Messenger RNA contains only

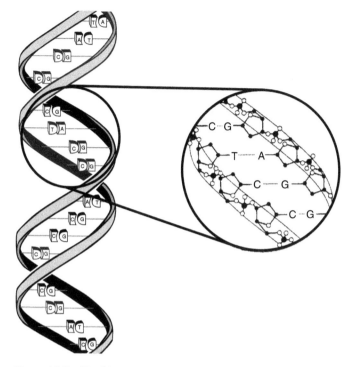

Figure 12.7 The DNA double helix. The two strands are held together by hydrogen bonding between complementary base pairs (C to G; T to A).

four bases A, G, C, and U. It is enzymatically synthesized in the cell nucleus in such a way that the base sequence of the mRNA molecule is complementary to the base sequence of one of the strands of the DNA molecule. Every mRNA molecule carries the code for one or more protein molecules. Transfer RNAs are relatively small molecules that act as carriers of specific amino acids during protein synthesis. Each of the 20 amino acids found in proteins has one or more corresponding tRNAs. The tRNA molecule may exist in its free form or attached to its specific amino acid. Finally, ribosomal RNA is the most abundant type of RNA. It plays an important role in the structure and biological function of ribosomes and constitutes up to 65% of their weight. Ribosomes are complexes of RNA and proteins, and undertake the job of translating DNA 'recipes' into proteins.

The information within individual genes resides in a coded form. The chemical structure of a protein can be written down in terms of its constituent amino acids. DNA can be represented as a sequence of base pairs linked by the sugar and phosphate components of nucleotides, which are simply part of the scaffolding. The sequence of base pairs within a gene on a DNA molecule can then represent a protein molecule in coded form. The parts of the gene sequence that are expressed in the protein are called exons, while the parts of the gene sequence that are not expressed in the protein are called introns.

A cell's full complement of DNA is called its genome. The average human cell is around 20 μm in diameter. Most of the DNA is stored in its nucleus, which is even smaller, 2–1 μm in diameter. Despite its small size, a nucleated human cell manages to include around 3 m of DNA. The generic information does not reside on one large single strand of DNA but is divided up into several sections, each of which is contained in a structure called a chromosome. In the case of the human genome, Ridley makes the analogy with a book [16]. If one supposes that the human genome is a book, then there are 23 chapters, called chromosomes. Each chapter contains several thousand stories, the genes. Each story is made up of paragraphs, called exons, which are interrupted by advertisements known as introns. A paragraph is made up of words called codons and, finally, every word is written in letters called bases.

There are 20 different amino acids that occur in proteins, but only four DNA bases. So a complete DNA-protein code can be established by taking groups of bases, the codons, to represent each amino acid. A code based on 'one base to one amino acid' clearly will not work. A DNA code in which the characters are groups of just two bases will only provide $4 \times 4 = 16$ different elements – insufficient to encode all the amino acids. But with groups of three bases, there are 64 possible characters, which is more than enough. The DNA-protein code must therefore use at least three bases to represent each amino acid.

All living organisms employ the same code. Because there is some redundancy in the code – there are 64 different base triplets available to represent 20 amino acids – some amino acids are encoded by more than one triplet. Furthermore, some triplets do not represent amino acids at all, but are control codes, which signify the end of the protein-coding sequence in a gene [13]. For example, the amino acid alanine can be represented by the codons GCU, GCC, GCA, or GCG, while lysine can be represented by AAA or AAG (Problem 12.1).

12.3.3 ATP, ADP

Adenosine triphosphate (ATP) is a nucleotide that performs many essential roles in the cell. This molecule may be considered as the energy 'currency' of life. The ATP molecule is composed of three components, as shown in Figure 12.8. At its centre is a sugar molecule, ribose (the same sugar that forms the basis of DNA). Attached to one side of this is the base adenine. The other side of the sugar is attached to a number of phosphate groups. ATP is remarkable for its ability to enter many coupled reactions, and to extract and provide energy. In animal systems, ATP is synthesized in the tiny energy factories called mitochondria.

When the third phosphate group of ATP is removed by hydrolysis, a substantial amount of free energy is released, the exact amount depending on the conditions:

$$ATP + H_2O \rightarrow ADP + phosphate \tag{12.1}$$

where ADP is adenosine diphosphate. The conversion from ATP to ADP is a key reaction for the energy supply for life processes. The process is facilitated by an enzyme ATPase. Living things can extract energy from ATP like a battery. The ATP can power reactions by losing one of its phosphate groups to form ADP, but then food energy in the mitochondria can be used to convert the ADP back to ATP, 'recharging' the battery (via another enzyme ATP synthase). In plants, sunlight can

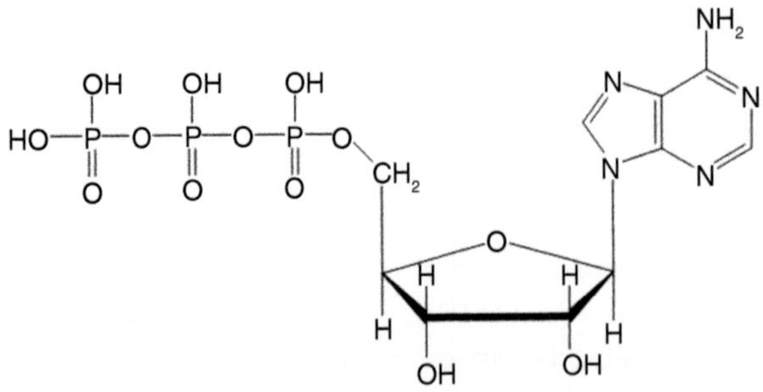

Figure 12.8 Chemical structure of adenosine triphosphate (ATP).

be used to convert the less active compound back to the highly energetic form (Section 12.7). For animals, the energy from high-energy storage molecules is used to maintain life; these are then recharged to their high-energy state. The oxidation of glucose operates in a cycle called the Krebs cycle in animal cells to provide energy for the conversion of ADP to ATP.

12.3.4 The Biological Membrane and Ion Transport

Membranes are amongst the most important biological structures. Many of the key functions of living systems, e.g. the ability to maintain steady-state conditions, are directly linked to the existence of shell-like entities, or membranes, surrounding the cells. Biological membranes have associated potential gradients, which regulate these processes. In 1972, Singer and Nicolson proposed the now widely accepted fluid mosaic model of their structure [13, 17, 18]. Phospholipids, or lipids, constitute about 50% of the mass of most animal cell membranes, with the remainder nearly all protein. Lipids are more complex versions of long-chain fatty acid molecules. Figure 12.9 contrasts the chemical structures of palmitic acid, a long-chain fatty acid (Figure 12.9a) with a typical phospholipid (Figure 12.9b). Both molecules are amphiphilic (or amphipathic), that is they possess a hydrophilic (water-loving) or polar head and a hydrophobic (water-hating) or non-polar tail. Phospholipids contain two non-polar hydrocarbon chains, one of which usually includes *cis* double bonds, which provide a kink(s) in the long chain [13]. The phospholipids are arranged in bilayers in biological membranes. These are dynamic, fluid structures and most of the molecules

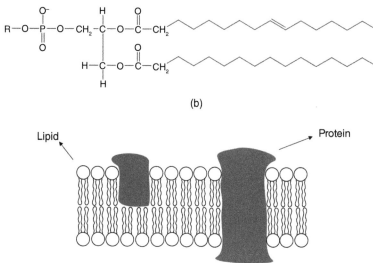

Figure 12.9 (a) Chemical structure for palmitic acid, a long-chain fatty acid. (b) General structure of a phospholipid. (c) Model of a biological membrane showing protein molecules embedded in and traversing a lipid bilayer.

can move around in the plane of the membrane. In contrast, movement from one monolayer to the other is much more restricted. Lipids in cell membranes are therefore more likely to be in a liquid-crystalline rather than a condensed solid state [9]. The hydrophilic polar groups associated with a phospholipid bilayer are on the outside, in contact with the aqueous media, while the hydrocarbon chains are in the interior. The biomembrane model also proposes that integral membrane proteins are embedded in the bilayer. Some of these proteins extend all the way through the bilayer, whilst others are only partially across it. Figure 12.9c depicts the fluid mosaic biological membrane model. Artificial hydrocarbon-layered structures, based on fatty acids, phospholipids, and other materials, and which exhibit some resemblance to biological membranes, can be built-up on solid surfaces using the Langmuir–Blodgett (LB) technique [19]. Protein molecule may also be incorporated in, or bound to, such thin films. These multilayer films invariably exhibit a condensed solid-state morphology rather than the fluidity of their naturally occurring biological counterparts.

The biological bilayer can be symmetric in terms of the polar head groups, but, more commonly, it exhibits asymmetry. For the red blood cell, the lipids sphingomyelin and phosphatidylcholine are disproportionately located in the outward-facing monolayer, while phosphatidylserine and phosphatidylethanolamine are mainly in the monolayer facing the inside of the cell. Lipid asymmetry is functionally important. Animals exploit the phospholipid asymmetry to distinguish between live and dead cells. Most membranes are electrically polarized with the negative inside the cell (typically −60 mV). Asymmetry in artificial membranes, such as LB films, will give rise to piezoelectric and pyroelectric behaviour (Section 7.2.3). However, pyroelectric coefficients found in alternating multilayer films of phospholipids are modest [20], and it is unclear if these are relevant in physiological processes.

Biological membranes are highly selective permeability barriers. The movement of molecules across the lipid bilayer consists of transfer from one aqueous environment to another and is restricted to solute molecules and water, illustrated in Figure 12.10. Generally, the smaller and less polar the molecule, the more easily it passes through the bilayer. Water is an exception; the small size of H_2O molecules, which offsets their large polarity, leads to a very rapid exchange of water across the bilayer structure. Gases, such as oxygen and carbon dioxide, important in cell metabolism, pass in or out of the cell in a dissolved state, and the rates of transfer are determined by the extent to which the gases are soluble in the aqueous environment. Carbon dioxide is very soluble in water and therefore passes freely through membranes. In contrast, oxygen is less soluble, and this becomes a limiting factor in cellular metabolism.

The existence of a concentration gradient of solute molecules across a membrane tends to cause a net movement of solute molecules in the direction of this concentration gradient. Transport occurs in both directions and the net flux is the sum of these two movements. In the simplest case, the rate of flow, the flux J ($mol\ m^{-2}\ s^{-1}$) of uncharged molecules in the direction of the gradient can be described by Fick's Law of diffusion (Eq. (2.58)).

Movements of ionized solutes are also influenced by electrical gradients. The flow of solute is given by the Nernst–Planck equation:

$$J = -\mu c \left(\frac{k_B T}{ce} \frac{dc}{dx} + z \frac{dV}{dx} \right) \tag{12.2}$$

where μ is the mobility of the ion ($m^2\ V^{-1}\ s^{-1}$), z is number of (electron) charges on the permeating molecule (its valence); dc/dx and dV/dx are gradients of the ion concentration and electrical potential across the membrane, respectively. The first term on the right-hand side of Eq. (12.2) is due to diffusion, while the second term originates from drift (electric field effect). The equation is very similar to those describing the currents due to electrons or holes under the influence of both an

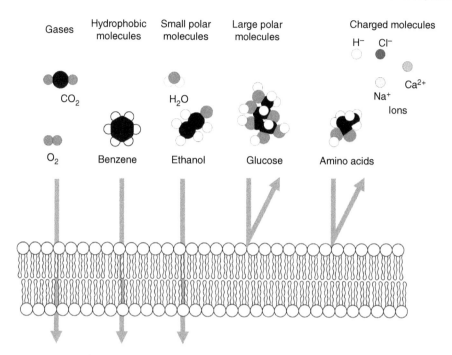

Figure 12.10 Schematic diagram showing the permeability of a lipid bilayer to different molecules. The smaller and less polar the molecule, the more easily it passes through the bilayer. Water is an exception.

electric field and a carrier concentration gradient (Eqs. (2.62) and (2.63)) as the underlying physics is the same. The mobility of the ion in Eq. (12.2) can also be expressed in terms of the diffusion coefficient, D, using the Einstein relation (Eq. (2.69)).

In equilibrium, when the net flux is zero, Eq. (12.2) can be rearranged:

$$\frac{dV}{dx} = -\frac{k_B T}{cez}\frac{dc}{dx} \tag{12.3}$$

Integrating, with respect to x across the width of the membrane:

$$V_i - V_o = V_m = -\frac{k_B T}{ez}\ln\left(\frac{c_o}{c_i}\right) \tag{12.4}$$

where V_i, V_o, and c_i, c_o refer to the potentials and ion concentrations, respectively, at the inner and outer surfaces of the membrane; and V_m is the voltage across the membrane or resting (equilibrium) potential. This expression is the Nernst equation and is often written in the format:

$$V_m = -2.30\frac{RT}{zF}\log_{10}\frac{c_o}{c_i} \tag{12.5}$$

where R is the universal gas constant ($=k_B N_A = 8.31 \times 10^3$ J kmol^{-1} K^{-1}), and F is Faraday's constant ($=eN_A = 9.65 \times 10^7$ C kmol^{-1}). An order of magnitude difference in the concentration of a univalent ion across a biological membrane will therefore result in potential difference magnitude of 58 mV at 20 °C (and 61.5 mV at 37 °C) between its inner and outer surfaces, with the inside negative with respect to the outside (Problem 12.2). The same figure is noted for the minimum sub-threshold swing of a field effect transistor (Section 8.2).

Osmosis is a special example of diffusion. This phenomenon is restricted to liquid media and is the movement of a solvent (normally water) through a semi-permeable membrane from a more

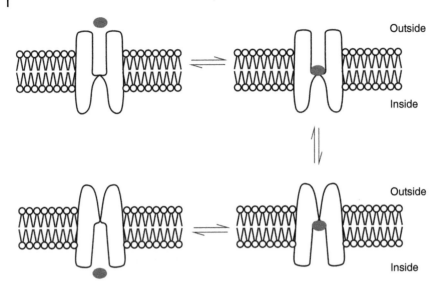

Figure 12.11 Mechanism for material transport across a bilayer. The non-covalent association of a molecule with the transporter protein triggers structural changes that result in transport to the other surface of the membrane. Only a small structural change may be required for this.

dilute solution to a more concentrated solution. A semi-permeable membrane is a barrier that permits the passage of some substances (by size) but not others. A cell membrane can be considered as being selectively permeable, i.e. the membrane 'chooses' what passes through. If a semipermeable membrane separates two solutions of different concentrations, then the solvent will tend to diffuse across the membrane from the less concentrated to the more concentrated solution. Osmosis is an essential process in the natural world; a good example is the absorption of water by plant roots.

Diffusion and osmosis are passive transport processes whereby ions or molecules, driven by thermal motion, move between solutions separated by biological membranes in living systems. In some instances, very rapid diffusion of material can take place across a membrane. Here, other constituents of the membrane, usually the proteins, play an important role. In a simple case, the transport (or carrier) protein possesses a specific binding site that recognizes the substance to be translocated. Non-covalent association of the substance triggers structural changes in the protein that effectively allow it to move to the other side of the membrane, illustrated in Figure 12.11. This process is facilitated diffusion. The degree of movement of the transported entity on the protein surface may be quite small – a few tenths of a nanometre – and it is probably not correct to envisage a permanent pore or hole through the protein. Where the flow is down a concentration gradient, no energy input is needed, but when the flow is up the gradient some form of energy is needed to produce the desired conformational state in the protein. Such 'uphill' transport systems are referred to as pumps; the energy input can be provided chemically by the downhill flow of another solute or by light. An example is the Na^+-K^+ pump, found in the plasma membrane of virtually all animal cells. This pump moves Na^+ out of the cell against its steep electrochemical gradient, while transporting K^+ in. This also plays a crucial role in regulating cell volume by controlling the solute concentration inside the cell.

Ion permeabilities of lipid bilayers and of natural membranes can be greatly increased by the incorporation of small molecules called ionophores. Such materials have been used as antibiotics (drugs that kill or slow the growth of bacteria). Many ionophores form stable complexes with cations. The nonpolar groups of the ionophore molecule are directed outward so that the ion becomes enclosed in a purse-like structure with a polar lining and a nonpolar exterior. One example is valinomycin, which is capable of selectively complexing with and transporting potassium ions across both biological and synthetic membranes. Complexation is associated with a change in conformation of the ionophore, facilitating this transport. Infrared spectroscopy has revealed that fatty acid LB films containing valinomycin exhibit structural changes on exposure to potassium containing solutions, suggesting the formation of a valinomycin-K^+ complex [21]. Dissociation of the complex did not occur, possibility the result of the condensed morphology of the fatty acid host matrix (compared to a biological membrane).

In other transport systems, ions moving by facilitated diffusion can traverse the cell membrane through channels created by proteins. These embedded transmembrane proteins allow the formation of a concentration gradient between the extracellular and intracellular contents. Ion channels are highly specific filters, even between ions of a similar character, e.g. Na^+ over K^+, allowing only desired ions through the membrane. For example, throughout the human body, sodium has a much higher concentration on the outside of cells (typically 10–20 times higher), while the reverse is true for potassium. The specificity of an ion channel is a well-researched topic. Over one hundred types of ion channels are known to exist [13]; a single neuron cell might typically contain 10 or more kinds of ion channels. Protein channels are highly efficient means of ion transport. Up to 10^8 ions s^{-1} can pass through one open channel (i.e. representing a maximum current of 16 pA for a monovalent ion), a rate that is 10^5 times greater than the fastest rate of transport mediated by any known carrier protein [13].

Ion channels are said to be 'gated' if they can be opened or closed. The main types of stimuli that are known to cause ion channels to open are a change of voltage across the membrane (voltage gated), a mechanical stress (mechanically gated), or the binding of a small signalling molecule or ligand (ligand gated). Some ion channels are gated by extracellular ligands and some by intracellular ligands. In both cases, the ligand is not the substance that is transported when the channel opens. Voltage gated channels are found in neurons (Section 12.3.6) and muscle cells; these open or close in response to voltage changes that occur across the membrane. For example, as an electrical impulse passes down a neuron, the reduction in the voltage opens sodium channels in the adjacent portion of the membrane. This allows the influx of Na^+ into the neuron and thus the continuation of the nerve impulse. Figure 12.12 shows an example of sodium and potassium ion channels embedded in a phospholipid bilayer. Here, Na^+ moves from the extracellular region of the cell to the intracellular region.

Proton movement through membranes deserves a special mention. Although protons resemble other positive ions, such as Na^+ and K^+, in their membrane transport properties, in some respects these are unique. In water, protons are highly mobile and travel through the hydrogen-bonded network of water molecules by rapidly dissociating from one molecule and then linking up with a neighbour. Protons are thought to move across a proton pump embedded in a lipid bilayer in a similar manner, transferring from one amino acid side chain to another.

Proton transport is intimately linked to electron transport. Whenever a molecule is reduced by acquiring an electron, it becomes negatively charged. In many cases, this charge is rapidly neutralized by the addition of a proton from water, so the net effect of the reduction is to transfer an entire hydrogen atom. Similarly, when a molecule is oxidized, a hydrogen atom removed from it can be readily dissociated into its constituent electron and proton – allowing the electron to be transferred separately to a molecule that accepts electrons, while the proton is passed to water. Therefore, in a membrane in which electrons are being passed along an electron-transport chain, pumping protons from one side of the membrane to the other can be relatively simple. The electron carrier merely needs to be arranged in the membrane in a manner that allows it to pick up a proton from one

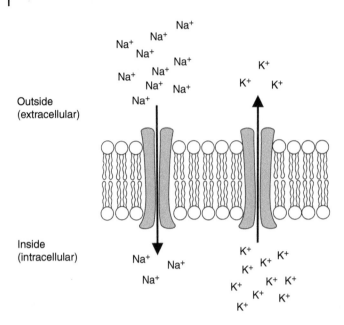

Figure 12.12 Example of the transport of sodium and potassium ions across a biological membrane facilitated by ion-selective channels in protein molecules. Sodium and potassium are shown to move in opposite directions in the example shown.

side when it accepts an electron, and release the proton on the other side of the membrane as the electron is passed to the next carrier molecule in the chain.

The electrical behaviour of a particular biological membrane can be modelled by an equivalent circuit consisting of basic electrical elements. For example, Figure 12.13 depicts the equivalent circuit of a two-ion (Na^+ and K^+) transport system. The bilayer membrane itself will have an associated capacitance, C_m. The membrane resting potentials V_{mK}, V_{mNa} are due to sodium and potassium ions, respectively, while G_K and G_{Na} are the variable conductances of the potassium and sodium channels, respectively.

The relatively small thickness of the biological membrane (6–10 nm) leads to large values of C_m. A figure for the membrane capacitance per unit area of $1\,\mu F\,cm^{-2}$ is typical. This is essentially the geometrical capacitance of a phospholipid bilayer and is little influenced by all the complexities of biology. In contrast, the values of conductance vary by many orders of magnitude across different cells and different transport systems. A pure phospholipid bilayer will be insulating, with a conductance per unit area as low as $10^{-17}\,S\,cm^{-2}$ [22]. Conductance values per unit area for potassium and sodium channels in human nerve cells have been given as 36 and $120\,mS\,cm^{-2}$, respectively [23]. The product of the cell's capacitance and resistance (or C/G) will determine its electrical response time [24].

12.3.5 Electron Transport

As noted earlier, nature favours ions as charge carriers as these come in many chemical forms and, in the biological environment, can carry more 'information' than electrons. The latter may be considered simply as point charges, whereas ions possess different sizes and charges. The previous section illustrates that biological membranes with their associated channel-forming proteins

Figure 12.13 Electrical equivalent circuit of the two-ion membrane transport system shown in Figure 12.12. C_m = membrane capacitance. V_{mK}, V_{mNa} are membrane resting potentials due to sodium and potassium ions, respectively. G_K and G_{Na} are the variable conductances of the potassium and sodium channels, respectively.

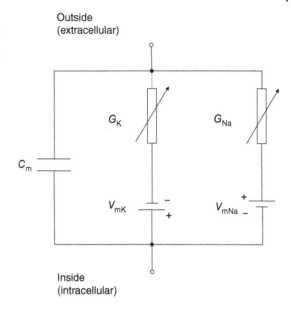

offer considerable scope for regulating biological processes by controlling the movement of ions. However, certain processes in nature also require the efficient transport of electrons.

In biochemical reactions, any electrons removed from one molecule are always passed to another, so that whenever one molecule is oxidized, another is reduced. The tendency for such redox reactions to proceed spontaneously will depend on the relative electron affinities of the two molecules. Very often this occurs in conjunction with, but separate from, the transport of protons. Most of the electron transport systems in biology are involved in some way with bioenergetics, the best characterized structurally being the photosynthetic system (Section 12.7).

Important biological molecules involved in redox processes are nicotinamide adenine dinucleotide (NAD) and its derivate nicotinamide adenine dinucleotide phosphate (NADP). Found in all living cells, NAD is called a dinucleotide because it consists of two nucleotides joined through their phosphate groups. One nucleotide contains an adenine nucleobase and the other nicotinamide. NAD exists in two forms: an oxidized and reduced form, abbreviated as NAD$^+$ and NADH (H for hydrogen), respectively. NADH and NAD$^+$ are redox pairs, since NADH is converted to NAD$^+$ by the loss of electrons in the reaction:

$$\text{NADH} \rightleftharpoons \text{NAD}^+ + \text{H}^+ + 2\text{e}^- \tag{12.6}$$

NADH is a strong electron donor because its electrons are held in a high-energy linkage. The transfer of electrons from NADH to O_2 is energetically very favourable, but most of the energy would be released as heat. Nature has evolved processes to make this reaction proceed more gradually, by transferring the high-energy electrons via many electron carriers in an electron-transport chain. The reaction $2\text{H}^+ + 2\text{e}^- + 1/2O_2 \rightarrow H_2O$ therefore occurs in many small steps. This enables almost half of the released energy to be stored, instead of being dissipated.

In metabolism, NAD is involved in redox reactions, carrying electrons from one reaction to another. These electron transfer reactions are the main function of NAD. However, the molecule is also used in other cellular processes, most notably as a substrate of enzymes in the addition or removal of chemical groups to or from proteins.

The cytochromes are a group of iron-containing, electron-transferring proteins that act sequentially in the transport of electrons. These contain iron porphyrin groups and resemble haemoglobin

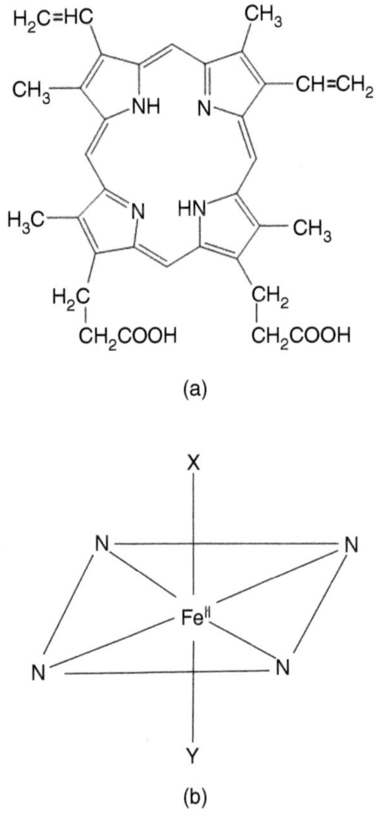

(a)

Figure 12.14 (a) Chemical structure of protoporphyrin IX. (b) Binding of an iron atom in cytochrome *c*. The four nitrogen atoms of the porphyrin ring bind to the iron in a planar arrangement. X and Y represent binding groups contributed by the protein.

(b)

and myoglobin; all are members of the class of heme proteins. Porphyrin rings usually absorb visible light (e.g. these are responsible for the red colour of blood) and change colour when they are oxidized or reduced. Porphyrins are very similar to the phthalocyanine molecules, which are stable organic semiconductors described elsewhere in this book (e.g. Section 1.5.2 and Section 6.6.1). These molecules are named and classified on the basis of their side-chain substituents. Protoporphyrin IX (Figure 12.14a) is the most abundant and contains four methyl groups, two vinyl groups, and two propionic acid groups. This is the porphyrin that is present in haemoglobin, myoglobin, and most of the cytochromes. Protoporphyrin forms very stable complexes with di- and trivalent metal ions. Such a complex of protoporphyrin with Fe(II) is called hemin, or hematin. The cytochromes undergo Fe(II) → Fe(III) valence changes during their function as electron carriers. In cytochrome *c*, the single iron-protoporphyrin group is covalently linked to the protein, as illustrated in Figure 12.14b. A related porphyrin is chlorophyll, which binds magnesium rather than iron, is a key molecule in the processes of photosynthesis (Section 12.7). This is responsible for the green colour of leaves.

12.3.6 Neurons

Neurons (also spelled neurones or called nerve cells) are the chemically and electrically excitable primary cells of the nervous system. A human brain contains around 10^{11} neurons. These register information from the environment, integrate and evaluate these data and then decide whether electrical signals are further transmitted. Many specialized types of neurons exist, and these differ

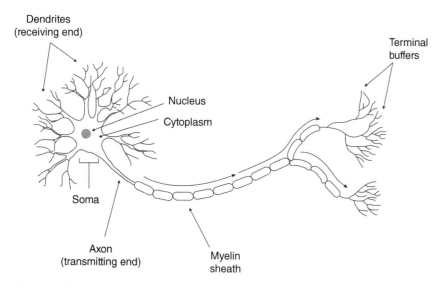

Figure 12.15 Schematic diagram showing the typical structure of a neuron. See text for details.

widely in appearance. Neurons are highly asymmetric, and a typical architecture is depicted in Figure 12.15. The soma, or cell-body, is the relatively large central part of the cell (on the left in the figure) between the dendrites and the axon. This is the metabolic centre of the cell and the site of protein synthesis and production of energy. The cell body gives rise to two kinds of cellular extensions: several short dendrites and a single long axon. The dendrites (or projections) branch out in a tree-like fashion and receive incoming signals from other neurons. Certain neurons in mammals have over 1000 dendrites each, enabling connections with tens of thousands of other cells. The axon is a much finer, wire-like part of the neuron, which may extend tens, hundreds, or even tens of thousands of times the diameter of the soma. This is the structure that carries nerve signals away from the neuron. Each neuron has only one axon, but this may undergo extensive branching and thereby enable communication with many target cells. The axon and dendrites are typically only about a 1 µm thick, while the soma is usually about 25 µm in diameter and not much larger than the cell nucleus it contains. The axon of a human motor neuron (or motoneuron) can be over 1 m long, reaching from the base of the spine to the toes.

Surrounding the axon is an electrically insulating layer called the myelin sheath. This is made up of protein and lipid. The purpose of the myelin sheath is to allow rapid and efficient transmission of impulses along the nerve cells. This occurs in two ways, by decreasing the capacitance of the nerve cells and by increasing their membrane resistance [25]. If the myelin is damaged, the impulses are disrupted. This can cause diseases like spinal cord injury, multiple sclerosis, and cerebral palsy.

As is the case with every cell in the body, neurons are surrounded by a phospholipid membrane, described above. At rest, neurons maintain a difference in the electrical potential on either side of the membrane, determined by ionic concentrations in the intracellular and extracellular regions (Section 12.3.4). Potassium and organic anions are typically found at higher concentration within the cell, whereas Na^+ and Cl^- are usually found outside the cell. The equilibrium potential is referred to as the resting potential. Receipt of a signal from an adjacent neuron can result in a positive current flowing into the cell, which subsequently depolarizes it. The resulting membrane potential is called an action potential. These are the direct consequences of voltage-gated cation channels and are brief ($\sim$1 ms duration) and relatively large amplitude ($\sim$100 mV) electrical pulses

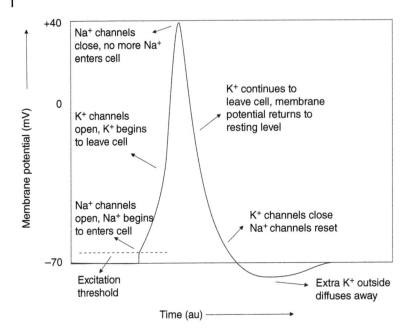

Figure 12.16 Biological membrane action potential – membrane potential versus time diagram. The action potential begins when voltage-gated sodium ion channels open due to an electrical stimulus. As the sodium ions enter the cell, the membrane potential becomes more positive. Once the cell has been depolarized, the voltage-gated sodium ion channels close. Potassium channels then open and K$^+$ moves out of the cell and the membrane potential falls. Typically, the repolarization overshoots the resting membrane potential, making the membrane potential more negative.

that make up the electrical signals by which the brain receives, analyses, and conveys information. The action potential is either present or not. While stimuli that do not reach a certain threshold value of the membrane potential produce no action potential, all stimuli above the threshold generate the same signal. The basic model for ionic processes involved in the generation of action potentials was formulated by John Eccles, Alan Hodgkin, and Andrew Huxley, for which they received the Nobel Prize for Medicine in 1963 [26].

The action potential is illustrated in Figure 12.16, in the form of a membrane potential versus time graph. At rest, typical values of a neuron membrane potential range from −55 to −70 mV. The stimulation of the cell partially opens protein channels in a patch of the membrane. Sodium diffuses into the cell, changing the potential in that region of the membrane towards a less-negative polarization. If this local potential reaches a critical state called the threshold potential (measuring about −60 mV), then sodium channels open completely. Sodium floods that part of the cell, which instantly depolarizes to an action potential of about +55 mV.

Following depolarization, the voltage-gated sodium ion channels close. The raised positive charge inside the cell causes potassium channels to open; K$^+$ ions now move down their electrochemical gradient. As the K$^+$ moves out of the cell, the membrane potential falls and starts to approach the resting potential. This repolarization typically overshoots the resting membrane potential, making the membrane potential more negative. This is known as hyperpolarization. When a neuron has just generated an action potential, it is unable to generate another within a certain time span, known as the refractory period. This phase lasts a few milliseconds and results from the time taken for the ion channels to recover. The refractory period therefore limits the number of action potentials that a given neuron can produce per unit time. As might be expected,

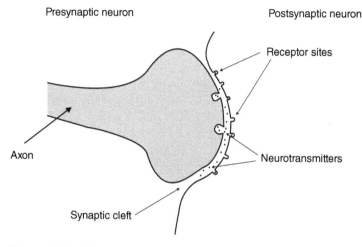

Presynaptic neuron

Postsynaptic neuron

Receptor sites

Axon

Neurotransmitters

Synaptic cleft

Figure 12.17 Schematic diagram of a synapse.

different types of neurons have different maximum rates of action potential firing due to the different natures and densities of ion channels. Many types of neurons emit action potentials constantly at rates of 10–$100\,s^{-1}$. However, some types are much 'quieter', and may go for minutes or longer without emitting any action potentials.

The above description of an action potential applies to one patch of the membrane. However, the self-amplifying depolarization of this region is sufficient to depolarize neighbouring areas of the membrane, which then go through the same cycle. The action potential spreads as a travelling wave from its initial site of depolarization (the base of the axon, called the axon hillock) and propagates without decay along the axon to its tip at a speed of up to $150\,m\,s^{-1}$. The action potential travels only one way along the stoma because of the Na^+ channel inactivation (refractory period). Near to its end, the axon divides into fine branches that make contact to neighbouring neurons. The point of contact between two communicating cells is called the synapse, shown schematically in Figure 12.17. The nerve cell transmitting the signal is called the presynaptic cell, while that receiving the signal is the postsynaptic cell. There are two types of synapses, chemical and electrical. In chemical synapses, action potentials that arrive at the presynaptic neuron trigger the release of neurotransmitter chemicals from the pre-synaptic to the post-synaptic neuron. These chemicals bind to specific receptors of the post-synaptic terminal and thereby modulate the voltage of the post-synaptic neuron. Neurotransmitters can either stimulate or suppress (inhibit) the electrical excitability of a target cell. An action potential will only be triggered in the target cell if neurotransmitter molecules acting on their post-synaptic receptors cause the cell to reach its threshold potential. In electrical synapses, the voltage of the post-synaptic neuron is modulated with ionic currents through the synapse when the action potentials arrive at the pre-synaptic neuron. For both electrical and chemical synapses, the connection strength between the pre- and post-synaptic neurons can be quantified and is known as the synaptic weight.

The shape of action potentials differs considerably among various types of neurons in the mammalian brain [26]. This results from the diversity of membrane channels, which allows neurons to encode information with a wide range of patterns and frequencies. Single neurons are essentially complex computational devices.

12.4 Computing Strategies

The processing of information has become an indispensable part of everyday life. A computer might be described simply as a device that automatically performs routine calculations. The earliest-known calculating device is probably the abacus, which dates back over 3 000 years. The abacus is a digital device; a bead is either in one pre-defined position or another, representing unambiguously, say, 1 or 0. Over the centuries, various mechanical calculating devices were built. A milestone was Charles Babbage's Difference Engine, developed in the 1820s and 1830s. This was more than a simple calculator. It mechanized not just a single calculation but also a whole series of computations on several variables to solve a complex problem. Like modern computers, the Difference Engine possessed a means to store data, and it was designed to stamp its output into soft metal, which could later be used to produce a printing plate.

It is generally believed that the first electronic digital computers were the Colossus (UK) and Electronic Numerical Integrator and Computer (ENIAC, USA), both built in the 1940s. Designed specifically for computing values for artillery range tables, ENIAC occupied a space of $15 \text{ m} \times 9 \text{ m}$ and contained more than 17 000 valves (vacuum tubes). The computer ran continuously generating 174 kW of heat. It could execute up to 5000 additions per second, several orders of magnitude faster than its electromechanical predecessors. The invention of the junction transistor in 1947 and the integrated circuit about a decade later paved the way for the microprocessor and the compact computing architectures of today [9]. The performance of computers can be quantified using several parameters. A simple measure is the number of instructions that the computer processor carries out per second (IPS). In respect of the ability to perform arithmetic operations, an appropriate factor is the FLOPS, an acronym for floating-point operations per second. Supercomputers can perform well over 10^{17} FLOPS, or 100 PetaFLOPS (PFLOPS).

12.4.1 Von Neumann Computer

Most computers use the stored-program concept designed by John von Neumann. The architecture, described in 1945, is a model for a computing machine that uses a single storage structure to hold both the set of instructions on how to perform the computation and the data required or generated by the computation. The essential features are a processing unit containing an arithmetic logic unit (ALU) and registers, a control unit with an instruction register, memory to store both data and instructions, external mass storage, and input and output mechanisms. A schematic diagram is shown in Figure 12.18. Such machines are also known as stored-program computers. The separation of storage from the processing unit is implicit in this model. Programs and data are contained in a slow-to-access storage medium (such as a hard disk) and work on these is undertaken in a fast-access, volatile storage medium (Section 9.2); the programs and data share the same memory. A program for a von Neumann machine consists of a set of instructions that are examined sequentially; a program counter in the control unit indicates the next location in the memory from which an instruction is to be taken. The data on which the program operates may include variables. Storage locations can be named so that the saved value can subsequently be referenced or changed during execution of the program.

The key advantage of the von Neumann architecture is its simplicity. The stored program concept was a considerable improvement over previous computing systems that were essentially hard-wired to perform a specific task. However, drawbacks are that memory for instructions and data are unified and shared between processor and memory, with one data bus (set of connection lines) and

Figure 12.18 Von Neumann computer architecture.

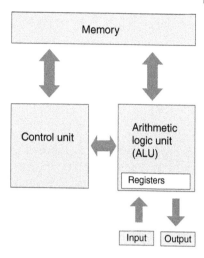

one address bus. The instructions and data must be fetched in sequential order, which limits the bandwidth for memory accesses.

The von Neumann computer has some fundamental limitations concerning its operating speed and switching energy, which are common to most computing models. As assertion, put forward by von Neumann in a 1949 lecture, is that a computer must dissipate an energy per reversible operation (or per bit in a binary computer), ΔE, given by

$$\Delta E = k_{\mathrm{B}} T \ln 2 \tag{12.7}$$

which is approximately 3×10^{-21} J at room temperature. A further restriction for a binary computer element is that its switching energy must be greater than the thermal energy, otherwise, the device will turn on and off randomly. This requires

$$\Delta E > k_{\mathrm{B}} T \tag{12.8}$$

At room temperature, $k_{\mathrm{B}} T = 4 \times 10^{-21}$ J, a similar figure to that provided by Eq. (12.7).

Quantum mechanics provides a further constraint. The Heisenberg Uncertainty Principle (Section 2.5) links the energy associated with a physical measurement (and the change of the bit state of a switch may be considered as a measurement) and time:

$$\Delta E \Delta t \geq \frac{\hbar}{2} \tag{12.9}$$

This equation predicts a minimum energy dissipated in a nanosecond switching device (i.e. operating at 1 GHz) is of the order of 10^{-25} J, many orders of magnitude below the actual switching energies of transistors in computer processors.

12.4.2 Biological Information Processing

The brain (nature's computer) utilizes parallel processing, instead of the serial approach in von Neumann systems. This means that the brain can send a signal to hundreds of thousands of other neurons in less than 20 ms, even though it takes a million times longer to send a signal than a semiconductor computer switch. Neural logic may be either analogue or digital. In the latter form, the neuron is designed to respond to the sum of its N inputs (E of which may be excitory and I, which may be inhibitory). Figure 12.19 shows a schematic diagram of such a gate. Provided that the

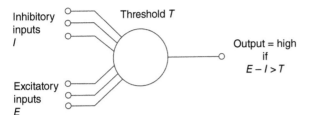

Inhibitory inputs *I*

Threshold *T*

Excitatory inputs *E*

Output = high
if
$E - I > T$

Figure 12.19 Representation of neural logic. An output from the neuron is produced when the difference between the excitatory inputs *E* and inhibitory inputs *I* exceeds a threshold value *T*.

sum exceeds a given threshold T ($E - I > T$) the neuron will output logic 1, otherwise, it produces logic 0.

The brain works mainly by non-linear computation using the rate of pulse production by a neuron as the information signal being sent to another cell. There are about 10^{11} neurons in the human brain, and each is connected to 10^3–10^4 others. This gives a crude 'bit-count' of 10^{11}–10^{15}. An equivalent artificial 'brain' might therefore be built from $10^5 \times 8$ Gbit chips, with a power dissipation of many MW. It is important to recognize that information is stored in the brain in very different manner to that in a computer [27]. For example, long-term memories are not located in just one part of the brain but are widely distributed throughout the cortex (brain's information processing centre). Such memories are stored as groups of neurons, which are promoted to fire together in the same pattern that created the original experience. It is difficult to compare directly the performance of the human brain with that of a supercomputer [28, 29]. If a calculation is simply based on the firing of neurons, then 10^{18} FLOPS (1 exaFLOPS) may be needed to emulate the computing operations (Problem 12.3). However, neurons are not strictly binary. The firing of a single neuron will depend on many factors (e.g. type of neuron, neuron connectivity, and nature of the neurotransmitters). For a simulation at the metabolic level (considering the concentrations of neurotransmitters in the various biological compartments), it is possible that up to 10^{25} FLOPS (10 yottaFLOPS) might actually be required [28].

12.4.3 Artificial Neural Networks

Artificial neural networks (ANNs) can be developed on traditional computer architectures and are the basis of machine learning and artificial intelligence [30]. The simplest form of an ANN is a unit called a perceptron, which consists of one artificial neuron, as illustrated in the schematic diagram of Figure 12.20. This single-layer ANN consists of four main parts: inputs (fixed and variable), weights, a weighted sum, and an activation function. The weights might be considered to be the synthetic equivalents of the synaptic weights that describe natural neuron connectivity. The computation process begins by taking all the input values, x_i, and multiplying these by their weights, w_i. Subsequently, the values obtained are added together to create the weighted sum. The weighted sum is then applied to the activation function, f, producing the perceptron's output, y, which can be described mathematically:

$$y = f\left(\sum_i w_i x_i\right) \tag{12.10}$$

The activation function is used to map the input between required values such as (0, 1) or (−1, 1). An example of f is a step function, e.g. $f(x) = 0$ for $x < 0$ and $f(x) = 1$ for $x \geq 0$. The weight of an input is indicative of the strength of a node. Similarly, an input's bias value (Figure 12.19) gives the ability to shift the activation function curve up or down. A perceptron is usually used to group data into two classes. Therefore, it is also known as a linear binary classifier (or linear threshold unit). More

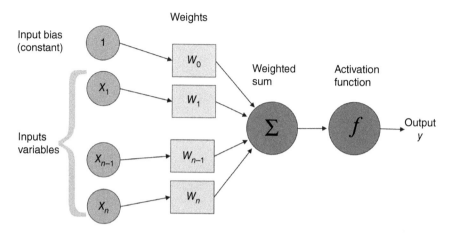

Figure 12.20 Schematic diagram of perceptron – a simple artificial neuron – showing inputs x_i and output y. Weights associated with the inputs are designated w_i.

complex ANNs consist of multiple artificial neurons arranged in several layers. The technique of back-propagation is the essence of neural network training. This involves fine-tuning the weights of a neural net based on the error rate obtained in the previous iteration. ANNs have widespread use in many disciplines, including data mining, object/pattern recognition (e.g. in devices such as the electronic nose, as described later in Section 12.9), autonomous driving and decision-making. However, these systems remain limited with respect to biological intelligence. This stems from the fact that ANNs are a very simple (and abstract) representation of the mammalian nervous system.

Spiking neural networks (SNNs) aim to bridge the gap between neuroscience and machine learning by using biologically realistic models of neurons to carry out computation. An SNN is fundamentally different from the ANN described above as the model incorporates the concept of time. The artificial neurons in the SNN do not transmit a voltage response at each propagation cycle (as in the case of with typical multilayer perceptron networks) but rather transmit information, in the form of spikes, only when a 'membrane' potential reaches a threshold. The occurrence of a spike is determined by differential equations that represent various biological processes, the most important of which is the membrane potential of the neuron. When a neuron reaches its threshold potential it spikes, and the potential of that neuron is reset (as for a natural membrane). Computing networks using SNNs can be developed using both software and hardware approaches and is generally referred to as neuromorphic computing. Prominent examples of neuromorphic computers using contemporary silicon technology include IBM's TrueNorth and Intel's Loihi chip [31, 32].

Most computers can be described as finite-state machines [12, 33, 34]. This is a model of computation based on a hypothetical machine comprising of one or more states. Only a single state can be active at a time, so the machine must make a transition from one state to another in order to perform different actions. A cellular automaton is a regular array of identical finite-state machines, each connected to a fixed number of neighbours using the same interconnection net. Von Neumann first investigated such systems as models for completely discrete physical dynamical systems such as the brain. Figure 12.21 shows an example of a two-dimensional cellular automaton layout. It consists of a regular spatial array of identical cells, and each characterized by a finite discrete-valued state function. At successive discrete intervals of time, each cell makes a transition to a new state which depends on the previous state values of the connected neighbours.

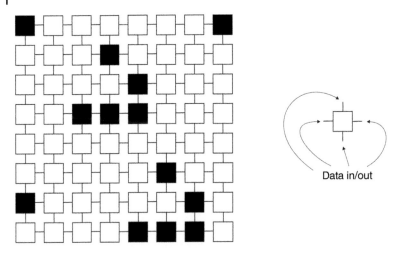

Figure 12.21 A cellular automaton. An individual cell, shown on the right, is a finite-state machine.

The best-known example of a cellular automaton is the Game of Life, devised by the mathematician John Conway [see, for example [35]]. This is played on a field of cells, as in Figure 12.21, each of which has eight neighbours (adjacent cells). In the standard Game of Life, a cell is 'born' if it has exactly three neighbours, stays alive (survival) if it has two or three alive neighbours but otherwise dies. Ever since its publication, the game has attracted considerable fascination with the surprising ways the patterns can evolve. I recommend readers to obtain a copy of the Game of Life (widely available to download) and have fun exploring the variety of patterns that can be produced.

12.4.4 Organic Neuromorphic Devices

Since the 1980s, neuromorphic digital computers have been built using silicon technology using field-programmable gate arrays (FPGAs) as an appropriate and cost-effective technology platform. Interest is now focused on dedicated circuitry fabricated from non-CMOS technologies such as the nonlinear resistive memory elements described in Section 9.3 [36–38]. A range of neuromorphic functions has been demonstrated with organic devices exploiting electrochemical, electronic and ferroelectric phenomena [37].

An example of an organic artificial synapse is shown in Figure 12.22 [39, 40]. This depicts an electrochemical device in the form of a field effect transistor, but one operating on rather different principles to the structures described in Chapters 8 and 9. The device is fabricated by sandwiching a layer of Nafion (a perfluorinated polymer electrolyte) between two poly(3,4-ethylene dioxythiophene):polystyrene sulfonate (PEDOT:PSS) films; one of these is permeated with polyethylenimine (PEI). On application of a voltage pulse to the PEDOT:PSS electrode (the gate in Figure 12.22), electrons are injected into (or removed from) this film, which, in order to keep its electroneutrality, absorbs cations from (or releases these to) the Nafion electrolyte. Such changes are reflected in the proton concentration in the PEDOT:PSS/PEI layer (the channel) and in the reduction/oxidation state of PEDOT, which ultimately determines the channel conductance. The introduction of the gate electrode decouples the write and read operations to achieve both switching at low energy (projected to be lower than a biological synapse following device scaling to the nanometre scale) and a long retention time (25 hours).

The artificial synapse depicted in Figure 12.22 exhibits many non-volatile and reproducible states (>500) and operates at small switching voltages (≈0.5 mV) and with low power consumption

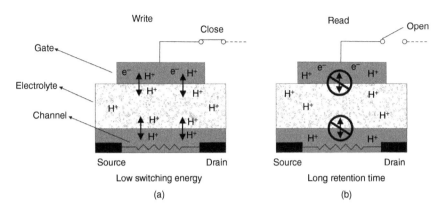

Figure 12.22 Schematic depiction on an artificial synapse. During the Write operation (a), an external switch is closed, allowing electrons (e$^-$) to flow in and out of the gate. These charge carriers trigger proton (H$^+$) exchange between the gate, electrolyte and channel, resulting in channel conductance changes between the source and drain. During the Read operation (b), the external switch is opened to prevent electron flow. As a result, the protons in each layer are frozen, leading to a stable channel conductance with a long retention time. Source: Yang and Xia [39].

($<10\,$pJ for $10^3\,\mu$m^2 cells). Electrochemical transistors can be combined into circuits that emulate Boolean and reversible logic gates [38]. Although such devices are promising candidates for computation devices, technical challenges such as reliability, device-to-device variation, and large-scale integration still hinder the transfer of these emerging neuromorphic computing technologies to the domain of consumer electronics.

Another approach to organic neuromorphic computing is to manipulate directly biological neurons that have been immobilized on support structures. A means to achieve this is to exploit stem cell technology, whereby neurons can be grown directly onto nanoprinted microscaffolds [41]. This so-called brain-on-a-chip technology might mimic the complex interconnectivity of the human brain and offer new approaches in human disease research. A much longer-term strategy is to integrate these biological architectures with silicon MOSFETs to achieve hybrid computational structures.

12.4.5 DNA and Microtubule Electronics

The study of the electronic behaviour of organic compounds has led unsurprisingly to interest in the electrical properties of biological materials. DNA, described in Section 12.3.2, is arguably the most significant molecule in nature. It can be considered as a potential candidate for nanowires and may also be an important material for molecular electronics applications. Reports into the electronic properties of DNA have already generated controversy in the literature. According to some, DNA is a molecular wire of very small resistance. Others, however, find that DNA behaves as an insulator or semiconductor [42–48]. These seemingly contradictory findings can probably be explained by the different experimental conditions used to monitor the conductivity [42]. The preservation of the double helix structure during the electrical measurements is a key issue. Methods to achieve this have included the attachment of anchoring chemical groups such as fullerenes to secure the integrity of the DNA structure [46, 48]. DNA in its natural hydrated state and in a dried form will almost certainly exhibit different conductivities (not least because of the ionic contribution to the conductivity). The DC resistivity of the DNA double helix over long length scales ($<10\,\mu$m) is very high ($\rho > 10^6\,\Omega$cm). However, an appreciable AC conductivity can arise from the polarization of the surrounding water molecules [44]. Computer simulations indicate that charge transport along the long axis of the molecule is strongly dependent on DNA's instantaneous conformation, varying over many orders of magnitude. It is also suggested that the charge transport can be active over longer length scales than the commonly accepted 2–3 base pairs [46], giving rise to a hopping conductivity process (Section 6.4) along the helix axis [48].

Chemical or biological reactions can also be exploited to perform computing operations [34]. For example, if a string of DNA is assembled in the right sequence, it can be used to solve combinatorial problems. The 'calculations' are performed in test tubes filled with strands of DNA. Gene sequencing is used to obtain the result. Solutions to the 'travelling salesman' and other challenges demonstrate the power of this technique [49]. Further examples include the application of a DNA-mediated multitasking processor to route planning and an analogue processor that can add, subtract and multiply [50–52]. The parallel processing offered by DNA potentially provides 10^{14} MIPS (millions of instructions per second) and uses less energy and space than conventional supercomputers. While CMOS supercomputers operate at about 10^9 operations J^{-1}, a DNA computer could perform about 10^{19} operations J^{-1}. The von Neumann restriction (Section 12.4.1) predicts a limit of about 3×10^{20} operations J^{-1}. Data could potentially be stored on DNA in a density of approximately 1 bit nm^{-3} while existing storage media such as DRAMs require 10^{12} nm^3 to store 1 bit.

Microtubules (MTs) are other cylindrically shaped biological structures that have attracted interest for information processing. These are polymers of the protein tubulin that form part of the cytoskeleton, a structure that helps cells maintain their shape and internal organization [53–57]. The MTs typically consist of 13 linear rows of tubulin dimers, known as protofilaments, which together form long hollow pseudo-helical cylinders with internal and external diameters of about 16 and 26 nm, respectively, depending on the precise number of protofilaments, and lengths commonly in the range 5–10 μm. Each protofilament within the MT comprises end-to-end negatively charged α- and β-tubulin dimers. A strong electric dipole is present along the MT axis and this can be aligned by an applied electric field. Although the structural functions of MTs in some physiological processes have been widely accepted (e.g. cell mitosis, cell motility and motor protein transport), their precise role in others (e.g. neuron activity) is unclear.

Some intriguing experimental data have been published on the electrical behaviour of MTs. For example, it has been demonstrated that MTs can work as biomolecular transistors capable of amplifying electrical information, as well as functioning as nonlinear electrical transmission lines that generate spontaneous electrical oscillations by changes in electric field and/or ionic gradients [53, 54]. The voltage-dependent electrical resistance observed for MTs also suggests that these resemble memristor devices (Section 9.3), offering scope for use as memory structures [55].

12.4.6 Quantum Computing

The unit of conventional information is the bit, which takes one of the two possible values, 0 or 1. Any amount of information can be expressed as a sequence of bits. A classical computer executes a series of simple operations with gates (e.g. AND gate in Figure 12.1), each of which acts on a number of inputs. By executing many gates in succession, the computer can perform complex calculations.

A qubit (or quantum bit) is the quantum mechanical analogue of a classical bit. A qubit is a two-level quantum system, such as the spin of an electron, where the two basis qubit states are usually written as $|0\rangle$ and $|1\rangle$. A qubit can be in state $|0\rangle$, $|1\rangle$ or, unlike a classical bit, in a linear combination of both states. The essential difference between a bit and a qubit is that whereas in a classical computer a bit of information will encode either a 0 or a 1, the nature of quantum

Figure 12.23 Representation of classical and quantum bits. Left: classical digital bits can occupy one of two states, 0 and 1. Right: qubits (quantum bits) are the quantum superpositions of two states, |0⟩ and |1⟩ on the surface of a Bloch sphere.

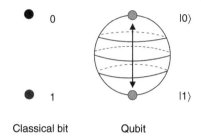

Classical bit   Qubit

mechanics allows the qubit to be in a superposition of both states at the same time. The state of a qubit is not known until it is measured.

The concept behind quantum computing is to use an array of such quantum 'particles' to perform mathematical operations [58–61]. In theory, a quantum computer would process all the states of a qubit at the same time, and with every qubit added to its memory size its computational power increases exponentially. For instance, eight bits are required for a classical computer to represent any number between 0 and 255. But eight qubits are sufficient for a quantum computer to represent every number between 0 and 255 at the same time. It does not take a lot of qubits to quickly surpass the performance of the most powerful supercomputers.

Figure 12.23 contrasts the classical bit with the qubit. A classical bit possesses just two states, 1 and 0, as indicated on the left of Figure 12.23. Qubits are conveniently represented using a Bloch sphere, which is a unit-radius sphere where the top half represents binary digit 0, and the bottom half represents binary digit 1, as shown on the right of Figure 12.23 [59]. Within the Bloch sphere, there is a vector originating from its centre and pointing outwards in any direction. If the vector points above the horizon of the sphere, the qubit represents a |0⟩; if it points below the horizon, the qubit represents a |1⟩. When a qubit is measured, the direction of the vector will determine its state.

A further property that distinguishes quantum computers from conventional computers is entanglement. If two qubits are 'entangled' there is a correlation between these two qubits. If one qubit is in one state, the other has to be in a different state. If two electrons become entangled, their spin states are correlated such that if one of the electrons has a spin-up, then the other one has a spin-down after measurement. The creation and manipulation of entangled states plays a central role in quantum information processing.

Qubits can be realized using different solid-state materials, including semiconducting quantum dots and superconductors. Semiconductor spin qubits are popular for several reasons. These possess a relatively long lifetime before decohering (interacting with their environment). This type of qubit can also be controlled electrically and be integrated at a high density on a chip. In 2019, the company Google published data claiming 'quantum supremacy' for a 53-qubit computer, named Sycamore, using superconducting qubits operating at a temperature of 15 mK [61]. The machine was able to perform a target computation in 200 seconds, which it was asserted would take the world's fastest supercomputer 10 000 years to produce a similar output.

Despite such remarkable progress, quantum computers are exceedingly difficult to engineer, build and program. One issue is the low temperature of operation. Other problems are the errors resulting from noise, manufacturing defects and loss of quantum coherence. There currently remains a gap between theory and practice that may take some years to resolve [62].

The manipulation of electron spins on organic molecules offers considerable promise for the development of organic qubits. Their structural variability and synthetic accessibility should offer potential for engineering organized molecular architectures suited to quantum computing.

Qubits have been formed from metal–organic frameworks of spin-paired nitroxides [63]. There has also been a suggestion that the microtubule structures described in the previous section might be exploited as qubits [56]. However, there are very few other reports on organic qubits and a quantum computer based on organic materials has yet to emerge.

The possible links between quantum mechanics and the biological world are topics of much fascination and have spawned the relatively new field of quantum biology [64]. Orchestrated objective reduction (Orch OR) is a biological theory, introduced by Stuart Hameroff and Roger Penrose, proposing that consciousness originates at the quantum level inside neurons, rather than the conventional view that it is a product of connections between neurons [65]. The idea is that computations in microtubules influence the firing of neurons and, by extension, consciousness. Intriguing as it is, there does not seem to be a simple way to test the Orch OR theory.

12.4.7 Evolvable Electronics

Living systems can achieve incredible feats of computation with remarkable speed and efficiency. Many of these tasks have not yet been adequately solved using algorithms running on the most powerful computers. Natural evolution is a bottom-up design process. Living systems assemble themselves from molecules and are extremely energetically efficient when compared to man-made computational devices. The technological drive to produce ever-smaller devices is leading to the construction of machines at the molecular level (Section 12.2). However, the basic computational paradigm is still von Neumann.

In a rather different approach to emulating nature, *evolution-in-materio* (EiM) uses computer-controlled evolution to create information processors [66–69]. Disordered material is trained via external stimuli, usually electric fields, to perform a specific computational task. The EiM process consists of two stages. In the first instance, training, the material is configured, using an iterative process, to perform a specific task. In the subsequent step, verification, the configured material is tested with unseen data. Most research to date has used the training phase to optimize the electrical connections to the material, as for example in FPGAs. Functions such as tone discriminators and maze solving can be realized within a liquid crystal display [66]. Moreover, the operation of logic gates and the solution to a variety of computational problems may be solved using carbon nanotube-polymer composites [70, 71]. There is also interest in exploiting memristors in unconventional computing applications [72]. All these materials/devices may be considered as 'static'. The application of electric field does not significantly perturb their morphology and the process of evolution is essentially that of optimization of the magnitudes and distribution of the applied voltages.

Dynamic EiM processors allow the morphology of the material to be altered during the training process. An example is that of a thin film composite in which single-walled carbon nanotubes (SWNT) are suspended in a liquid crystal (LC). The nanotubes act as a conductive network while the liquid crystal provides a host medium to allow the conductive network to reorganize when voltages are applied [73]. A schematic diagram of such an EiM processor is shown in Figure 12.24. The experimental arrangement allows input and configuration voltages to be applied to the computational material, which is in the form of a thin film deposited on a microelectrode array, and output currents to be measured. The morphology of the SWNT network will determine its initial electrical behaviour. 'Programming' is then achieved by an evolutionary algorithm that adjusts the nanotube configuration for a target application. The algorithm assesses the suitability of a population of candidate configurations for the task at hand. Each configuration represents a particular set of connections to the nanotube network and configuration voltage levels.

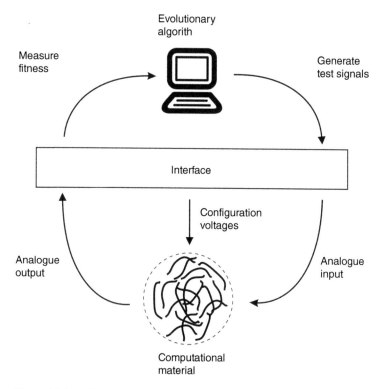

Figure 12.24 EiM (*evolution-in-materio*) processing system. Configuration voltages are applied to a material to train it to perform a particular task.

Figure 12.25 shows the results obtained with an EiM processor, based on a SWNT-LC blend, and working on a binary classification problem (i.e. deciding to which of two groupings the variables in a particular dataset belong) [74]. This problem has several applications within the domain of machine learning. The data in Figure 12.25 show the error (as a percentage) achieved for each iteration of the training phase for both a SWNT-LC sample and a control (set of reference electrodes with no computational material). The result for the control remained around 50% during the training, demonstrating that the system was unable to provide solutions to the classification problem better than a random (50:50) guess. In contrast, the error obtained using the nanotube material converged to zero with increasing iterations of the evolutionary algorithm. Furthermore, following the training phase, the application of the optimum configuration voltages allowed the material to act as a binary classifier with no ill-classified data points from the verification set [74]. The data used in this experiment were generated artificially, and intentionally simple. However, the EiM approach can also be applied to datasets obtained from real life issues, including medical problems [75]. Other approaches to dynamic EiM processors exploit processes in biological materials, including the microtubules described earlier [76].

Such computational systems are unlikely to compete with silicon-based processors for speed but may provide a complementary technology suited to particular tasks. One advantage of evolvable electronics is its potential ability to deal with damage occurring during operation. For example, if the binary classifier described above develops a fault (e.g. as the result of disruption of part of the nanotube network), then the system can be retrained, and new connections established. This might be time-consuming but provides a more effective solution for electronic systems that are

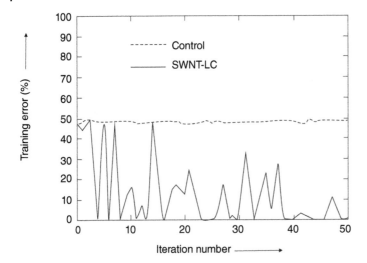

Figure 12.25 Training errors versus iteration number for the training phase of an evolutionary algorithm working on a binary classification problem. Data are shown for a control sample (dashed line) and for a carbon nanotube-liquid crystal blend (full line). Source: Vissol-Gaudin et al. [74].

embedded in hostile, or not easily accessible, environments. Such issues are explored further in the next section.

12.5 Fault Tolerance and Self Repair

A major obstacle associated with the manufacture of any molecular electronic device or system will almost certainly be the relatively poor fabrication yield resulting from problems such as the influence of background charge, difficulties of making reliable contacts and lithographic inaccuracies at the nanometre level. Although technical solutions to such challenges may be found, these will almost certainly increase the fabrication cost. Therefore, the search for an alternative to the expensive but reliable CMOS-based electronics in the form of a much less expensive but less robust technology should be accompanied by fault-tolerant schemes, permitting the functionality of the chips even if some of the individual devices do not work. An early example of such an approach was the 'Teramac' computer, which was designed and fabricated by the Hewlett Packard company in the late 1990s [77]. This system was based on a high redundancy of wires connecting 'unreliable' switching devices, so that even if some of the paths contained non-operational devices there was always one functional path. Once built, the system was checked by software and the operational paths identified. Consequently, the problem of expensive fabrication costs to make perfect devices was transferred to the time-consuming procedure of identifying the operational paths by software. The above scheme was mainly designed with wires and switching devices that permitted a reconfiguration of the system when a non-functional path is found.

Digital electronics, particularly communications networks, commonly includes a coding system that can detect and/or correct errors in a binary coded message. In every case, some redundancy is necessary – a similar strategy used by nature that resulted in the evolution of the two strands in the DNA helix. The simplest method of error detection is the addition of one extra binary digit, or bit, to a string of 1s and 0s – a parity bit [78]. Any group of bits contains an even or an odd number of 1s. A parity bit is attached to a string of bits to make the total number of 1s in the group always even

(even parity) or always odd (odd parity). A given digital electronics system works with even or odd parity, but not both. For example, for even parity, a check is made on each group of bits received to ensure that the total number of 1s in that group is even. If there is an odd number of 1s, an error has occurred. The parity bit can be attached to the code at either the beginning (most significant bit or MSB) or at the end (least significant bit or LSB), depending on the system design. There are two disadvantages of the simple parity bit approach to error detection. First, the system will fail if two errors (or an even number of errors) have occurred in the binary group. But, more importantly, there is no means to correct the error.

Hamming code is a more sophisticated coding system used in digital electronics systems that includes an error correcting provision. However, this requires the addition of more redundant information to the original data. If the number of data bits is d, then the number of parity bits p needed to locate and correct one error is determined using:

$$2^p \geq d + p + 1 \tag{12.11}$$

For example, with four data bits $p = 3$ satisfies Eq. (12.11), so three parity bits are required to provide single-error correction; for $d = 5$, $p = 4$. The positions and values of the parity bits are chosen so that they provide error checks on other digits in the binary message. The parity bits are in the positions corresponding to ascending powers of two. If the left-most bit is designated bit 1, this gives the locations of the parity, P_n, and data, D_n, bits as: $P_1, P_2, D_1, P_3, D_2, D_3, D_4$. If the data are represented by the 4-bit binary number 1001 (decimal 9), the locations of the data and binary bits are shown in Table 12.1. Each parity bit provides a check on certain other bits in the binary string [78]. The first parity bit checks the least significant positions of 1s in the binary set of numbers 1 to 10, i.e. positions 1, 3, 5, 7, etc. The second digit checks the next significant positions of 1s, i.e. 2, 3, 6, 7, etc. and so on. The procedure is summarized in Table 12.2. It should be noted that error

Table 12.1 Positions of parity bits, P_n, and data bits, D_n, for Hamming even parity code based on a four-bit binary message (1001).

Bit position	1	2	3	4	5	6	7
Bit Designation	P_1	P_2	D_1	P_3	D_2	D_3	D_4
Binary position number	001	010	011	100	101	110	111
Data bits (D_n)			1		0	0	1
Parity bits (P_n)	0	0		1			

The binary numbers corresponding to the bit positions are also shown, e.g. 001 (decimal 1) for bit number 1 and 111 (decimal 7) for bit number 7.

Table 12.2 Positions checked by each of the first four parity bits in Hamming code.

Check digit	Positions checked
First	1, 3, 5, 7, 9, ...
Second	2, 3, 6, 7, 10, ...
Third	4, 5, 6, 7, ...
Fourth	8, 9, 10, ...

detection and correction are provided for all bits, both parity and data, i.e. the parity bits also check themselves. So, for example, for even parity, the first parity bit P_1 for the data string 1001 must take on the value 0 to provide an even number of 1s in bit positions 1, 3, 5, and 7; the second parity bit P_2 must also be 0 to provide an even number of 1s in bit positions 2, 3, 6, and 7. The transmitted binary message for even parity Hamming code for the data string 1001 is therefore 0011001. A check is then made on the 1s received. First, the group of bits checked by P_1 is verified. If there is an even number of 1s, a 0 is noted. The parity check is repeated with the bits checked by P_2 and then by P_3. The parity checks form a binary number, known as the error position code. The first parity check generates the LSB of this code and the final parity check generates the MSB (it is important to note the order in which the bits of this error code are generated). In the above example, the result is a three-bit code 000, indicating that there is no error in the received message. However, if an error in bit position 4 has occurred, with the 1 replaced by a 0 – the received message would then read 0010001. The parity checks on the received message would generate the error position code 100 (decimal 4). This indicates that the error in the fourth bit position, which can subsequently be corrected – from 0 back to 1 (problem 12.4). The addition of an extra parity bit will allow for two errors to be detected. But, as in nature, the general rule is that for more error detection/correction to be introduced, the redundancy must be increased.

An alternative approach to building an electronic system with fault tolerance is to construct it from devices that are inherently capable of self-repair. Nature already provides examples of self-repairing processes. Regeneration is the renewal, restoration, and tissue growth that makes genomes, cells, organisms, and ecosystems robust to natural fluctuations that cause disturbance or damage. Every species, from bacteria to humans, is capable of some kind of regeneration. Certain animals, such as types of starfish, lizards, and salamanders, can regrow entire limbs. Other organisms, including worms and sponges, can grow a whole new body from a small surviving piece. Although humans have some regenerative capacity – for example, skin can repair itself from wounds – this ability pales in comparison to that of these resilient species. (There are speculations, however, that we have inherited the ability to regenerate limbs, but that the relevant bits of genetic code may be switched off.)

Within the cell, an inbuilt repair system (DNA repair) eliminates spontaneous changes in DNA, created by heat, metabolic accidents, radiation and exposure to substances in the environment. A large set of DNA repair enzymes continuously scans the DNA and replaces any damaged nucleotides. The double helix structure of DNA is ideally designed for repair because it carries two separate copies of all the genetic information – one in each of the two strands. Consequently, when one strand is damaged, the complementary strand retains an intact copy of the same information and this can be used to restore the correct nucleotide sequences to the damaged strand. A potentially dangerous type of DNA damage occurs when both strands of the double helix are broken.

The forces that drive phospholipids to form bilayers, with the outers surfaces hydrophilic and an inner hydrophobic region also provide self-healing. A small tear in the bilayer creates a free edge with water. Because this is energetically unfavourable, the lipids rearrange spontaneously to eliminate the free edge. Large tears can be repaired by cell fusion. The only way in which a bilayer can avoid free edges is by closing in on itself to form a sealed compartment – the cell. This remarkable feature is a direct consequence of the shape and amphiphilic nature of the phospholipid molecule (Figure 12.9b). The fluid nature of the bilayer membrane is key to this self-healing ability. Individual phospholipid molecules can rapidly change places with neighbouring molecules with the same monolayer, approximately 10^7 times per second, which means that an average molecule may diffuse the length of a large bacterial cell (2 μm) in about one second [13]. The hydrocarbon

Figure 12.26 Phospholipid mobility. The types of movement possible for phospholipid molecules in a lipid bilayer. The flip-flop process between lipid layers rarely occurs.

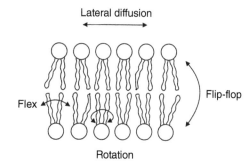

chains in the lipid molecules allow rapid rotation of individual lipid molecules about their long axes and the long chains to be flexible. However, the molecules in a lipid bilayer rarely migrate from one side of the bilayer to the other – a process called flip-flop and that occurs less once per month. Figure 12.26 provides an indication of the possible movement of phospholipid molecules in a bilayer membrane.

A key question is how do integrate the above ideas into current or future organic or molecular electronic devices to confer these with a degree of fault tolerance? The use of EiM has been noted in Section 12.4.7. Many other approaches to the self-repair of materials and devices are actively being pursued [79–83]. Current self-healing material systems can be broadly classified as autonomous and nonautonomous, according to the trigger requirements and the nature of the self-healing process [81]. Nonautonomous systems require external triggers, e.g. light, heat or chemicals, for healing. In contrast, autonomous (or autonomic) systems initiate self-healing process upon damage. Self-healing in materials can further be divided into intrinsic and extrinsic processes. The former exploits a healing mechanism based on the reversible formation of weak chemical bonds of polymer chains. Following damage, chemical bonds such as S—S bonds and hydrogen bonds are broken but can then reform at the fractured interfaces because of the dynamic characteristics of polymer chains [81, 82]. Intrinsic self-healing materials do not require the addition of healing reagents, and damaged regions are able to repeatedly heal (in some cases in ambient conditions) through the reorganization of the polymer matrix. These intrinsically healable materials are often flexible and may be important in the development 'soft' electronics.

Extrinsic self-healing materials contain external healing agents contained in capsules or in a vascular network. These agents typically consist of reactive precursors, which are released after damage. The material damage is then be repaired through autonomous polymerization reactions and the reconstruction of crosslinked networks. Figure 12.27 illustrates the self-healing process using microcapsules [84]. Following the formation of a crack due to damage (a), the growing crack ruptures microcapsules in its path (b), thereby releasing the healing agent into the crack plane (c), polymerization is then triggered by contact with an embedded catalyst in the polymer matrix, and the faces of the crack are bonded. This approach unfortunately only allows the material to heal a limited number of times and usually cannot repeat healing at the same locations.

The above healing strategies have been applied, with different degrees of success, to the various types of materials used in organic electronics. Conferring organic conductors and semiconductors, such as polymers, with self-healing properties is generally achieved by exploiting composite materials systems, although in some cases annealing at elevated temperatures and/or in solvent environments is required. Both intrinsic and extrinsic repair strategies can be exploited. Self-healing has been achieved in organic devices such sensors, capacitors and solar cells [79]. However, applying such strategies to the nanometre material dimensions and multilayer architectures found in

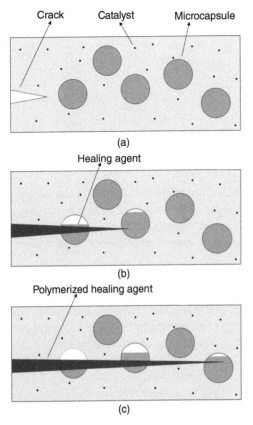

(a)

(b)

(c)

Figure 12.27 Self-healing using microcapsules. Following the formation of a crack due to damage (a), the growing crack ruptures microcapsules in its path (b), thereby releasing the healing agent into the crack plane (c); polymerization is then triggered by contact with an embedded catalyst in the polymer matrix, and the crack faces are bonded. Sources: Latif et al. [82]; McDonald et al. [84].

devices such as OFETs is challenging [83]. Self-healing can take place in thin polymer semiconducting layers that have hydrogen-bonding moieties. Nanosized cracks formed by fatigue in such a polymer semiconducting film have been found to heal by solvent annealing and thermal annealing, allowing efficient recovery of its charge carrier mobility [83].

Organic monolayers of phospholipids have also been shown to exhibit self-healing at room temperature [85]. Following electrical breakdown, such dielectrics heal autonomously, and the initial value of their electrical resistance can be recovered after one hour. The physical processes involved may mimic the self-healing processes in natural lipid membrane via lipid mobility (Figure 12.26) and suggest one way ahead for the development of individual molecular electronic devices. However, complex molecular electronic computing architectures will almost certainly require the integration of fault-tolerant processes like those described above, and perhaps device redundancy engineered at material level.

12.6 Bacteriorhodopsin – A Light-Driven Proton Pump

Bacteriorhodopsin is one of the most studied proteins for use in bioelectronics and deserves a special section in this chapter. It is a compact molecular machine that pumps protons across a membrane powered by green sunlight. The molecule is the light harvesting protein in the purple membrane of a micro-organism *Halobacterium salinarum* (formerly *Halobacterium halobium*), which lives in extreme conditions, in salt marches [86, 87]. The bacteriorhodopsin found in halobacteria

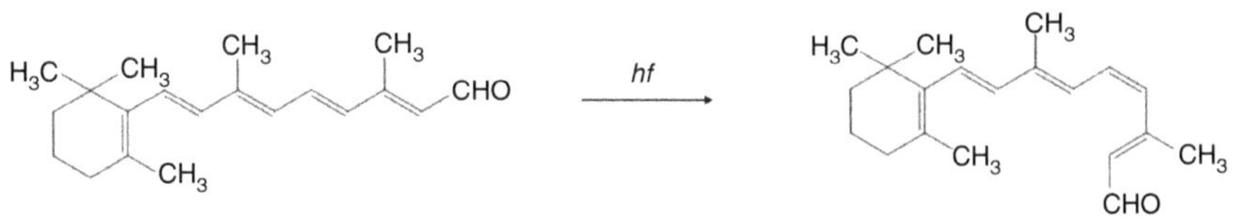

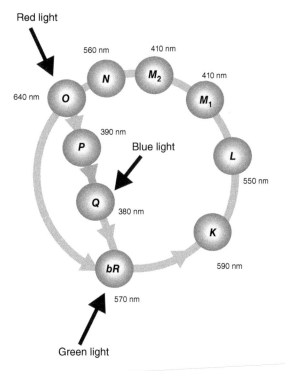

trans-Retinal   **cis-Retinal**

Figure 12.28   Light-induced transformation of the retinal by photons of energy *hf* from a *trans* to *cis* configuration.

is in the form of a two-dimensional crystal integrated into their cell membranes. This results in excellent chemical and thermal stability (up to 140 °C for dry layers).

Sunlight interacts with this protein to pump protons outwards across the cell membranes, making the inside more alkaline than the outside. By this process, light energy is converted to chemical energy and a proton gradient across the membrane is established, which may be used for energy storage.

Bacteriorhodopsin consists of 248 amino acids, arranged in seven α-helical bundles inside the lipid membrane, which form a cage. At the heart of the bacteriorhodopsin is a molecule of all-*trans* retinal (Vitamin A aldehyde), which is bound deep inside the protein and connected through a lysine amino acid. Retinal is a chromophore, containing a delocalized carbon atom chain that strongly absorbs light. When a photon is absorbed, it causes a change in the conformation of the molecule from its *trans* to its *cis* form (photoisomerization) as shown in Figure 12.28. This change in molecular shape drives the pumping of protons.

Following photon absorption and proton transfer, the bacteriorhodopsin recovers to its initial conformation. The entire sequence of transitions is called the photocycle. The precise details of this are a matter of some debate; a simplified version is given in Figure 12.29 [86, 87]. Each photocycle

Figure 12.29   Photocycle of bacteriorhodopsin. Green light transforms the resisting *bR* state to the intermediate *K*. Next *K* relaxes, forming *L*, M_1, M_2, *N*, and then *O*. If the *O* intermediate is exposed to red light, a so-called branching reaction occurs. Structure *O* converts to the *P* state, which quickly relaxes to the *Q* state – a form that remains stable almost indefinitely. Blue light, however, will convert *Q* back to *B*. Sources: Wagner et al. [86]; Wickstrand et al. [87].

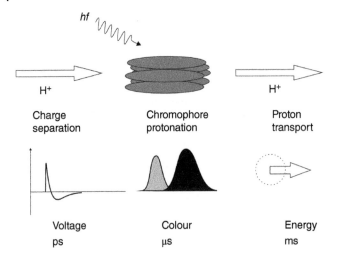

Figure 12.30 Basic molecular functions of bacteriorhodopsin. The proton transport is initialized by photon (of energy *hf*) absorption and a charge separation step on the picosecond time scale. After about 50 μs, deprotonation leads to the main photochromic shift during the photocycle. After about 10 ms the proton transport is completed. Source: Hampp [88].

lasts 10–15 ms and results in the net translocation of a single proton across the membrane. There are several spectrally distinct intermediates, labelled bR, K, L, M_1, M_2, N and O; Figure 12.29 indicates the absorption maximum for each state. The initial resting state of the molecule is known as bR. Green light transforms this initial state to the intermediate state K. Next K relaxes, forming other intermediate states L, M_1, M_2, N, and O and finally back to bR. If the intermediate state O is irradiated with red light, it converts to a further state P (a branching reaction) which then relaxes to a very stable state Q. Blue light converts Q back to the initial bR state.

The basic molecular functions of bacteriorhodopsin and their corresponding physical effects are depicted in Figure 12.30 [88]. The proton transport is initialized by proton absorption and a charge separation step on the picosecond timescale. After about 50 μs, the main colour change occurs and after 10–15 ms the proton transport is completed.

The various photochemical changes in bacteriorhodopsin are under study for exploitation in optical and opto-electronic devices [89, 90]. Different technologies, such as LB deposition and electrostatic adsorption, have been used to reconstitute bacteriorhodopsin layers on solid substrates. The cyclicity of the protein (i.e. the number of times that the protein can be photochemically cycled before denaturing) exceeds 10^6, a result of its evolution in the harsh environment of a salt marsh. For optical storage applications, the principle is to assign any two long-lasting states of the protein to the binary values of 0 and 1. The prospects for three-dimensional data storage seem more promising than for two-dimensional systems. There are three different types of volume storage under investigation. The first is page-oriented holographic storage; the second is based on the branched photocycle scheme and the third uses two-photon excitation of individual data points in the volume of the material. The storage capacity of such devices is very high (10 GB), although it remains unclear whether this type of memory will complete with hard disks or semiconductor memory. The limitation of capacity is mainly connected with problems of lens system and quality of protein.

12.7   Photosynthesis and Artificial Molecular Architectures

Food and fossil fuel are the products of photosynthesis, the process that converts the energy in sunlight to chemical forms, which can then be used by biological systems. Photosynthesis takes place in many different organisms, from plants to bacteria. The best-known example of photosynthesis is that carried out by higher plants and algae, which are responsible for a major part of photosynthesis in oceans. All these organisms convert carbon dioxide to organic material by reducing the gas to carbohydrates in a rather complex set of reactions.

Sunlight is absorbed and converted initially to electronic excitation energy. This starts a chain of electron-transfer events leading to charge separation across a photosynthetic membrane. The resulting potential energy is then used to pump protons across the membrane, generating an osmotic and charge imbalance, which in turn powers the synthesis of ATP. For the sugar glucose ($C_6H_{12}O_6$ – one of the most abundant products of photosynthesis) the relevant equation is:

$$6CO_2 + 12H_2O + light \rightarrow C_6H_{12}O_6 + 6H_2O + 6O_2 \tag{12.12}$$

Light provides the energy to transfer electrons from water to $NADP^+$ forming NADPH (Section 12.3.5); and to generate ATP (Section 12.3.3). ATP and NADPH provide the energy and electrons to reduce CO_2 to organic molecules. Electrons for this reduction reaction ultimately come from water, which is then converted to oxygen and protons.

The first step in photosynthesis is the absorption of light by pigments such as primarily chlorophylls and carotenoids. Chlorophylls (such as chlorophyll *a*) absorb blue and red light and carotenoids (such as β-carotene) absorb blue-green light. However, green and yellow photons are not effectively absorbed by photosynthetic pigments in plants; therefore, light of these colours is either reflected by the leaves or passes through (why plants are green, and carrots are orange).

Other photosynthetic organisms have additional pigments that absorb the colours of visible light which are not absorbed by chlorophyll and carotenoids. Some contain bacteriochlorophyll, absorbing in the infrared in addition to the blue part of the spectrum. These bacteria do not evolve oxygen but perform photosynthesis under anaerobic (oxygen-less) conditions using the infrared light.

Photosynthetic pigments are normally bound to proteins, which provide the pigment molecules with the appropriate orientation and positioning with respect to each other. Light energy is absorbed by individual pigments but is not used immediately for energy conversion. Instead, the light energy is transferred to chlorophylls that are in a special protein environment where the actual energy conversion event occurs. The pigments and proteins involved with this primary electron transfer event are together called the reaction centre. Many pigment molecules (100–5000), collectively referred to as antenna, 'harvest' light and transfer the light energy to the same reaction centre. The purpose is to maintain a high rate of electron transfer in the reaction centre, even at lower light intensities. Antennas permit an organism to increase significantly the absorption cross-section for light without having to build an entire reaction centre and associated electron transfer system for each pigment, which would be very costly in terms of cellular resources. The photosynthetic antenna system is organized to collect and deliver excited state energy by means of transfer to the reaction centre complexes where photochemistry takes place. The process is illustrated in Figure 12.31.

Many antenna pigments transfer their light energy to a single reaction centre by transmitting the energy to another antenna pigment, and then to another, until the energy is finally trapped in the reaction centre. Each step of this energy transfer must be very efficient to avoid a large loss in the overall transfer process. The association of the various pigments with proteins ensures that

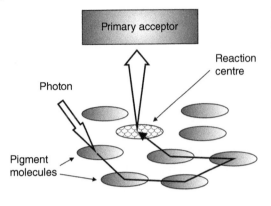

Figure 12.31 Schematic diagram of a photosynthetic reaction centre. A large number of pigment molecules 'harvest' the light and transfer the energy to the same reaction centre.

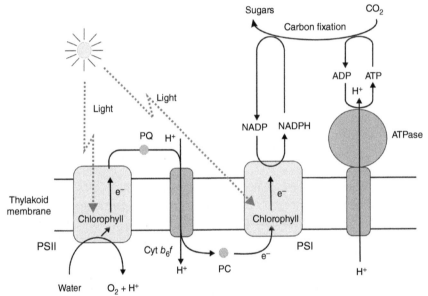

Figure 12.32 Process of photosynthesis. Two reaction centres, PSI and PSII, are involved, which are located in the thylakoid membrane. PQ = plastoquinone. PC = plastocyanin. b_6f = cytochrome complex. ATPase = enzyme that catalyses the decomposition of ATP into ADP.

transfer efficiencies are high by having the pigments in close proximity, and by providing an appropriate molecular geometry of the pigments with respect to each other.

The overall process of photosynthesis is complex. Figure 12.32 shows the important reaction steps, which are summarized in the following section. A detailed discussion can be found elsewhere [91].

All photosynthetic organisms that produce oxygen have two types of reaction centres, named photosystem II and photosystem I (PSII and PSI, shown in Figure 12.32), both of which are pigment/protein complexes located in membranes called thylakoids. In eukaryotes (plants and algae) these thylakoids are in specialized entities called chloroplasts and often are found in membrane stacks. Photosystem I, which is activated by light at 680 nm, is associated with chlorophyll *a* and is not involved in oxygen evolution. Photosystem II, which is activated by shorter wavelengths of light, between 500 and 600 nm, appears to be involved in oxygen evolution; it uses a second type of chlorophyll as well as accessory pigments.

Upon oxidation of the reaction centre chlorophyll in PSII, an electron is transferred from a nearby amino acid (tyrosine), part of the surrounding protein, which in turn acquires an electron from the water-splitting complex. From the PSII reaction centre, electrons flow to free electron-carrying molecules in the thylakoid membrane (plastoquinione – PQ in Figure 12.32) and from there to a cytochrome complex (Cyt $b_6 f$ in Figure 12.32). Finally, the electrons are transported to the PSI centre by a small protein (plastocyanin – PC in the figure).

The other photosystem, PSI, also catalyses light-induced charge separation in a fashion similar to PSII: an antenna harvests light, and light energy is transferred to a chlorophyll reaction centre, where light-induced charge separation is initiated. In PSI, electrons are transferred eventually to NADP. The oxidized chlorophyll reaction centre receives the electron from the cytochrome complex. Therefore, electron transfer through PSII and PSI results in water oxidation (producing oxygen) and NADP reduction, with the energy for this process provided by light (two quanta for each electron transported through the whole chain).

The conversion of carbon dioxide into organic compounds during photosynthesis is called carbon fixation. The electron flow from water to NADP requires light and is coupled to generation of a proton gradient across the thylakoid membrane. This proton gradient is used for the synthesis of ATP. ATP and reduced NADP, which result from the light reactions, are used for CO_2 fixation in a process that is independent of light.

In summary, the process of photosynthesis converts solar energy into chemical energy and stores this in the form of carbohydrates. Photosynthesis occurs through the cooperation of two photosystems, PSI and PSII. On exposure to sunlight, water is oxidized in PSII into oxygen, releasing protons and electrons. These are transferred via a cytochrome complex to PSI, in which the separated electrons and protons are finally consumed by CO_2 reduction to produce carbohydrates.

Plants use sunlight to make carbohydrates from carbon dioxide and water. Most work on artificial photosynthesis seeks to use the same inputs – solar energy, water, and carbon dioxide – to produce energy-dense liquid fuels. The estimated maximum efficiency for photosynthesis (solar energy to biomass) is only a few percent [92–94] (Problem 12.5). However, the efficiencies of the initial photochemical and chemical stages of photosynthesis, which are not directly involved in biomass production, are significantly higher and are the subject of much of the research on artificial photosynthesis [95–97]. The Artificial Leaf accomplishes the solar process of natural photosynthesis – the splitting of water to hydrogen and oxygen using sunlight – under ambient conditions [94].

There has also been much interest in developing synthetic reaction centres, which may find use in photovoltaic cells. The initial work, in the late 1970s, consisted of linking molecular analogues of components involved in the natural photosynthetic reaction centre with covalent bonds [98]. Porphyrins (e.g. Figure 12.14a) and carotenoids (β-carotene is a type of carotenoid), the main chromophores of natural photosynthesis, are candidates for the design of artificial systems. Antenna architectures can be based on arrays of such molecules [99]. Examples of more sophisticated systems include two chlorophyll-type molecules linked together (one serves as the electron donor, the other as the acceptor) with the electron-accepting molecule linked to two quinones, which serve as electron acceptors in the natural system. The electron donating chlorophyll analogue is linked covalently to a carotenoid, and this can donate an electron to the oxidized chlorophyll. Upon excitation of the chlorophyll, a charge separation occurs resulting in an oxidized carotenoid and a reduced quinone. This charge-separated state is formed with high efficiency. In other artificial reaction centres, porphyrin moieties have been used in place of chlorophyll and fullerenes have been exploited as the electron acceptor. More complex molecular architectures based on covalently linked carotenoid-porphyrin-quinone triads and carotenoid-porphyrin-diquinone tetrads may be used to yield long-lived charge-separated states upon excitation [98]. Such systems can be incorporated into bilayer lipid membranes, separated by appropriate solutions, and used to generate a steady-state photocurrent upon illumination.

Artificial photosynthetic solar-to-fuels cycles typically terminate at hydrogen, with no process installed to complete the cycle via carbon fixation. A development uses a pair of catalysts to split water into oxygen and hydrogen and feeds the hydrogen to bacteria along with carbon dioxide. The bacteria, micro-organisms that have been bioengineered to specific characteristics, convert the carbon dioxide and hydrogen into liquid fuels [100]. It is suggested that coupling this hybrid device to existing photovoltaic systems would yield a CO_2 reduction efficiency of about 10%, exceeding that of natural photosynthetic systems.

The realization of an artificial photosynthetic molecular system that mimics nature (solar energy to biomass) remains elusive. However, there have been some notable advances, such as the development of a synthetic structure (liposome) incorporating functional ATP synthase and bacteriorhodopsin to establish a proton gradient, and thus driving ATP synthesis [101]. This system encompasses the basic function of photosynthesis: using light to synthesize an energy carrier molecule. The creation of a fully functional artificial photosynthetic cell would certainly have many technological implications for molecular electronic and nanoelectronic devices. More significantly, this would signify major progress in the understanding of life and the development of synthetic organisms.

12.8 Bio-Chemical Sensors

The exploitation of organic and biological materials as chemical sensors has been noted in Section 8.8. Biological molecules offer scope for highly selective and sensitive devices [9, 102–104]. Evolution of species by means of natural selection has led to extremely sensitive organs that can respond to the presence of just a few molecules (the sense of smell of canines is renowned). Artificial sensors exploit biologically active materials in combination with different physical sensing elements. The bio-recognition element works like a bioreactor on the top of the conventional sensor. The response of the biosensor will be determined by the diffusion of the analyte, by the reaction products, by the co-reactants or interfering species and/or by the kinetics of the recognition process. Organisms, tissues, cells, membranes, enzymes, antibodies and nucleic acid can all be detected by means of a biosensor. Biosensors find applications in medicine, food and process control, environmental monitoring, and defence and security. However, the market is driven mainly by medical diagnosis and, in particular, glucose sensors for people with diabetes. The most significant trend likely to impact on development of biosensors is the emergence of personalized medicine, in particular biosensors as wearable devices [105, 106].

One of the key issues for biological sensing systems is the immobilization of the active element on the physical transducer. The biologically active material must be confined to the sensing element and kept from 'leaking' while allowing contact with the analyte solution. Furthermore, the reaction products must readily diffuse out of the sensing layer so as not to denature its biologically active characteristics. Many biosensing materials are proteins or contain proteins in their chemical structure. Two techniques employed to immobilize the proteins are binding (adsorption or covalent binding) and retention. Physical retention involves separating the biologically active material from the analyte solution with a layer on the surface of the sensor, which is permeable to the analyte and to any products of the recognition reaction, but not to the biologically active materials.

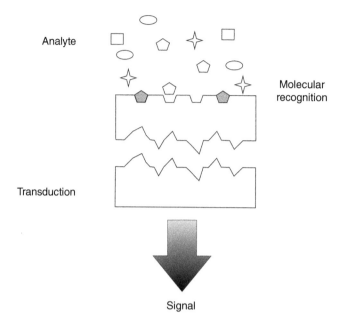

Figure 12.33 Molecular recognition and signal transduction parts of a biosensor.

A biosensor usually has two functional parts, as illustrated in Figure 12.33: one for molecular recognition and the other for signal transduction (e.g. conversion to a voltage signal). To achieve a high selectivity, either a biocatalyst or bioaffinity (immunological) material can be exploited as the molecular recognition element.

12.8.1 Biocatalytic Sensors

A biocatalyst recognizes the corresponding substrate (analyte) and immediately generates products by a specific reaction. The complex of the catalyst and the analyte remain stable in a transition state. A change in either the analyte or product is detected in the signal transducing device of the biosensor. Redox (oxidation–reduction) enzymes are important materials for both biocatalytic and bioaffinity sensors. Biocatalytic sensors for glucose, lactate and alcohol utilize glucose oxidase, lactate dehydrogenase (lactate oxidase) and alcohol dehydrogenase (alcohol oxidase), respectively as the molecular recognizable material. Since these redox enzymes are mostly associated with the generation of electrochemically active substances, many electrochemical enzyme sensors have been developed by linking redox enzymes for molecular recognition with electrochemical devices for signal transduction.

Nanotechnology is now making an impact on biosensor development. For example, carbon nanotubes can be blended into several sensing layer formulations to improve current densities and performance of enzyme electrodes. However, the most widely used nanomaterial for commercial biosensors are silver nanoparticles. These have been exploited as a simple electrochemical label in a highly sensitive and inexpensive amperometric immunoassay intended for distributed diagnostics [102].

12.8.2 Bioaffinity Sensors

A bioaffinity sensor involves an antibody, binding protein or receptor protein that forms a stable complex with a corresponding functional group (a ligand). The bioaffinity protein-ligand complex formation is sufficiently stable to result in signal transduction. Immunosensors take advantage of the high selectivity provided by the molecular recognition of antibodies. Because of significant differences in affinity constants, antibodies may confer an extremely high sensitivity to immunosensors in comparison to enzyme sensors. Furthermore, antibodies may be obtained (in principle) for an unlimited number of determinants. Immunosensors are therefore characterized by high selectivity, sensitivity and versatility.

Immunosensors can be divided in principle into two categories: nonlabelled and labelled immunosensors. Nonlabelled immunosensors are designed so that the immunocomplex, i.e. the antigen–antibody complex, is directly determined by measuring the physical changes (e.g. thickness, mass, refractive index) induced by the formation of the complex.

Nonlabelled immunosensors are based on several principles, as illustrated in Figure 12.34. Either the antibody or the antigen can be immobilized on the solid matrix to form a sensing device. The

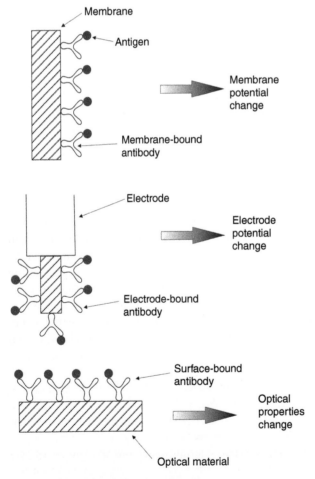

Figure 12.34 Principles of nonlabelled immunosensors. The highly selective antigen–antibody reaction can give rise to electrical or optical signals.

solid matrix must be sensitive enough, in its surface characteristics, to detect the immunocomplex formation. Electrodes, membranes, piezoelectric materials, or optically active surfaces may be used to construct nonlabelled immunosensors. The antigen or the antibody to be determined is dissolved in a solution and reacts with the complementary matrix-bound antibody or antigen to form an immunocomplex. This formation changes the physical properties of the surface, such as the electrode potential or the transmembrane potential, or the refractive index. A sufficiently high selectivity may be obtained with nonlabelled immunosensors, although such problems as non-specific adsorption onto the matrix-bound antibody surface remain unresolved.

A common configuration for a labelled immunosensors is the lateral flow test, as used to detect the SARS-CoV-2 virus. This uses a paper strip typically coated with immune-system antibody molecules, which recognize specific antigen chemical groups of viral proteins [107, 108]. A patient's sample is mixed with a small amount of liquid, which is applied to one end of the strip and then flows, via capillary action, towards the other end. Along the way, the sample passes through the antibodies (or similar binding proteins), which are seized by any viral antigens in the sample. This antigen–antibody complex migrates to a test zone and initiates a chemical reaction that causes a colour change, indicating a positive result. Excess antibodies will travel the length of the strip to a control zone, and again cause a colour change. This second change provides reassurance that the test is working correctly.

An obvious target for electrochemical affinity sensors is DNA. The advent of the DNA microarray, or DNA chip, has focused attention on alternatives to fluorescence detection. These exploit the fact that short strands of DNA will bind to other segments of DNA that have complementary sequences and can therefore be used to probe whether certain genetic codes are present in a given specimen of DNA. Different types of micro/nano-fluidic technologies have facilitated DNA purification, amplification and detection to be integrated into one chip, which combine the advantages of automation, small sizes, much shorter reaction times and reduced cost [109–111].

The feasibility of single molecule detection is currently a hot topic [112–114]. Many approaches exploit optical detection (e.g. fluorescence). However, electrical methods can offer advantages [114]. Detection platforms consisting of single-molecule electronic devices, for example, graphene/molecule/graphene junctions and silicon-nanowire-based single-molecule electrical nanocircuits are also under study. The experimental configuration here is essentially that described for single molecule electrical measurements in Section 4.5. Host molecules, including DNA, can be assembled between nanogap electrodes (e.g. break junctions) and the change in electrical conductivity measured. Subsequently, interactions with guest molecular species can then be monitored. Another electrically based sensing system is based on nanopore technology for DNA sequencing [112]. The concept is to observe the ion current across a nanopore, which is extremely sensitive to changes in the shape and size caused by single nucleotides passing through it. Other work in this area has explored the possibility of sequencing DNA by passing the molecule through nanopores in a sheet of graphene [113].

12.9 Electronic Olfaction and Gustation

Many chemical sensors exploiting organic materials (e.g. resistance-based devices) do not possess the specificity to be of practical use [9]. For example, a semiconductive polymer (chemiresistor) may show a similar change in electrical resistance to a range of oxidizing (or reducing) gases. Fortunately, nature suggests a way forward.

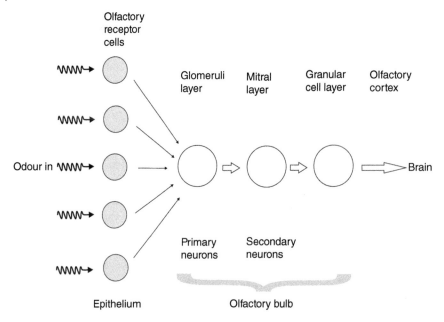

Figure 12.35 Schematic representation of the human olfactory system. See text for details.

The human olfactory system has many receptors cells (sensors) that are individually non-specific; signals from these are fed to the brain via a network of primary and secondary neurons for processing. It is generally believed that the selectivity of the olfactory system is a result of a high degree of parallelism in the neural architecture. In humans, the process of olfaction begins with the inhalation of molecules and their transport to the olfactory receptor cells, located within a specialized membrane, of approximate dimensions 2 cm by 5 cm, called the epithelium, in the nose. The subsequent interactions generate electrical signals, which propagate down the axon (Section 12.3.6) of the olfactory receptor cells to the olfactory bulb, as depicted in Figure 12.35. Further layers, referred to as glomeruli, process the signals. Here, the nature of the odorous stimulus is encoded as a specific combination of activated glomeruli, which takes the form of a two-dimensional map. Particular odorants are associated with specific topographic patterns of activity. Although mammals possess in the order of 1000 different receptor proteins, they can detect more than this number of different odorants. This suggests that any given odorant can interact with more than one receptor protein and, conversely, that any single receptor is able to interact with more than one odorant. It is worth noting that olfactory receptor cells are replaced every 30–60 days, so at any one time the population of olfactory receptor cells contributing to a particular glomerulus will possess a range of ages. This will provide the system with a mechanism to overcome the problem of a loss of sensitivity of the receptor cells with time. Although the sensitivity of any given receptor is likely to decrease with time, the summed response from a population of cells of all ages will be more stable [115, 116]. The molecular basis for the discrimination of odours in mammals is still a matter of some debate. Mechanisms based on receptor-ligand (lock-key) docking and a bond vibration model (i.e. olfaction involves the detection of the vibrational frequencies of odorant molecules) have been proposed. Current evidence favours the former idea [117].

The map of glomerular activity, as detected by the mitral cells, then feeds into the next level of information processing, the granular cell level, before passing via the olfactory cortex to the brain. Olfactory information travels not only to the limbic system, primitive brain structures that govern emotions, behaviour, and memory storage, but also to the brain's cortex, where conscious thought occurs. In addition, the data combine with taste information in the brain to create the sensation of flavour.

The electronic nose is an attempt to imitate the human olfactory system [118–120]. Emulation of gustation, the sense of taste, would similarly lead to an electronic tongue [118]. Individual sensors can be based on semiconductive polymer films. Each element is treated in a slightly different way during deposition so that it responds uniquely on exposure to a particular gas or vapour. The pattern of resistance changes in the sensor array can then be used to fingerprint the vapour. Alternatively, data for a single sensor, but measured under different conditions, such as the conductivity of a chemiresistor at different frequencies, may be used to provide a suitable fingerprint [9]. Electronic tongues might exploit ISFETs (Section 8.8.1) sensitized with appropriate organic or biological thin films to provide the individual sensing elements.

Figure 12.36 shows a schematic diagram of an electronic nose. If more than two linear sensors are used to characterize a sample, a multivariate evaluation is usually applied to visualize and evaluate the data [115]. In electronic noses, pattern recognition approaches are generally used. Unsupervized pattern recognition algorithms try to cluster the results, which are usually visualized in a two-dimensional plot. A common technique that is applied is principal component analysis. This involves the use of a mathematical procedure to transform several (possibly) correlated variables into a (smaller) number of uncorrelated variables called principal components. Another approach to multivariate analysis uses an ANN, as described above in Section 12.4.3. The synaptic weights are determined either during a training (or learning) phase for supervized neural networks, or by an algorithm for unsupervised neural networks. Electronic noses and tongues have been in the marketplace for several years and find many applications, particularly quality control in the food and drink industry, and in flavour and fragrance testing [119, 120]. Nonetheless, the electronic olfactory equipment possesses some limitations regarding sensitivity and specificity as compared to the biological system. The weak links are often the individual sensors that make up the array, which can degrade with extended use. Frequent recalibration of sensors arrays is often required, which can be time consuming. Nature has a distinct advantage here as individual receptor cells are continuously replaced.

Problems

12.1 Undertake research to suggest answers to the following:
   a) In RNA, the base thymine, found in DNA, is replaced by uracil (Figure 12.6). Why do you think this is? Why do not DNA and RNA possess the same four bases?
   b) The genetic code, in which three bases are used to encode the 20 amino acids, is degenerate – the same amino acids are encoded by different base sequences. For example, alanine can be represented by the codons CGU, GCC, GCA or GCG and lysine can be represented by AAA or AAG. However, two amino acids are represented by only one codon: methionine is denoted by AUG and tryptophan by UGG. Suggest some reasons for this.

12.2 a) The concentrations of Na^+ and K^+ ions across human neuron cells are as follows: Potassium – 5 mM outside, 140 mM inside. Sodium – 12 mM inside, 140 mM outside. Calculate the equilibrium potentials for these ionic species.
   b) A spherical human cell of diameter 10 μm contains a concentration of potassium ions of 100 mM inside and 10 mM outside. Assume that the cell membrane has a thickness of

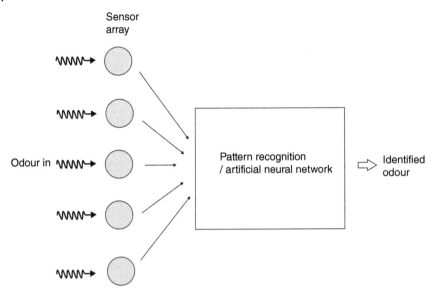

Figure 12.36 Components of an electronic nose.

4.4 nm and a relative permittivity of 5.0. Calculate the fraction of the ionic charges contained within the cell to maintain its equilibrium potential. (Hint: a molar [M] solution contains N_A molecules of a substance per 1 l of solution).

12.3 In computing, the term FLOPS refers to the number of FLoating-point Operations Per Second (or bit operations per second) and is a measure of computer performance. The computing power of the human brain is based on the 'firing' of neurons. If there are about 10^{11} neurons in the brain and a neuron can fire about once every 5 ms (approximately 200 times per second) and each neuron is connected to 10^3–10^4 other neurons, estimate the computation power of the brain in FLOPS. Compare your estimate to the power of modern supercomputers.

12.4 a) Look up the 7-bit ASCII code (American Standard Code for Information Interchange) for the letter 'K' and the symbol '='. In each case, add an even parity bit to the beginning of the code.

b) Correct any error in the following Hamming code that is received: 101101010. Odd parity is used.

12.5 In photosynthesis, the energy of at least eight 'red' photons is required per O_2 molecule released or CO_2 molecule fixed. A typical product of carbon fixation is glucose ($C_6H_{12}O_6$), the energy content of which is 2805 kJ mol^{-1} (the enthalpy of combustion or energy generated when glucose is burnt). Estimate the efficiency of the photosynthetic process. Why is the practical efficiency likely to be less than this figure?

References

1 Balzani, V., Credi, A., and Venturi, M. (2008). *Molecular Devices and Machines: Concepts and Perspectives for the Nanoworld*, 2e. Weinheim: Wiley-VCH.

2 Lorente, N. and Joachim, C. (eds.) (2013). *Architecture and Design of Molecule Logic Gates and Atom Circuits*. Heidelberg: Springer-Verlag.

3 Erbas-Cakmak, S., Kolemen, S., Sedgwick, A.C. et al. (2018). Molecular logic gates: the past, present and future. *Chem. Soc. Rev.* 47: 2228–2248.

4 Horowitz, P. and Hill, W. (2015). *The Art of Electronics*, 3e. Cambridge: Cambridge University Press.

5 De Silva, A.P., De Silva, S.A., Dissanayake, A.S., and Sandanayake, K.R.A.S. (1989). Compartmental fluorescent pH indicators with nearly complete predictability of indicator parameters – molecular engineering of pH sensors. *J. Chem. Soc., Chem. Commun.*: 1054–1056.

6 De Silva, A.P., Gunaratne, H.Q.N., and McCoy, C.P. (1997). Molecular photoionic AND logic gates with bright fluorescence and 'off-on' digital action. *J. Am. Chem. Soc.* 119: 7891–7892.

7 Carter, F.L. (ed.). (1981). Proceedings of the Molecular Electronic Devices Workshop. *Naval Research Laboratory Memorandum Report 4662*. Washington, DC: Naval Research Laboratories.

8 Roth, S. (1995). *One-Dimensional Metals*. Weinheim: VCH.

9 Petty, M.C. (2019). *Organic and Molecular Electronics: From Principles to Practice*, 2e. Chichester: Wiley.

10 Aviram, A., Seiden, P.E., and Ratner, M.A. (1981). Theoretical and experimental studies of hemiquinones and comments on their suitability for molecular storage elements. In: *Proceedings of the Molecular Electronic Devices Workshop*, Naval Research Laboratory Memorandum Report 4662 (ed. F.L. Carter), 3–16. Washington, DC: Naval Research Laboratories.

11 Echegoyen, L. and Echegoyen, L.E. (1998). Electrochemistry of fullerenes and their derivatives. *Acc. Chem. Res.* 31: 593–601.

12 Barker, J.R. (1995). Molecular electronic logic and architectures. In: *An Introduction to Molecular Electronics* (eds. M.C. Petty, M.R. Bryce and D. Bloor), 345–376. London: Edward Arnold.

13 Alberts, B., Johnson, A., Lewis, J. et al. (2014). *Molecular Biology of the Cell*, 6e. New York: Garland Science.

14 Swart, M., Snijders, J.G., and Van Duijnen, P.T. (2004). Polarizabilities of amino acid residues. *J. Comput. Methods Sci. Eng.* 4: 419–425.

15 DeepMind (2021) AlfaFold: a solution to a 50-year-old grand challenge in biology. https://deepmind.com/blog/article/alphafold-a-solution-to-a-50-year-old-grand-challenge-in-biology (accessed 25 March 2021).

16 Ridley, M. (1999). *Genome*. London: Harper Perennial.

17 Singer, S.J. and Nicolson, G.L. (1972). The fluid mosaic model of the structure of cell membranes. *Science* 175: 720–731.

18 Findlay, J.B.F. (1995). Molecular electronic logic and architectures. In: *An Introduction to Molecular Electronics* (eds. M.C. Petty, M.R. Bryce and D. Bloor), 279–294. London: Edward Arnold.

19 Petty, M.C. (1996). *Langmuir-Blodgett Films*. Cambridge: Cambridge University Press.

20 Petty, M., Tsibouklis, J., Petty, M.C., and Feast, W.J. (1992). Pyroelectric behaviour of synthetic biomembrane structures. *Thin Solid Films* 220–211: 320–323.

21 Howarth, V.A., Petty, M.C., Davies, G.H., and Yarwood, J. (1989). Infrared studies of valinomycin-containing Langmuir-Blodgett films. *Langmuir* 5: 330–332.

22 Goldup, A., Ohki, S., and Danielli, J.F. (1970). Black lipid films. *Rec. Prog. Surf. Sci.* 3: 193–260.

23 Feher, J.F. (2016). *Quantitative Human Physiology: An Introduction*, 2e. London: Academic Press.

24 Dubey, A.K., Dutta-Gupta, S., Kumar, R. et al. (2009). Time constant determination for electrical equivalent of biological cells. *J. Appl. Phys.* 105: 084705.

25 Bakiri, Y., Káradóttir, R., Cossell, L., and Attwell, D. (2011). Morphological and electrical properties of oligodendrocytes in the white matter of the corpus callosum and cerebellum. *J. Physiol.* 589 (3): 559–573.

26 Bean, B.P. (2007). The action potential in mammalian central neurons. *Nat. Rev. Neurosci.* 8: 451–465.

27 Kandel, E.R., Dudai, Y., and Mayford, M.R. (2014). The molecular and systems biology of memory. *Cell* 157: 163–186.

28 Eth, D., Foust, J.-C., and Whale, B. (2013). The prospects of whole-brain emulation within the next half-century. *J. Artif. Gen. Intell.* 4: 130–152.

29 Strukov, D., Indiveri, G., Grollier, J., and Fusi, S. (2019). Building brain-inspired computing. *Nat. Commun.* (Q & A) 10: 4838.

30 Aggarwal, C.C. (2018). *Neural Networks and Deep Learning: A Textbook*. Cham: Springer.

31 Merolla, P.A., Arthur, J.V., Alvarez-Icaza, R. et al. (2014). A million spiking-neuron integrated circuit with a scalable communication network and interface. *Science* 345: 668–673.

32 Jeong, D.S. (2018). Tutorial: neuromorphic spiking neural networks for temporal learning. *J. Appl. Phys.* 124: 152002.

33 Donzellini, G., Oneto, L., Ponta, D., and Anguita, D. (2018). *Introduction to Digital Systems Design*. Cham: Springer.

34 Amos, M. (2006). *Genesis Machines*. London: Atlantic Books.

35 Adamatzky, A. (ed.) (2010). *Game of Life Cellular Automata*. London: Springer-Verlag.

36 Tuchman, Y., Mangoma, T.N., Gkoupidenis, P. et al. (2020). Organic neuromorphic devices: past, present, and future challenges. *MRS Bull.* 45: 619–630.

37 Van de Burgt, Y. and Gkoupidenis, P. (2020). Organic materials and devices for brain-inspired computing: from artificial implementation to biophysical realism. *MRS Bull.* 45: 631–640.

38 Perez, J.C. and Shaheen, S.E. (2020). Neuromorphic-based Boolean and reversible logic circuits from organic electrochemical transistors. *MRS Bull.* 45: 649–654.

39 Yang, J.J. and Xia, Q. (2017). Battery-like artificial synapses. *Nat. Mater.* 16: 396–397.

40 Van de Burgt, Y., Lubberman, E., Fuller, E.J. et al. (2017). A non-volatile organic electrochemical device as a low-voltage artificial synapse for neuromorphic computing. *Nat. Mater.* 16: 414–419.

41 Harberts, J., Fendler, C., Teuber, J. et al. (2020). Toward brain-on-a-chip: human induced pluripotent system cell-derived guided neuronal networks in tailor-made 3D nanoprinted microscaffolds. *ACS Nano* 14: 13091–13102.

42 Dekker, C. and Ratner, M.A. (2001). Electronic properties of DNA. *Phys. World* 14: 29–33.

43 Seeman, N.C. (2003). DNA in a material world. *Nature* 421: 427–431.

44 Briman, M., Armitage, N.P., Helgren, E., and Grüner, G. (2004). Dipole relaxation losses in DNA. *Nano Lett.* 4: 733–736.

45 Jin, J.-I. and Grote, J. (eds.) (2012). *Materials Science of DNA*. Boca Raton, FL: CRC Press.

46 Tan, B., Hodak, M., Lu, W., and Bernholc, J. (2015). Charge transport in DNA nanowires connected to carbon nanotubes. *Phys. Rev. B* 92: 075429.

47 Sharipov, T.I. and Bakhtizin, R.Z. (2017). The study of electrical conductivity of DNA molecules by scanning tunnelling spectroscopy. *Scanning Probe Microscopy 2017. IOP Conference Series: Materials Science and Engineering* 256, 012009.

48 Jiménez-Monroy, K.L., Renaud, N., Drijkoningen, J. et al. (2017). High electronic conductance through double-helix DNA molecules with fullerene anchoring groups. *J. Phys. Chem. A* 121: 1182–1188.

49 Adleman, L.M. (1994). Molecular computation of solutions to combinatorial problems. *Science* 266: 1021–1024.

50 Shu, J.-J., Wang, Q.-W., Yong, K.-Y. et al. (2015). Programmable DNA-mediated multitasking processor. *J. Phys. Chem. B* 119: 5639–5644.

51 Song, T., Garg, S., Mokhtar, R. et al. (2016). Analog computation by DNA strand displacement circuits. *ACS Synth. Biol.* 5: 898–912.

52 Zhou, C., Geng, H., Wang, P., and Guo, C. (2019). Programmable DNA nanoindicator-based platform for large-scale square root logic biocomputing. *Small* 15: 1903489.

53 Priel, A., Ramos, A.J., Tuszynski, J.A., and Cantiello, H.F. (2006). A biopolymer transistor: electrical amplification by microtubules. *Biophys. J.* 90: 4639–4643.

54 Cantero, M.R., Villa Etchegoyen, C., Perez, P.L. et al. (2018). Bundles of brain microtubules generate electrical oscillations. *Sci. Rep.* 8: 11899.

55 Cantero, M.R., Perez, P.L., Scarinci, N., and Cantiello, H.F. (2019). Two-dimensional brain microtubule structures behave as memristive devices. *Sci. Rep.* 9: 12398.

56 Hameroff, S., Nip, A., Porter, M., and Tuszynski, J. (2002). Conduction pathways in microtubules, biological quantum computation, and consciousness. *BioSystems* 64: 149–168.

57 Kalra, A.P., Patel, S.D., Bhuiyan, A.F. et al. (2020). Investigation of the electrical properties of microtubule ensembles under cell-like conditions. *Nanomaterials* 10: 265.

58 Ladd, T.D., Jelezko, F., Laflamme, R. et al. (2010). Quantum computers. *Nature* 464: 45–53.

59 Debnath, S., Linke, N.M., Figgatt, C. et al. (2016). Demonstration of a small programmable quantum computer with atomic qubits. *Nature* 536: 63–70.

60 Jazaeri, F., Beckers, A., Tajalli, A., and Sallese, J.-M. (2019). A review on quantum computing: from qubits to front-end electronics and cryogenic MOSFET physics. *Proceedings of the 26th IEEE International Conference on Mixed Design of Integrated Circuits and Systems (MIXDES)*, pp. 15–25.

61 Arute, F., Arya, K., Babbush, R. et al. (2019). Quantum supremacy using a programmable superconducting processor. *Nature* 574: 505–510.

62 Dyakonov, M. (2019). The case against quantum computing. *IEEE Spectr.* 56: 24–29.

63 Jellen, M.J., Ayodele, M.J., Cantu, A. et al. (2020). 2D arrays of organic qubit candidates embedded into a pillared-paddlewheel metal-organic framework. *J. Am. Chem. Soc.* 142: 18513–18521.

64 Al-Khalili, J. and McFadden, J. (2015). *Life on the Edge: The Coming Age of Quantum Biology*. Cambridge: Black Swan.

65 Hameroff, S. and Penrose, R. (2014). Consciousness in the universe: a review of the 'Orch OR' theory. *Phys. Life Rev.* 11: 39–78.

66 Miller, J., Harding, S., and Tufte, G. (2014). Evolution-in-materio: evolving computation in materials. *Evol. Intell.* 7: 49–67.

67 Eiben, A.E. and Smith, J. (2015). From evolutionary computation to the evolution of things. *Nature* 521: 476–482.

68 Chen, C., Van Gelder, J., Van der Ven, B. et al. (2020). Classification with a disordered dopant-atom network. *Nature* 577: 341–345.

69 Miller, J.F. (2019). The alchemy of computation: designing with the unknown. *Nat. Comput.* 18: 515–526.

70 Massey, M.K., Kotsialos, A., Qaiser, F. et al. (2015). Computing with carbon nanotubes: optimization of threshold logic gates using disordered nanotube/polymer composites. *J. Appl. Phys.* 117: 134903.

71 Mohid, M., Miller, J.F., Harding, S.L. et al. Evolution-in-materio: solving computational problems using carbon nanotube-polymer composites. *Soft Comput.* 20: 3007–3022.

72 Vourkas, I. and Sirakoulis, G.C. (2016). *Memristor-Based Nanoelectronic Computing Circuits and Architectures*. Cham: Springer.

73 Massey, M.K., Kotsialos, A., Volpati, D. et al. (2016). Evolution of electronic circuits using carbon nanotube composites. *Sci. Rep.* 6: 32197.

74 Vissol-Gaudin, E., Kotsialos, A., Massey, M.K. et al. (2017). Solving binary classification problems with carbon nanotube/liquid crystal composites and evolutionary algorithms. *2017 IEEE Congress on Evolutionary Computation (CEC)*, San Sebastian, pp. 1924–1931.

75 Vissol-Gaudin, E., Kotsialos, A., Groves, C. et al. (2017). Computing based on material training: application to binary classification problems. *2017 IEEE International Conference on Rebooting Computing (ICRC)*, Washington, DC, pp. 1–8.

76 Vissol-Gaudin, E., Pearson, C., Groves, C. et al. (2021). Electrical behaviour and evolutionary computation in thin films of bovine brain microtubules. *Sci. Rep.* 11: 10776.

77 Heath, J.R., Kuekes, P.J., Snider, G.S., and Williams, R.S. (1998). A defect-tolerant computer architecture: opportunities for nanotechnology. *Science* 280: 1716–1721.

78 Floyd, T.L. (2017). *Digital Fundamentals*, 11e. Upper Saddle River, NJ: Pearson.

79 Chen, D., Wang, D., Yang, Y. et al. (2017). Self-healing materials for next-generation energy harvesting and storage devices. *Adv. Energy Mater.* 7: 1700890.

80 Tan, Y.J., Wu, J., Li, H., and Tee, B.C.K. (2018). Self-healing electronic materials for a smart and sustainable future. *ACS Appl. Mater. Interfaces* 10: 15331–15345.

81 Aïssa, B., Haddad, E., and Jamroz, W. (2019). *Self-Healing Materials*, 2e. London: IET Digital Library.

82 Latif, S., Amin, S., Haroon, S.S., and Sajjad, I.L. (2019). Self-healing materials for electronic applications: an overview. *Mater. Res. Express* 6: 062001.

83 Kang, J., Tok, J.B.-H., and Bao, Z. (2019). Self-healing soft electronics. *Nat. Electron.* 2: 144–150.

84 McDonald, S.A., Coban, S.B., Sottos, N.R., and Withers, P.J. (2019). Tracking capsule activation and crack healing in a microcapsule-based self-healing polymer. *Sci. Rep.* 9: 17773.

85 Dumas, C., El Zein, R., Dallaporta, H., and Charrier, A.M. (2011). Autonomic self-healing lipid monolayer: a new class of ultrathin dielectric. *Langmuir* 27: 13643–13647.

86 Wagner, N.L., Greco, J.A., Ranaghan, M.J., and Birge, R.R. (2013). Directed evolution of bacteriorhodopsin for applications in bioelectronics. *J. R. Soc. Interface* 10: 20130197.

87 Wickstrand, C., Dods, R., Royant, A., and Neutze, R. (2015). Bacteriorhodopsin: would the real structural intermediates please stand up? *Biochim. Biophys. Acta* 1850 (3): 536–553.

88 Hampp, N. (2000). Bacteriorhodopsin as a photochromic retinal protein for optical memories. *Chem. Rev.* 100: 1755–1776.

89 Ashwini, R., Vijayanand, S., and Hemapriya, J. (2017). Photonic potential of haloarchaeal pigment bacteriorhodopsin for future electronics: a review. *Curr. Microbiol.* 74: 996–1002.

90 Li, Y.-T., Tian, Y., Tian, H. et al. (2018). A review on bacteriorhodopsin-based bioelectronic devices. *Sensors* 18: 1368.

91 Blankenship, R.E. (2014). *Molecular Mechanisms of Photosynthesis*, 2e. Oxford: Wiley Blackwell.

92 Barber, J. and Tran, P.D. (2013). From natural to artificial photosynthesis. *J. R. Soc. Interface* 10: 20120948.

93 Zhang, J.Z. and Reisner, E. (2020). Advancing photosystem II photoelectrochemistry for semi-artificial photosynthesis. *Nat. Rev. Chem.* 4: 61–21.

94 Dogutan, D.K. and Nocera, D.G. (2019). Artificial photosynthesis at efficiencies greatly exceeding that of natural photosynthesis. *Acc. Chem. Res.* 52: 3143–3148.

95 Whang, D.R. and Apaydin, D.H. (2018). Artificial photosynthesis: learning from nature. *ChemPhotoChem* 2: 148–160.

96 Zhang, B. and Sun, L. (2019). Artificial photosynthesis: opportunities and challenges of molecular catalysts. *Chem. Soc. Rev.* 48: 2216–2264.

97 Gaut, N.J. and Adamala, K.P. (2020). Towards artificial photosynthesis. *Science* 368: 587–588.

98 Llansola-Portoles, M.J., Gust, D., Moore, T.A., and Moore, A.L. (2017). Artificial photosynthetic antennas and reaction centres. *C.R. Chim.* 20: 296–313.

99 Prathapan, S., Johnson, T.E., and Lindsey, J.S. (1993). Building-block synthesis of porphyrin light-harvesting arrays. *J. Am. Chem. Soc.* 115: 7519–7520.

100 Liu, C., Colón, B.C., Ziesack, M. et al. (2016). Water splitting-biosynthesis system with CO_2 reduction efficiencies exceeding photosynthesis. *Science* 352: 1210–1213.

101 Berhanu, S., Ueda, T., and Kuruma, Y. (2019). Artificial photosynthetic cell producing energy for protein synthesis. *Nat. Commun.* 10: 1325.

102 Turner, A.P.F. (2013). Biosensors: sense and sensibility. *Chem. Soc. Rev.* 42: 3184–3196.

103 Karunakaran, C., Bhargava, K., and Benjamin, R. (eds.) (2015). *Biosensors and Bioelectronics*. Amsterdam: Elsevier.

104 Swager, T.M. and Mirica, K.A. (eds.) (2019). Chemical sensors. *Chem. Rev.* 119: 1–726.

105 Kim, J., Campbell, A.S., De Ávila, B.E.-F., and Wang, J. (2019). Wearable biosensors for healthcare monitoring. *Nat. Biotechnol.* 37: 389–406.

106 Tu, J., Torrente-Rodríguez, R.M., Wang, M., and Gao, W. (2020). The era of digital health: a review of portable and wearable affinity biosensors. *Adv. Funct. Mater.* 30: 1906713.

107 De Puig, H., Bosch, I., Gehrke, L., and Hamad-Schifferli, K. (2017). Challenges of the nano-bio interface in lateral flow and dipstick immunoassays. *Trends Biotechnol.* 35: 1169–1180.

108 Wu, Y., Zhou, Y., Leng, Y. et al. (2020). Emerging design strategies for constructing multiplex lateral flow test strip sensors. *Biosens. Bioelectron.* 157: 112168.

109 Wu, K.R., Cao, W., and Wen, W. (2014). Extraction, amplification and detection of DNA in microfluidic chip-based assays. *Microchim. Acta* 181: 1611–1631.

110 Missoum, A. (2018). DNA microarray and bioinformatics technologies: a mini-review. *Proc. Nat. Res. Soc.* 2: 02010.

111 Wöhrle, J., Krämer, S.D., Meyer, P.A. et al. (2020). Digital DNA microassay generation on glass substrates. *Sci. Rep.* 10: 5770.

112 Howorka, S., Cheley, S., and Bayley, H. (2001). Sequence-specific detection of individual DNA strands using engineered nanopores. *Nat. Biotechnol.* 19: 636–639.

113 Merchant, C.A., Healy, K., Wanunu, M. et al. (2010). DNA translocation through graphene nanopores. *Nano Lett.* 10: 2915–2921.

114 Li, Y., Yang, C., and Guo, X. (2020). Single-molecule electrical detection: a promising route toward the fundamental limits of chemistry and life science. *Acc. Chem. Res.* 53: 159–169.

115 Gardner, J.W. and Bartlett, P.N. (1999). *Electronic Noses: Principles and Applications*. Oxford: Oxford University Press.

116 Persaud, K.C., Marco, S., and Gutierrez-Galvez, A. (eds.) (2013). *Neuromorphic Olfaction*. Boca Raton, FL: CRC Press.

117 Block, E. (2018). Molecular basis of mammalian odor discrimination: a status report. *J. Agric. Food Chem.* 66: 13346–13366.

118 Méndez, M.L.R. (2016). *Electronic Noses and Tongues in Food Science*. London: Academic Press.

119 Cipriano, D. and Capelli, L. (2019). Evolution of electronic noses from research objects to engineered environmental odour monitoring systems: a review of standardization approaches. *Biosensors* 9: 75.

120 Wasilewski, T., Migoń, D., Gębicki, J., and Kamysz, W. (2019). Critical review of electronic nose and tongue instruments prospects in pharmaceutical analysis. *Anal. Chim. Acta* 1077: 14–29.

Further Reading

Adamatzky, A. (ed.) (2016). *Advances in Physarum Machines: Sensing and Computing with Slime Mould*. Cham: Springer.

Ball, P. (1994). *Designing the Molecular World*. Princeton, NJ: Princeton University Press.

Bar-Cohen, Y. (ed.) (2011). *Biomimetics: Nature-Based Innovation*. Boca Raton, FL: Taylor and Francis.

Birge, R.R. (ed.) (1994). *Molecular and Biomolecular Electronics*. Washington, DC: American Chemical Society.

Contero, S. (2019). *Nano Comes to Life*. Princeton, NJ: Princeton University Press.

Drexler, K.E. (1990). *Engines of Creation: The Coming Era of Nanotechnology*. New York: Anchor Books.

Extance, A. (2016). Digital DNA. *Nature* 537: 22–24.

Fruk, L. and Kerbs, A. (2021). *Bionanotechnology: Concepts and Applications*. Cambridge: Cambridge University Press.

Jones, R.A.L. (2004). *Soft Machines: Nanotechnology and Life*. Oxford: Oxford University Press.

Karunakaran, C., Bhargava, K., and Benjamin, R. (2015). *Biosensors and Bioelectronics*. Amsterdam: Elsevier.

Nicolini, C. (2016). *Molecular Bioelectronics: The 19 Years of Progress*, 2e. Singapore: World Scientific.

Shen, J.-R., Satoh, K., and Allakhverdiev, S.I. (eds.) (2021). *Photosynthesis: Molecular Approaches to Solar Energy Conversion*, Advances in Photosynthesis and Respiration, vol. 47. New York: Springer.

Wright, R.H. (1982). *The Sense of Smell*. Boca Raton, FL: CRC Press.

Index

Electrical Processes in Organic Thin Film Devices: From Bulk Materials to Nanoscale Architectures, First Edition. Michael C. Petty.
© 2022 John Wiley & Sons Ltd. Published 2022 by John Wiley & Sons Ltd.
Companion Website: www.wiley.com/go/petty/organic_thin_film_devices